The Butterflies *and* Moths
of NORTHERN IRELAND

The Butterflies and Moths of NORTHERN IRELAND

Robert Thompson   Brian Nelson

First published 2006

© National Museums Northern Ireland 2006

Photographs © Robert Thompson 2006, except where stated
Book design © Robert Thompson 2006

ISBN 0 900761 47 4
Publication No. 018

All rights reserved. No part of this publication may be reproduced, stored in a retrieval system, or transmitted in any form or by any means, electronic, mechanical, photocopying, recording or otherwise, without the prior permission of the publisher and the various copyright holders.

Published by
National Museums Northern Ireland
Botanic Gardens
Belfast BT9 5AB
www.ulstermuseum.org.uk

Printed by Nicholson & Bass Ltd

Contents

Foreword		vi
Acknowledgements		ix
Introduction		x
Chapter One	*Notable Lepidoptera recorders of Northern Ireland*	1
Chapter Two	*The composition and conservation of the fauna*	9
Chapter Three	*Introduction to species accounts*	23
	Butterfly accounts	33
	Moth accounts	57
Chapter Four	*Habitat gallery*	355
Chapter Five	*Studying butterflies and moths in the field*	377
Photography		397
Checklist of Northern Ireland's fauna		399
Bibliography		412
Appendices		414
List of plants mentioned in the text		414
List of places mentioned in the text		417
List of recorders		421
Index		422

Richard Lewington '04

Foreword

THIS BOOK DESCRIBES what we know of the occurrence and distribution of the butterflies and larger moths of Northern Ireland.

Taken together, butterflies and moths make up the order of insects known as Lepidoptera. In scientific terms, the distinction between them is arbitrary. Indeed, there is no single feature which reliably distinguishes one from the other; we rely on a combination of features to tell them apart, and yet their popular images could scarcely be more different. Butterflies are seen as bright and beautiful creatures of the sunshine, if sometimes a trifle frivolous; moths are murky and mysterious denizens of the night, sinister and destructive devourers of clothes and carpets.

Only the good parts of these images are justified. I have studied Lepidoptera on and off since, at the age of nine, my sister stopped me trampling on a Northern Eggar caterpillar and I picked it up and took it home to rear. Growing familiarity quickly dispelled any notions that butterflies are frivolous or that moths are sinister. But even after fifty years, the fascination and sense of mystery have not waned. Some of our butterflies, although they may not be all that colourful, have an extraordinarily interesting life history. When you look at them closely, many moths have exquisitely pretty patterns and delicate colouring. Notable among these are the carpet moths, so called because of the resemblance of their wing patterns to oriental carpets, rather than indicating the dietary preferences of their larvae. The only two kinds of Lepidoptera which are a regular nuisance in my house and garden are not moths but butterflies, namely the Large and Small Whites which eat my cabbages. Very few moths actually eat clothes or other textiles, and the few that do are members of what we call the microlepidoptera, and so are not to be found in the pages of this book.

One of the advantages of an interest in butterflies and moths is that your observations need not be confined to choice habitats or rural areas. They are all around us, occurring everywhere there is vegetation, in town and country alike, though the more specialised kinds may have very strict habitat needs. Likewise, your activities are not dictated by the seasons. There are some species on the wing in Northern Ireland in favourable weather at all times of year, but the numbers and diversity are admittedly much reduced in winter. One or two of those which hibernate as adults, such as the Small Tortoiseshell and Herald Moth, may be seen in almost any month of the year. As if it wished to prove the point, I recall being distracted by a Small Tortoiseshell flying round inside church during a watchnight service.

One of the most intriguing aspects of butterflies and moths is their English names. Unlike many other insects, there are long-established English (but not Irish) names, some of them plainly descriptive, while others are delightfully imaginative. What is there in the name to suggest to the uninitiated that the Powdered Quaker and Setaceous Hebrew Character are, in fact, moths? At the start of each species account, the authors have quoted the scientific names, but have preferred to use the English ones elsewhere in the text. Wonderful! Savour these treasures.

Being in a position to produce a book of this standard about the butterflies and moths of Northern Ireland is no mean achievement. We have almost 500 species in total, a little more than half of all the species present in the British Isles as a whole. The list on page 2 of twenty main recorders who were once active in Northern Ireland but are now deceased is relatively short. In recent years, interest in these insects has grown, and a total of around 700 recorders is listed in Appendix 3, many of them still active. Species have been discovered that were previously unknown, either because they had never been seen before or because they had not been recognised. Others have been found to be more widespread than formerly believed. Some species, though, have apparently declined, in a few cases to the extent that there have been no recent sightings despite, generally, an increased effort in searching for moths.

The main purpose of the book is to document the distribution and known flight periods of each of the species, and to provide an outline of their habitat requirements. It does not set out to describe their variation or to assess their ecological needs in detail. To do so would require a great deal more space; perhaps more to the point is the fact that our knowledge of these aspects of Lepidoptera in Northern Ireland remains very sketchy. In practice, it is usually assumed that Irish populations belong to the same species as the most similar looking populations in Great Britain. This assumption is usually correct, but a close examination has shown that it is sometimes not. For years it was believed that our Wood White butterflies were the same as the English *Leptidea sinapis*, though we now know that all of them are in fact Réal's Wood White, *L. reali*.

While there is a wealth of information in the pages that follow, scrutiny of the maps and the text will soon reveal that there are huge gaps. Recording has been uneven, both geographically and over time; the north-west in particular is under-recorded. In all probability, any species which fly at night, but are not attracted to light, are also under-recorded. We know little about the importance of Lepidoptera in ecological relationships, beyond the generalisation that they are a significant part of the diet of bats and some other mammals and birds. Our understanding of their conservation requirements is largely based on the principle that conserving habitats will conserve the species that live there. Experience of managing nature reserves for butterflies or moths has demonstrated that this principle alone is not necessarily enough.

This brings me to a second objective of the book, and arguably the one of greater long-term value. This is to give momentum to the growing interest in butterflies and moths, to prompt the making and reporting of observations, and to encourage active concern for the well-being of these beautiful and seductive members of the natural world. There is no need to be a trained biologist to contribute. When you have perused the pages that follow, consider what more you might do to join in the action. Chapter Five gives a lot of practical advice, and most experienced lepidopterists will be more than pleased to help.

John Faulkner
Loughgall, County Armagh

Richard Lewington

Acknowledgements

THE PROCESS OF researching and writing this book through editing to final production has involved many people who have given their time freely to ensure its successful completion. The book has benefited immeasurably from their advice, help and knowledge and we are extremely grateful to them for their commitment and support. In particular, the authors would like to acknowledge Environment and Heritage Service for their generous support in funding this publication and the invaluable contribution of the following people:

Damian McFerran, the Records Manager of the Centre for Environmental and Data Recording (CEDaR), for his constant support and input throughout the preparation and publication of this book; Dave Allen and Kenny Murphy for information on species and localities; Trevor Boyd and Ted Rolston, respectively the butterfly and moth recorders for the Northern Ireland Branch of Butterfly Conservation, who have provided data in their care and answered many queries; Maurice Hughes and Vincent McLoughlin for access to their records and information about species; Peter Crowther, Marianne McKeown and Robert Nash at the Ulster Museum and Michael Meharg and Mark Wright of Environment and Heritage Service for assistance and support; John Faulkner for writing the foreword and Richard Lewington for preparing the artwork; Paul Roper of Johnsons Photopia for photographic equipment; David Anderson and Robert South of Nicholson & Bass Ltd; Stephen Dalton, Maurice Hughes and Roy Leverton for permission to use their photographs; the Irish Naturalists' Journal for permission to reproduce images; Bernard Skinner and Paul Waring for species queries and identification; Alan Morton, the creator of the DMAP software used to generate the distribution maps; Dave Allen, John Faulkner, Maurice Hughes, Kenny Murphy, Ian Rippey and Ted Rolston who suggested improvements to the manuscript – any errors remaining are entirely our responsibility; our wives and families, Catherine, Jonathan, Ethan, Shirley, Peter and Calum for their continuous support; and last, all the butterfly and moth recorders, living and deceased, who have ever contributed records or information on these insects in Northern Ireland. We owe them our sincere thanks and hope this publication meets their expectations, both in its presentation and content.

Introduction

BUTTERFLIES HOLD A special place in the affections of many people and there are few naturalists who are unable to identify one instantly. They are the most flamboyant of the insects, gaudily proclaiming their identity. Moths are much more understated and secretive, most only revealing themselves after dark; however, many moths, as the photographs in this volume confirm, are highly attractive insects. These insects are the subject of this book and included in it are written accounts describing the distribution, habitat and flight period of the almost 500 species of butterfly and larger moth recorded in Northern Ireland. Each species account is complemented by a map showing the distribution of recent and past records. Additional chapters provide brief biographies of deceased naturalists who have contributed to the study of Northern Ireland's butterflies and moths, the composition and conservation of the fauna, and field techniques for studying Lepidoptera. Included within the text are over 260 photographs illustrating a wide selection of the species and a habitat gallery showing a representative sample of the local habitats. A complete checklist and a bibliography of the relevant literature is included.

The text is written for the general naturalist and in particular for those who wish to develop an interest in these captivating and often beautiful insects. It aims to answer two of the basic questions of natural history – what species have been recorded and where have they been seen. This information has not previously been presented in one book as no complete account of the butterfly and moth fauna of Northern Ireland exists. The last Irish catalogue was written in 1964, but is now very out of date and often sketchy in detail. Our primary aims, therefore, have been to inventory the fauna and to give a current assessment of the status of each species.

Butterflies and moths belong to the same order of insects, the Lepidoptera. Whilst the identity of both groups appears obvious, in reality no single character exists that strictly defines a butterfly from a moth; most of the features thought characteristic of a butterfly can also be found in some moths. The division of the moths into the larger or macro species and the micromoths also needs explanation. By a long-standing convention the term 'larger moths' has become synonymous with the large and relatively easy to identify species in the more advanced families of Lepidoptera that are represented in Britain and Ireland. The micromoths, which contain the largest number of species, are inconspicuous, difficult to identify insects in the more primitive families. They tend not to come to light, therefore finding them requires different techniques and so they remain the realm of the specialist. However this division, like the butterfly/moth distinction, is not a rigid one. There is a significant number of species amongst the macros that are challenging to identify and would be defined by many people as small moths, the pugs being a prime example. Equally, some of the moths within the micro grouping are striking and obvious species and in practice are as easy to identify as many of the larger moths. A major inconsistency is apparent in the inclusion of the Hepialidae, the swift moths, within the larger moth grouping, as it is considered the most primitive family of all the Lepidoptera. Nevertheless, despite its inconsistencies, this distinction between the larger and the micromoths has survived to the present with minor amendment. We have followed it and so the reader will not find mention of any species of moth that is not included within the most popular identification books.

It has been common in book titles to give butterflies prominence greater than or equal to that of moths, even though a simple count shows that the larger moths outnumber the butterflies by about twenty to one in Northern Ireland. Increasingly, larger moths are getting more attention. There are several possible explanations for this. One may be related to the unfortunate circumstance that butterflies are becoming a rare sight, whereas many moths remain common, even in urban areas. Another reason may be that the basic tools for detecting moths – light traps, and the means to power them – are much more available than in the past. The allure that light has for these nocturnal insects means that they can easily be attracted to a place of our choosing, where they may be identified at leisure. Just a single night's light trapping may reveal that an urban garden is home to a few dozen species of moths, even though it is never visited by a single butterfly. Those living in suburban and rural gardens will find the species diversity and abundance of moths in light traps even greater. Many naturalists, however, simply find moths fascinating in their own right, providing many challenges to their identification skills and deserving of further study.

Butterflies and moths are primarily plant-feeding insects and only a few species deviate from this habit. Adults feed mainly on energy-rich liquids such as nectar from flowers. Attracting butterflies by growing appropriate flowers is an easy way to see them, but flower-feeding by moths generally escapes attention as it happens at night. Nevertheless, it is significant behaviour and the availability of flowers is probably as important to moths as it is to the diurnal butterflies. Some species of moths including, surprisingly, several of our largest moths – for example, the Poplar Hawkmoth and Emperor – are physiologically incapable of feeding as they lack mouthparts. However, it is in the larval stage that the dietary preference of butterflies and moths is most apparent as it is a caterpillar's function to eat, grow and survive to adulthood. It is the needs of the larvae that primarily dictate where butterflies and moths live and the larvae of many species can only complete development successfully under very specific conditions of light, heat and food availability. The species are therefore extremely sensitive to changes in the countryside and, increasingly, it is being recognised that Lepidoptera have considerable potential as indicators of environmental change.

Species of Lepidoptera have been present in the Irish countryside for thousands of years. In the millennia since the end of the last Ice Age, the composition of the fauna has undoubtedly changed in response to natural ecological and climatic trends. Additionally, they have had to survive the transformation to the landscape brought about by human activity including the creation of agricultural land, development of towns and infrastructure, drainage of wetlands and forestry. It is widely documented that the pace of change in the Irish countryside has been increasing over the last 60 years. Northern Ireland is also becoming much more urbanized with an increasing acreage of land under housing and infrastructure for transport, retail and leisure activities. The present-day landscape in most of lowland Northern Ireland is dominated by agricultural land and only small, often isolated, fragments of natural habitats survive. It is in these refuges that the rarer species tend to be concentrated. The pressure on these precious sites seems bound to increase, creating greater problems for our coexisting wildlife. In the next 100 years, if the predictions are correct, the impact of global climate change may be even more profound.

Several recent analyses have revealed the fortunes of the butterflies and moths of Britain and Ireland over the last 40 years. Foremost amongst these are the *Butterflies for the New Millennium* (BNM) project and *The state of Britain's larger moths*. The former project mapped the distribution of all species in Britain and Ireland for the first time since 1982 and analysed population trends from regularly monitored sites; the latter reported an analysis of the catches of 337 moth species by the network of Rothamsted traps over the last 35 years. Both these studies have reported significant increases in abundance and range for some species, but these are outnumbered by those that have declined. The largest declines amongst the butterflies have been seen in habitat specialist species, those whose ecological requirements are the most specific. Two-thirds (226 species) of the moths showed a decreasing population trend over the 35-year study. The largest declines reported were more than 90 per cent. Another worrying finding was that the total population of all species had declined by 30 per cent in 35 years. Whether the declines reported in Britain have been paralleled in Ireland is unknown, but given the similar trends in the countryside in both islands, it would seem likely to be so. Despite this gloomy scenario we have to remain positive about the prospects for the species. There are many species which, although less common than in previous eras, still have robust populations in many sites, and given the chance could recover lost ground. There is also hope in the continuing discovery of new sites for rare species.

This lack of information on the health of the populations of the butterflies and particularly the moths in Northern Ireland provides the major justification for writing this book, namely, to foster a greater interest in the fauna and to promote recording and monitoring of the various species. There has been a very welcome increase in recording since 1980 and much of this can be attributed to the efforts of the Northern Ireland branch of Butterfly Conservation. The branch was an active participant in the BNM and also the successor mapping project. It has also been very active in moth recording, coordinating data collection from all the active moth recorders in Northern Ireland. As a result, butterfly and moth records form a significant proportion of the total number of biological records held by CEDaR at the Ulster Museum. There were over 154,000 individual butterfly and moth records in the database at the end of 2004. Support for recording by Butterfly Conservation has been provided by CEDaR and Environment and Heritage Service. The results of this collaborative work can be seen in the maps produced in this book. However, the maps also show that more recording needs to be done, especially in the north and west of Northern Ireland; Londonderry and Tyrone, in particular, are poorly known. This is ironic as east Tyrone was the home territory of Thomas Greer, one of the most accomplished entomologists ever resident in Northern Ireland. It is our hope that this book will stimulate a new generation of recorders to emulate Greer and all the other eminent lepidopterists, as well as encourage existing recorders to continue their work.

Robert Thompson
September 2006

Brian Nelson

Richard Lewington

CHAPTER ONE

Notable Lepidoptera recorders of Northern Ireland

Preceding page

Henry Heal on Brackagh Moss National Nature Reserve, County Armagh, 1980.

Drawn by Raymond Piper.

A KNOWLEDGE OF the lives of the naturalists who have studied the Lepidoptera of Northern Ireland is needed to understand the past history of species. When and where they lived and where they were active is required information in interpreting the pattern of past recording. Here we provide short biographies of the deceased entomologists who have contributed most to the documenting of our fauna.

The scientific study of Irish butterflies and moths began in the mid-nineteenth century. More naturalists have been involved in their study and the literature is more extensive than for any other insect order. Five catalogues of the Irish fauna have been produced, whereas other groups have been catalogued just once or, in some cases, not at all. Ireland has produced a number of prominent lepidopterists, some of whom were resident or active in Northern Ireland. Few, however, have gone on to international prominence, seemingly content with studying the Irish species. The Irish fauna has received scant attention from continental European entomologists as our fauna is depauperate, lacking endemic species; it does, however, have a sufficiently distinctive character to have attracted the notice of several British entomologists who have made a major contribution to its documentation.

Little information survives about the activities of the earliest naturalists in the early nineteenth century, but evidently some lepidopterists were active as the first Irish catalogue was produced in 1854. Progress since then has been intermittent. The number of entomologists active in Ireland in any era has always been limited, so periods of intense activity have been sporadic. This is especially frustrating to anyone seeking an understanding of changing distributions and abundance of species. The apparent absence of a species is almost certainly due to inadequate recording effort in particular periods of time and geographical areas.

The history of Irish entomology between 1850 and 1950 was written by Beirne and his account refers to many of the entomologists active in Northern Ireland during that time; it is an important source of information, especially as Beirne knew many of the people personally. Other sources are the *Bibliography of Irish Entomology* by Ryan *et al.*, the brief biographies in Praeger's *Some Irish Naturalists* and the pages of the *Irish Naturalist* and *Irish Naturalists' Journal.* From these sources we have distilled the following accounts. We have omitted several figures about whom we could find no information. The biographies are presented in approximate chronological order of birth and only deceased entomologists and naturalists are included. The maps and the records mentioned in the individual species accounts testify to the contribution of the living lepidopterists, but the recounting of their stories will have to wait for the future.

SAMUEL ARTHUR BRENAN (1837–1908)

The long-time rector of Cushendun and a general naturalist who collected widely throughout Northern Ireland. He published little, but was listed as a contributor to the 1901 catalogue by Kane.

WILLIAM FRANCIS DE VISMES KANE (1840–1918)

Kane lived at Drumreaske House in Monaghan and collected in that area, and also at Favour Royal in Tyrone. He is best remembered as the author of the third (and first comprehensive) catalogue of the Irish macrolepidoptera in 1901. This replaced and updated two nineteenth-century catalogues. The first, written by Joseph Greene (1824–1906) and Arthur Riky Hogan (fl. 1850–1860), was published in 1854. Greene's and Hogan's catalogue is not relevant to us as it was largely a list of the macrolepidoptera of the Dublin area. The second, published in 1866 and authored by Edwin Birchall (1819–1884), is more complete but also now very outdated. Birchall was trusting of the information supplied to him and his catalogue contains records that later cataloguers, including Kane, considered doubtful. Kane's catalogue was published in 1901 and was the first to have a significant number of contributions from naturalists living in the counties that comprise Northern Ireland. However, like Birchall's, the catalogue is tarnished by the inclusion of many records of moths falsely claimed by Robert Edward Dillon (later Lord Clonbrock 1869–1926). This, however, does not affect the veracity of the northern records. After 1901, Kane suddenly gave up entomology and left Ireland, leaving his collection to the National Museum in Dublin. Beirne speculates that this was because of his unwitting connection with the Dillon records, but, equally, it could have been for family reasons.

William Francis de Vismes Kane.

Author of the first comprehensive catalogue of the Irish macrolepidoptera, published in 1901.

CHARLES W. WATTS (fl. 1890–1902)
Very little is known of Watts. He collaborated with Kane, contributing records from Antrim and Down to the 1901 catalogue and published a small number of papers including one on the Lepidoptera of the Mournes. Arguably, his most significant records were those of the Small Blue from south Antrim.

WILLIAM FREDERICK JOHNSON (1852–1934)
A clergyman and teacher who lived most of his life in County Armagh, Johnson was one of the most significant collectors of Irish insects and a prolific author of entomological notes and records. Specimens collected by Johnson are prominent in the collections of the Ulster Museum and the National Museum in Dublin, surpassed only by those of Arthur Wilson Stelfox (1883–1972). Johnson's bibliography includes some 100 notes and papers on the Lepidoptera. He lived at several places in Armagh, including Mullinure and Acton Glebe, which appear frequently on his specimen labels. He also collected at Churchill (Peatlands Park) and Poyntzpass in Armagh and Rostrevor and Warrenpoint in Down. Johnson was undoubtedly a consummate entomologist and Praeger's comment in *Some Irish Naturalists* that 'his eye was scarcely sufficiently critical for the discrimination of some of the more difficult species' belittles his immense contribution to Irish entomology. Johnson had wide contacts (especially in the Entomological Society of London) and sent difficult species to specialists for accurate determination.

DAVID CALLENDER CAMPBELL (1860–1926) and W.H. CAMPBELL (fl. 1878–1894)
These two brothers, who ran a flour-importing business, were a rarity, being virtually the only lepidopterists of any era ever resident in Derry. David Campbell appears to have been the most active, publishing notes on birds and insects over a thirty-year period from 1892. He collected widely in Counties Londonderry and Donegal and also Antrim, but often no locality is given for his records. David Campbell discovered the Belted Beauty at Ballycastle, sending larvae to Kane. His brother W.H. was also a naturalist, publishing a few notes between 1878 and 1894 on birds and insects, but nothing is written about him. Kane acknowledged both for their assistance.

S.L. BRAKEY (fl. 1872)
Kane acknowledged 'my friend' the Rev. S.L. Brakey of Trory Glebe in Fermanagh in his catalogue, but nothing more is known of him. Brakey appears never to have published any of his own records and the only surviving ones attributed to him are in Kane. Amongst these is a sighting of Camberwell Beauty at Trillick (Tyrone) 'settled on the roadside, but not

captured, it being Sunday.' It was also reported by Kane that Brakey 'set free' 20 adults of Privet Hawkmoth *Sphinx ligustri* (not an Irish species) at Trillick in 1872, that he had bred from English pupae.

JAMES BRISTOW (fl. 1872–1902)

Little information is available about Bristow, a clergyman from Belfast who published a few notes on moths, including one on the Belted Beauty in Antrim. Kane acknowledged Bristow especially in the introduction to his 1901 catalogue.

C.E. PARTRIDGE and E.W. BROWN (fl. 1893–1897)

Two little-known naturalists who collected in Fermanagh at the end of the nineteenth century whilst serving in the army. Both served in the same regiment and were presumably garrisoned at Enniskillen. Partridge had the rank of Lieutenant-Colonel, whilst Brown was a Captain. Partridge is the author of four papers published between 1893 and 1897 of which the first, entitled 'Lepidoptera at Enniskillen' is the most significant. Whilst Enniskillen is given as the location for the records, it is likely that he recorded throughout Fermanagh. Partridge's most significant records were those of the Purple Hairstreak and the only Northern Ireland record of the Lackey. E.W. Brown also had an interest in Lepidoptera and he published five short papers between 1893 and 1897. However, like Partridge, nothing more is known of him.

CHARLES DONOVAN (1863–1951)

Donovan was author of the fourth catalogue of the Irish macrolepidoptera which was published privately in 1936. He was a rigorous compiler of records and he firmly rejected all of Dillon's false claims that had appeared in Kane. Donovan did not collect in Northern Ireland as far as is known, but he received records from the active recorders. He was born in Calcutta, but his family home was near Timoleague, in County Cork. After a distinguished career as a doctor in India, he returned to live in England from 1919 onwards. During his retirement he visited Ireland frequently.

CHARLES LANGHAM (1870–1951)

Charles Langham, a resident of Fermanagh, published just a few notes and brief reports on his collecting activities, but it is clear from his collection, which is now in the Ulster Museum, that he was a significant entomologist. The bulk of the collection comprises Irish, British and continental European butterflies, but there are also many moth specimens. Langham lived at Tempo Manor in Fermanagh and most of his specimens are labelled simply 'Tempo', but from the little data given in his papers it is clear he also collected elsewhere in the county. For example, there are Dingy Skipper specimens labelled Tempo, but it is extremely unlikely that they came from there, as the limestone grassland habitat required by this species does not exist in this part of Fermanagh. A few of his records came from Portaferry (Down), including the only Northern Ireland occurrence of Ringed Carpet. Langham was mentioned frequently by Kane, but his name rarely appeared in the subsequent catalogues.

J.E.R. ALLEN (fl. 1897–1930)

Another resident of Fermanagh (he taught Classics at Portora Royal School in Enniskillen), Allen published notes and papers on moths between 1897 and 1930, but mostly before 1915. Nothing more appears to be known about him. One of his most significant finds was the Slender-striped Rufous on Benaughlin, County Fermanagh.

GEORGE FOSTER (1870–1935)

A Kerry-born clergyman but long-time resident of Down, Foster rose to become a canon of Down Cathedral. He authored a list of the macrolepidoptera of Down, the only list ever produced for one of the Northen Irish counties. He collected also in Antrim and Armagh. Many of his records appeared in Donovan's catalogue. His publications on moths often had lyrical titles including 'Moth hunting among the reeds on a September night.'

WILLIAM MONOD CRAWFORD (1872–1941)

The son of one of Belfast's most prominent industrialists, Crawford was born in Paris. The family remained there until 1888 when they returned to Belfast, where Crawford completed his education. After leaving university he joined the Indian Civil Service in 1895, returning to Belfast in 1919. Between 1921 and his death, Crawford was a prolific author of insect notes, principally about Coleoptera and Lepidoptera. Many of his notes concerned migrant moths and butterflies and he documented the occurrences of several rare hawkmoths, including the only Northern Ireland record of Silver-striped Hawkmoth. One of his most significant finds was the Small Eggar in Fermanagh in 1928.

Above
William Monod Crawford.

Crawford had a particular interest in migrant Lepidoptera. He also discovered the Small Eggar in County Fermanagh.

THOMAS GREER (1874–1949)

Thomas Greer lived all his life near Cookstown in Tyrone. He was one of the most important entomologists ever resident in Northern Ireland. Greer amassed what Beirne described as 'probably the best collection of Irish Lepidoptera ever made.' However, after his death, the collection was broken up and sold through dealers to private collectors. Beirne states in Greer's obituary that the collection was remarkable for the many rarities and the series of varieties and aberrations of butterflies. Some of Greer's specimens were purchased by Edward Alfred Cockayne (1880–1956), an inveterate collector of oddities, and they are now preserved in the Rothschild-Cockayne-Kettlewell collection in the Natural History Museum, London. Greer also sent Cockayne the specimens of the Bordered Grey which were described as a new subspecies *tyronensis*. Greer was the author of almost 120 papers and notes on Lepidoptera and he was a regular correspondent with Henry Jerome Turner (1856–1951), the English editor of *Entomologist's Record and Journal of Variation*, the journal in which many of these papers were published.

Greer lived a long, colourful and sometimes charmed life. Beirne described him as brusque and often tactless but, despite this, immensely likeable. Outside entomology, which was no doubt viewed as a strange hobby by his family and acquaintances, he is best remembered as an ardent motorcyclist. This ultimately caused his death after a collision between his bike and a lorry. He raced motorbikes and was a founder member of the Cookstown and District Motor Cycle Club and its president until his death. He was the first Northern Ireland rider to compete in the T.T. Races on the Isle of Man in 1909. At the age of 75, he came out of retirement to race again in the Cookstown 100 road race, although a mechanical problem caused his failure to finish. He had competed in the same race when aged 63 and it is recorded that he received minor injuries when his 598 c.c. Scott seized at 70 m.p.h.

Left
Thomas Greer.

One of the most accomplished entomologists of his generation. His contribution to the study of the Lepidoptera of Northern Ireland, and Tyrone in particular, is unsurpassed.

EDWARD STUART AUGUSTUS BAYNES (1889–1972)

Baynes was an Englishman posted to Ireland as a Trade Commissioner after the Second World War. He lived and trapped moths at Glenageary, near Dublin. Following Donovan, Baynes was the next cataloguer of the Irish macrolepidoptera. His catalogue appeared in 1964 with a supplement in 1970. Many Northern Ireland records were mentioned in both parts, but more especially in the supplement; however, details of the records are frustratingly sparse. Throughout the catalogue, Baynes compared his experience of species with Donovan's assessment, but there was no attempt at explanation of the different views of the two authors. Whilst it is a useful publication, it lacks many important details about the source of the records, rendering it less useful than it might have been. Baynes also compiled for many years the migrant reports in the *Irish Naturalists' Journal*.

PHYLLIS ISMAY NELSON (1907–1979)

The most significant recorder of butterflies and moths in Armagh between 1940 and her death, Phyllis Nelson was a stalwart member of the Armagh Field Naturalist's Society and she published some records in the proceedings of the society. Her collection of moths and butterflies was given to the Armagh County Museum. She lived first at Drumbanagher, where her father was rector and, later, near Drumman More Lough.

DENIS HENDERSON RANKIN (1919–1944) and MATTHEW NEAL RANKIN (1918–1999)

The Rankin brothers were active naturalists in the late 1930s and early 1940s. Together they published several notes and papers mainly on birds, but also, moths. However, this promising partnership did not survive the Second World War and it would be many years before any other naturalists became interested in moths in Northern Ireland. Denis, a pilot in the RAF, was killed in 1944 and Neal, whilst maintaining his interest in natural history, moved to Scotland. Their records were mainly from the hills above north Belfast near their home and also in north Antrim at Portballintrae and on Rathlin Island.

Denis (left) and Neal (right) Rankin.

Both brothers were interested primarily in birds but they also recorded Lepidoptera in the Belfast Hills, on Copeland Bird Observatory and on Rathlin Island before and during the Second World War.

BRYAN PATRICK BEIRNE (1918–1998)

A distinguished professional entomologist and specialist in the management of agricultural pests and the microlepidoptera, Beirne lived in Canada from 1947 to his death. He was probably the most internationally renowned entomologist of Irish origin. Beirne's contributions to Irish entomology were varied and mostly highly significant. They include the first bibliography of Irish entomological literature, books and papers, especially on microlepidoptera. One of his papers concerned the maritime moths found living around Lough Neagh by Greer. He presented this as evidence for some of his theories on the postglacial colonisation and origins of the Irish fauna. However, these theories have not been substantiated and are no longer credible. The interesting fauna on Lough Neagh that Beirne documented, although real, is probably no longer present today.

Bryan Patrick Beirne.

An eminent entomologist, widely published, with an international reputation. His interest lay primarily with the microlepidoptera and professionally with agricultural pests.

WILLIAM STUART WRIGHT (fl. 1946–1976)

Captain W.S. Wright lived in south-east Antrim near Aghalee, an area, then as now, of meadows, quiet lanes lined with tall hedges and the incomparable wetland of the Montiaghs. He trapped in this area, recording much of its rich fauna and was the first to record the moths on the dunes at Murlough NNR on the coast of Down. Tollymore Forest (Down) was another of his trapping sites. Wright was the of the author of the *Catalogue of Macro-Lepidoptera in Northern Ireland*, published in 1964 by the Ulster Museum. It simply records the presence of each species in each county, so it is of limited value. He published a few other papers on the moths of Antrim and Down and, for a short period in the 1970s, he compiled the annual summary of migrant Lepidoptera for the *Irish Naturalists' Journal*. Part of Wright's collection of moths is in the Ulster Museum, but it is very poorly labelled. Wright was neglectful in recording this information. Despite these shortcomings, he was virtually the only active recorder in Northern Ireland during the 1950s and 1960s. His most significant discovery was of the Burren Green *Calamia tridens* in Clare in 1949, which encouraged visits by many British entomologists to this area.

HENRY GEORGE HEAL (1920–1986)

Reader in Chemistry at Queen's University, Henry Heal is largely known for his knowledge of the Irish microlepidoptera, unsurpassed until recently. Fluent in German and French, Henry Heal used continental literature, unusual even among professional entomologists in England where he had many contacts. He travelled extensively in Northern Ireland and had an intimate knowledge of the landscape and habitats. Henry Heal was widely known for his dedication to conservation and was a member of the Nature Reserves Committee and an active participant in setting up the Ulster Trust for Nature Conservation (now the Ulster Wildlife Trust). He also studied the butterflies of Northern Ireland and was the first to note the spread of the Wood White along old railway lines and the conservation value of these redundant areas as potential wildlife refuges. He was instrumental, along with Joe Furphy, in establishing Peatlands Country Park. Among his many other accomplishments was the creation of the first wild Butterfly Garden at Drum Manor (Tyrone). Some of his other activities would perhaps be considered controversial today, especially his attempt at introducing the Large Copper *Lycaena dispar* to Downpatrick marshes (Down). However, he was held in very high regard both as an academic and ardent conservationist as shown by the number of personal appreciations published after his death.

ROY MORELL (d. 1991)

Roy Morrell was a lecturer in English at the University of Ulster at Coleraine in the 1960s and early 1970s. He previously lived in East Africa and Malaysia and seems to have had an interest in tropical butterflies. While resident here, he recorded butterflies and moths, especially on the Londonderry coast. He never published these, but some of his records appear in Baynes's supplement and others in the distribution maps of the early volumes of the *Moths and Butterflies of Great Britain and Ireland*.

Silver-striped Hawkmoth.

An exceptionally rare immigrant which has been recorded just once in Northern Ireland.

CHAPTER TWO

The composition and conservation of the fauna

The Butterflies and Moths of Northern Ireland

THERE ARE 25 SPECIES of resident butterfly and 462 species of larger moth currently on the Northern Irish list. The composition of the fauna is summarised in Table 1 which shows the number of species in each family present in Northern Ireland, the whole of Ireland and Great Britain. The totals for each butterfly family include only resident species, but those for each moth family include both resident and migrant species, as it is impossible to determine the status of some of the rarer species.

The figures in Table 1 reveal two main facts. First, the butterfly and larger moth fauna of Ireland is depauperate compared to Great Britain; only 49 per cent of the British butterfly species and some 62 per cent of the species of moth are found in Ireland. The average figure for Irish insects as a whole is 65 per cent representation of the British species. Second, the proportion of the Irish fauna present in Northern Ireland is 88 per cent which indicates a rather homogeneous fauna across the whole island, a reflection of the relative uniformity of climate and habitats throughout Ireland.

Preceding page

*Orange Tip.
Brackagh Moss National Nature Reserve, County Armagh.*

Above
Réal's Wood White.

Brackagh Moss National Nature Reserve, County Armagh.

A recent study has found that two species of 'wood white' occur in Ireland. In Northern Ireland only Réal's Wood White has been recorded.

The absence of so many species of British Lepidoptera from Ireland is a mystery as suitable conditions would appear to exist. One obvious factor is the absence of the foodplant of the species from Ireland, which may be related to the suitability of habitat. This is exemplified by the Adonis Blue butterfly *Polyommatus bellargus*, whose sole foodplant, Horseshoe Vetch, is entirely absent from Ireland. Other species are absent for reasons of climate, the conditions in Ireland being either not warm enough for thermophilic 'southern species' or conversely, too mild and unsuitable for cold-tolerant 'northern species'. Examples of 'southern species' are the Honeysuckle-feeding Broad-bordered Bee Hawkmoth *Hemaris fuciformis* and the Ribwort Plantain-feeding Glanville Fritillary *Melitaea cinxia*, which are both restricted in Great Britain to southern England. There are fewer examples of the latter group, the 'northern species', but two are the Scotch Argus *Erebia aethiops* (main foodplant, Purple Moor-grass) and the Rannoch Brindled Beauty *Lycia lapponaria* (main foodplants, Bog Myrtle and Heather).

There are two species of larger moth (Irish Annulet *Odontognophos dumeteta* and Burren Green *Calamia tridens*) and one species of butterfly (Réal's Wood White *Leptidea reali*) present in Ireland but absent from Great Britain. Of these, only Réal's Wood White is found in Northern Ireland, the other two being confined to the Burren area of Clare. Réal's Wood White is a cryptic species, only recognised as distinct in 1988, and the reason for its apparent absence from Britain is unknown. It is found throughout Ireland in sheltered areas of flower-rich grassland including sand dunes, limestone grassland and abandoned quarries.

The only Irish records of eight species of larger moth, Alder Kitten *Furcula bicuspis*, Barred Hook-tip *Watsonalla cultraria*, Four-dotted Footman *Cybosia mesomella*, Juniper Carpet *Thera juniperata*, Northern Winter Moth *Operophtera fagata*, Slender Brindle *Apamea scolopacina*, Smoky Wave *Scopula ternata* and Yellow-ringed Carpet *Entephria flavicinctata*, are from Northern Ireland. All except the last are recent additions to the Irish list and their

The composition and conservation of the fauna

Barred Hook-tip.

Ballykinler Dunes, County Down.

This is a photograph of the first adult recorded in Northern Ireland in 1999.

Juniper Carpet.

Finaghy, County Antrim.

This is the first adult recorded in Ireland.

Yellow-ringed Carpet.

Murlough Bay, County Antrim.

Adults of this elusive species have only been found by day, disturbed from their resting places under overhanging rocks.

status is uncertain. They may be overlooked residents or recent colonists. The Yellow-ringed Carpet has been known in Northern Ireland for almost 100 years, but has only ever been recorded from Murlough Bay in north-east Antrim and one locality in Fermanagh. Its foodplants are species of saxifrage and stonecrop and, in Northern Ireland, larvae have been found on Mossy Saxifrage. Yellow-ringed Carpet is still extant at the Antrim site, but there have been no records from Fermanagh since 1914.

Table 1: The numbers of species recorded from Northern Ireland, Ireland and Great Britain in the families of butterfly and larger moth present in both islands and resident in the last 200 years. Figures from Great Britain are taken from Asher *et al.* (2001) for the butterflies and Waring and Townsend (2003) for the moths. The numbers of species in each butterfly family refer only to those generally considered as resident.

Family	Northern Ireland	Ireland	Great Britain
Butterflies			
Hesperiidae	1	1	8
Papilionidae	0	0	1
Pieridae	6	7	7
Lycaenidae	6	7	18
Nymphalidae	12	14	25
Totals	25	29	59
Moths			
Hepialidae	4	4	5
Cossidae	0	2	3
Zygaenidae	3	4	10
Sesiidae	3	6	14
Lasiocampidae	7	7	10
Saturniidae	1	1	1
Drepanidae	4	4	6
Thyatiridae	5	6	9
Geometridae	185	204	305
Sphingidae	11	12	17
Notodontidae	14	17	23
Lymantriidae	4	6	11
Arctiidae	16	19	29
Nolidae	1	1	4
Noctuidae	204	235	c. 400
Totals	462	528	c. 847

The changing fortunes of the species

It would seem likely that the abundance and distribution of the butterflies and moths of Northern Ireland will have varied over the millennia, but firm evidence for this is lacking. The status of species prior to 1850 is simply unknown as there is no information from this period, whilst the data from 1850 to the present is variable in quantity and quality and unsuitable for objective analysis. An additional problem in interpreting the information is the changed behaviour of butterfly and moth recorders over the past 150 years. The entomologists in the nineteenth and early twentieth centuries used a variety of collecting techniques to gather specimens for their collections that are rarely employed today, such as sugaring and searching for larvae and pupae. Today, in contrast, moth recorders rely almost exclusively on light traps and the emphasis is on regular trapping of a single site or inventory surveys. These gross differences in recording effort and behaviour make it impossible to state with any confidence whether species have increased or decreased in this period.

Small Blue.

Ballintra, County Donegal.

The Small Blue, our smallest and rarest resident butterfly, is feared extinct at its last-known colony in Fermanagh. There have been no records since 2001.

Since 1980, standard techniques of recording have been adopted and consistent and interpretable data have become available. The two main methods in use are butterfly transects and Rothamsted moth traps. Both methods impose consistency in the collection of data, aiding objective analysis. Any trends revealed will reflect those of the whole population rather than changes in methodology.

Butterfly transects are regular weekly counts of butterflies throughout the flight season along a set route. In the analysis of transect data, the weekly counts for each species are summed to calculate annual indices of abundance, which are estimates of population size. Site indices can be combined to calculate regional and national indices. Plotting these indices over time reveals trends in butterfly populations. Butterfly transects were initiated on approximately twenty sites in Northern Ireland in the 1980s, but few were continued for more than a few years and the results of most have never been analysed. The data from three Northern Ireland transects – Hillsborough Forest, Killard Point and Murlough NNR (Down) – are included in the UK Butterfly Monitoring Scheme. This scheme has been collating data for over 30 years and over this period it has revealed changes in abundance of species on a national and regional scale, the response of individual species to climate change and the impact of management and site designation on species of conservation concern.

Rothamsted traps are fixed moth traps of a uniform design which are operated continuously. There is just one of these traps operating in Northern Ireland at Hillsborough Forest, but Butterfly Conservation is working to get more installed so that there will be more effective monitoring of the local moth population. A recent analysis of trends in catches of moths in the network of Rothamsted traps across the UK has shown the value of the information gathered by these long-running traps. This study revealed an alarming picture of decline in the total moth population and the numbers of some species previously considered common. It is likely, given the similarities in the two islands, that the Irish moth population has undergone a similar decline.

Threatened species

Species continue to be found in Northern Ireland for the first time, but there have been few documented examples of extinction. The extinction of a species is difficult to prove as there are many reasons why species may remain undetected. They may persist in low numbers making them difficult to detect, or they may be of such obscure habits that they can only be found by special effort. One of the few species that appears to have become extinct in Northern Ireland is the Small Blue. This tiny butterfly is searched for annually at its sole site, Monawilkin, in Fermanagh. No adults have been seen since 2001, raising fears that the colony has died out. Its extinction is difficult to prove, but given the interest in this site and the species, it is unlikely to have been overlooked. The reasons for its demise are unclear as the site appears to have been managed appropriately, so it may simply have died out for natural reasons.

Table 2 lists 17 species of larger moth that have not been recorded in Northern Ireland for more than 25 years. The status of some of the species on this list is uncertain and they may never have had established populations in Northern Ireland. This applies to Currant Clearwing, Small Emerald and V-Moth, which all feed on non-native or cultivated plants. All three may have been accidentally introduced. Others, for example, Light Feathered Rustic and Slender-striped Rufous, are species of remote habitats that have seldom been visited by moth recorders and they may well survive undetected. There are some species which have been specifically looked for, but have not been rediscovered and are likely to be extinct; Belted Beauty, Crescent Striped and White Colon fall into this category. These are all coastal species and the habitats and sites they occurred in have all been affected by habitat loss. The remaining species are mainly found in woodland and scrub habitats that appear to be still present in Northern Ireland, so again there is a reasonable possibility that they are still extant. The Sword-grass is a species

Above
Belted Beauty.

Belmullet, County Mayo.

Adults of this attractive moth were recorded at Ballycastle, County Antrim for a few years up to 1900. Although suitable sites along the north coast have been searched recently, no larvae or adults have been found. Large colonies still exist on the machair and dunes at Belmullet, County Mayo.

Right
Currant Clearwing.

Kent.

There is only one early twentieth-century record for this wasp-mimicking moth in Northern Ireland. There is no evidence the species was ever a resident in the region.

which has seemingly declined in southern Britain for unknown reasons. There is a recent suggestion that the species may not be as rare as previously thought but that, reluctant as it is to come to light, it is being overlooked. At a site in Scotland, adults were only seen regularly at sugar and in much larger numbers than ever appeared at light traps. The decreasing use of this technique may therefore have exaggerated its apparent decline.

Table 2: Species of larger moth which have not been recorded for more than 25 years in Northern Ireland with the date of the last confirmed record (if known) and the main habitat of the species.

Species	Main Habitat
Sesiidae	
Currant Clearwing (1892)	Gardens, fruit fields, allotments, woodland
Lasiocampidae	
Lackey (pre-1900)	Sunny patches of scrub and woodland edge
Geometridae	
Small Emerald (1918)	Hedges, scrub on limestone
Cream Wave (1935)	Broadleaved woodland
Beech-green Carpet (1942)	Woodland and moorland
Slender-striped Rufous (1914)	Wet flushes in upland grassland and moorland
Marsh Pug (1973)	Wetlands and sand dunes
V-Moth (1937)	Gardens and allotments
Belted Beauty (1893)	Sand dunes
Ringed Carpet (1914)	Damp heath and moorland
Grey Birch (1936)	Birch woodland
Noctuidae	
Light Feathered Rustic (1897)	Herb-rich grassland on rocky ground
Crescent Dart (1973)	Coastal; cliffs and rocky shorelines
White Colon (1978)	Sand dunes, saltmarsh
Sword-grass (1953)	Moorland, rough grassland, open woods
Suspected (pre-1949)	Heaths, wetlands, woods
Crescent Striped (1978)	Salt and estuarine marshes

Six species of butterfly and nineteen species of larger moth considered threatened or declining rapidly in Northern Ireland have been included in the Northern Ireland Biodiversity Strategy. These are listed in Table 3 and indicated by the abbreviation SOCC in the species accounts. Regional action plans have been compiled for the Marsh Fritillary and are in preparation for the other priority species.

Seven species of butterfly are given legal protection in Northern Ireland under The Wildlife (Northern Ireland) Order 1985. Schedule 5 of this Order gives the species protection at all times. They are also listed on Schedule 7 which makes it an offence to sell live or dead specimens of any of the seven species. The listed species are the Brimstone, Dingy Skipper, Holly Blue, Large Heath, Marsh Fritillary, Purple Hairstreak and Small Blue. It is arguable whether giving these species this legal protection will ensure their survival as collecting is much less of a threat than habitat loss. It certainly has not prevented the disappearance of the

Table 3: Species of Conservation Concern (SOCC) and Priority species listed in the Northern Ireland Biodiversity Plan. Priority species are indicated by (P). The main habitat used by each species in Northern Ireland is described.

Butterflies	Main Habitat
Dingy Skipper (P)	Unimproved limestone grassland
Green Hairstreak	Heaths and margins of bogs
Marsh Fritillary (P)	Fens, cutover bogs, sand dunes
Réal's Wood White (P)	Unimproved grassland with scrub
Small Blue (P)	Unimproved limestone grassland
Wall (P)	Sparse dry grassland with bare ground
Moths	
Argent and Sable (P)	Heaths, flushed margins of bogs
Belted Beauty (P)	Sand dunes
Bordered Grey	Bogs and coastal heaths
Cloaked Pug (P)	Mature conifer plantations
Confused	Rocky coasts, moorland
Dark Tussock	Heaths and bogs
Forester	Open, often damp, habitats
Narrow-bordered Bee Hawkmoth (P)	Unimproved grassland and cutover bogs
Olive	Wet woodland
Pale Eggar (P)	Open woodland and heathland
Pretty Pinion (P)	Unimproved grassland
Red-tipped Clearwing (P)	Fens
Small Eggar (P)	Hedges, scattered bushes
Sprawler	Broadleaved woodland
Square-spot Dart	Coastal grassland
Sword-grass (P)	Moorland, upland rough grassland
White-line Snout (P)	Moorland and damp woodland
Wood Tiger (P)	Moorland, wet heaths, sand dunes
Yellow-ringed Carpet (P)	Rocky areas on coasts and hills

Marsh Fritillary.

Montiaghs Moss National Nature Reserve, County Antrim.

Although recorded from all counties, the Marsh Fritillary has disappeared from many of its former haunts. It is protected by Northern Ireland and European legislation.

Brimstone in the 1980s or the more recent loss of the Small Blue. Both these extinctions are more likely to have been caused by natural factors. The highest level of protection given any species of Lepidoptera in Northern Ireland is afforded to the Marsh Fritillary. Not only is the species legally protected here, but it is the sole insect species found in Northern Ireland that is listed in Annex II of the Habitats Directive. It is a requirement of the Habitats Directive that national governments in the European Union (EU) must designate sites for Marsh Fritillary as Special Areas of Conservation (SAC). In Northern Ireland, a number of SACs have Marsh Fritillary as the qualifying feature including Aughnadarragh Lough (Down), Montiaghs Moss (Antrim) and Murlough NNR (Down).

Habitats

A diligent naturalist will probably be able to find at least one species of Lepidoptera anywhere in Northern Ireland wherever there is vegetation. However, it will quickly become apparent that some habitats support many more species than others. The richest habitat in Northern Ireland for moths is broadleaved woodland, whereas butterfly diversity is greatest in areas of unimproved grassland.

Broadleaved woodland is a widespread but relatively uncommon habitat in Northern Ireland. The percentage of Northern Ireland that is covered by woodland is much lower than in Great Britain and the woods also tend to be smaller and younger. Many types of broadleaved

Annagarriff Wood National Nature Reserve, Peatlands Park, County Armagh

Peatlands Park is one of the most important sites in Northern Ireland for moths. Two national nature reserves are included in the park, Annagarriff Wood and Mullenakill. Annagarriff is a drumlin covered in oak and birch woodland.

woodland exist but the two most significant types for Lepidoptera are the oak-dominated plantation or native woodland and the secondary woodland that is typically dominated by birch and found on acid, peaty soils. Less common types include woods dominated by Ash, Alder or Hazel. Mature woodland is a rich habitat for moths but it is less important for butterflies and there is only a single tree-feeding butterfly in Northern Ireland, the Purple Hairstreak. Amongst the best examples of mature woodland habitat are Banagher Glen (Londonderry), Crom (Fermanagh) and Rostrevor Wood (Down); all these are known to support a rich and notable assemblage of moths. Many of the woodland species can also be found in suburban gardens and hedgerows that possess some mature trees.

Secondary woodland is another very important habitat for many moths and birch (usually the dominant tree species) is the foodplant of a high proportion of our tree-feeding species. Secondary woods of birch, oak and Holly are common throughout Northern Ireland, growing on relict areas of cutover bog and at the edge of conifer plantations. These woods are often small and difficult to conserve and few have any statutory protection. They are threatened by neglect leading to abandonment and reclamation. Many of the boggy areas are used as dumps causing permanent loss of habitat.

Grassland is the commonest type of habitat in Northern Ireland, but most of the resource is intensively managed for agriculture, resulting in a sterile habitat for most Lepidoptera. The grassland fauna is concentrated in the small areas of semi-natural habitat which have survived agricultural intensification and development. These remnants exist mainly on the coast, the edges of upland areas and in abandoned quarries and undeveloped land, so-called brownfield sites. Many sites are small and suffer from neglect and inappropriate management. Some of the richest sites are the brownfield sites but these are often seen as prime sites for development. However, they do provide some of the last inland refuges for butterfly species like the Common Blue and Wall. Good examples of this habitat can be found at the Craigavon Lakes (Armagh), where it has developed along the line of a planned road, the old railway line at Dundrum (Down) and in the abandoned limestone quarry at Navan Fort (Armagh).

Bogs, heaths and moorland are other rich habitats for butterflies and moths. Collectively, these habitats are relatively common in Northern Ireland, but like semi-natural grassland, the surviving examples are becoming fragmented and are often suffering from habitat change brought about by neglect. The best sites are those with a varied structure with stands of birch and willow, areas of grassland and a varied surface topography. The species diversity of butterflies and moths is greatest on lowland bogs. The large expanses of bogs found in upland areas are not without interest, but they support fewer species. Their fauna is also less well known because of the difficulties in working in these remote areas. Peatlands Park (Armagh) is one of the most intact and largest surviving examples of lowland bog habitat and home to many of the specialist bog species.

Conservation of habitats

It is generally recognised that conservation of species is best achieved by conserving habitat. There are simply too many species for the needs of each to be addressed, so it is more effective to protect the most representative and the best examples of habitats. The threats to the habitats of our species of butterfly and moth are diverse and hierarchical ranging from global issues such as climate change to more local impacts on whole landscapes, the habitats within them and individual sites. Conservation strategies are similarly tiered from supranational designations such as SACs, through national statutory designations to the lowest level, represented by some form of non-statutory listing of sites of local importance and local nature reserves. The main objective of these designations is to protect representative examples of the range of habitats found in Northern Ireland.

Conservation in Northern Ireland follows the model applied in Britain. The emphasis is on conserving habitats through site designation and in Northern Ireland the main

The Umbra, County Londonderry.

The Umbra is one of the few localities on the north Londonderry coast that has been regularly trapped for moths. The site comprises flower-rich dune grassland with some wet slacks, scrub and woodland. Species diversity is high, including some of Northern Ireland's rarer species, such as Small Eggar.

Brackagh Moss National Nature Reserve, County Armagh.

This wetland site has been actively trapped for many years producing a long species list. It supports many scarce and rare species, such as Red-tipped Clearwing (at its only known site) and Valerian Pug.

instrument is the statutory designation of Area of Special Scientific Interest (ASSI) by Environment and Heritage Service (EHS). EHS is mandated to declare all areas of land as an ASSI where they need protection 'by reason of their flora, fauna or geological, physiographical or other features'. The sites are evaluated using objective scientific criteria. When complete, the ASSI network is intended to encompass the best and most representative examples of all the freshwater and terrestrial habitats in Northern Ireland.

Site selection of ASSIs is first and foremost based on the characteristics of the vegetation and to date no sites have been declared primarily on the basis of their butterfly and moth populations. However, many of the existing ASSIs are known to support important assemblages of butterflies and moths and rare species. Some examples include Montiaghs Moss and Rathlin Island (Antrim), Derryleckagh (Down) and Monawilkin (Fermanagh). The dry grassland at Monawilkin is the habitat of many uncommon species of moth. It was once the richest butterfly site in Northern Ireland but it has lost two species, the Small Blue and Wall, in recent years. Derryleckagh is a site complex comprising a broadleaved woodland and a fen. Both habitats have interesting species and the fen is the location of the only known colony of the Marsh Fritillary in south Down. The latter species is also found at the Montiaghs Moss, a valuable wetland with an outstanding suite of grassland, bog and wet woodland species. The diverse habitats on Rathlin Island support many uncommon species. The island is especially valuable for coastal and heathland species of butterfly and moth.

Whilst ASSI designation should safeguard the best examples of habitats in Northern Ireland, the butterflies and moths living in the general countryside lack any statutory protection. It is clear that the current ASSI network, by concentrating on examples of single habitats, does not adequately cover complex sites that have a mosaic of habitats. Many species require different conditions at specific times in their life cycle and if these are not provided

The composition and conservation of the fauna

at a site, the species will be absent. For example, adult butterflies and moths require access to flowers to feed in order to produce their eggs; if these are not available, the species will not be able to survive. Moths in particular appear to be most common in sites that have an intimate mixture of trees, scrub and flowery areas. Such sites may be difficult to categorise and therefore difficult to fit into the ASSI system. One good example is Rehaghy Mountain, a site in Tyrone that has been trapped intensively since 1995. The site comprises a mosaic of open woodland, wet and dry heath and wetland. The moth fauna is exceptional for such a small site and several rare species (including Sprawler, one of the Northern Ireland priority species) occur along with a good range of common and generalist species.

Other species may occur only at the margins of two habitats. For example, Argent and Sable is found on the edges of bogs where it meets birch scrub. If the ASSI is designated to protect the bog habitat, the birch scrub may be removed as it is perceived as a threat, and the microhabitat needed by the moth may be lost. Similarly, many species considered typical of fens feed, not on the wetland plants, but on willow and birch trees. These are often seen as a threat to the habitat and are frequently removed by management.

Several of the most important sites for butterflies and moths in Northern Ireland are owned and managed directly by EHS as National Nature Reserves (NNRs). Forest Service and a number of non-governmental conservation bodies also own sites with important species, including the National Trust (NT), Royal Society for the Protection of Birds (RSPB), Ulster Wildlife Trust (UWT) and Woodland Trust (WT). Some of these have also been declared as NNRs or ASSIs and a few have qualified as SACs.

Argent and Sable.

Lough Alaban, County Fermanagh.

This conspicuous day-flying moth appears always to have been rare in Northern Ireland. All of the recent records emanate from County Fermanagh, although historically, it has been recorded from a few other counties.

Pre-eminent amongst the National Trust properties is the Crom Estate in south Fermanagh and Murlough NNR near Dundrum (Down). Crom, a varied and large site, supports the most complete woodland fauna in Northern Ireland as well as significant populations of wetland species. Many of the recent additions to the Northern Ireland list have been discovered at Crom. Several parts of the estate have been declared as NNRs. Murlough NNR is the prime site in the region for coastal and dry heathland Lepidoptera and also the best-monitored butterfly site. Other National Trust properties of importance are the Argory (Armagh) for woodland and bog species; Fair Head and Murlough Bay (Antrim) for wet heath and upland grassland fauna and Portstewart Dunes (Londonderry) for coastal species. The newly acquired property at Divis (Antrim) in the hills above Belfast is likely to have a notable fauna of moorland and perhaps upland species.

The UWT reserve at the Umbra (Londonderry) is noted for its coastal and grassland species. Its recent acquisition at Slievenacloy (Antrim) on the north side of the Belfast Hills has not been fully surveyed but the habitat quality suggests it is probably a rich site for grassland species. The Inishargy reserve on the Ards Peninsula (Down) has a colony of Marsh Fritillary and the Glenarm reserve (Antrim) is an important woodland site.

The network of nature reserves and country parks owned by EHS includes many important species and habitats. Peatlands Park (Armagh) is the most important site in Northern Ireland for species of lowland bog and wet heath; the woodland fauna is also notable. The coastal NNRs at Ballymaclary (Londonderry) and at Killard Point (Down) are exceptionally important for their grassland species. Killard Point boasts one of the largest colonies of Common Blue in Britain or Ireland and the Small Eggar population at Ballymaclary is considered one of the largest in Ireland. The oak woodland at Rostrevor NNR (Down) supports a species-rich assemblage of woodland moths, surpassed only by Crom. Brackagh Moss (Armagh) and the Montiaghs Moss are probably the best sites in Northern Ireland for wetland and wet woodland species. Other notable EHS sites are Banagher Glen (Londonderry), Breen Wood (Antrim), Crawfordsburn Country Park (Down) and Kebble on Rathlin Island (Antrim).

Ultimately the long-term survival of many of our resident butterflies and moths rests with us. If we are to pass on a fauna as rich as the one we have inherited, the support of the public at every level is required. The trends apparent in the countryside have to be reversed if even our most familiar species are to remain common. It will be a sad indictment of our current stewardship if the next generation is denied the opportunity to see the species we enjoy today.

CHAPTER THREE

Species accounts

The Butterflies and Moths of Northern Ireland

Illustrations depicting anatomical terms commonly used in the text.

Butterfly

Noctuid moth

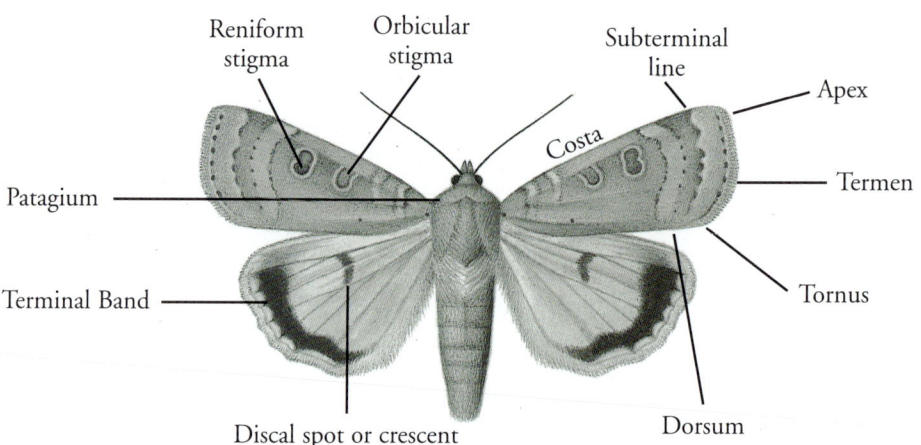

Geometrid moth

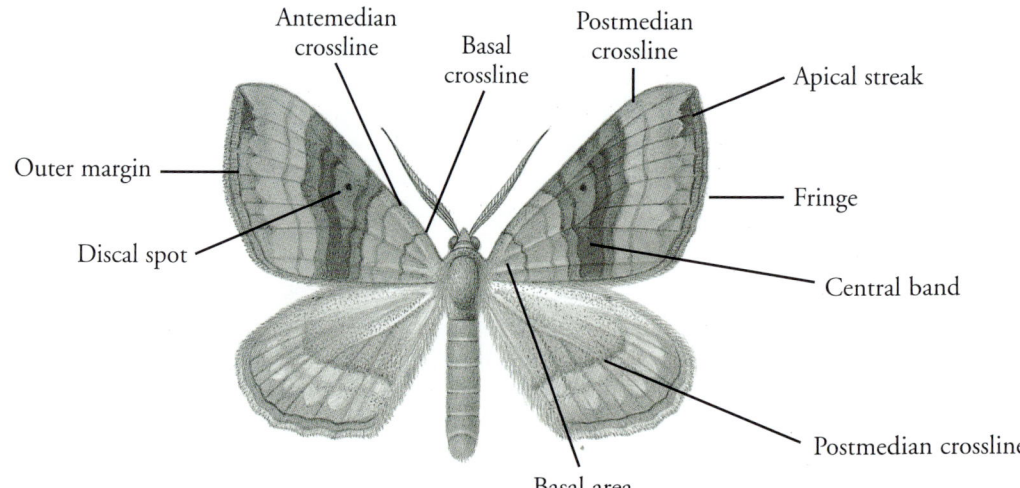

Preceding page

Poplar Hawkmoth.

Selshion Moss, County Armagh.

Format of the species accounts

THIS CHAPTER PRESENTS accounts for the species of butterfly and larger moth recorded in Northern Ireland up to the end of 2004. The butterfly accounts precede those of the moths and each account is accompanied by a distribution map.

Taxonomic order of the butterfly species follows the *Millennium Atlas* (Asher *et al.* 2001) whilst the moth species are ordered using the Bradley checklist (Bradley 2000) also followed by Waring and Townsend (2003) and Bond *et al.* (2006). In all accepted classifications of the Lepidoptera, the butterfly families are placed between the Sphingidae and the Drepanidae, but for ease of use in this publication, we follow convention and deal with the two groups separately.

Each account is headed with the vernacular name, which is retained throughout the text, and is followed on the next line by the scientific name. For the butterflies, both names are taken from the *Millennium Atlas*; for the moths, the vernacular names are taken from Waring and Townsend (2003) and the scientific names from Bradley (2000). Owing to the uncertainty of taxonomy and identification, with regard to subspecies, their names have been omitted in all cases, thus avoiding over-interpretation.

The Northern Ireland Lepidoptera Records Committee (NILRC) has devised a system of recording categories for all species of moth as an aid to recording. The same categories have been applied to the butterflies for this publication. These are defined as follows:

Category one includes species that are widespread, abundant and readily identifiable.

Category two includes species which have a restricted distribution or occurrence, even though they may be readily identifiable.

Category three covers species that are not readily identifiable and may be confused with other species. It also includes species that are considered sufficiently scarce for the NILRC to monitor the records received. Species which can only be identified by examination of the genitalia have 'G' appended.

Category four covers everything else – species which are not on the current Northern Ireland checklist plus a small number for which there are few records. These species may be hard or easy to identify, localised or rare. Voucher specimens are normally required for the acceptance of these records but, in some cases, high-quality photographs may suffice.

The categorisation of each species is shown following the scientific name. The abbreviation SOCC is included on this line for the six species of butterfly and nineteen species of larger moth that qualify as Species of Conservation Concern in the Northern Ireland Biodiversity Strategy. The Priority species are indicated by (P).

The species account text follows, commencing with a one-word assessment of the status of each species in Northern Ireland. This evaluation is based on the number of 10km square records for each species, applying the formula given below. The square count for the butterflies is obtained from the recording period for the BNM project (1995-2004). The square count for the larger moths is derived from the maps presented here, taking the period from 1980 to 2004 (all the coloured dots).

Unknown	no valid records of butterflies between 1995 and 2004 or moths since 1980
Rare	present in **1 to 5** 10km grid squares
Notable	present in **6 to 10** 10km grid squares
Scarce	present in **11 to 30** 10km grid squares
Widespread	present in **31 to 80** 10km grid squares
Ubiquitous	present in **81 or more** 10km grid squares

The first sentences of each account describe the general distribution pattern of the species in Northern Ireland. In many accounts, mention is made of historical records, especially if there have been no recent records from a county. Site references are usually accompanied by their respective county, in brackets. The grid references for the sites mentioned in the text are given in Appendix 2. We follow the convention of using Armagh City and Derry when referring to these cities, but Armagh and Londonderry for the respective counties.

A brief description of the adult insect follows with information on the habitat, flight period, larval behaviour and foodplants completing the account. Foodplant information is derived largely from published sources, but with reference to any specific knowledge from Northern Ireland. The flight period is derived from the butterfly and moth databases compiled by the Northern Ireland branch of Butterfly Conservation (BCNI) and held at the Centre for Environmental Data and Recording (CEDaR) in the Ulster Museum. For most species this is expressed as an actual date rather than a vague term such as the 'beginning of May.' The dates given are the first and last dates that the species was seen between 1980 and 2004. This is done deliberately as it provides recorders with precise information as to the accuracy of a record, especially of a scarce species. A common flaw in the BCNI databases was date error; many such were found and corrected during the compilation of the flight period data, vindicating our approach. While occasional early and late individuals occur (and with changing climate, these can be expected to become more common), we believe these to be exceptional.

Interpreting the data for double-brooded species was more difficult than it was for species with single generations. Even within the small latitudinal range of Northern Ireland, the phenology of species is not consistent from south to north and it appears that some species may be double-brooded in southern areas, but single-brooded in the north. Phenology may also be changing with warming climate. There is anecdotal evidence that some single-brooded species are starting to produce a second annual generation. A detailed analysis of the flight period data may be possible, but is not within the remit of this publication. We have therefore been conservative with the analysis and interpretation of the data.

Mapping and interpreting the data
The data for producing the maps is taken from the BCNI butterfly and moth databases. Application to view or obtain any of the data should be made to BCNI. The maps only show records of butterflies and moths that have been accepted and validated by BCNI and NILRC. The databases are incomplete for historical pre-1980 records for many of the common species, but also for some scarcer ones. Information on these is taken from a number of publications, principally the catalogues of Donovan and Baynes; however, much of the data included in these publications is vague and impossible to map. Reference therefore may be made in the text to some records which have not been mapped.

The validity of historical records is a continuing problem plaguing the assessment of all Lepidoptera species in Ireland. It is beyond the scope of this publication to attempt a complete review of these, but it is highly desirable that an updated catalogue of the Irish fauna is produced, to provide recorders with reliable and current information. The new Irish checklist (Bond *et al.* 2006) is a welcome first step in this process.

The distribution of each species is shown using a standard dot map, the symbols indicating the presence of the species in each of the 10km squares that lie partly or wholly in Northern Ireland. Dots are conventionally shown centrally in each square. Only Northern Ireland records are mapped in the squares that straddle the border with the Republic of Ireland. The dots are colour coded to indicate the date of the most recent record from each square according to the following formula:

- Grey   most recent record on or before 31 December 1979
- Yellow   most recent record between 1 January 1980 and 31 December 1999
- Orange   most recent record between 1 January 2000 and 31 December 2004

Recording of the butterflies and moths has been uneven and limited, geographically and historically, thus the maps only indicate the general distribution of each species in Northern Ireland. There is insufficient data to determine from the maps any but the broadest of trends in the population of species. Those statistics and patterns that do emerge from the data should be appreciated when examining any of the maps.

The total number of individual records in the BCNI database at the end of 2004 was 154,122, consisting of 66,551 records of butterflies and 87,571 records of larger moths (57%). The statistics are further summarised in Figure 1 which shows the number of individual records of butterfly and moth from each county. The graph reveals the discrepancy in coverage between both groups in the counties. Recording of both groups has been most intense in Down and just over 41% of all the records have originated from this county. The numbers of records of butterfly and moth from Antrim and Armagh are remarkably similar. Fermanagh is unusual as the number of moth records is almost four times greater than the number of butterfly records, which, despite the rich fauna, is the second lowest of any of the counties. The main feature of this graph is, however, the low numbers of records, especially of moths, from both Londonderry and Tyrone. This always has to be kept in mind when viewing the distribution maps.

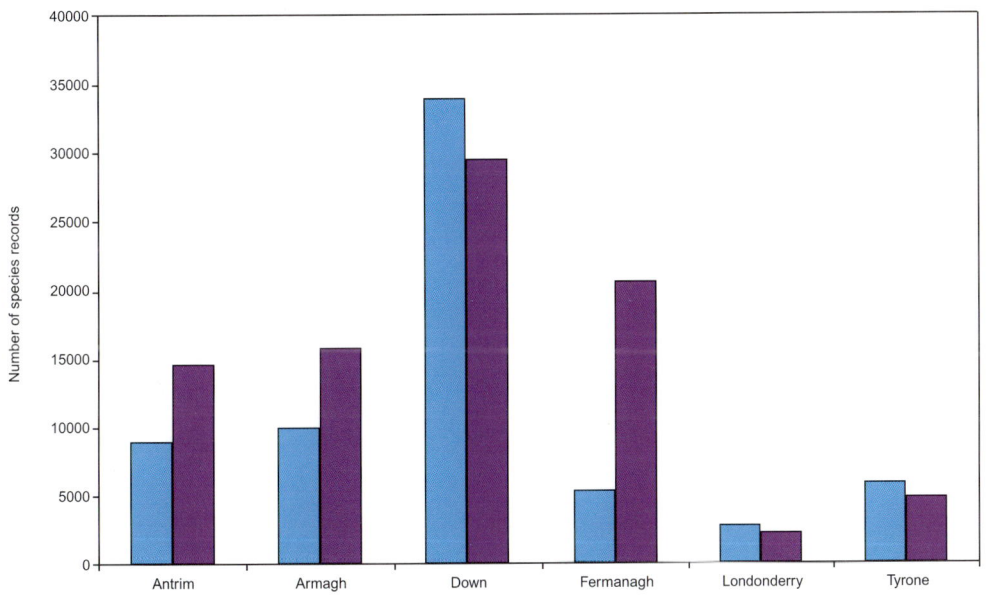

Figure 1: *The number of species records of butterflies (blue) and larger moths (purple) contained in the Northern Ireland Lepidoptera database from each county up to the end of 2004.*

The vast majority of the records (88%) have been gathered since 1988. This is shown in Figure 2. 1988 was the year the Northern Ireland branch of Butterfly Conservation was formed. CEDaR was created in 1995, which was also the first year of recording for the *Millennium Atlas of Butterflies*.

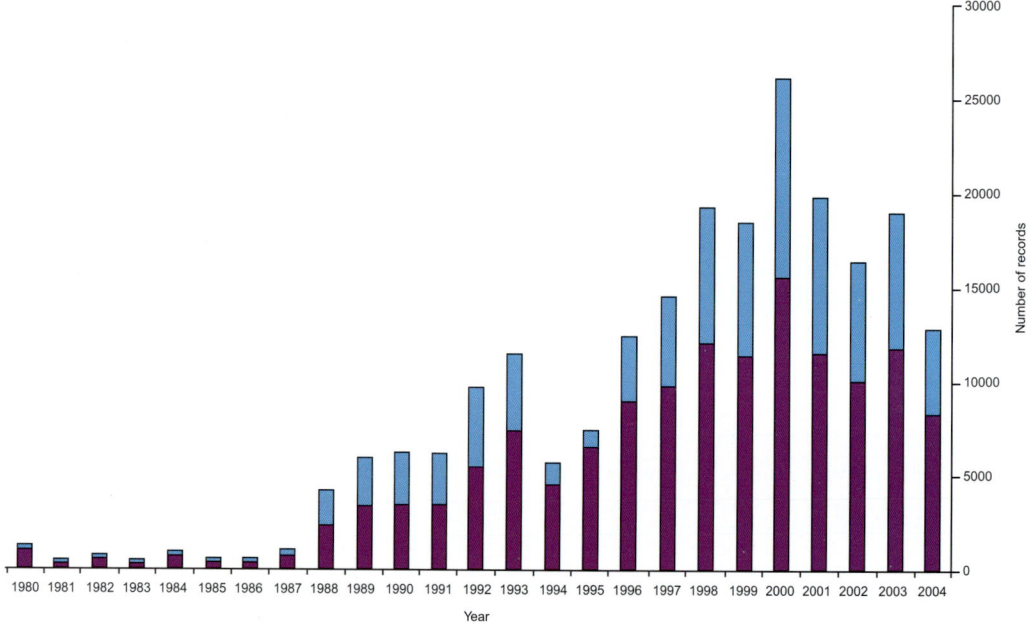

Figure 2:
The number of species records of butterflies (blue) and larger moths (purple) contained in the Northern Ireland Lepidoptera database for each year from 1980 to 2004.

Figures 3 to 7 which follow, show aspects of the level and extent of recording of butterfly and moths in Northern Ireland and the patterns in species diversity. Overall, the maps show that while coverage has been complete (there has been at least one record of a butterfly or moth from each of the 187 10km squares that lie wholly or partly within Northern Ireland), it has not been uniform. The variation in coverage is most apparent in the moths. Figure 7 shows that the squares in the north and west of Northern Ireland have been poorly recorded for moths compared to those in the southern counties. This is reflected in the species richness map (Figure 6) by the preponderance of green and grey squares in Antrim, Londonderry and Tyrone.

Figures 3, 4 and 5 reveal the pattern of species richness in the butterflies. Figure 3 shows the species diversity per 10km of the complete fauna of 31 species. This map shows that a minimum of 15 species has been recorded from most of the grid squares in Northern Ireland. Figure 4 is a refinement, showing the pattern of species richness of just the resident butterflies, excluding records of the Clouded Yellow, Brimstone, Red Admiral, Painted Lady, Camberwell Beauty, Comma and Monarch. The pattern revealed is very similar to Figure 3, showing that species richness of the resident butterflies is highest in certain areas of Northern Ireland and especially the coastal region of south Down, the shores of Belfast Lough, Fermanagh and the north coast of Londonderry. Areas of low diversity can clearly be distinguished covering the high ground of Antrim and the Sperrins (Londonderry and Tyrone). Figure 5 refines the data further by mapping only the number of habitat specialist butterflies and Species of Conservation Concern recorded in each 10km square since 1995. The category 'habitat specialist' was defined by Butterfly Conservation and explained in the *Millennium Atlas* and there are nine such species present in Northern Ireland — Dingy Skipper, Réal's Wood White, Green Hairstreak, Small Blue, Dark Green Fritillary, Silver-washed Fritillary, Marsh Fritillary, Grayling and Large Heath. The Wall is also included in this analysis as it is a Priority Species in the Northern Ireland Biodiversity Strategy. There are two main concentrations of these rare and threatened species in north-west Fermanagh and on the coast of south Down, particularly around Dundrum Bay. There is also a conspicuous hotspot on the north coast of Londonderry at Magilligan.

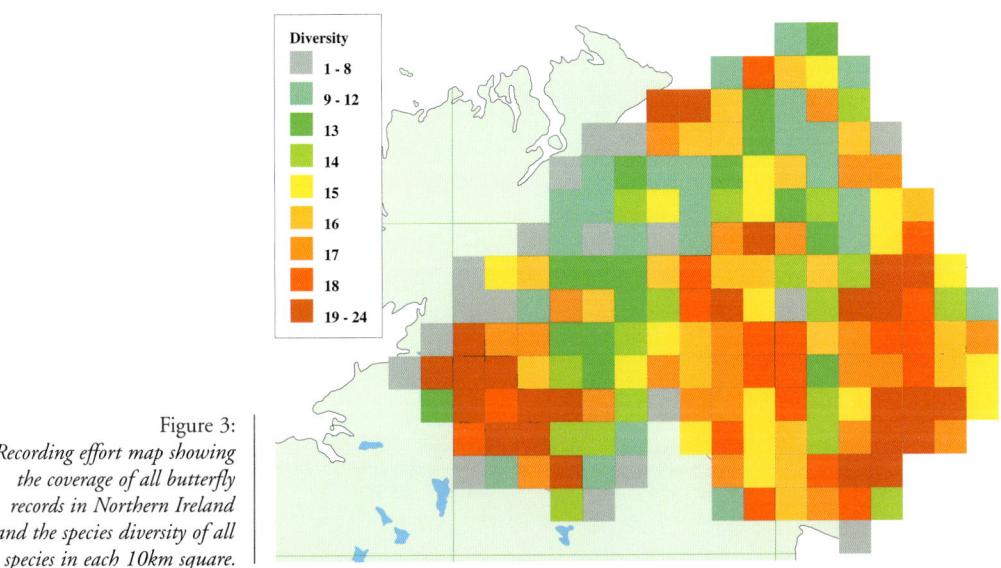

Figure 3:
Recording effort map showing the coverage of all butterfly records in Northern Ireland and the species diversity of all species in each 10km square.

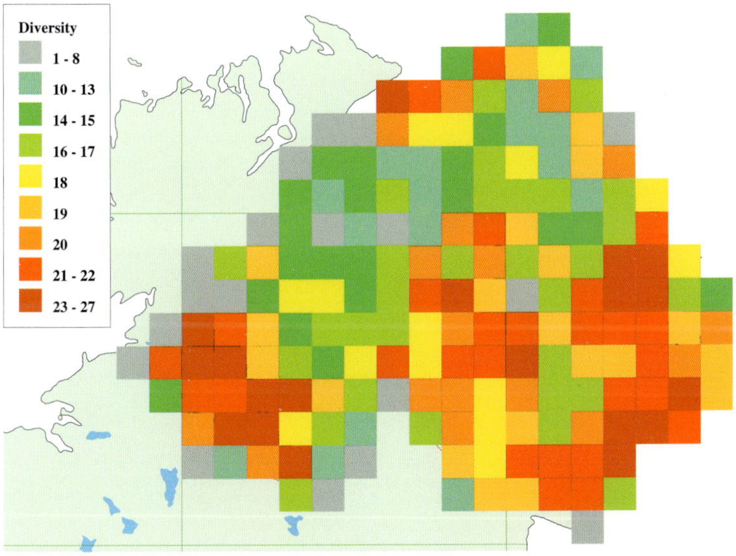

Figure 4:
The number of resident butterfly species recorded in each 10km square in Northern Ireland.

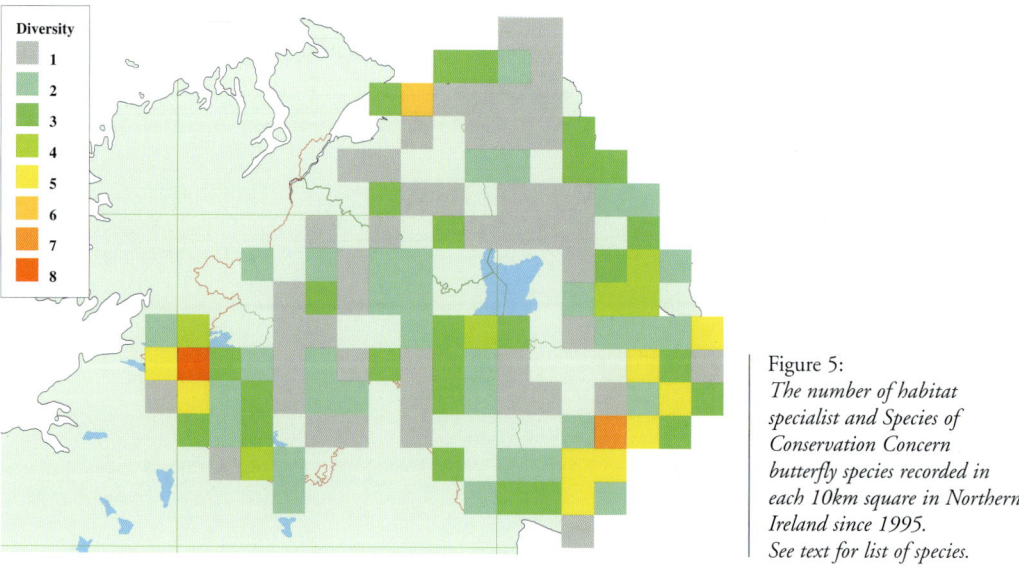

Figure 5:
The number of habitat specialist and Species of Conservation Concern butterfly species recorded in each 10km square in Northern Ireland since 1995. See text for list of species.

Figure 6 shows the pattern of coverage and species richness in the moths. Coverage is almost complete but there have been records of 33 species or fewer from 87 (47%) of the 10km squares in Northern Ireland. Figure 7 shows that there is a large variation in the recording effort for moths across Northern Ireland. It reveals that coverage across most of the north and west is low and that this corresponds to the areas of lowest species diversity shown in Figure 6. The conclusion from this is that whilst the larger moth data set is comprehensive in terms of coverage, there is need for much more recording in the north and west and that many of the gaps in distribution may be explained by lack of recording effort.

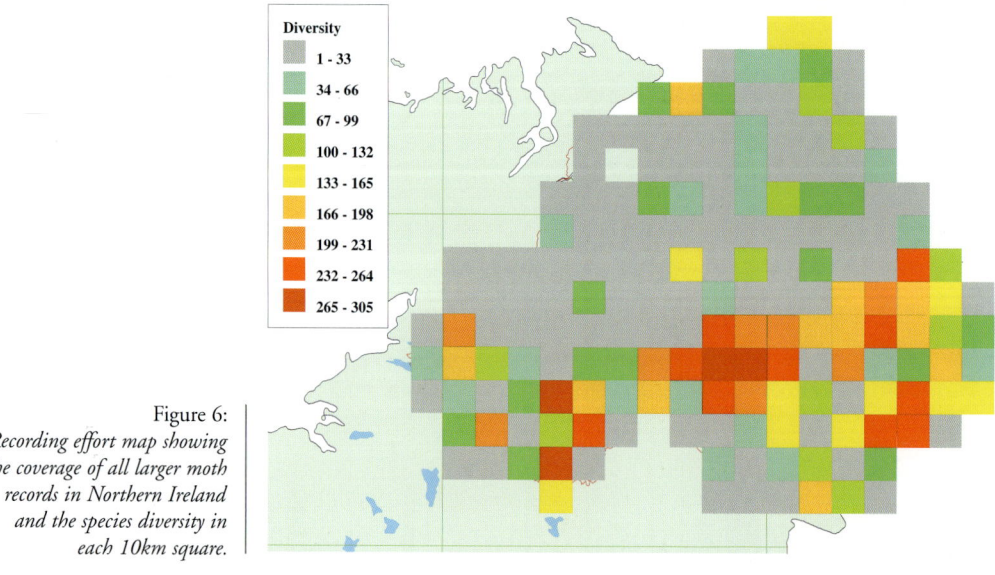

Figure 6:
Recording effort map showing the coverage of all larger moth records in Northern Ireland and the species diversity in each 10km square.

Species accounts

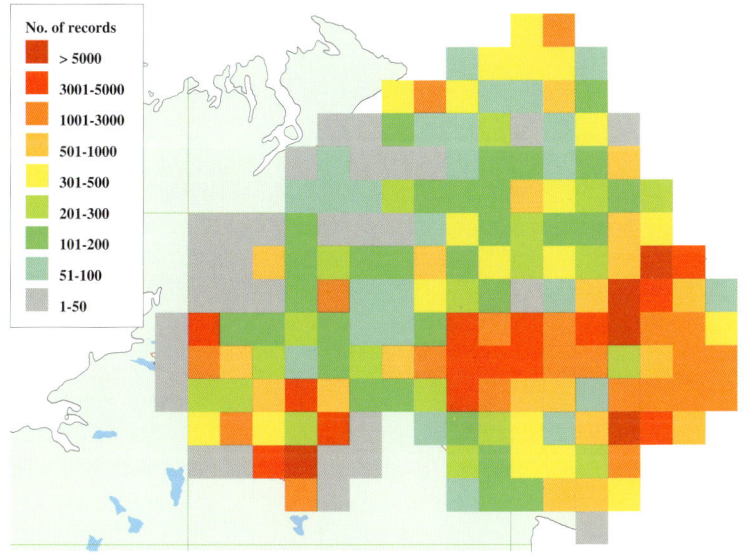

Figure 7:
The number of individual records of larger moths in each of the 10km squares in Northern Ireland.

Clouded Magpie.

Rostrevor Wood National Nature Reserve, County Down.

Overleaf
Oak Beauty.

Rostrevor Wood National Nature Reserve, County Down.

31

The Butterflies and Moths of Northern Ireland

Butterfly accounts

The Butterflies and Moths of Northern Ireland

HESPERIIDAE

*S*kippers *are small, rather moth-like butterflies, so called because of their fast direct flight, 'skipping' over the vegetation. Adult skippers can be told by their large-headed appearance, the well-separated antennae and the short, dumpy wings. The larvae live within spun leaves of their foodplant. There is just one Irish representative of this family, the Dingy Skipper, compared to eight in Britain.*

Dingy Skipper
Erynnis tages (Linnaeus) SOCC (P) Category two.
Notable. Only certainly recorded from the limestone regions of Fermanagh. A minimum of sixteen colonies has been documented, mainly in the north-west of the county. The colony at Monawilkin has been the best documented and recorded. Other colonies in the Marlbank area, on Belmore Mountain, and on Knockninny have been visited less frequently and their current status is uncertain. A colony exists at Kinarla near Enniskillen and there are three more close to Belleek. East of Lower Lough Erne, it is known from just one site, Crockanaver, but it has not been seen here since 1987. Colonies appear small as the peak counts of adults rarely exceed ten. Dingy Skipper is a small, rather moth-like butterfly which can be confused with day-flying moths, particularly Mother Shipton and Burnet Companion, which share its habitat. Adults are soberly coloured with mottled brown and grey wings. They only fly during sunshine, moving fast and low over the ground and so are hard to track. They perch frequently to bask on bare ground where, likewise, they can be difficult to detect. Colonies are found in

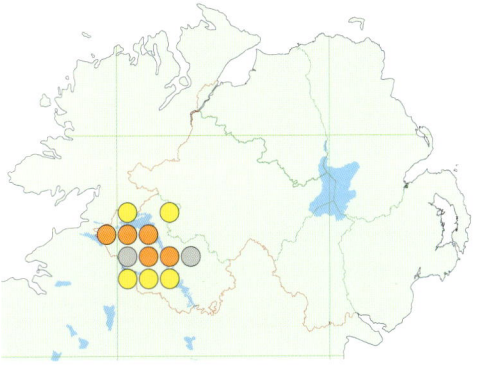

Preceding page
Green-veined White.

Clare Glen, County Armagh.

Opposite
Common Blue.

Thompson's Quarry, County Armagh.

Left
Dingy Skipper.

Monawilkin, County Fermanagh.

sunny and sheltered areas of unimproved limestone grassland, road cuttings and in abandoned quarries. One of the Belleek sites is on limestone waste beside the River Erne. The recorded flight period is 3 May to 7 July, peaking around the beginning of June. The larval foodplant is Bird's-foot Trefoil. It overwinters as a fully-grown larva.

PIERIDAE

Most members of this family of medium to large butterflies are white or yellow, often with black markings. Of the seven species found in Northern Ireland, five are resident, one is a migrant and one a vagrant. There is an additional species resident elsewhere in Ireland, the Wood White Leptidea sinapis. *Several other species are occasional vagrants to Britain and Ireland. The life history of the species varies considerably. The Large and Small Whites are regarded as pest species of cultivated cabbages and are closely associated with farmland and gardens.*

Réal's Wood White

Leptidea reali Reissinger SOCC (P) Category three G.
Widespread. There are two species of 'wood white' butterfly in Ireland, *Leptidea reali* and *L. sinapis*. They cannot be distinguished in the field and certain identification requires examination of the genitalia. Only Réal's Wood White has been confirmed in Northern Ireland whilst the other species is confined to Clare and Galway. The map shows the distribution of all 'wood white' records in Northern Ireland. The vast majority are unconfirmed but are presumed to be Réal's Wood White. The species is most common in southern counties Armagh and Down and the west and north of Fermanagh. It is much more local in the north in Antrim, Londonderry and Tyrone. Adults have rounded wings with grey wing tips and conspicuously white-tipped antennae. The underside of the hindwing has a greenish tinge with variable patterning of darker patches. They fly with loose, slow flaps and perch distinctively with the wings closed. Despite its name this species is found in open, flower-rich grassland such as on dunes, in abandoned quarries and limestone grassland. Patches of scrub are a feature of many of its sites. The recorded flight period is 9 April to 21 September but most are in May and June. The records in August and September indicate that occasionally there is a partial second generation. The larvae feed during June and July on Meadow Vetchling and Bird's-foot Trefoil. It overwinters as a pupa.

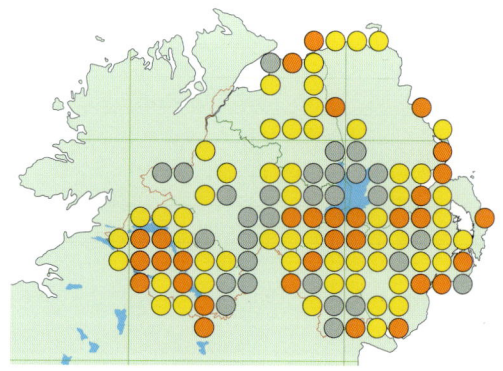

Clouded Yellow

Colias croceus (Geoffroy) Category three.
Widespread. Since 1990 Clouded Yellow has been of almost annual occurrence in Northern Ireland, but formerly it was a rare visitor. Up to the mid-1980s adults were reported on average only once every eight years. In most years only a few are seen but many were recorded in 1998 and 2000. None were seen in 2001. It has been seen in all counties, though like all migrant species it is most commonly recorded in the south and east, especially on the coast of Down. Adults are distinctive and unlike any of our resident species. The uppersides are orange-yellow with wide black borders. The underside, which is always shown when it is perched, is yellow and green with a silver figure 8 in the centre of the hindwing. A proportion of the females are of f. *helice*

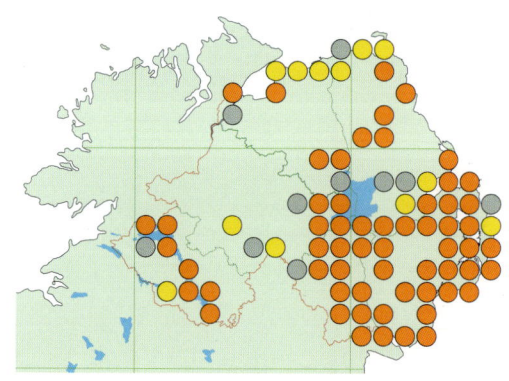

in which the yellow is replaced by creamy white, but these can still be easily told by the wide black borders. Clouded Yellow has been reported in every month between May and November. Most records have been in August and September. Numbers are greatest in years when migrants arrive in early summer as these will breed and produce a new generation of adults in late summer. It can be found in any open grassy habitat but tends to be seen in flower-rich habitats, especially on the coast. The larvae feed on clovers and related legumes.

Brimstone

Gonepteryx rhamni (Linnaeus) Category three.

Scarce. This species is a former resident and all the recent sightings are believed to have been wandering adults. Where these individuals originate is unknown, but as most of the accepted reports have been in the east, Britain is considered the most likely source. Up to the mid-1980s, Brimstone was resident in Fermanagh. The reasons why it became extinct there are unclear. The large adults are highly distinctive in colour, shape and behaviour. The wings are yellow or greenish without any black markings and the antennae are pink. Brimstone adults are large butterflies with a slow, floppy flight. They perch with their wings held tightly together and they bear a remarkable resemblance to a leaf. The adults, which are the longest lived of our

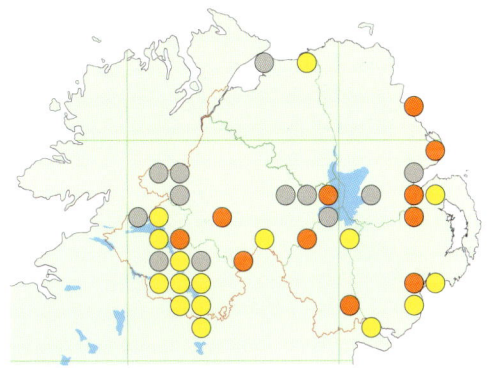

butterflies, emerge in late summer and quickly enter hibernation. They appear again in the following spring when they feed and search for a mate. Adults wander widely but are generally seen in scrubby woods, especially on lake shores and limestone pavement. When it was resident in Fermanagh, Brimstone could be seen on the shores and islands on Upper Lough Erne, especially around Crom. There were reports of larvae from Lower Lough Erne, but there is no evidence it ever bred elsewhere. The larval foodplants are species of buckthorn. In Northern Ireland the Purging Buckthorn is likely to be the sole foodplant, as the Alder Buckthorn is extremely rare and not present in Fermanagh.

Brimstone.

The Burren, County Clare.

This distinctive and charismatic butterfly was resident in County Fermanagh for a few years in the mid-1980s.

The Butterflies and Moths of Northern Ireland

Large White.

Banbridge, County Down.

Large White
Pieris brassicae (Linnaeus) Category one.

Ubiquitous. This is a resident species, but the local population is supplemented by immigrants and its abundance varies considerably from year to year. It is recorded from all lowland parts of Northern Ireland, but is most common in the south and east, reflecting the influence of migration on the local population and the availability of foodplants. In the north and west and in upland areas the species is much more local or completely absent. This is the largest of the three *Pieris* species in Northern Ireland. It can be told by the broad black margins to the wings and black spots on the undersides. Large White is a highly mobile non-colonial species which can be seen in virtually any habitat. The females have an uncanny ability to find even small patches of their foodplant. It is a frequent garden visitor but is most common in arable areas with crops of brassicas. Cultivated brassicas are the principal foodplants, but it will readily use Nasturtium. The yellow and black larvae feed communally and conspicuously on the foodplant. Adults can be seen between April and September. There are two generations a year which overlap in late June/early July. The second brood is larger than the first, especially in years when a large immigration has occurred. It overwinters as a pupa.

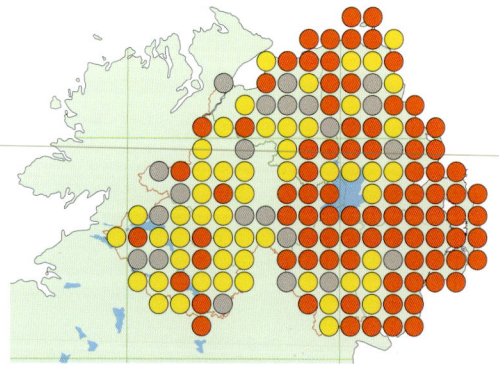

Small White
Pieris rapae (Linnaeus) Category one.

Ubiquitous. This is a resident butterfly but it is thought the local population is augmented by migrants. It is commonest in the south and east, reflecting its habitat requirements and partial migrant status. It is rare in upland areas and patchily distributed in much of the north and west. Small White has white forewings with black tips and a small, central black spot. The underside of the hindwing is a uniform yellow colour. It can easily be mistaken for Green-veined White, especially in flight. They are best distinguished by the patterning on the underside of the hindwing. This is a common butterfly of gardens, urban areas and farmland where the larval foodplants (cabbages and related crops) are cultivated. The larvae are green and well camouflaged and occur singly on the foodplant. Adults have been recorded in all months between March and October. There are two generations each year, one in spring and early summer and a second, usually more numerous one, in midsummer through to autumn. The spring brood peaks in late April and early May and the summer brood in late July and early August, and the two overlap in late June and early July. The species overwinters as a pupa.

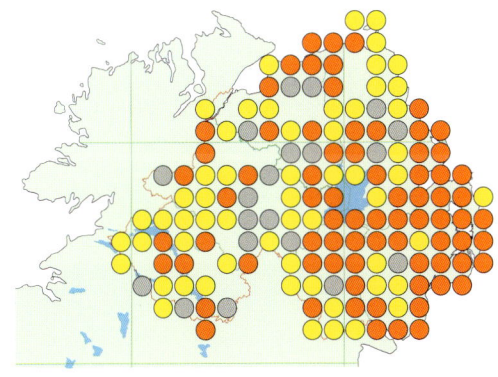

Green-veined White
Pieris napi (Linnaeus) Category one.
Ubiquitous. Green-veined White is perhaps the commonest butterfly in Northern Ireland. It certainly is the most widely distributed and absent only from the highest ground, heavily built-up areas and dense woodland. It is common on Rathlin Island (Antrim). The adults are medium-sized, mainly white with black wing tips. It is very similar to Small White from which it is distinguished by the green mottling to the veins on the underside of the hindwing. The extent of the black mark on the apex of the forewing also differs in the two species. In Green-veined White, the black tip is as extensive on the costal as on the side margin. In Small White, the black extends more along the costal margin. The female Orange Tip is also similar, but the green mottling on the underside of the hindwing is more extensive and not confined to the veins. The adults can be seen in most habitats in Northern Ireland, including gardens. The largest populations are found in damp grassland. Adults have been recorded in all months between March and October. In lowland sites there are two broods each year, but upland colonies may have a single generation in midsummer. The spring brood peaks in late April and early May and the summer brood in late July and early August. The larvae feed on the leaves of small crucifers including Cuckoo-flower. It overwinters as a pupa.

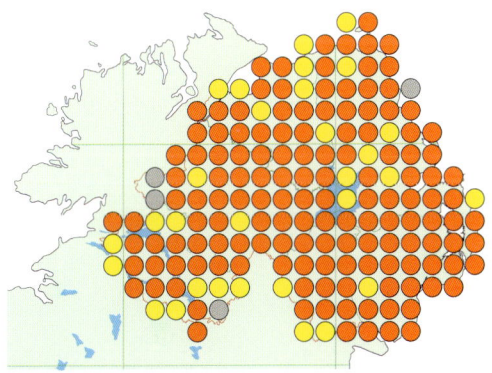

Orange Tip
Anthocaris cardamines (Linnaeus) Category one.
Ubiquitous. Present in most lowland parts of Northern Ireland. The only gaps in the distribution are in upland areas. It has been recorded on Rathlin Island (Antrim) and Copeland Bird Observatory (Down). The males with their bright orange and black tips to the forewings are unmistakable and their sighting is one of the signs of spring. Females are less conspicuous and may pass undetected amongst the more common Green-veined White. The marbling on the underside of the hindwing, readily seen on perched individuals, distinguishes both sexes from the other whites. Orange Tip occurs in open, uncultivated and usually damp habitats including grassland, open glades in woods, gardens, fens and lake shores. Adults have been recorded between

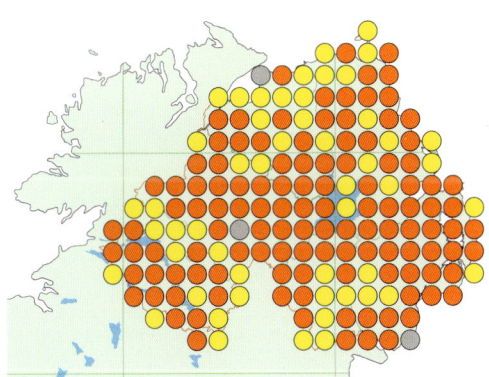

31 March and 19 July. The peak period for sightings is late April to the end of May. The most frequently used foodplant in damp habitats is Cuckoo-flower, but it can utilise other foodplants in drier habitats, especially Garlic Mustard; however, the incidence of this is unknown in Northern Ireland. The eggs, which are probably the most easily found of any of the Irish butterflies, are laid singly on the flower stalk of the foodplant and turn bright orange a few days after they are laid. The larvae feed on the flowers and developing seed heads of the foodplant, rendering the species unable to tolerate intensive cutting and grazing.

Orange Tip.

Montiaghs Moss National Nature Reserve, County Antrim.

LYCAENIDAE

Members of this family are small to medium-sized butterflies. There are six species resident in Northern Ireland and there is one other species on the Irish list, Brown Hairstreak Thecla betulae, but it has never been seen here. The 'blues' and 'coppers' are characterised by their vibrant, often metallic colouring. Hairstreaks are more sombre, but many species have patches of bright colours. Most species in this family exhibit sexual dimorphism. Many have eyespots on the wings, often associated with a short tail on the edge of the hindwing. The larvae are small and dumpy and most live a solitary life. An association with ants is apparent in many lycaenids, but none of our species have an obligate relationship with ants.

Green Hairstreak
Callophrys rubi (Linnaeus) SOCC Category two.

Ubiquitous. Recorded in all counties but absent from much of east Antrim, central Down, central Armagh and east Tyrone. It is probably commonest in the marginal land in the Sperrins, the Antrim hills and the remnant areas of bog in north Armagh. This is a small, inconspicuous but unmistakable butterfly which can easily be overlooked. The upperside is a uniform red brown, but this is only visible in flight as the wings are always held tightly closed when the insect is perched. The underside is bright green with a variable number of small white spots forming a fine line across both wings. This camouflages the butterfly superbly when it perches amongst the young leaves of birch and other plants. It is found on sheltered lowland raised 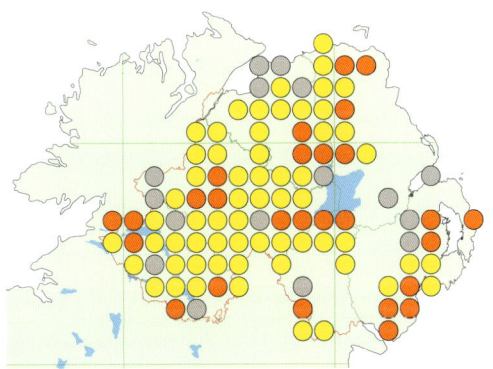 bogs, heathland and areas of moorland with patches of scrub and scattered trees. The males are territorial, defending a prominent perch on a bush from rivals, behaviour that gets the butterfly noticed. There is a single generation annually from late spring to midsummer. The recorded flight period is 5 April to 5 July. Throughout its range the Green Hairstreak utilises many different foodplants; Gorse and Bilberry are probably the most important in Northern Ireland. It overwinters as a pupa.

Purple Hairstreak
Neozephyrus quercus (Linnaeus) Category two.

Rare. Confined to a few localities in Fermanagh and Londonderry. A record from Gosford Forest (Armagh) has never been confirmed. Otherwise unknown from other counties, but undetected colonies may exist, especially in Tyrone. The species was first reported in Northern Ireland in 1893 'near Enniskillen'. It was not reported again until 1983 when it was found at Reilly Wood, part of the Crom estate (Fermanagh). It has since been found to occur throughout the extensive woodlands on this National Trust property. In north Fermanagh it was reported from Correl Glen in 1988, although it has not been seen since. A sighting on Inish Doney in 1995, a wooded island on Lower Lough Erne (Fermanagh), has similarly not been repeated. In 1997 it was 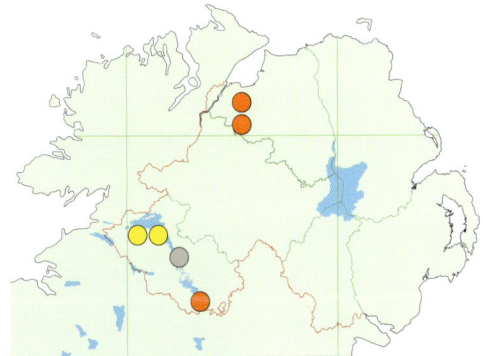 discovered at Ness Wood (Londonderry) and subsequently found in two nearby woods. Adults are dark above and light grey beneath. The male has glossy, purple patches on both wings, the female only on the forewing. Purple Hairstreak is found only in mature oak woodland. It is an elusive butterfly as it flies only during sunshine and mainly stays around the crowns of the trees.

Green Hairstreak.

Brackagh Moss National Nature Reserve, County Armagh.

An attractive, often overlooked species of bogs and heaths. Males are territorial, defending the airspace around prominent bushes and small trees.

Purple Hairstreak.

Crom, County Fermanagh.

The most elusive of our resident butterflies and seldom seen at close range.

However, females may descend lower, especially in the late afternoon, when they can be seen on Bramble and the lower branches of trees. The recorded flight period is 15 July to 5 September. The larvae feed on the young leaves of oak. The egg is the overwintering stage.

Small Copper
Lycaena phlaeas (Linnaeus) Category one.

Ubiquitous. This is found throughout Northern Ireland, although it is scarce in the north and west and in upland areas. It has been recorded once on both Rathlin Island (Antrim) and Copeland Bird Observatory (Down). Adults are small but unmistakable and easily visible as they are fond of basking on low vegetation with their wings held open. Males and females are similarly marked with brilliant copper forewings with black edges. The hindwings are grey with a copper band close to the rear edge. The underside of the forewing is orange with black spots. Small Copper is found in many open habitats including unimproved grassland, the drier margins of bogs and fens, dunes and, rarely, gardens. It prefers sunny, sheltered localities. The numbers seen 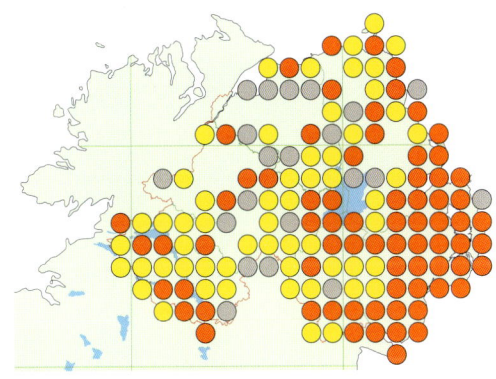 at colonies tend to be small. Males especially are very active insects, rarely stationary for any length of time. They are territorial and will defend the airspace around a perch against rivals and other insects. This is a double-brooded species. The first brood peaks in mid-May to early June, the second, usually more numerous one, in mid- to late August. The recorded flight period is 17 April to 21 October. The principal foodplants are species of sorrel and mainly Sheep's Sorrel and Common Sorrel. It overwinters as a larva.

Small Blue
Cupido minimus (Fuessly) SOCC (P) Category four.

Rare. This is the rarest resident butterfly in Northern Ireland. It has been long confined to just a single colony at Monawilkin (Fermanagh), but it is feared that it may even be extinct here as the last adult sighting was in June 2001. A second Fermanagh site was lost in the 1970s. The only other Northern Ireland populations existed on the north shore of Belfast Lough and on Islandmagee (Antrim), but these became extinct for unknown reasons around 1900. Records from Armagh and Down are considered erroneous. The Monawilkin colony was discovered in the 1970s and adults were recorded annually between 1980 and 2001. Small Blue is the smallest Irish butterfly and easily identified. The uppersides are slate-blue or dark brown with a complete white fringe.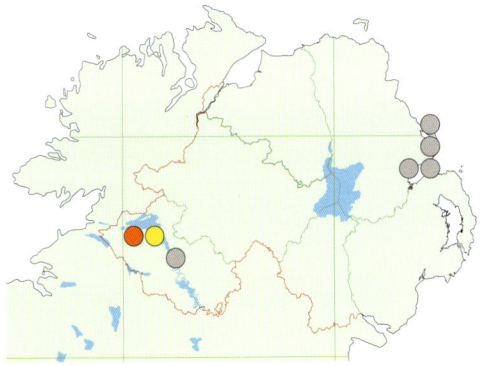
The hindwings are silvery-grey with fine black dots. The Monawilkin colony was restricted to a south-facing area of short calcareous grassland. Elsewhere in Ireland colonies exist on limestone pavement, abandoned quarries and dunes. The recorded flight period in Northern Ireland is 17 May to 18 July with the peak period in the first three weeks of June. The larvae feed on the flowers and seed heads of Kidney Vetch. The overwintering stage is the pupa.

Common Blue
Polyommatus icarus (Rottemburg) Category one.

Ubiquitous. Common Blue is easily the most widespread and abundant blue butterfly in the region. Colonies can be found around the entire coastline and on Rathlin Island (Antrim). Inland it is much more locally distributed and only found in areas with good drainage and thin

The Butterflies and Moths of Northern Ireland

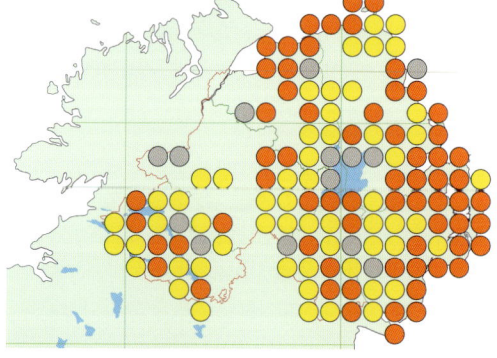

Common Blue.

Thompson's Quarry, County Armagh.

The males of this species are easily recognised in flight, with their bright, iridescent blue forewings.

soils such as on the basalt and chalk in Antrim, the limestone of Armagh and Fermanagh and the granite in the Mournes. It is absent from large parts of Antrim, Londonderry and Tyrone with cold and wet clay or peaty soils. Adult males have brilliant blue uppersides with conspicuous white outer fringes. The underside is mainly brown with black-centred white spots and orange lunules. Females are browner than the males on the uppersides, although some may be mainly blue, but all have a row of orange spots on the outer edge of both wings. This species is found in open grassy habitats that are dry and sunny. The largest populations are found on coastal sites on dunes and rocky coasts. Inland it is found on well-drained sites on scarps and abandoned quarries. Adults have been reported in Northern Ireland in all months between May and October. Most colonies are considered to have just a single generation of adults, but the late records indicate that a partial second generation is occasionally produced. The main foodplant is Bird's-foot Trefoil. A few colonies exist on the fringes of wetlands and these presumably use Large Bird's-foot Trefoil.

Holly Blue
Celastrina argiolus (Linnaeus) Category two.
Scarce. This has been recorded in all counties except Armagh. Most sites are in Down, in an area bounded by Comber, Carryduff and Rostrevor. Within this area it has been recorded in many places, but is especially frequent around Saintfield, between Newcastle and Castlewellan and inland of Rostrevor. Elsewhere in Northern Ireland, the species has been recorded only sporadically and mainly on single occasions. The one exception is in the Garrison and Correl Glen area in west Fermanagh where it is clearly resident. This is a bright blue butterfly with silvery-blue undersides with small dark spots. The fringes of the wings are chequered black and

Holly Blue.

Rostrevor Wood National Nature Reserve, County Down.

When weather conditions are favourable, adults can be seen fluttering around mature Holly trees in the more open parts of woodland.

white. Females have a wide dark border on the outer edge of the forewings. The male resembles the male Common Blue but can be told by the lack of orange spots and the black chequers on the white fringe. Holly Blue is found in open woods, parks and large gardens. In southern Britain and southern Ireland this species can have two or even three generations a year. In Northern Ireland there is evidence that a second brood is developing, but it is still overwhelmingly single-brooded. First-brood adults have been recorded between 9 April and 26 June, but sightings are unusual after late May. There have been four records in August of second-brood adults, all but one since 2000. The larvae of the first brood feed on the flowers and developing fruits of female Holly trees. The second brood develops on the flowers of Ivy. It overwinters as a pupa.

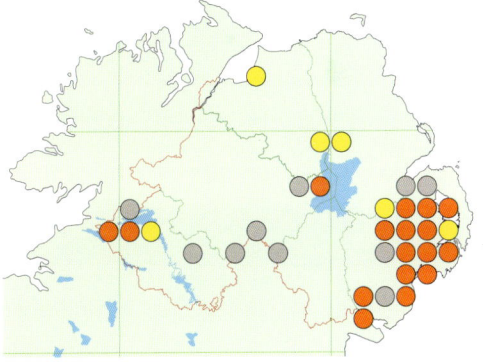

NYMPHALIDAE

This large family of medium to large butterflies contains some of the most familiar resident and migrant butterflies. In the latest checklist the former families of the Satyridae ('browns') and the Danaidae ('monarchs') are now included with the Nymphalidae. The predominant colours of the adults are browns, reds and oranges. Many of the 'browns' have eyespots on the underside of the hindwing or upperside of the forewing. The habits and life cycle of the Irish species are very varied. Sixteen of the 21 species on the Irish list have been recorded in Northern Ireland. Two, Pearl-bordered Fritillary Boloria euphrosyne *and Gatekeeper* Pyronia tithonus, *are resident. Mountain Ringlet* Erebia epiphron *is extinct and Queen of Spain Fritillary* Issoria lathonia *and American Painted Lady* Vanessa virginiensis *are both rare vagrants.*

The Butterflies and Moths of Northern Ireland

Red Admiral.

Banbridge, County Down.

The missing section on the hindwing is probably the result of an unsuccessful bird attack.

Red Admiral
Vanessa atalanta (Linnaeus) Category one.

Ubiquitous. This is the most commonly and regularly occurring of the migrant butterflies in Northern Ireland and there are few areas where it has not been seen. The numbers reported each year vary enormously due to the interaction of weather patterns and the success of breeding in southern Europe. The largest numbers are usually seen in years when there is a substantial arrival in early summer as these will breed and produce a new generation of adults in August and September. These may be augmented throughout the summer by further arrival of immigrants when weather conditions are favourable. The large adults, with their striking red bands and white spotting on the velvety black wings, are unmistakable. Adults can be seen anywhere, but are especially drawn to gardens and other sources of nectar-rich flowers. The flowers of Buddleia and the flowers and fruit of Bramble are amongst the most highly sought-after sources of nourishment. In autumn, adults will commonly feed on fruit. The adults have been seen in every month of the year. The adults hibernate, but winters here are not conducive to this and successful overwintering this far north is exceptional. There is increasing evidence that adults in northern Europe move south as summer declines, but each year's adults are the result of fresh influxes. The larval foodplant is Common Nettle.

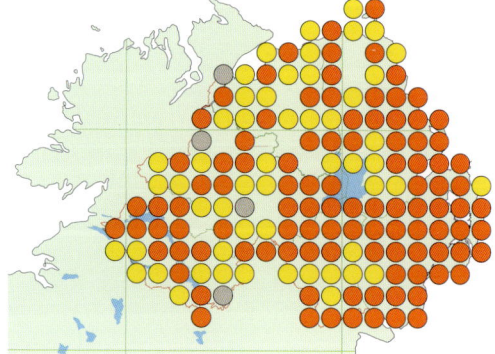

Painted Lady
Vanessa cardui (Linnaeus) Category one.

Ubiquitous. This is an irregular, occasionally common, migrant from southern Europe. It has been recorded in all parts of Northern Ireland. Most sightings have been in late summer and early autumn, but there have been many reports of adults in spring and early summer. Successful breeding only occurs in Northern Ireland following these early influxes. Painted Lady is a large, long-winged butterfly. The upperside is mainly orange, but fresh individuals are often salmon pink. The tips of the forewings are black with conspicuous white spots. This is a migrant, non-colonial species and it can be seen in any habitat. Adults tend to congregate in areas with abundant sources of nectar, often in the company of Small Tortoiseshell and Red Admiral.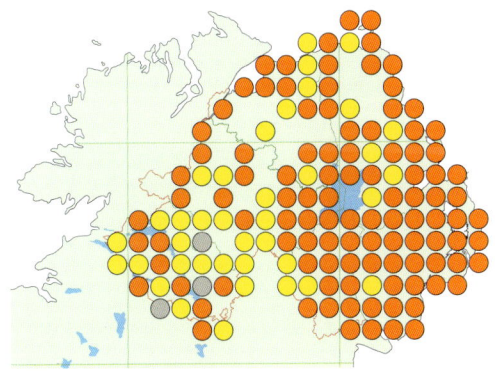
It is particularly attracted to patches of flowering thistles, so it is often seen on weedy ground both inland and on the coast and in derelict urban sites. Adults have been recorded in all months in Northern Ireland between March and November, but the peak period for sightings is in August and September. The larvae feed on species of thistle. Adult Painted Lady do not hibernate and cannot overwinter in northern Europe.

Small Tortoiseshell
Aglais urticae (Linnaeus) Category one.

Ubiquitous. The far-ranging adults of this well-known species have been recorded in virtually all of Northern Ireland. There are no large gaps in the distribution apparent even on fine-scale maps; however, the abundance of this species fluctuates greatly from year to year. This is perhaps the most familiar of our resident butterflies and the adults are easily recognised by the patterning of yellow, orange, black and white on the upper wings. They can be found in most habitats but are most commonly seen in open locations with concentrations of nectar-rich flowers, including gardens, or around beds of the foodplant, Common Nettle. The adults are long-lived with many surviving for eight months and they have been recorded on the wing in every month of the year.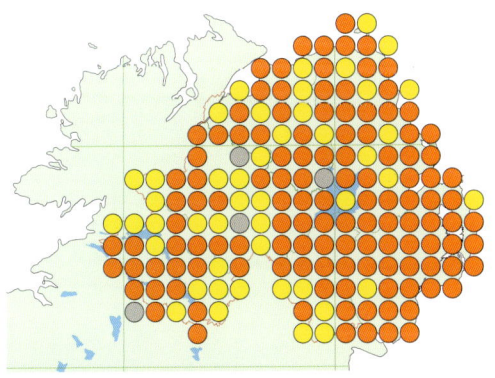
The new generation of adults emerges in July and the largest numbers are usually seen each year in late summer. These feed for a while, then enter hibernation. They often enter houses and outbuildings during this period as they seek a dark, cool place to sit out the winter. Adults emerge the following spring, although some appear precociously in mild winter spells. In spring and early summer, the adults mate and the females lay batches of eggs on Common Nettle. Most of the population in Northern Ireland has a single annual generation, but in some years a partial second brood produces adults in early autumn.

Camberwell Beauty
Nymphalis antiopa (Linnaeus) Category four.

Rare. This has been seen on only a few occasions in Northern Ireland. There have been sightings in every county except Fermanagh. The most recent records were in 2000 when singletons were seen on 17 June at Killard Point and 27 June at Murlough NNR on the south Down coast. One was seen in a small urban garden in south Belfast (Antrim) on 11 August 1997. In 1995 – a year when there was a large influx into eastern Britain – three were reported between 4 and 12 August at Holywood (Down), Dunmurry (Antrim) and Oxford Island

(Armagh). A fourth individual was seen during August at Bessbrook (Armagh). There are five earlier accepted records from Traad Point (Londonderry) in 1975; Stormont (Down) on 4 August 1969; Cookstown (Tyrone) in 1902; Greencastle (Antrim) in 1872 and Trillick (Tyrone) sometime before 1894, but the exact date is unknown. Camberwell Beauty is found in eastern and northern Europe. The pattern and timing of influxes to Britain and Ireland reflect the range and phenology of the species. Most individuals have been seen in late summer and on the eastern side of both islands, nearest to the continental populations. This was clearly evident in the pattern of records following the large influx of 1995. This is a large majestic butterfly and the adults are unmistakable, with large burgundy wings edged in creamy yellow. The larvae feed on species of willow and poplar and the adults overwinter.

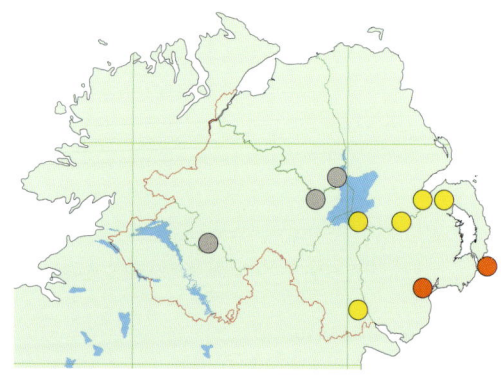

Peacock

Inachis io (Linnaeus) Category one.
Ubiquitous. Recorded in all parts of Northern Ireland, but the range and abundance have fluctuated considerably over the last 25 years. In periods of rarity, such as during the early 1990s, the distribution contracts to western counties, particularly Fermanagh. Since the mid-1990s the species has reoccupied much of its range in the east. This is a large butterfly with unmistakable markings. There are four prominent eyes, one on each wing, which give it its common name. The underside is very dark and subtly patterned. Peacock can turn up in any habitat but, in spring, they are especially frequent along the edge of woodland and in damp rides and clearings. Adults of the new generation are less tied to woodland and wander widely in search of sources of nectar, so they are often attracted into gardens in late summer. They are frequent visitors to Buddleia bushes, but the largest numbers are seen on Devil's-bit Scabious and Water Mint on the margins of bogs and fens or in woodland rides. Adults hibernate and can be seen virtually all year, but there are few records in the winter months from November to February. The larval foodplant is Common Nettle. Females preferentially select stands growing in woodland.

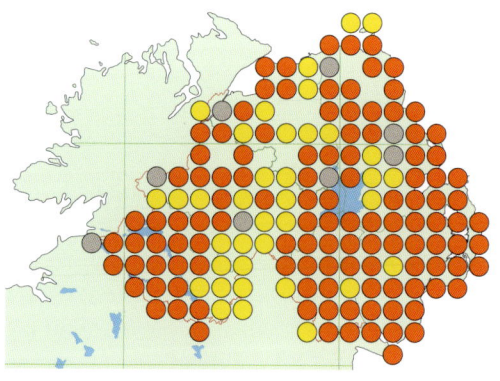

Comma

Polygonia c-album (Linnaeus) Category four.
Rare. There have been two confirmed reports of the Comma in Northern Ireland in 1997 and 1998, both from the Portaferry area on the southern tip of the Ards Peninsula (Down). It has not been reliably reported since, but there have been genuine records from elsewhere in Ireland. It is considered the sightings are wandering adults from the expanding populations in northern Britain and the possible forerunners of full colonization. Given adequate views, this butterfly (which is noticeably larger than a Small Tortoiseshell) is unmistakable on account of the orange and black patterned wings which have a ragged appearance at the edges. The underside is dark with obscure patterning and a small white mark shaped like a comma. Like other nymphalids, the Comma is a wide-ranging non-colonial butterfly. It does tend to stay close to broadleaved woodland, the habitat in which it breeds and hibernates. The main larval

Peacock.

Brackagh Moss National Nature Reserve, County Armagh.

foodplant is Common Nettle, but in the past Hop was commonly used. Successful breeding has so far not been proven in Ireland. The species overwinters as an adult. In Britain, the Comma is single-brooded in the north although a partial second brood occurs frequently in southern England.

Dark Green Fritillary
Argynnis aglaja (Linnaeus) Category three.
Widespread. Colonies of this large butterfly can be seen round the coast of Northern Ireland and on Rathlin Island (Antrim). It is much more local inland, but extant populations still exist in the Mourne Mountains (Down) and on Slieve Gullion (Armagh). There have also been records of, but no conclusive evidence for, established colonies from the Belfast Hills, east Tyrone, north Armagh and the hills of west Fermanagh. The most recent record from any of these areas was in 1996. The coastal colonies appear stable, with no reported losses in recent years. This is a large orange butterfly with a chequered pattern of black lines and spots on the upper side. The outer margins of the forewing are straight or, especially in the female, slightly convex. The

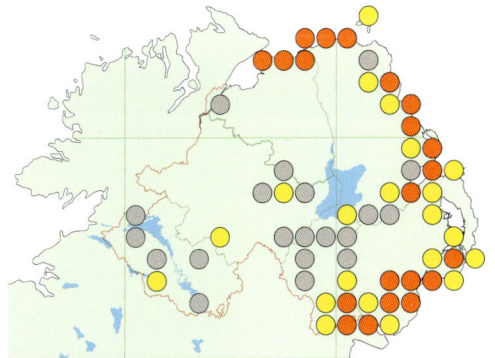

underside of the hindwing has a silvery-green background with silver spots. The only species it is liable to be confused with is the larger Silver-washed Fritillary. The two species rarely overlap in habitat, but can be told by the different underwing pattern and the shape of the wings. Dark Green Fritillary is found on dunes and on coastal and inland heaths. The Fermanagh records were from limestone grassland. Adults are powerful fliers, ranging widely over their habitat and occasionally visiting flowers like Wild Thyme for nectar. The flight period is 13 June to 9 September. The larval foodplants are species of violets such as Marsh Violet growing in short vegetation on open ground. It overwinters as an early instar larva.

The Butterflies and Moths of Northern Ireland

Dark Green Fritillary.

Murlough National Nature Reserve, County Down.

Adults of this colourful species can often be seen basking on bare sand in early morning sunlight, or feeding at flowers, particularly Wild Thyme.

Silver-washed Fritillary.

Rostrevor Wood National Nature Reserve, County Down.

This large, distinctive butterfly is frequently observed in midsummer, feeding on flowers in dappled sunlight along woodland rides and glades.

Silver-washed Fritillary
Argynnis paphia (Linnaeus) Category three.

Widespread. This, the largest of our resident butterflies, has been recorded in all parts of Northern Ireland where suitable habitat exists. It is commonest in west Fermanagh, in the woodland on the fringes of the Mourne Mountains (Down) and in the Glens of Antrim, especially Glenarm. It is absent from upland, treeless areas and offshore islands. This is a large, very striking butterfly with orange wings that are marked with black spots and lines. The outer margin of the forewing is concave. The underside of the hindwing is green with a pattern of silvery marks. It is closely tied to woodland, especially mature broadleaved woods with sunny rides and glades. However, the adults do wander and may appear in other habitats and, rarely, gardens. The 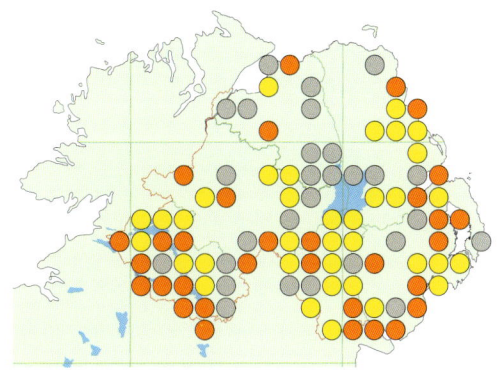 largest numbers are seen in sunny glades where it can find food, particularly flowering patches of Bramble. There is a single generation annually, flying from midsummer. The recorded flight period is 22 June to 15 September. The larval foodplants are woodland species of violet. The female lays her eggs on the trunks of trees adjacent to large clumps of the foodplant. The larvae hatch after a few weeks, then hibernate.

Marsh Fritillary
Euphydryas aurinia (Rottemburg) SOCC (P) Category three.

Scarce. This has been recorded from all counties, but few of the records refer to colonies. Most of the confirmed breeding records have come from sites in east and south Down. There has been no instance of successful breeding since the early 1990s from anywhere in Armagh, Londonderry or Tyrone. The last confirmed breeding record was in the early 1990s. The latest survey found that there were eight colonies in Down, two in Antrim and one in Fermanagh. This is a medium-sized butterfly with a variable pattern of bright orange, yellow and white chequers on a dark brown background. Males and females differ only in size, the latter being the larger. Colonies have been found on the drier margins of cutover bogs and fens, in wet heaths and on sand 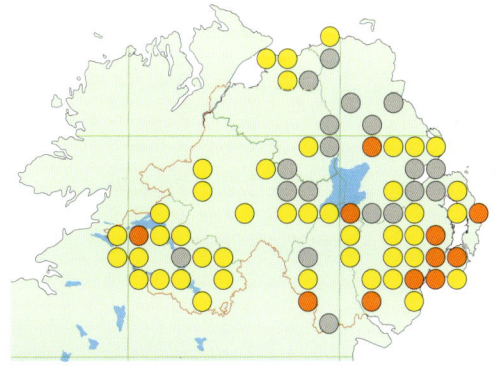 dunes. Wandering adults have appeared in other grassland habitats. Adults are usually seen basking on the dead bleached leaves of grasses or flying for a short distance low over the ground. They visit flowers only occasionally. There is just a single annual generation in late spring and early summer. Adults have been recorded between 14 May and 22 July and the peak period for sightings is early to mid-June. The only larval foodplant in Northern Ireland is Devil's-bit Scabious. The black, spiky larvae live in a communal web spun over the foodplant. These webs are very conspicuous and easily found in late summer and provide an easy method of both confirming breeding and monitoring each population. It overwinters as a third instar larva in a specially constructed web, low down at the base of the foodplant.

Speckled Wood
Pararge aegeria (Linnaeus) Category one.

Ubiquitous. Recorded from all counties. There are few large gaps apparent in the distribution, but, like many butterflies, it is less frequently recorded from the north and west. It has been recorded a number of times on Copeland Island and Copeland Bird Observatory (Down), but only once in the 1970s on Rathlin Island (Antrim). The adults are distinctive and easily recognised. The upper wings are brown with creamy spots, some containing white-pupilled black eyespots. The wings are narrowly edged in white. The undersides are more intricately patterned in shades of brown and cream. Speckled Wood occurs in woodland and scrub, along hedges, parks and large gardens. It tolerates more shade than any other butterfly in the region. Adults are most often seen perched in patches of sunlight on the ground or foliage. It is an infrequent flower visitor, preferring to feed on honeydew in the crowns of trees. The life cycle is complex, involving a series of overlapping generations. The adults can be seen continuously from March to early November, with the greatest numbers usually in late summer and early autumn. The larvae feed on grasses growing at the edge of woodland. It overwinters either as a larva or a pupa and this variation results in the complicated pattern of broods.

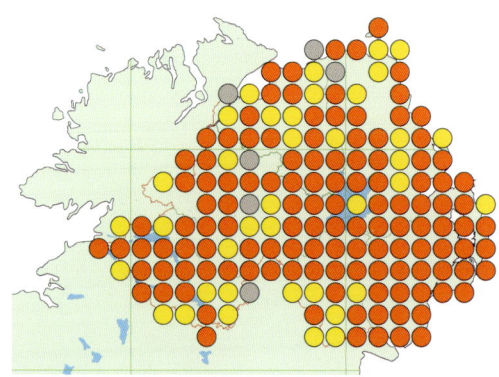

Wall

Lasiommata megera (Linnaeus) SOCC (P) Category two.
Scarce. Historically widespread, but following a major range contraction the Wall is currently scarce and, since 2000, not reported reliably away from the coast of Down. It has been recorded from all counties, but it was always scarce and local in Fermanagh and Tyrone. The last records from these counties were in 1982 and 1997, respectively. In Londonderry it was mainly found on the coast around Castlerock where it was last reported in 1997. Similarly, in Antrim the distribution was overwhelmingly coastal. It was lost from sites on the north coast in the early 1990s and further south in Belfast Lough and Islandmagee in 2000. It was formerly present throughout Armagh, but the last record was in 2000 from a quarry at Carganamuck. A major decline has been evident in inland parts of Down and, since 2000, most reports have come from the coast between Killard Point and Ardglass. It also persists at a few places on the coast between Newcastle and Kilkeel. The easily recognised adults are brown with bright orange patches on the upper wings and the underside of the forewing. The underside of the hindwing has a series of eyespots and is intricately patterned in greys and browns. Adults characteristically perch on bare ground and rocks, holding their wings flat to the ground. Wall is found in warm, sunny sites with sparse vegetation and areas of bare ground. It is found especially on low rocky coasts and similar places inland, including abandoned quarries. There are two generations a year, one from the end of April to mid-June and the second from July to September, rarely October. The recorded flight period is 20 April to 30 October. The larvae feed on fine-leaved grasses. It overwinters as a larva.

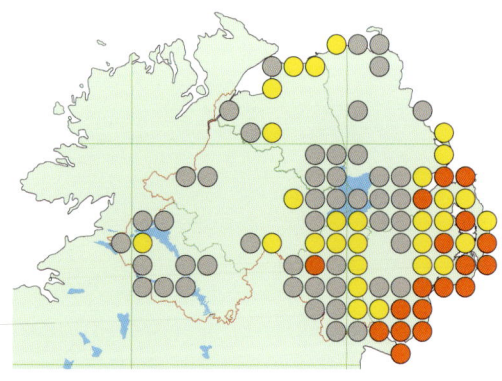

Grayling

Hipparchia semele (Linnaeus) Category three.
Widespread. The Grayling is found around the coast of Northern Ireland, in the Mourne Mountains (Down) and in west Fermanagh. It is also found on Rathlin Island at Kebble (Antrim). On the north coast it is present on most areas of dune and rocky coastline between Magilligan (Londonderry) and Fair Head (Antrim). In east Antrim there are colonies at Torr Head, and between Garron Point and Ballygalley. It also is present on Muck Island, off Islandmagee. On the north shore of Belfast Lough it is recorded from Knockagh and Cave Hill. It is absent from most of north Down and the Ards Peninsula, but is common on coastal sites in south Down between

Butterfly accounts

Wall.

Killard Point National Nature Reserve, County Down.

Once widespread inland, the Wall has declined dramatically and is now confined to a few sites on the coast of County Down.

Grayling.

Murlough National Nature Reserve, County Down.

This species regularly rests on bare ground or rock, with its wings angled to reduce its shadow.

53

Killard Point and Ardglass, and at the Murlough NNR dunes. There are also colonies on the low-lying heaths on the eastern edge of the Mourne Mountains. In Fermanagh it is found on the limestone hills in the west of the county as far south as Knockninny. In behaviour this is a distinctive species, but it appears rather drab. The adults rest frequently on the ground, always with their wings held closed so that only the underside of the hindwing is visible. This is intricately patterned in greys and brown that camouflage the butterfly to its background. The upperside is grey with lighter brown bands. The underside of the forewing has a large orange patch that appears like a flash as the butterfly takes off. Grayling can be found in dry grassy habitat where the vegetation is sparse with areas of bare ground including dunes, heaths, limestone outcrops. Sites are always sunny and well drained. The flight period is from 22 June to 26 September, peaking in August. The larvae feed on fine-leaved grasses. It overwinters as an early instar larva.

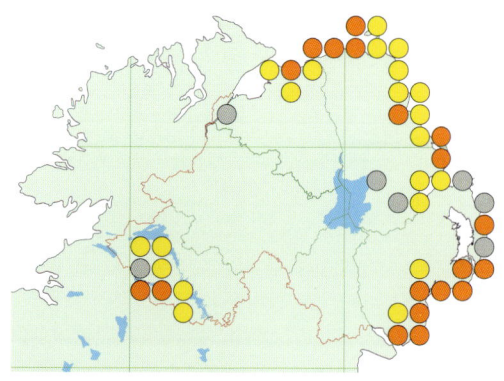

Meadow Brown

Maniola jurtina (Linnaeus) Category one.

Ubiquitous. The Meadow Brown is one of the commonest butterflies in Northern Ireland. It is generally distributed in all counties and it is widespread also on Rathlin Island (Antrim). The species can be very abundant and large patches of suitable habitat may support colonies producing hundreds or thousands of adults. Adult males are brown above with orange patches on the underwing. The female is larger and lighter-coloured than the male and, additionally, has a large orange patch on the upperside of the forewing. This is a grassland butterfly, preferring dry, unimproved types. It is absent from improved agricultural land and other intensively managed sites. Colonies may be found along wide woodland rides, but the species is absent from heavily-shaded woods, upland areas above 300m and wetland habitats. Adults generally remain within their colonies, flying, when conditions are suitable, with lazy flaps just above the vegetation. They are also attracted to flowers especially thistles, Knapweed and Bramble. There is a single generation a year from June to August. A few early individuals have been seen in May and a few adults may persist to October or early November in large, coastal colonies. The larvae feed on grasses. It overwinters as an early instar larva.

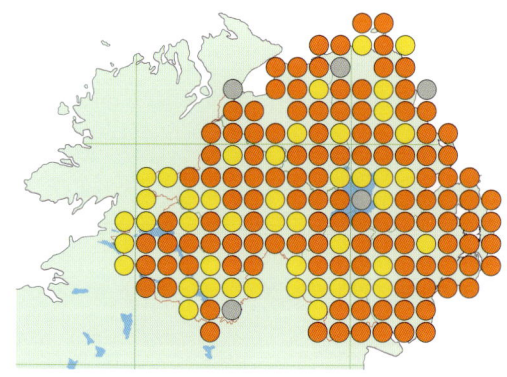

Ringlet

Aphantopus hyperantus (Linnaeus) Category one.

Ubiquitous. Recorded from all parts of Northern Ireland including Rathlin Island (Antrim) and absent only from the highest ground. Adult Ringlet are small and dark velvety brown above with a narrow white fringe to the wings. As the insect ages, the upperwing colour fades to a lighter brown. The underside is greyish-brown with a series of pale-centred, dark eyespots surrounded by a golden halo. The contrast between the dark upperside and the paler underside is conspicuous on a flying individual. It is possible to confuse this species only with Meadow Brown but, unlike that species, Ringlet is a smaller, darker butterfly that never shows any orange on the wings. Colonies of the Ringlet exist in open, grassy habitats and it

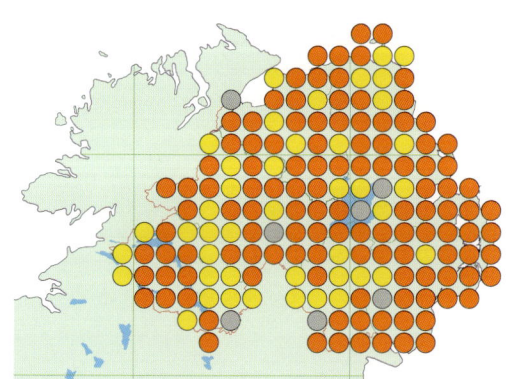

frequently co-exists with Meadow Brown. However, it prefers damper grassland than does the Meadow Brown. It can also be found in quite shady sites including lush grassy rides in woodland. Adults fly low over the vegetation usually in sunshine, but it is tolerant of cloudy and damp conditions as long as it is not too windy or cold. This is a midsummer butterfly with a single annual generation from mid-June to mid-August. There have been exceptional sightings in May and September. The larvae, which overwinter in an early instar, feed on grasses.

Small Heath
Coenonympha pamphilus (Linnaeus) Category one.
Ubiquitous. A widespread species recorded from all counties and on Rathlin Island (Antrim). In Britain this has undergone a severe recent decline, but the trend in Northern Ireland is unknown. This is a small butterfly with tawny upper wings. The underside of the forewing is tawny with a conspicuous black eyespot. The hindwing is a dusky grey-brown with a whitish patch. The wings are always held closed when the insect is perched. Large Heath is the only similar species, but the adults of the Small Heath are slightly smaller and brighter in appearance. Small Heath is found in short to medium height, dry, unimproved grassland and, here, the largest populations exist on the coast. It avoids very lush and rank grasslands and inland it is only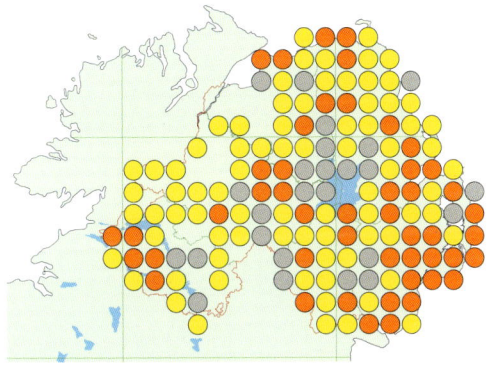
found on well-drained ground. Colonies can disappear quickly if sites are unmanaged and invaded by scrub. The recorded flight period is prolonged, from 29 April to 11 October. The peak of the flight period is between mid-May and the end of June, but in some large colonies, especially those on the coast, adults can be seen well into autumn. The phenology of this species in Britain and Ireland is complex. It is generally double-brooded in southern Britain and single-brooded in the north, but this is complicated by variation caused by larvae developing at different rates. At the monitored colonies of Killard Point and Murlough NNR (Down), two peaks in the flight period are apparent, but whether these represent separate broods is unclear. The larval foodplants are species of fine-leaved grasses. The species overwinters as an early instar larva.

Large Heath
Coenonympha tullia (Müller) Category three.
Scarce. Confined to the north and west of Northern Ireland from the north Antrim hills through the Sperrins in Londonderry and Tyrone to the north and west of Fermanagh. It has never been recorded in Down and the only Armagh population is found on Mullenakill Bog within Peatlands Park in the extreme north-west of the county. This is a readily identified butterfly which does not share its habitat with any other species. The upperside is a dull orange and the underside is grey with a distinctive, slightly greenish cast and cream and orange patches. Some individuals have white-pupilled, black eyespots on the underside of the hindwing. The adults always perch with their wings closed. The adults are very similar to the Small Heath, but they are generally larger and less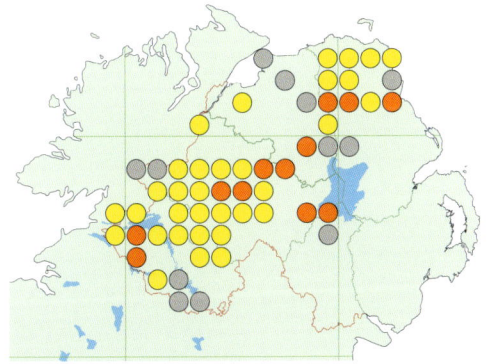
brightly coloured on the upperside. Large Heath is found on lowland raised bogs. A few colonies exist on blanket bogs, but it generally avoids large bleak expanses of moorland. It is also rare above 300m. The species has a single generation annually and the recorded flight period is

Large Heath.

Mullenakill National Nature Reserve, Peatlands Park, County Armagh.

29 May to 26 July, apart from two exceptional records in August and September. The foodplant is Hare's-tail Cottongrass, possibly other species. It overwinters as an early instar larva.

Monarch

Danaus plexippus (Linnaeus) Category four.

Rare. This North American and southern European butterfly has been seen five times in Northern Ireland. The first sighting was in Lurgan (Armagh) in July 1964. The other four were all recorded between 12 and 15 October 1995 in Belfast and east Down. This coincided with a well-documented and large influx of Monarch to western Britain and Ireland. The timing of the arrival, coincident both with the main southward migration of Monarchs along the eastern coast of North America and the eastward passage of a fast-moving Atlantic depression, strongly suggests these were blown across the Atlantic. A second large influx occurred in October 1999, but whilst some were seen along the south coast of Ireland, none were reported in Northern Ireland. The

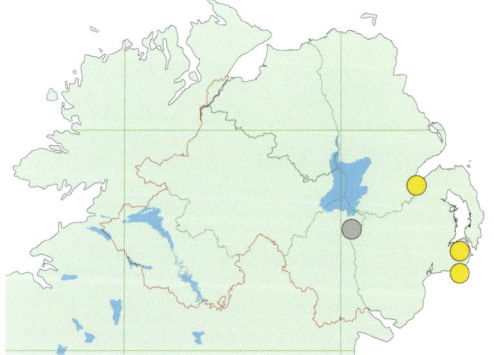

Monarch cannot be confused with any other Irish butterfly by virtue of its huge size and striking orange, black and white colouring. The species has colonised parts of southern Europe where its foodplants (species of milkweed) have been introduced, but there is no evidence that these populations are the source of the individuals seen in Britain and Ireland. As these plants cannot survive outside in Ireland, the species is only ever going to be a transient visitor. The possibility of individuals being escapees should be borne in mind, but autumn individuals like those in 1995 are most likely genuine migrants.

Moth accounts

HEPIALIDAE

The swifts are considered the most primitive of the Lepidoptera. Four of the five resident species in Britain and Ireland occur in Northern Ireland. Swift moths are readily identified by their elongated wings, short antennae and their characteristic resting posture of wings held tightly against the body. The adults lack a functional proboscis and are therefore unable to feed. Activity commences before dusk and continues after dark. Most species exhibit a weak attraction to light, both sexes appearing in small numbers in traps. Eggs are dispersed by the female as she flies low over the foodplant. The virtually hairless and unpigmented larvae are subterranean and feed on the roots of plants; pupation also takes place underground. Larval development is prolonged and most species have a two-year life cycle, with the larvae overwintering twice.

Preceding page
Wood Tiger.

Leitrim Lodge, Mourne Mountains, County Down.

Ghost Moth

Hepialus humuli (Linnaeus) Category one.

Widespread. A commonly recorded species in Antrim, Armagh, Down and south Fermanagh. It appears much more localised in the north and west although under-recording may account for its apparent absence from many areas. It has also been recorded from several sites on Rathlin Island (Antrim). The males are distinctive with their white forewings and characteristic flight behaviour commonly referred to as 'lekking'. This involves groups of males flying in a rhythmic manner, up and down, above long grassy vegetation at dusk on warm evenings. Ghost Moth requires areas of tall grassland and is found on embankments and wide verges, in rides and glades in woodland, coastal grassland and the margins of wetlands. Adults appear sparingly at light, the females more commonly than the males. The recorded flight period is 9 June to 20 August. The larva is subterranean and seldom seen, feeding on the roots of grasses and probably other low-growing herbaceous plants. Development probably takes two years, the larva overwintering twice.

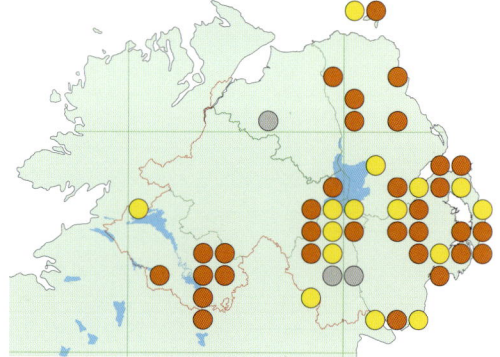

Gold Swift

Hepialus hecta (Linnaeus) Category one.

Widespread. Recorded from all counties but more frequently reported from north Armagh, Down, and parts of Fermanagh than elsewhere. Records are more scattered in Tyrone and the northern counties. It has been reported from Rathlin Island (Antrim), once, in the late 1980s. In 2003 it was found at Copeland Bird Observatory (Down). There have been recent records from Breen Wood, Glenariff, Lagan Meadows and Montiaghs Moss (Antrim); Hillsborough Forest and Killard Point (Down) and Peatlands Park (Armagh). Fermanagh and Tyrone localities include Altaveedan, Brookeborough, Crom and Monmurry. This is the smallest of the four swifts found in Northern Ireland. The males are brightly marked with golden-brown forewings and white markings which can vary in both size and shape. Females are slightly larger and much duller than the males. Gold Swift is mainly found where Bracken is common in lowland heaths and bogs, and on the margins of woods and patches of scrub, but it has also been recorded from some mature gardens where bracken does not occur. The males are frequently seen at dusk, often in large numbers, hovering above Bracken and tall grass. Paired adults can be found

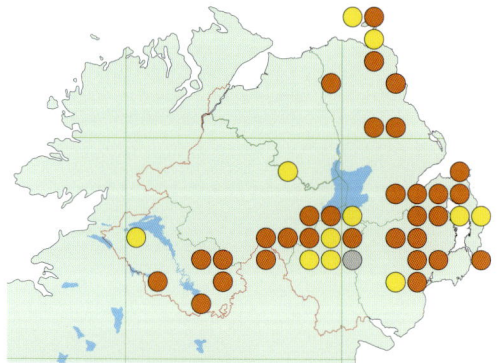

easily by searching the underside of the leaves and amongst low-growing vegetation. The recorded flight period is 19 May to 8 August. The larvae feed on the roots and inside the stems of Bracken and probably other low-growing plants which have not been identified in Northern Ireland. In Britain it is said to overwinter twice in the larval stage.

Above left
Ghost Moth (male).
The Argory, County Armagh.

Above right
Gold Swift.
Lackan Bog, County Down.

Common Swift

Hepialus lupulinus (Linnaeus) Category four.

Rare. Baynes described this species as being 'not generally abundant … but distributed from north to south.' The vernacular name of this species is inappropriate in Northern Ireland as it is by a considerable margin the rarest member of the family. There appear to have been only three reliable records since 1980 – two from Milford (Armagh) on 15 and 16 June 1984 and Hillsborough Forest (Down) on 10 June 1997 – and its present status needs to be carefully investigated. There are historical records for Belfast; Armagh City, Drumbanagher, Lurgan, Mullynure and Poyntzpass (Armagh); Lower Lough Erne (Fermanagh) and Killycolpy Wood and Tullyhogue (Tyrone). The adults are smaller than Ghost Moth and the forewings of the male are light brown with a series of white dashes which are variable and less apparent in some individuals. The few Northern Ireland records have been between the end of May and mid-June. Common Swift is found in many open habitats in Britain including agricultural land, gardens, moorland and rough, uncultivated grassland including roadside verges and woodland

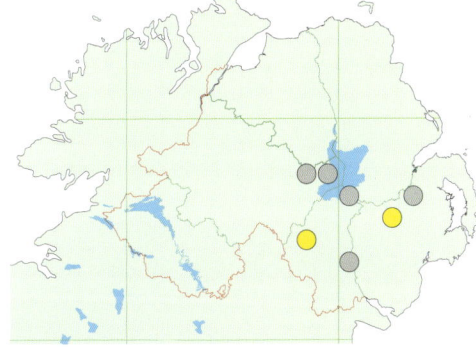

rides. In Britain the main flight period lasts from May to July. The larvae feed on the roots of grasses and other cultivated herbaceous plants. It overwinters as a larva.

Map-winged Swift
Hepialus fusconebulosa (DeGeer) Category one.
Widespread. This is the most frequently encountered species of swift in Northern Ireland. It has been recorded in all counties including suburban areas and on both Rathlin Island (Antrim) and Copeland Bird Observatory (Down). It derives its name from the irregular pale markings on the forewings which are said to resemble patterns on a map. There is much individual variation in the colour shape and pattern of these markings but, despite this, the moth is distinctive and easily identified. A variety known as f. *gallicus* with uniform reddish-brown forewings and a small, pale discal spot, exists in small numbers throughout most populations in Northern Ireland. Unlike other swifts this species is frequently seen at light, often in good numbers. Map-winged Swift occurs in heaths, bogs, rough grassland and open woodland. The recorded flight period is 21 May to 20 August. The larva is subterranean, feeding on Bracken and the roots of grasses. It has a two-year life cycle, overwintering twice in the larval stage.

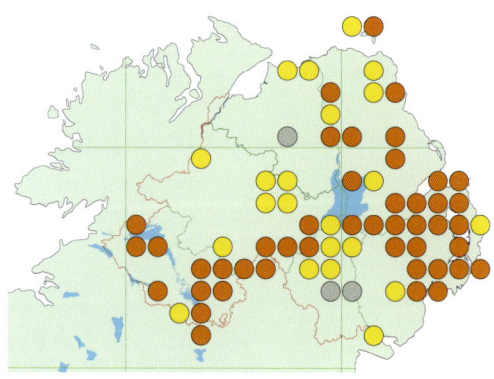

ZYGAENIDAE

This is a family of colourful medium-sized day-flying moths, most active in bright sunshine. Ten species occur in Britain and Ireland, three of which are resident in Northern Ireland. All species have bright metallic colouring, narrow forewings which are held tightly to the body and long, conspicuous antennae, which are clubbed in some species. Their vibrant colours serve as a warning to predatory birds and mammals that they are toxic. The burnets are colonial species typically of open, flowery habitats (dry or damp), especially in coastal sites where populations can be very large. The larvae, which are also toxic, feed on herbaceous plants. They pupate in conspicuous papery cocoons on grass stems.

Forester
Adscita statices (Linnaeus). SOCC Category three.
Scarce. This attractive species is confined to scattered colonies in southern and western counties. Colonies have been reported from the Argory, Ballynacor and Peatlands Park in north Armagh and at several localities around Belcoo, Florencecourt and Garrison in west Fermanagh. There are historical records for Aghalee (Antrim), Maghery (Armagh), Lisburn (Down) and Cookstown and Tamnaghmore (Tyrone). Forester is considered a declining species, but as it is never seen in large numbers, it is impossible to assess the strength of the colonies. The progressive loss of suitable habitat has no doubt contributed to its decline. The most stable population seems to be in north Armagh where one or two individuals were seen in most years up to 2002. There have been no recent records from many of the sites in the west. The adults are bluish-green with an iridescent sheen but, despite this, the adult moth can be quite elusive and difficult to see. They only fly in sunshine and can be seen visiting flowers such as Ragged Robin, clovers and buttercups. Colonies are found in fens and wet flowery meadows.

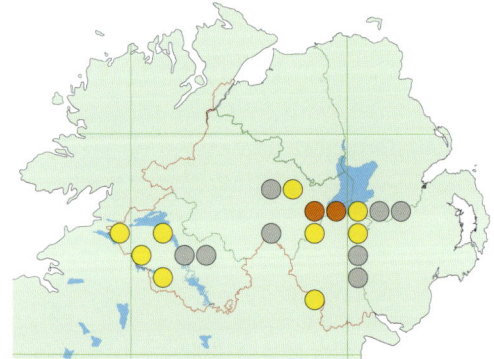

Moth accounts

Forester.

Peatlands Park, County Armagh.

Peatlands Park has been a reliable site for this declining moth in recent times; however, even here, no adults have been reported for a few years.

The recorded flight period is 5 June to 11 July. Its larval habits and foodplants in Northern Ireland are not known. In Britain, it feeds on Common Sorrel and Sheep's Sorrel and overwinters as a young larva concealed deep among the vegetation.

Six-spot Burnet
Zygaena filipendulae (Linnaeus). Category one.
Widespread. Generally distributed on the sandy coasts of Down, Antrim and Londonderry. Inland it is more local and mainly found in Armagh, Down and west Fermanagh. There have been no recent records from Tyrone. Six-spot Burnet is a colourful day-flying species, colonial in its habits and preferring mainly light well-drained soils such as coastal grassland and sand dunes. It can be seen abundantly on many of its coastal localities during the summer. It is much less common inland having a preference for disused quarries and roadside verges. Adults are easily recognised by the pairs of bright red spots on the dark bluish-black forewings. The shape and size of the red spots are variable and in some individuals the paired spots are merged into blotches. The adults are most conspicuous in bright sunshine visiting a wide variety of grassland flowers including Viper's Bugloss, thistles and knapweeds. On dull days they retreat low down among the vegetation and are often difficult to detect. The recorded flight period is 26 May to 9 September, peaking in late June and early July. The larval foodplant is Bird's-foot

The Butterflies and Moths of Northern Ireland

Six-spot Burnet.

Murlough National Nature Reserve, County Down.

Trefoil. The papery cocoons are formed conspicuously on the stems of Marram and other tall grasses in early summer. It overwinters in the larval stage.

Narrow-bordered Five-spot Burnet
Zygaena lonicerae (Scheven) Category two.
Scarce. Recorded from all counties but mainly found in Armagh where colonies are widespread especially around Armagh City and in the north of the county at Brackagh Moss, Tannaghmore and Selshion Moss. One of the largest populations is at Craigavon Lakes. It is very local in the other counties. Three extant colonies exist in Fermanagh at Derryvore (where

large numbers were seen in 2004), Legatillida and Monawilkin. There are also old records from Belleisle and Portora. In Tyrone, it has been recorded most recently at Wood Lough and Crilly in the south-east of the county, but there are early twentieth-century records for Favour Royal, Stewartstown and Tullylagan. There are populations in Antrim at Larne, Megaberry, the Montiaghs Moss and Whitehead. It is found at a few sites in Down on the coast at Ballyrobert, Helen's Bay and Murlough NNR and inland at Belvoir Park and Clandeboye. The only recent Londonderry colonies are found at Crockcor Quarry and Ranaghan Bridge, but Greer reported it from Coagh in the early twentieth century. Adults closely resemble Six-spot Burnet but differ in the number of spots on the forewings. The wings are slightly longer and narrower in this species with a stronger bluish tinge and the antennae are slightly longer. Unlike Six-spot Burnet, this species has a preference for inland sites particularly roadside verges, fens, damp meadows and marshy areas, but it has also been recorded in dry coastal localities. Adults can be seen taking nectar from flowers, but they are never as common or abundant as Six-spot Burnet. The recorded flight period is 12 June to 16 August. The larval foodplants are Bird's-foot Trefoil, Red Clover, White Clover and Bitter Vetch. It overwinters in the larval stage.

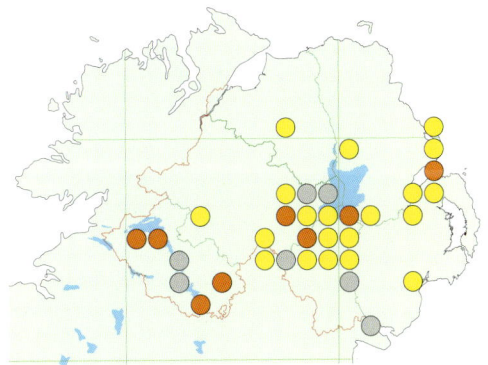

SESIIDAE

These rather unusual day-flying moths, of which there are three species recorded in Northern Ireland, are mimetic, resembling hymenopteran wasps; however, they are easily distinguished from wasps by the smaller head and eyes and the absence of a waisted abdomen. They are called clearwings as the wings lose most of their scales after emergence, leaving a transparent window. Emergence usually takes place in early morning. The newly-emerged adults of most species tend to rest on the trunks and branches of the host tree but disperse quickly when conditions warm up; they are most active in bright sunshine. Clearwings are by nature elusive and seldom seen and best looked for in the larval or pupal stages.

The larvae infest the stems, crowns and roots of plants and the trunks and branches of trees. They reach maturity in one or two years depending on species. Successive generations of adults will use the same trees year after year. Prior to pupation, the larva will prepare an exit hole through to the surface of the trunk or stem leaving a thin wall for protection before emergence. The empty pupal cases often remain protruding from the exit holes for several weeks following emergence and this can be a reliable method of recording the species. In general, however, clearwings are not well recorded in Northern Ireland and the current map is only indicative of their true distribution.

Lunar Hornet Moth
Sesia bembeciformis (Hübner) Category three.
Scarce. Recorded from all counties, but seems to have been more common in the nineteenth and early twentieth centuries; the larvae were often described as common and abundant in old willows and poplars. In recent times, it has declined in all counties and is known from just a few sites in each. The majority of records are from east Tyrone at Stillago and Terryscollop and in north Armagh at Brackagh Moss and close to Loughgall. Here, specific effort has been made to record it by searching the lower part of mature willows in the appropriate habitat for empty pupal cases or exit holes. Some sites have been found by the fortuitous discovery of feeding

tunnels in felled trees. In Fermanagh it is known from Crom, Gorteen, Monawilkin and Monmurry. It occurs at three sites in Down – Ballylesson, Derryleckagh, and Seaforde. In Antrim it is only known from the Aghalee area, where it was recorded by Wright, but it has not been seen for many years. Likewise in Londonderry there have been no recent records and it has been found only at Culcrow and Traad Point. Like all clearwings, this species is elusive in the adult stage and it is likely that it has simply been overlooked in some areas as suitable habitat still exists. As the name implies, adults resemble a large hornet and have distinctive yellow banding on the abdomen and a yellow collar at the back of the head. Lunar Hornet is found 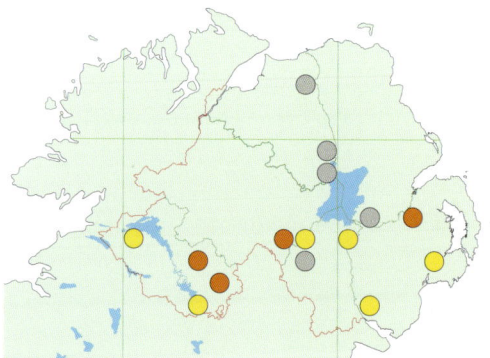 where mature willows grow in damp woodland, lake shores, fens and bogs. Adults seem to prefer old isolated trees in which to lay. They are most active in sunshine but are seldom seen except by accident or at rest on the tree trunk after emergence in the early morning. It is not attracted to pheromone lures. The pupal exuviae protrude from the exit holes on the trunk of the host tree and provide perhaps the easiest method of recording the species. Only a small number of adults have ever been seen in Northern Ireland, in late June and July. The larvae feed in the roots and trunks of old willows and spend two winter periods in the larval stage.

Lunar Hornet Moth.

Brackagh Moss National Nature Reserve, County Armagh.

This wasp-like mimic is elusive and seldom encountered by day.

Moth accounts

Above left
A Lunar Hornet Moth pupa close to emergence.

Brackagh Moss National Nature Reserve, County Armagh.

Above right
Lunar Hornet Moth exit holes in old willow trunk.

Brackagh Moss National Nature Reserve, County Armagh.

One of the easiest ways of detecting the presence of Lunar Hornet Moth is to examine mature willows for exit holes.

Left
Red-tipped Clearwing (female).

Brackagh Moss National Nature Reserve, County Armagh.

This photograph, taken in 1985, documents the first ever Northern Ireland record of this clearwing. The moth has been seen on only a few occasions since. The larvae probably develop in the willows growing in the wettest, most inaccessible parts of the fen.

Currant Clearwing
Synanthedon tipuliformis (Clerck) Category four.
Unknown. There is a single record of this species from Londonderry dating back to 1892. It is rare in Ireland, the only records being one from Cork and several from the Dublin City area. Its present status here is unknown. Adults are small and wasp-like, with three or four narrow yellow bands on a black abdomen. Like the other species in this family the wings have a transparent window with dark wing tips. In Britain, this species is seen most often in suburban gardens and allotments with established currant bushes, but it can also be found in damp woods and wetlands where the foodplants are growing wild or have become naturalised. All the Irish records appear to be from gardens. Like all clearwings, its secretive habits make it difficult to detect. Adults fly during the day in sunshine but they can be attracted to pheromone lures and this appears to be the best way of detecting its presence. This may be worth investigating here in suitable localities. The stems of mature plants can also be examined for exit holes and pupal exuviae. The larvae feed internally in the branches and main stems of Black Currant, Red Currant and occasionally Gooseberry. In Britain the main flight period is June and July. It overwinters in the larval stage within the foodplant.

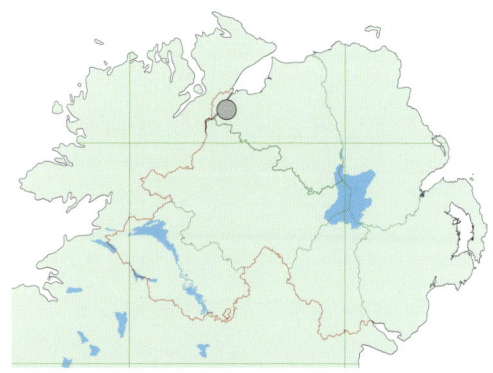

Red-tipped Clearwing
Synanthedon formicaeformis (Esper) SOCC (P) Category four.
Rare. Confined to a single site, Brackagh Moss, in north Armagh. It was first discovered in 1985 and has been seen on several occasions since, the last time in 1999. It was not listed from Northern Ireland by Baynes or previous authors. Despite frequent searching in suitable habitat, Brackagh Moss remains its only known locality in Northern Ireland. Adults have a blackish abdomen with a distinctive red band; the tips to the forewings are also red. Like other species of clearwings, the wings are transparent with dark veining. The few adults which have been observed here have a noticeable white tip on the antennae, a feature only found in the female. The Red-tipped Clearwing is found in wetlands with mature willows. Adults seem to be most active during sunshine, especially in the early morning and late afternoon, when they have been observed resting on Bramble or visiting flowers, particularly Meadowsweet. They are skittish by nature and, when disturbed, they either fly off or drop into the vegetation where they are impossible to find. In Britain, adults have been found resting on the trunks and branches of willows in early morning, or beaten from the foliage in dull overcast conditions. The recorded flight period is 11 June to 31 July which conforms with the main flight period in Britain. The larvae overwinter once, feeding inside the stems and trunks of willows with little outward sign to betray their presence.

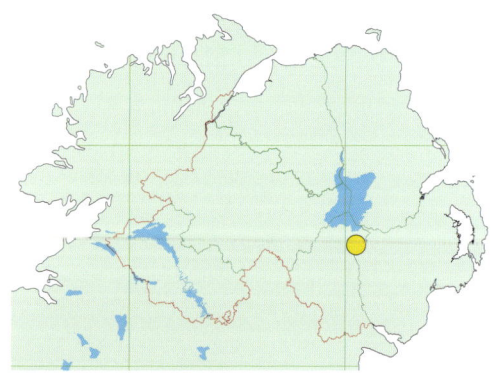

LASIOCAMPIDAE

Eggars are well-built, medium-sized or large moths. Of the ten resident species found in Britain and Ireland, seven occur in Northern Ireland. The adults are predominantly brown or yellowish in colour. The males have large feathered antennae which they use to pick up the scent emitted by the freshly-emerged females. This they can detect from a considerable distance. The mouthparts are only partially developed and therefore the adults do not feed. Two species have day-flying males that are active in bright sunshine. Female eggars are nocturnal, flying from dusk onwards and are infrequently attracted to light in large numbers.

The larvae are generally hairy, heavy-bodied and often conspicuous (many of the submitted records are of larvae), feeding mostly on the leaves of trees and shrubs. Two of the species are communal, living within a silken web until the final instar. Prior to pupation the larvae spin egg-like cocoons (hence the name eggars) in which to pupate, usually at ground level or among the foodplant. Two species fly in winter, the remainder in mid- to late summer.

December Moth
Poecilocampa populi (Linnaeus) Category two.

Widespread. This species has been recorded from all counties. Its main distribution is in a narrow band running west from Belfast Lough across north Down and south Antrim to the south shore of Lough Neagh and into south Tyrone. It is otherwise known from a scattering of localities in Antrim, south Down, Fermanagh and Londonderry. Frequently recorded sites include Shane's Castle (Antrim); Aughinlig, Brackagh Moss and Loughgall (Armagh); Belvoir Park, Helen's Bay, Hillsborough Forest and Seaforde (Down); Garvary Wood and Monmurry Bog (Fermanagh); Traad Point (Londonderry) and Rehaghy Mountain (Tyrone). Adult males have dark greyish-brown forewings with a distinctive, cream postmedian line. The forewings 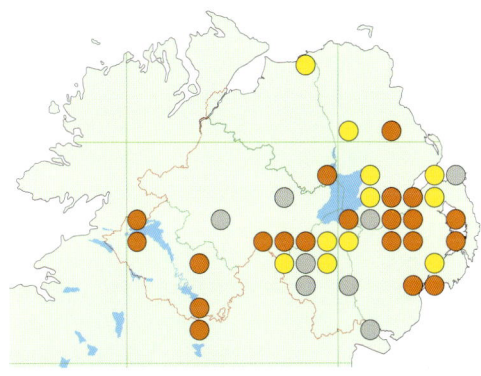 tend to lose their scales quite quickly and become semi-transparent, especially at the tips. The body is quite hairy and the males have large feathery antennae. Both sexes show little variation in colour. The December Moth is a species of broadleaved woodland but it also occurs in other habitats with mature trees, though rarely in large gardens. Adults are attracted to light in small numbers. This is one of the late-flying moths, a feature which undoubtedly contributes to under-recording. The recorded flight period is 22 October to 4 January, with November to mid-December the peak period for records. The larvae are apparently nocturnal and feed on the leaves of broadleaved trees and shrubs including oak, birch, willow, Blackthorn and Hawthorn. It overwinters as an egg, hatching in April.

Pale Eggar
Trichiura crataegi (Linnaeus) SOCC (P) Category four.

Rare. Restricted to three localities in west Fermanagh. It was discovered in Northern Ireland in 1990, when adults were trapped at Correl Glen and Monawilkin on 26 August. It has been recorded subsequently at Correl Glen in 1990, 1992, 1998 and 2002. The only other record is from Gortatole where a single adult was trapped in August 1995. Despite increased fieldwork in west Fermanagh, no other colonies have been detected; however, small isolated populations may await discovery in other parts of the county. There is an old published record in Donovan for Magilligan (Londonderry); the recorder was W. Savage who claimed to have found larvae of this species on Blackthorn. This is generally regarded as doubtful and most 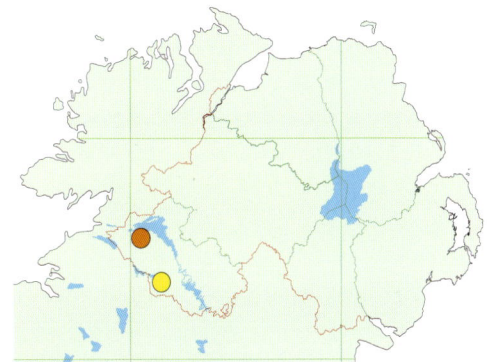 certainly a misidentification of the Small Eggar which is known from this locality. Elsewhere in Ireland this species is reported only from the Burren (Clare and Galway) and Kerry. Adult males have pale greyish forewings with a distinctive, darker central band. Females are slightly

larger and are browner in colour. The Pale Eggar has been taken in scrub at the edge of mature broadleaved woodland. Adults are attracted to light, but appear sporadically and usually only in ones and twos. The records in Northern Ireland have been between 18 August and 18 September, which conforms to the main flight period in Britain. The main foodplants used in Britain are birch, Blackthorn and Hawthorn. Birches are common at the main Fermanagh site, together with two other recorded foodplants, Heather and Bilberry. It overwinters as an egg.

Above left
Pale Eggar.

Correl Glen National Nature Reserve, County Fermanagh.

Above right
Small Eggar.

Monmurry, County Fermanagh.

Small Eggar
Eriogaster lanestris (Linnaeus) SOCC (P) Category four.
Rare. This species is unusual as the majority of records refer to the communal web-forming larvae and it was only in 2000 that an adult was found in the wild in Northern Ireland. It has been recorded from two locations in Down, Six Road Ends and Bohill Forest, both in 1972, but has not been reported in this county since. Crawford found two larval nests on the shore of Lower Lough Erne at Blaney (Fermanagh) 'in the early part of July 1926.' Currently the only extant populations exist on the extensive dunes on the north coast of Londonderry at Magilligan and at Monmurry (Fermanagh). The Londonderry population has been recorded almost continuously since 1990 (as larvae), although not always from the same site, whilst all the records from the Monmurry locality, discovered in 2000, have involved adults. The adult moths have reddish-brown forewings with a distinctive white spot located centrally in the forewing and a pale postmedian line. The species occurs in Northern Ireland in dune scrub and hedgerows. The few adult records here have been trapped between 23 February and 4 April. The main flight

period in Britain is February to March. The adults are nocturnal and rarely visit light and this combined with their early flight period makes them difficult to detect. In contrast, the communal larvae live conspicuously in a web spun over the branches of the foodplant from spring to midsummer. The main foodplant at the north coast sites appears to be Blackthorn, but they have also been reported on birch and Hawthorn. Crawford found the 1926 larval nests on willow, but he reared them on Hawthorn. It overwinters as a pupa protected in a brown cocoon. Captive larvae reared by Crawford remained in the pupal stage for two winters.

Lackey
Malacosoma neustria (Linnaeus) Category three.
Unknown. The status of the Lackey in Northern Ireland is uncertain as there have been no recent records. Donovan gives an nineteenth-century larval record from the Enniskillen area (Fermanagh). There is no recent evidence to suggest that this moth currently survives in Northern Ireland. Apart from this vague Fermanagh record, this species is confined to the southern half of Ireland. It is especially common in the Burren region of Clare. Adults vary from light yellow to reddish-brown; females tend to be darker than males. There are two pale median lines on the forewings but these occasionally may be absent. This species reportedly prefers sunny, sheltered positions in open woodland, along hedgerows and established scrub. The main

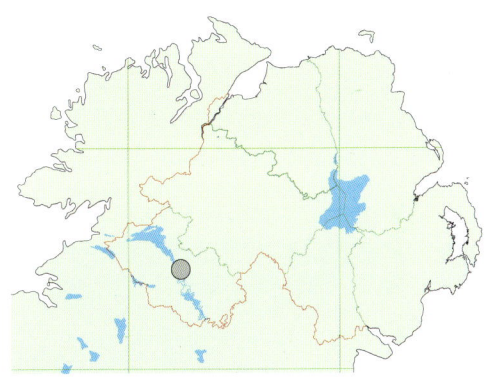

flight period in Britain is July and August. The larvae are communal, living inside a silken web for most of their development. In Britain the larval foodplants include Blackthorn, Hawthorn, willow, apple and plum. It overwinters as an egg.

Oak and Northern Eggar
Lasiocampa quercus (Linnaeus) Category one.
Widespread. Recorded in all counties, its distribution largely reflecting that of bogland, its principal habitat. It is found throughout Armagh, Fermanagh and Tyrone. In Londonderry it is confined to sites on the north coast and strangely absent from the Sperrins. In Antrim, it occurs along the coast and glens as far south as Larne, in relict bogs along the Lower Bann valley and on Rathlin Island. Most of the known sites in Down are in the east of the county, close to Strangford Lough. It is also found on the coast at Dundrum and along the northern edge of the Mourne Mountains, inland from Newcastle. Adult males are a dark reddish-brown with a small white discal spot on the forewings and a diffused yellow median band on all wings. There is a yellow basal patch

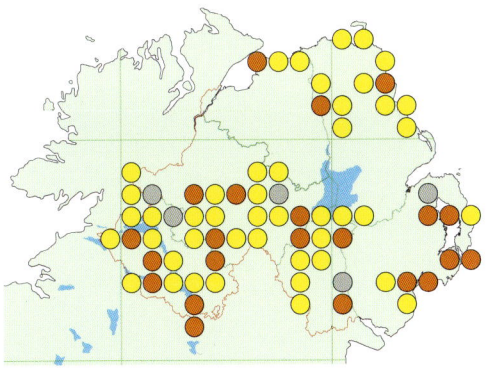

on the forewings, which can be variable. The large heavy females are yellowish-brown with similar markings to the male. The Northern Eggar is found on cutover bogs, low-level blanket bogs, heaths, sand dunes, and rarely, woodland rides and clearings. Males are often seen flying swiftly and erratically over the habitat during the day in sunshine. The females are nocturnal and are attracted to light in small numbers. Two subspecies have traditionally been recognised – *quercus* (Oak Eggar) with a one-year life cycle flying in May and June and *callunae* (Northern Eggar) with a two-year life cycle flying in July and August. The only subspecies considered to be present in Ireland is *callunae*; however, the validity of these subspecies has been questioned as intermediate populations exist. It is also evident that the flight period in Ireland is longer than previously recorded; adults have been reported in Northern Ireland from 30 April to 31 August. The larvae

The Butterflies and Moths of Northern Ireland

Oak Eggar (male).

Mullenakill National Nature Reserve, Peatlands Park, County Armagh.

are often seen resting openly on vegetation in the autumn prior to hibernation and throughout the spring and summer of the following year. The pupa is protected inside a tough silken cocoon spun among ground vegetation and can be found occasionally during the winter months. The larvae have been found in Northern Ireland feeding on Heather, birch, willow and Hawthorn.

Fox Moth
Macrothylacia rubi (Linnaeus) Category one.

Scarce. Recorded in all counties and also on Rathlin Island (Antrim) but there are large gaps in the distribution, especially inland in the north and west and in much of both Antrim and Armagh away from Lough Neagh. Frequently recorded sites include Peatlands Park (Armagh), Murlough NNR (Down), Eshywulligan, Garvary Wood and Glennasheevar (Fermanagh) and Magilligan and the Umbra (Londonderry). Adult males are reddish-brown with two medial cross-lines on the forewings. The female has similar markings but the overall colour is greyish-brown. The absence of a white spot on the forewing distinguishes it from other species in this family. The Fox Moth is mainly associated with cutover raised bogs, low-

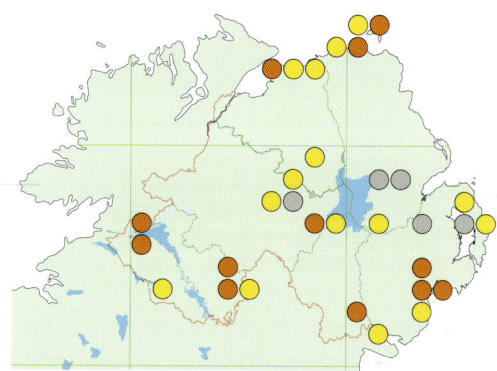

Drinker (male).

Rehaghy Mountain, County Tyrone.

The larva of this attractive moth is more commonly encountered than the adult, especially in early spring after hibernation.

level blanket bog, sand dunes and coastal heaths. It is still probably under-recorded in some areas where suitable habitat exists. Adult males fly by day, particularly in the late afternoon and early evening, whereas the female is nocturnal. Their flight is rapid, similar to that of Northern Eggar and Emperor which often shares its habitat, but the Fox Moth is a more elusive species. Adults have been recorded from 18 May to 3 July. The larvae have been found on Heather in Northern Ireland and are often seen exposed on open ground especially in coastal habitats, prior to hibernation in the autumn. Other recorded foodplants in Britain include Bilberry, Creeping Willow, Bramble and Meadowsweet. The larvae emerge in early spring and bask in sunshine for a short period before pupation.

Drinker
Euthrix potatoria (Linnaeus) Category one.
Widespread. Generally distributed in north Armagh, Fermanagh and Tyrone. It is very local in Antrim with records for Breen Wood and Garry Bog being the most recent. In Londonderry there are records for Ballymaclary, Magilligan and the Umbra whilst in Down, it is found mainly in the east of the county and in the Mourne Mountains. It is commonly encountered at the Argory, Oxford Island and Peatlands Park (Armagh), at Crom and Garvary Wood

(Fermanagh) and Rehaghy Mountain (Tyrone). Surprisingly, it has not been recorded from Rathlin Island (Antrim) which has plenty of suitable habitat. Adult males are generally reddish-brown suffused with yellow with a white discal spot in the central area of the forewing. There is a distinctive thin line running diagonally across the forewing, terminating at the apex. The female is larger and yellow in colour but is otherwise like the male. This species is found typically in damp, open habitats such as boggy meadows, reed beds, heaths and moorland. Adults come frequently to light, the males appearing more commonly than females. The recorded flight period is 3 June to 17 August. The larvae feed on Common Reed, Cock's-foot, couches and probably other grasses. The young larvae enter hibernation in the autumn and reappear in early spring, when they are often found resting exposed on prominent grass stalks. Pupation takes place inside a yellowish paper-like cocoon that is prominently formed on a grass stem, and persists long after emergence.

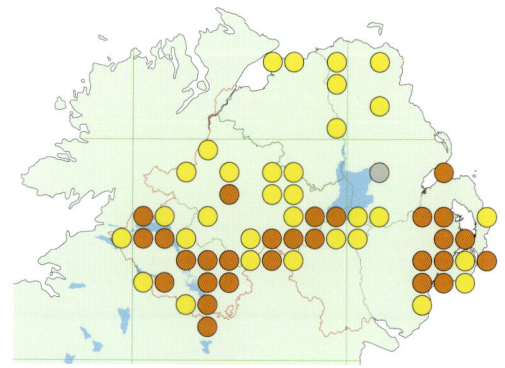

SATURNIIDAE

This family contains some of the most colourful moths in the world. They are essentially of tropical origin and renowned for their great size and beauty, with over 1300 species described worldwide. There is only one species resident in Britain and Ireland. The mouthparts are vestigial and therefore the adults do not feed. The males fly by day and are most active during sunshine. They are frequently observed flying swiftly across heaths and bogs during early spring in search of newly-emerged females. The females are nocturnal, flying from dusk onwards and may be attracted to light, but seldom more than one or two.

Emperor
Saturnia pavonia (Linnaeus) Category one.
Ubiquitous. This is one of the most generally distributed moths in the region found in all counties. It has also been reported from Rathlin Island (Antrim) in the late 1980s. The large colourful adults are distinctive and unlikely to be confused with any other species. The males are reddish-grey with two white patches on the forewings which contain prominent eyespots. There is a suffused white band at the termen of each forewing and a red apical streak. The dark eyespots on the tawny hindwings are concealed when at rest but exposed when the moth is alarmed or threatened. The female has similar markings to the male but is larger and predominantly grey. Many of the records of this moth are of the fully-grown larvae which are equally distinctive. These are very large, bright green and covered in rows of hairy, whitish tubercules. Emperor is found chiefly on wet heaths, bogs and moorland, but also in wet meadows and fens. It is scarce in parts of the east of Northern Ireland and absent from areas of intensive agriculture. The flight period is 23 March to 17 June, peaking in mid-April to mid-May. The larvae have been recorded in Northern Ireland on Heather, Meadowsweet, Bog Myrtle and willow. The fully-grown larvae overwinter in a tough silken cocoon attached to the stems of Heather and other low-growing plants. Cocoons which have suffered predation by small mammals have been found during the winter months.

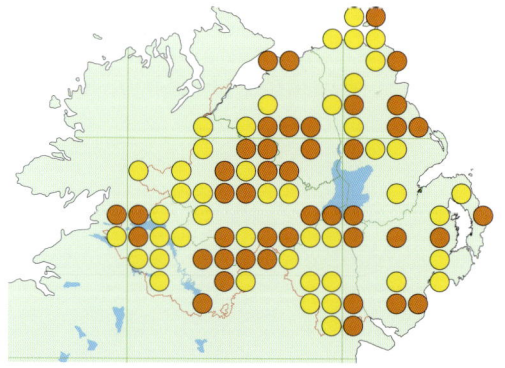

Moth accounts

Emergence sequence of a male Emperor.

Brackagh Moss National Nature Reserve, County Armagh.

Emergence is a vulnerable time for all Lepidoptera, when they are at risk from birds and other predators. For additional protection many moths undergo this transformation either in late evening or after dark.

DREPANIDAE

This is a small family of moths closely related to the geometrids. Six species occur in Britain and Ireland, four of which are resident in Northern Ireland. Occasional individuals can be disturbed by day, but most are encountered in light traps, though usually only in small numbers. The proboscis in this family is rudimentary and therefore the adults are not attracted to 'sugar' or flowers.

The hooked tip to the forewings is a characteristic feature of the Irish representatives of this family with the exception of the Chinese Character. The close relationship of this species to the others is apparent in the form of the larva. Drepanid larvae are characterised by a hump-backed appearance and projecting tips from the abdomen. The absence of the anal claspers on the last segment gives rise to a modified anal segment in the form of a raised tip. Pupation is carried out inside a flimsy cocoon constructed between leaves of the foodplant.

Scalloped Hook-tip
Falcaria lacertinaria (Linnaeus) Category two.

Scarce. This species has been recorded from all counties but the distribution appears to be mainly southern and western. There are old records for Armagh City; Kilkeel, Saintfield and Strangford (Down); Tempo (Fermanagh) and Favour Royal (Tyrone). Most of the recent records are from Fermanagh and along the southern shore of Lough Neagh. It is local and apparently absent from much of Antrim, Down and Londonderry. There are recent records from Derryola Bridge and Portmore Lough (Antrim); the Argory, Brackagh Moss, Drumnahavil and Peatlands Park (Armagh); Lackan Bog (Down) and Cuilcagh, Garvary Wood, Legatillida and Monmurry (Fermanagh). In Tyrone it has frequently been recorded from Rehaghy Mountain. Londonderry 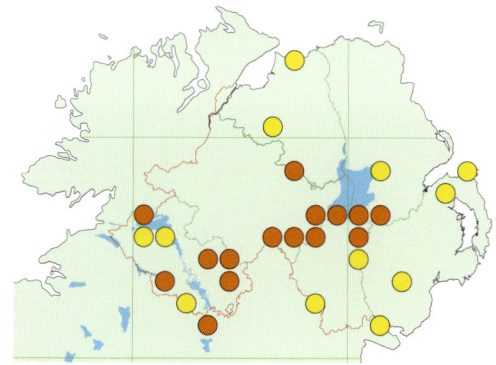 localities are Altnaheglish Forest, Curran Bog and the Umbra. Adults have pale brown forewings with two dark cross-lines and a small dark central spot; the edges are scalloped or ragged and often freckled with a hook-like tip at the apex. At rest, the adult closely resembles a dead birch leaf. The Scalloped Hook-tip has a preference for damp habitats where birches flourish, so is found on heaths, bogs and fens. Adults come frequently to light and have occasionally been found during the day in Northern Ireland, resting among the leaves of its foodplant. The recorded flight period is 1 May to 29 August. The Northern Ireland population would seem to be double-brooded as there are two clear peaks in the pattern of records, the first in late May and June and the second in early August. The larvae feed on birch and spend the winter months in the pupal stage protected in a cocoon.

Barred Hook-tip
Watsonalla cultraria (Fabricius) Category four.

Rare. This species was first recorded in Northern Ireland at Ballykinler Dunes (Down) on 4 September 1999. The single adult was discovered resting on the outer rim of an actinic trap. It was recorded for a second time on 4 August 2004 at Rostrevor Wood (Down). There appears to be no other published record for Ireland. It was previously thought that the first specimen may have been a vagrant since the habitat was atypical, and the foodplant, Beech, does not occur nearby. However, the discovery of a second adult in a woodland locality, suggests that it is an established, but rare, species. Whether it is a recent colonist is as yet unclear. Further investigation is needed in the Rostrevor area of south Down to establish the current status of this species. Adults are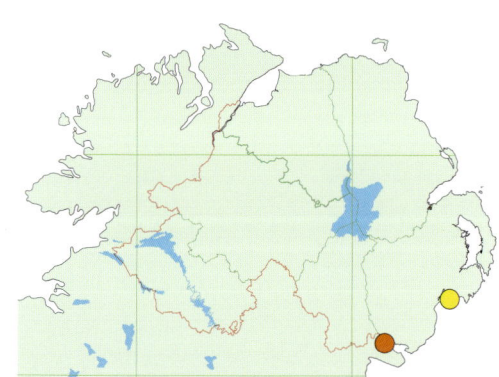

Moth accounts

Scalloped Hook-tip.

Lackan Bog, County Down.

This adult was discovered early in the morning at rest on the underside of a birch leaf beside an actinic trap.

ochreous brown with a hook-like tip to the apex of the forewings. There is a dark central band with a discal spot. Barred Hook-tip is a woodland moth found in calcareous beech-dominated woodland in southern England and also in woods where Beech has been introduced as far north as southern Scotland. Adults come to light, but males reputedly also fly during the day. Two generations are normal in Britain in May and June and again from mid-July to early September. In Britain, it feeds on Beech and spends the winter months in the pupal stage.

Pebble Hook-tip.

Lackan Bog, County Down.

Pebble Hook-tip
Drepana falcataria (Linnaeus) Category three.
Notable. This is a very local species that is known from just a few large bogs close to Lough Neagh. There are old records for Enniskillen (Fermanagh) and Altadavan, Favour Royal and Tamnamore (Tyrone). Most of the recent records have come from the Argory and Peatlands Park (Armagh), Lackan Bog (Down) and Rehaghy Mountain (Tyrone). Monmurry is the only recent site in Fermanagh. In Antrim, it has been recorded several times at the Montiaghs Moss and, once, in 1974, at Rea's Wood. Surprisingly, as they appear to have suitable habitat, it has not been found on the bogs of south and mid-Londonderry. Adults are generally pale brown with the characteristic hook-tip at the apex of the forewings. There is a small dark blotch near the centre of the forewing

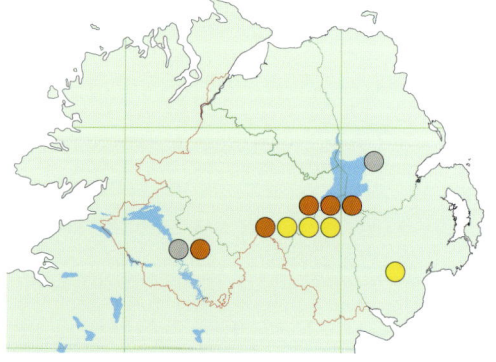

and a distinctive reddish-brown line that runs from the apex of the forewing to the dorsum. The Pebble Hook-tip is recorded almost exclusively from cutover and raised bogs in Northern Ireland. Adults are attracted to light but appear only sparingly. The recorded flight period is 28 May to 19 August. The range of dates would suggest that the Northern Ireland population has but a single generation, as it does in Scotland, but more study is required to ascertain if this is the case. The larval foodplants are birch and, possibly, Alder. It overwinters as a pupa.

Chinese Character
Cilix glaucata (Scopoli) Category two.
Widespread. Generally distributed across Armagh, Down and Fermanagh. Apparently absent from most of the north of Northern Ireland, north of Lough Neagh, apart from Magilligan and the Umbra (Londonderry) and Ballymena and Magheramorne (Antrim). Commonly recorded

sites in the south include the Argory, Loughgall and Peatlands Park (Armagh); Belvoir Park, Helen's Bay, Hillsborough Forest and Lackan Bog (Down); Crom, Garvary Wood and Monmurry (Fermanagh) and Rehaghy Mountain (Tyrone). There are also old records for Favour Royal (Tyrone) and Kilkeel, Saintfield and Strangford (Down). This small but distinctive moth is unmistakable due to its strange shape and patterning that bears a remarkable resemblance to a bird-dropping. The wings are predominantly white apart from a greyish blotch on the forewing. They are held folded tightly against the body in a high arched position. The Chinese Character can be seen in woods, hedgerows, scrubby areas and occasionally, gardens. Adults come to light in small

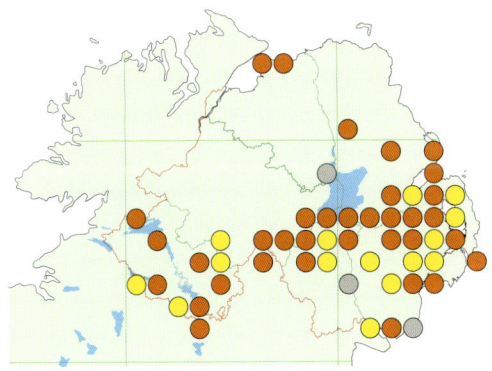

numbers but can also be found during the day especially by beating foliage. When disturbed they have a tendency to drop into the vegetation rather than fly. There are two annual generations, one in May and June and a second in late July and August, but with some overlap in early July. The recorded flight period is 4 May to 3 September. The larval foodplants are not known in Northern Ireland. In Britain, Hawthorn, Blackthorn and Crab Apple are stated to be the main foodplants, although it has been reported also on Bramble and Pear. It spends the winter months in the pupal stage among leaves, or cracks and crevices in bark.

THYATIRIDAE

*T*his is a small family of medium-sized moths. Nine species occur in Britain and Ireland and five of these are resident in Northern Ireland. The adults of lutestring moths resemble noctuids although they are generally slimmer with a prominent humped thorax. When at rest they hold their wings close to the body. Some species are attractively marked and most are easily identified. The adults have a well-developed proboscis and are attracted to flowers and, in some cases, sugar.

Many of the larvae are nocturnal and retreat to the base of the foodplant during the day making them difficult to find. Others construct a protective tent from rolled-up leaves in which to hide, emerging only at night to feed.

The larvae feed on a wide variety of trees and shrubs and produce a single brood each year. Some species are subterranean constructing chambers just below ground, while others overwinter in cocoons spun among leaves. The adults of all the Northern Ireland species come readily to light in reasonable numbers.

Peach Blossom
Thyatira batis (Linnaeus) Category one.
Widespread. A commonly recorded species particularly in southern counties. It is more locally distributed in the north and west, but it is probably under-recorded in many areas. Northern localities include Duncrun, Kilrea and Traad Point (Londonderry) and Glenariff Forest, Portglenone Wood and Rathlin Island (Antrim). It is regularly recorded from suburban areas of Belfast and surrounding towns; Loughgall and Peatlands Park (Armagh); Castlewellan Forest, Helen's Bay and Rostrevor Wood (Down); Brookeborough, Crom and Garvary Wood (Fermanagh) and Rehaghy Mountain (Tyrone). Adults are distinctive with olive-brown forewings and large colourful blotches of white, suffused with pink and brown. The pink colour varies in intensity and

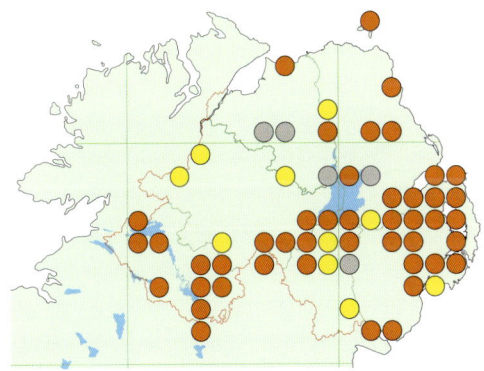

Peach Blossom.

Castlewellan Forest, County Down.

some individuals are more richly marked than others. Peach Blossom is mainly associated with broadleaved woodland, hedgerows, gardens and parkland where Bramble flourishes. Adults are rather secretive and seldom found during the day, preferring to hide in vegetation. They are attracted to light in moderate numbers, but often individuals are found resting outside the trap. They can also be attracted to sugar. The recorded flight period is 15 May to 13 August apart from an exceptional early September record. The larva feeds on Bramble after dark, remaining concealed at the base of the foodplant during the day. It overwinters as a pupa.

Buff Arches
Habrosyne pyritoides (Hufnagel) Category one.
Common. Widespread in the southern counties, but more local north of Lough Neagh with the most recent records from Breen Wood and Glenariff Forest (Antrim) and a few suburban gardens. In the south it is widespread across north Armagh, Down and Fermanagh. This is a distinctive and beautifully marked moth that is unlikely to be confused with any other species.

The forewings are pale brownish-grey with a broad buff band edged with white. There is also a conspicuous whitish linear blotch along the costal area and a well-defined white line running across the outer edge of the forewing. The thorax has similar colouring. Buff Arches is found in open woodland, hedgerows, bogs and occasionally, gardens. It is probably more widespread in the north than current records suggest. Adults are attracted to light in small numbers but are seldom seen during the daytime, preferring to hide among leaves and other ground vegetation. The flight period is 2 June to 3 November. Mid-June to late July is the peak period for records. Occasional adults have been reported through late summer to early autumn, but the significance of this is unknown. The larvae feed at night on Bramble. The species overwinters as a pupa.

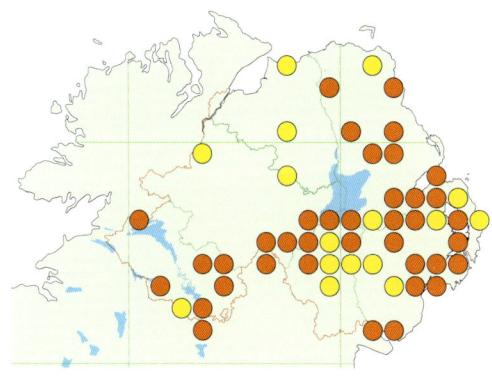

Poplar Lutestring
Tethea or (Denis & Schiffermüller) Category three.
Rare. Reported mainly from Fermanagh where it has been recorded annually since its rediscovery in 1998. However, its recent detection at Crilly in Tyrone on 19 May 2004 may indicate a more widespread distribution that was previously thought. Donovan listed it from Upper Lough Erne (Fermanagh), recorded by Langham. Recently recorded sites in Fermanagh include Garvary Wood and Lusty More Island in the north of the county and Altaclanabryan, Crom, Killynick and Legatillida in the south. Adults can be distinguished from the other lutestrings by the darker medial bands across the grey forewings and the yellow stigma. Poplar Lutestring is found in woodland with stands of its foodplant, Aspen. In Northern Ireland, this tree is widespread in rocky places especially on lake shores and the margins of semi-natural woodland in Fermanagh. Investigation of habitats where the foodplant exists may lead to the discovery of other populations. The recorded flight period is 14 May to 23 July. There is a nineteenth-century record from the Enniskillen area of Fermanagh of larvae having been found on Aspen. In Britain the larva feeds at night on the leaves of Aspen and possibly other related species. It spends the winter months in the pupal stage.

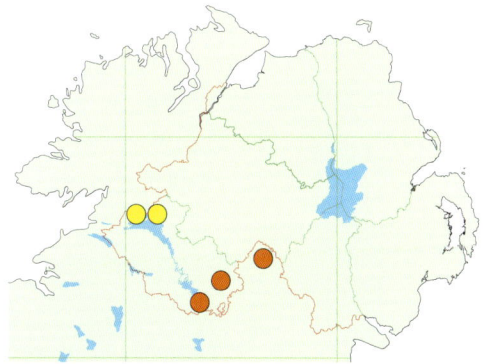

Common Lutestring
Ochropacha duplaris (Linnaeus) Category one.
Common. Widespread and frequently encountered in southern counties especially the woodlands, fens and bogs of north Armagh. Surprisingly there are virtually no records from north of Lough Neagh, the only ones being from Cromore (Londonderry) in 1980 and Rea's Wood (Antrim) in 1976. It seems likely to be under-recorded in many places in the north as suitable habitat exists in many localities. The forewings of the adult moth are pale grey with a darkish central band, although individuals vary slightly. A distinctive feature is the presence of two small black dots towards the centre of the forewing. It normally rests with its wings tightly closed along its body. Common Lutestring is found in a wide variety of habitats including amenity woodland and parkland, and secondary birch woodland on relict bogs and fens. Adults are attracted to light in small numbers. The recorded flight period is 19 May to 5 September. The larvae remain concealed by day utilising the leaves of the foodplant to hide in, emerging to feed at night.

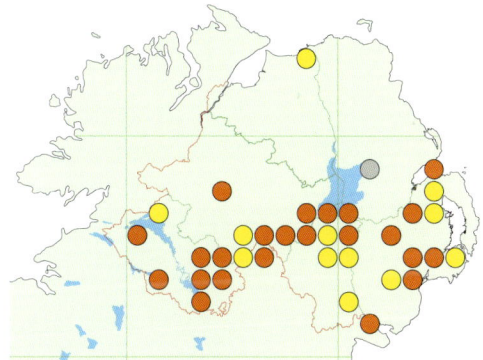

The Butterflies and Moths of Northern Ireland

Poplar Lutestring.

Crom, County Fermanagh.

A rare Aspen-feeding species which has its stronghold in County Fermanagh.

Larvae have not been found in Northern Ireland, but it seems probable that birch is the most likely foodplant here as it occurs at many of the recorded habitats. In Britain, other recorded foodplants include Alder, Hazel and oak. It overwinters as a pupa.

Yellow Horned

Achlya flavicornis (Linnaeus) Category two.

Scarce. This species was once considered a rarity in Northern Ireland, but with increased trapping in suitable habitat (birch woodland) during its early spring flight period, it has been found at sites in all counties. It is recorded from fens and bogs in the Lough Neagh basin including Brackagh Moss, Gosford Forest and Peatlands Park in north Armagh, and a scattering of sites (relict bogs and woodlands) in Down, Fermanagh and east Tyrone. Notable

Yellow Horned.

Lackan Bog, County Down.

An early spring species that was largely overlooked in the past due to its early flight period.

sites in these counties include Lackan Bog (Down), where it has been recorded in large numbers; Crom, Garvary Wood and Monmurry (Fermanagh) and several times at Rehaghy Mountain (Tyrone). There are a few records from Antrim including Breen Wood, Sharvogues Bog and Taylorstown. In Londonderry it has been recorded from Curran Bog and the woodland on the dunes at the Umbra on the north coast. Most of these sites have an abundance of birch and it is likely this moth is present in other birch woodlands especially in the west. Adults are generally grey although some specimens, particularly from Fermanagh, have a greenish-grey appearance. The forewings have a series of medial lines with a reniform stigma close to the costa. The

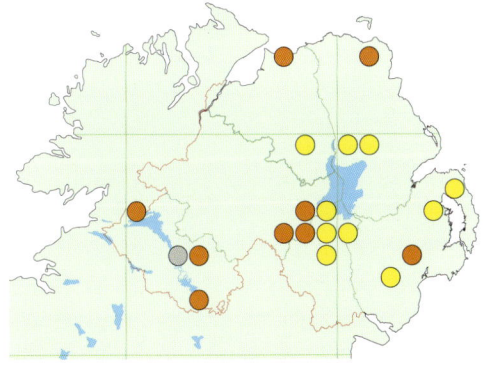

recorded flight period is 16 February to 22 April. It comes frequently to light in moderate numbers. The adults are said to rest on the branches and trunks of the foodplant during the day and have been reported to visit willow catkins after dark. The larvae have been recorded in Northern Ireland on birch. They remain concealed during the day between two leaves spun together. It spends the winter months in the pupal stage.

GEOMETRIDAE

This is the second largest family of larger moths in the Northern Irish fauna with 185 recorded species. The Irish checklist includes just over 200 species and another 100 species are known from Britain. The Geometridae are a varied group of moths, particularly in size and structure. The vast majority have a triangular butterfly-like appearance with slim, rather delicate bodies and long simple antennae. These are feathered in males and slender in females but are never clubbed as they are in butterflies. The females of some species have vestigial or rudimentary wings. The general shape and structure of geometrids means they are not normally noted for long-distance travel, but a few species are regular migrants.

Many geometrids have a skittish nature and readily take flight when disturbed. The majority of species become active from dusk onwards and most have an attraction to light. A few are day-flying; others are easily disturbed by day from low-growing vegetation or by shaking the branches of trees and bushes. Many species have functional mouthparts but do not visit flowers, although they will readily take moisture from the surface of leaves or branches when necessary.

The larvae of this family are quite diverse but all share one feature, the absence of three of the four pairs of prolegs normally found on larvae. They are commonly called loopers because of their distinctive form of locomotion. The larvae progress by moving the hind part of the body forward to just behind the front legs, so the middle of the body is temporarily arched in a loop, before it is straightened again by moving the front legs forward. The vast majority pupate in the ground or as a cocoon attached to the foodplant.

March Moth
Alsophila aescularia (Denis & Schiffermüller) Category one.
Widespread. This species has been recorded mainly in east and north Down and around Lough Neagh, particularly the wooded bogs and fens close to the southern shore of the lake; it is also known from several woods in Fermanagh. Sites in these areas include the Argory and Loughgall (Armagh); Belvoir Park, Helen's Bay and Lackan Bog (Down); Crom, Garvary Wood and Monmurry (Fermanagh) and Rehaghy Mountain (Tyrone). March Moth is also known from a scattering of sites along the north coast including Breen Wood and Portrush (Antrim) and the Umbra (Londonderry). The distribution is probably more widespread than the map indicates, especially in the north and west where recording effort is low in the early part of the year. Only the males of this species are fully winged, the females are completely wingless and spider-like. The males are generally a light greyish-brown with well-defined pale, jagged crossbands. The wings appear long and triangular and are overlapped when at rest. The records have chiefly been from woodland, wooded cutover bogs and fens along with some records from suburban gardens. The males are attracted to light in moderate numbers. Females remain hidden during the day, although the occasional specimen may be found in the early morning, resting on a tree trunk. The recorded flight period is 29 January to 27 April. The larva feeds

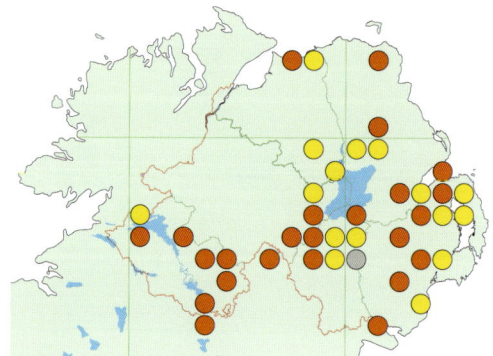

Moth accounts

Yellow Horned.

Lackan Bog, County Down.

An early spring species that was largely overlooked in the past due to its early flight period.

sites in these counties include Lackan Bog (Down), where it has been recorded in large numbers; Crom, Garvary Wood and Monmurry (Fermanagh) and several times at Rehaghy Mountain (Tyrone). There are a few records from Antrim including Breen Wood, Sharvogues Bog and Taylorstown. In Londonderry it has been recorded from Curran Bog and the woodland on the dunes at the Umbra on the north coast. Most of these sites have an abundance of birch and it is likely this moth is present in other birch woodlands especially in the west. Adults are generally grey although some specimens, particularly from Fermanagh, have a greenish-grey appearance. The forewings have a series of medial lines with a reniform stigma close to the costa. The

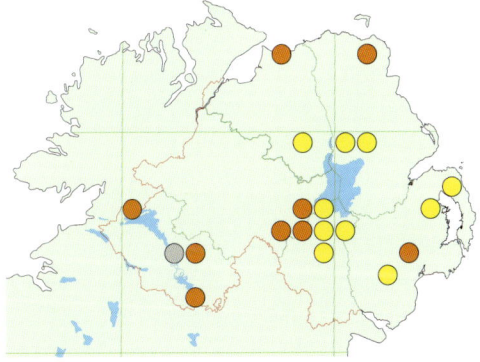

recorded flight period is 16 February to 22 April. It comes frequently to light in moderate numbers. The adults are said to rest on the branches and trunks of the foodplant during the day and have been reported to visit willow catkins after dark. The larvae have been recorded in Northern Ireland on birch. They remain concealed during the day between two leaves spun together. It spends the winter months in the pupal stage.

GEOMETRIDAE

This is the second largest family of larger moths in the Northern Irish fauna with 185 recorded species. The Irish checklist includes just over 200 species and another 100 species are known from Britain. The Geometridae are a varied group of moths, particularly in size and structure. The vast majority have a triangular butterfly-like appearance with slim, rather delicate bodies and long simple antennae. These are feathered in males and slender in females but are never clubbed as they are in butterflies. The females of some species have vestigial or rudimentary wings. The general shape and structure of geometrids means they are not normally noted for long-distance travel, but a few species are regular migrants.

Many geometrids have a skittish nature and readily take flight when disturbed. The majority of species become active from dusk onwards and most have an attraction to light. A few are day-flying; others are easily disturbed by day from low-growing vegetation or by shaking the branches of trees and bushes. Many species have functional mouthparts but do not visit flowers, although they will readily take moisture from the surface of leaves or branches when necessary.

The larvae of this family are quite diverse but all share one feature, the absence of three of the four pairs of prolegs normally found on larvae. They are commonly called loopers because of their distinctive form of locomotion. The larvae progress by moving the hind part of the body forward to just behind the front legs, so the middle of the body is temporarily arched in a loop, before it is straightened again by moving the front legs forward. The vast majority pupate in the ground or as a cocoon attached to the foodplant.

March Moth
Alsophila aescularia (Denis & Schiffermüller) Category one.
Widespread. This species has been recorded mainly in east and north Down and around Lough Neagh, particularly the wooded bogs and fens close to the southern shore of the lake; it is also known from several woods in Fermanagh. Sites in these areas include the Argory and Loughgall (Armagh); Belvoir Park, Helen's Bay and Lackan Bog (Down); Crom, Garvary Wood and Monmurry (Fermanagh) and Rehaghy Mountain (Tyrone). March Moth is also known from a scattering of sites along the north coast including Breen Wood and Portrush (Antrim) and the Umbra (Londonderry). The distribution is probably more widespread than the map indicates, especially in the north and west where recording effort is low in the early part of the year. 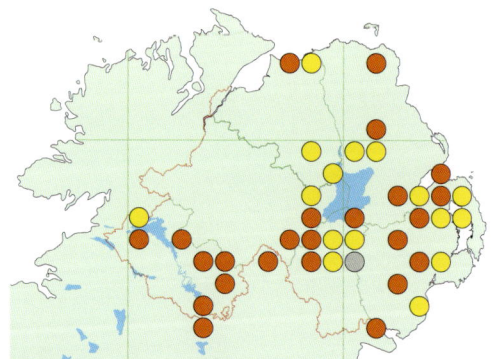 Only the males of this species are fully winged, the females are completely wingless and spider-like. The males are generally a light greyish-brown with well-defined pale, jagged crossbands. The wings appear long and triangular and are overlapped when at rest. The records have chiefly been from woodland, wooded cutover bogs and fens along with some records from suburban gardens. The males are attracted to light in moderate numbers. Females remain hidden during the day, although the occasional specimen may be found in the early morning, resting on a tree trunk. The recorded flight period is 29 January to 27 April. The larva feeds

March Moth.

Belvoir Park, County Down.

from May to June on the leaves of broadleaved trees including oak, birch, Blackthorn, Hawthorn and willow. It overwinters as a pupa.

Grass Emerald
Pseudoterpna pruinata (Hufnagel) Category two.
Widespread. Recorded throughout Armagh and Down, and frequent at several sites in both counties. Elsewhere, in the other four counties, this is a very local species, known from just a few sites. In Antrim it has been found at Dunmurry and White Mountain and once, in 1939, on Rathlin Island. In Londonderry, there have been records from Banagher Glen, Duncrun and, most recently, the Umbra. The Fermanagh sites are Altaclanabryan, Crom, Garvary Wood, Legatillida and Monmurry. It is only known from Dungannon, Ravella and Rehaghy Mountain Mountain in south Tyrone. Freshly-emerged adults are bluish-green with one white and two dark wavy cross-lines on the forewings. The green pigment is unstable and fades quickly, so most specimens taken at

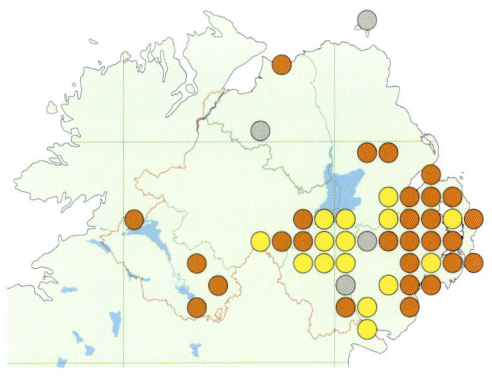

light show little or no trace of colour even though they appear to be in good condition. Grass Emerald occurs in heaths, open woods and other areas where Gorse is common. Adults come to light in small numbers and can, on occasion, be disturbed from vegetation during the day. The recorded flight period is 3 July to 1 September. The larva feeds at night and can be found from September to May of the following year on Gorse and Broom. It overwinters as a small larva.

Large Emerald
Geometra papilionaria (Linnaeus) Category one.
Widespread. Recorded in all counties and present as far north as Breen Wood (Antrim) and the Umbra on the north Londonderry coast. Known sites in the south and west include the

The Butterflies and Moths of Northern Ireland

Large Emerald.

Lackan Bog, County Down.

Argory, Aughinlig and Peatlands Park (Armagh); Belvoir Park (Down); Crom, Garvary Wood and Legatillida (Fermanagh) and Rehaghy Mountain (Tyrone). This is one of the largest of the geometrids and is instantly recognisable. The adult moth is green with white median cross-lines on both wings. The green colour pigment is more stable than in other species. They rest in a butterfly-like manner with the wings spread and held almost horizontally. Large Emerald is a woodland moth, found in open, broadleaved woodland and also wooded heaths and bogs. It is rarely seen in gardens. Adults are attracted to light in moderate numbers. They tend not to enter the traps, remaining outside in the vegetation. The recorded flight period is 20 June to 12 August.

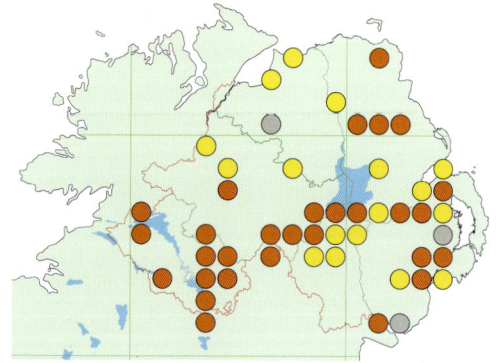

The larval foodplants are not known in Northern Ireland but, in Britain, it feeds on birch, Hazel and possibly Alder. It overwinters as a larva attached to its foodplant.

Common Emerald

Hemithea aestivaria (Hübner) Category two.

Scarce. Contrary to its vernacular name, the Common Emerald is less frequently recorded here than either the Grass Emerald or the Large Emerald. It is known chiefly from east Down, north Armagh and south Fermanagh. It has been found just three times in south Antrim at Belfast, Derryola Bridge and Lisburn. The most reliable sites for it (where it has been taken in most years) appear to be Aughinlig and Peatlands Park (Armagh), Aghalane and Crom (Fermanagh) and Mill Bay (Down). This species probably has a wider distribution than the current map indicates, especially in parts of Antrim, Tyrone and west Fermanagh where plenty of suitable habitat exists. Newly-emerged adults are dark green, but this colour fades rapidly. However, the adults remain easily recognised by the shape and the markings on the wings. The tip of the forewing is sharply pointed and there is also a sharp point on the rear edge of the hindwing. Both wings have a chequered outer fringe and there are two white cross-lines running roughly parallel to the hind margins. The median lines become less apparent as the colour fades. Common Emerald is found mainly in open woodland, scrub and bogs. The recorded flight period is 22 June to 16 August. The larva, which is the overwintering stage, can be found from August to May of the following year on Hawthorn, Blackthorn, birch, oak and willow.

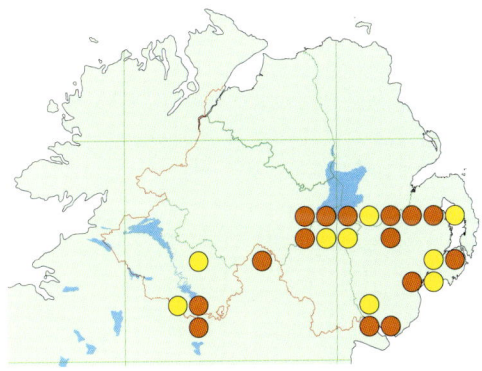

Small Emerald
Hemistola chrysoprasaria (Esper) Category four.
Unknown. There has been only one record of this species involving a few specimens recorded from Coalisland (Tyrone) in 1917. Donovan regarded these as 'probably introduced on imported plants of Clematis.' There have been no confirmed records since, so it appears that a population has not become established. The only other Irish record is an older one from Dublin which is supported by a voucher specimen in the collection of the National Museum, Dublin. Adults are a pale bluish-green colour, with two white curved cross-lines on the forewings. The hindwings are slightly angled and have a single white cross-line. The main flight period in Britain is late June to early August. The larval foodplants are Traveller's-joy and cultivated species and varieties of Clematis. Traveller's-joy is a naturalised species in Ireland and present in many parts of Northern Ireland, so it is possible the Small Emerald could become established if it were ever introduced. It overwinters as a small larva.

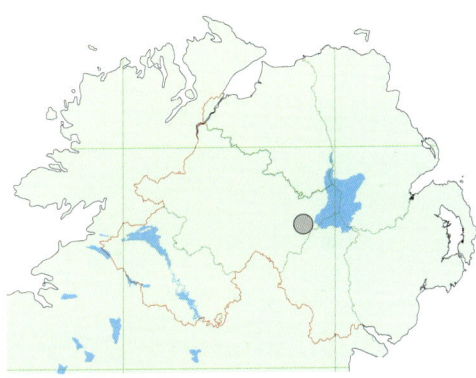

Little Emerald
Jodis lactearia (Linnaeus) Category one.
Scarce. This attractive species is known from a small number of sites in every county except Londonderry; most are in south Down and west Fermanagh. In Down, it occurs at Bohill Forest, Derryleckagh and Rostrevor Wood. The Fermanagh records are all from the west of the county at Correl Glen, Ely Lodge, Hanging Rock, Marble Arch and Monawilkin. It has been recorded just once in Antrim at Glenarm in 1993. In Armagh, it is also rare; there is an old 1960s' record from Richhill (not found there since) and a 2001 record from Carganamuck. There are also recent records from Altadavan and Rehaghy Mountain in Tyrone. This is the smallest of the emeralds found in Northern Ireland. The freshly-emerged adult is a very pale green, but this fades to white a short time afterwards and so its true colour is rarely observed on wild adults. Adults captured at light will often look remarkably fresh, even though their colour is usually white with a slight greenish tinge. Little Emerald is a woodland species and the records seem largely to be from Hazel and birch woodland in Northern Ireland. The moth rests by day among the foliage of trees where it has been beaten from the branches on several occasions. It flies at dusk along woodland clearings but rarely comes to light so is perhaps under-recorded in some localities. The recorded flight period is 17 May to 21 July. Larvae have not been found in Northern Ireland but it seems

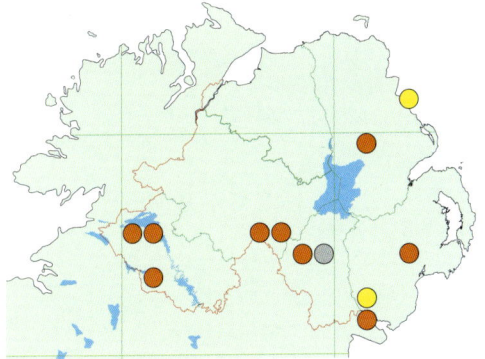

Little Emerald.

Derryleckagh Wood, County Down.

This is the smallest emerald found in Northern Ireland. It is generally associated with Hazel and birch woodlands.

probable that Hazel and birch are the likely foodplants here. In Britain other recorded foodplants include oak, Bilberry, Hawthorn and willow. It spends the winter months as a pupa.

Birch Mocha

Cyclophora albipunctata (Hufnagel) Category three.

Scarce. Since 1990 this has been recorded from a few sites in Armagh, Fermanagh and Tyrone. The majority of these records have come from Peatlands Park in north Armagh. Other localities in Armagh are the Argory, Brackagh Moss and Selshion Moss. In Fermanagh, there have been records from Aghalane, Enniskillen Agricultural College, Lisblake Bog, Monmurry, Mountgibbon Bog, Tullychurry and Tullyreagh Bridge. Recent Tyrone localities are Altaveedan and Rehaghy Mountain. There are historical records from Rostrevor Wood (Down) and Killymoon, Favour Royal and Altadiawan (Tyrone). The only record for Londonderry is from Tamniaran Bog in 1975. Adults are variable in colour but the majority of individuals taken here have pale greyish forewings with a diffuse reddish central band. There is a dark discal spot with a white centre on each wing (a distinctive characteristic of all mochas) and a broken medial cross-line that runs across both wings. Birch Mocha is recorded from heaths and bogs, especially where birches flourish. Adults are attracted to light but never appear commonly. It may be more widespread than records suggest in the west, where much suitable habitat exists. The recorded flight period is 19 May to 30 July. The larvae can be found on birch in Northern Ireland. It overwinters as a pupa.

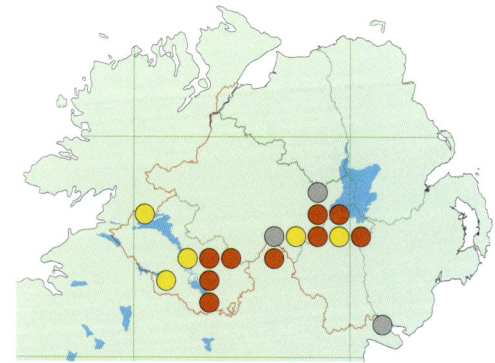

Clay Triple-lines

Cyclophora linearia (Hübner) Category four.

Rare. Currently this species is known only from Down and Fermanagh. It was recorded for the first time in Northern Ireland at Derryvore (Fermanagh) on 23 June 2001. There were no

Moth accounts

other records until 2004 when an adult was taken at Hillsborough Forest (Down) on 31 May. Baynes describes it as rare and local in Ireland and gave records only for Glengarriff (Cork) and Rathdrum (Wicklow). It has recently been recorded from Dromore Wood (Clare) and Newbridge (Dublin). There may be undetected populations, particularly in Fermanagh, where there are long-established habitats; ongoing fieldwork there may prove this to be the case. This is a fairly distinctive moth with brownish-yellow forewings which are slightly pointed at the apex. Most individuals have three conspicuous brown cross-lines on each wing. The median cross-line is broader and more evident than the other two and is the most constant and reliable feature. The head and thorax is a similar colour to the forewings. Clay Triple-lines has been recorded from mature broadleaved woodland in Northern Ireland. In Britain it is associated

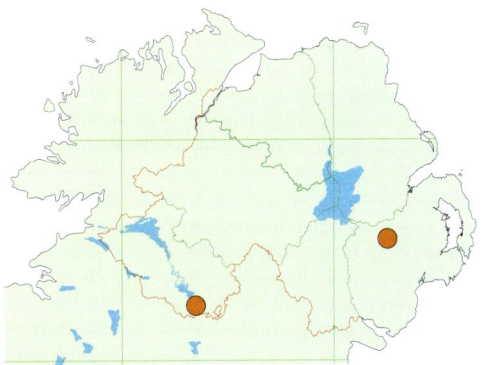

Birch Mocha.

Brackagh Moss National Nature Reserve, County Armagh.

with mature Beech woodland, hedgerows with established Beech trees, parkland and old estate woodland. Adults are attracted to light and can be found by day at rest on Bracken and other vegetation. There is insufficient data to determine its phenology in Northern Ireland. In Britain it is mainly single-brooded, flying from late May until early July, but a partial second generation from mid-August to mid-October is not uncommon. The larvae have not been found in Northern Ireland. In Britain, it feeds on Beech. It overwinters as a pupa.

Blood-vein

Timandra comae A. Schmidt Category four.

Rare. This distinctive species has been recorded once in Northern Ireland at Killard Point on the coast of Down on 19 August 1996. Despite repeated trapping at this site no other specimens have been reported. It is difficult to ascertain whether this individual was an immigrant or from an established population. Futher evidence is needed to prove its status here. The Irish distribution is mainly western and coastal in counties Cork, Kerry, Galway, Sligo and Louth. Blood-vein is not easily confused with any other species in the region. Adults are light creamy brown with a conspicuous pinkish-red stripe extending from the apex of the forewing to the inner margin of each hindwing. There is also a reddish fringe to each wing and a small discal spot on the forewing. In Britain it is found 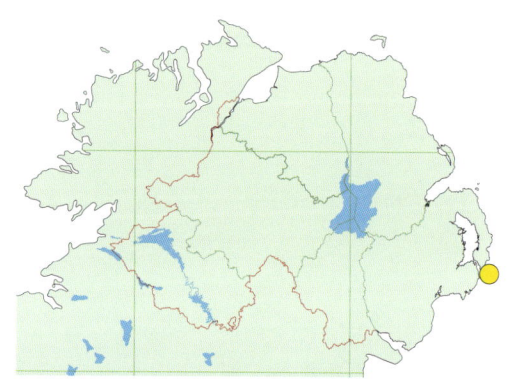 in weedy, ruderal habitats, and in rather damp, tall vegetation along ditches, hedgerows and wetland habitats. There is insufficient information to give an accurate flight period in Northern Ireland. It is mainly double-brooded in Britain from May to early July and early July to September. The recorded larval foodplants are docks, Common Sorrel and Knotgrass. Nothing is known about its larval habits in Northern Ireland. The Blood-vein overwinters as a larva.

Mullein Wave

Scopula marginepunctata (Goeze) Category four.

Rare. The Mullein Wave has been recorded seven times in Northern Ireland. Baynes listed it from Antrim, but gave no location details. The other records all originate from Down, from Kilkeel in 1894 and 1902, Helen's Bay in 1944, Hillsborough Forest on 25 July 1996 and, most recently, Cranfield on 11 July 1999. Elsewhere in Ireland it is known from the east and south coasts in Dublin, Wexford, Waterford, Cork and Kerry. Adults are mainly greyish-white with a cream fringe on all of the wings. The colour and the heavily freckled markings on the wings of the adult moth can be quite variable. This is a coastal species of rocky and sandy coasts and saltmarshes; it rarely strays inland. There are insufficient records to give an accurate flight period in Northern Ireland. It is double- 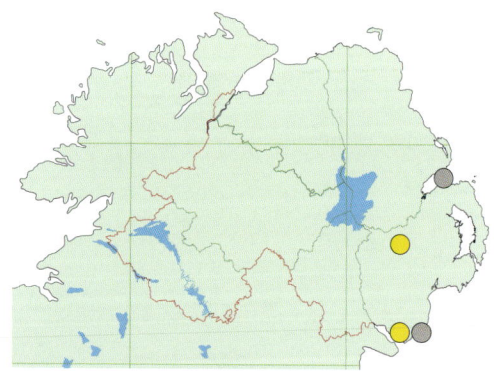 brooded in southern Britain, flying in June and July and again in August and September, but with a single generation in June and July in the northern part of its range. Its larval habits and foodplants in Northern Ireland are not known. The recorded foodplants in Britain include Mugwort, Wild Marjoram, Wood Sage, stonecrop, Yarrow and plantain. It overwinters as a larva.

Lesser Cream Wave

Scopula immutata (Linnaeus) Category three.

Rare. This species has been recorded at four sites in Down, and single sites in both Antrim and Armagh. The principal site here appears to be Brackagh Moss (Armagh) where it has been

Mullein Wave.

Disused quarry near Cranfield, County Down.

An elusive, mainly coastal, species which has rarely been seen in the wild.

recorded virtually annually since its discovery in 1989. The sites in Down are all in the south of the county at Derryleckagh, Lackan Bog, Maghery Bog and Mill Bay. Montiaghs Moss is the sole locality in Antrim. There is no evidence the species has increased in range, rather it was probably overlooked in the past. Adults are pale grey with a silky sheen to the wings. There is a series of ochreous medial lines across all four wings and a small black discal spot on the forewing. This spot may be difficult to see on worn specimens. It is similar in appearance to the rarer Cream Wave, which is usually larger with a more pointed forewing and no discal spot. Adults appear sparingly at light and their condition when captured is often poor, making it difficult to confirm identification. At some of its

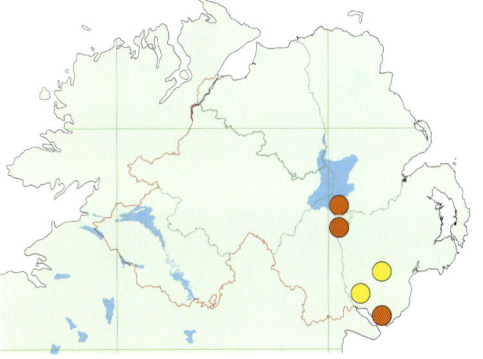

sites adults have been disturbed during the day from vegetation. Lesser Cream Wave occurs on bogs, fens and other damp, marshy localities. The recorded flight period is 23 June to 10 August. The larval habits and foodplants are not known in Northern Ireland. In Britain, the larvae feed on wetland herbs such as Meadowsweet and Common Valerian. It overwinters as a larva.

Cream Wave
Scopula floslactata (Haworth) Category four.
Unknown. There has been one acceptable record of this woodland moth from Rostrevor Wood (Down) in 1935. This record is given in Donovan's catalogue and ascribed to Foster. Recent claims have all proved to be incorrect or unsubstantiated. It is uncommon generally in Ireland, although widely distributed according to Baynes, and 'quite abundant' at Killarney (Kerry). Adults are ochreous white or smoky grey towards the front edge of the forewings. There is a series of continuous cross-lines on both wings, but sometimes these lines are faint and can be

difficult to see on slightly worn specimens. It is similar in appearance to the closely related Lesser Cream Wave, but adults of Cream Wave are larger and have a more pointed forewing. Any claimed specimens need to be examined carefully and records should be supported by a voucher specimen. The two species occur in different habitats, with Cream Wave in broadleaved woodland and hedgerows. Waring and Townsend state that it can often be disturbed from hedgerows and undergrowth by day. There is insufficient information to indicate the phenology of this species in Northern Ireland. The main flight period in Britain is May and June and into July in Scotland, but generally earlier than Lesser Cream Wave. The larval habits and foodplants are not known in Northern Ireland. Recorded foodplants in Britain include bedstraws and Woodruff. It overwinters as a larva.

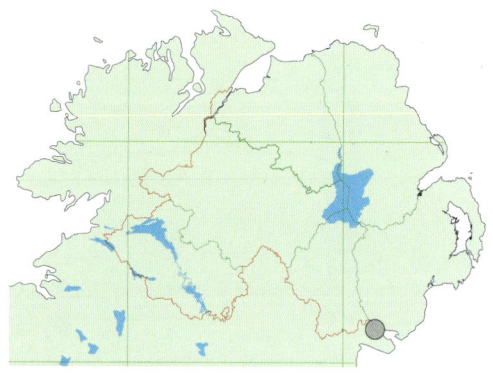

Smoky Wave
Scopula ternata (Schrank) Category four.
Rare. This species has only been recorded once, on 28 June 1997 at Murlough Bay in north-east Antrim, the first fully verified Irish record. Old records from Waterford and Kerry were considered doubtful by both Donovan and Baynes. The Antrim specimen was initially thought to be a Cream Wave, but was confirmed as this species by Bernard Skinner. The Smoky Wave is widely distributed in northern Britain, including south-west Scotland. Whether the Antrim specimen came from a resident population (perhaps recently established) or was a vagrant, is unknown. Vagrancy, however, has not been reported in this species, so it would appear more likely this specimen came from a resident population. The adults closely resemble both Cream Wave and Lesser Cream Wave and specimens of all three species require careful examination to confirm identification. Records from new sites ideally should be supported by a voucher specimen. In Britain, adults have been taken at light in small numbers and individuals have also been disturbed from the vegetation by day. The Antrim specimen was flushed from vegetation and netted. In Britain, the adults fly in June and July in upland heaths and moorland and, rarely, rocky calcareous grassland. There is insufficient information to give a flight period in Northern Ireland. Larvae have not been found here but, in Britain, they feed on Heather and Bilberry. It overwinters as a larva.

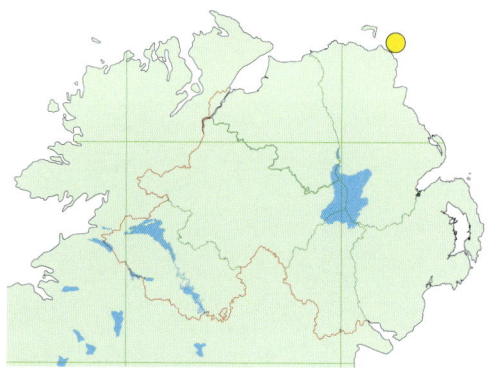

Small Fan-footed Wave
Idaea biselata (Hufnagel) Category one.
Widespread. Generally distributed across southern counties but more local north of Lough Neagh where it is mainly confined to coastal localities in Antrim and Londonderry; however, this distribution pattern probably reflects recording effort. Recently recorded sites in the north include Breen Wood, Clare Wood and Rathlin Island (Antrim) and Magilligan and the Umbra (Londonderry). Most adults have pale, ochreous wings, although this can vary between individuals. There is a distinctive dark greyish outer band and a small dark discal spot on each wing. Like many of the *Idaea* species, worn adults may be difficult to identify as the markings are obscured and, in this event, examination of the genitalia may be needed to correctly determine species. Small Fan-

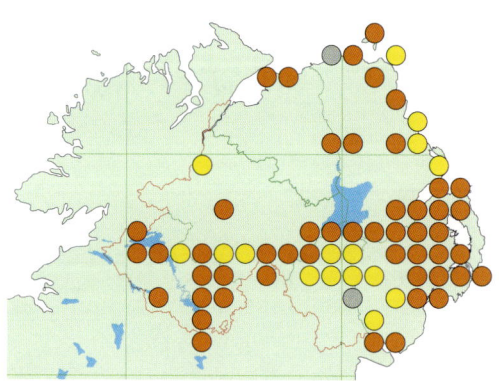

footed Wave is most commonly encountered in woodland, bogs, fens and mature suburban gardens. It is occasionally disturbed during the day from low vegetation and the branches of trees. Adults appear frequently at light in moderate numbers. The recorded flight period is 10 June to 1 September. The larval foodplants are unknown in the wild. It apparently overwinters as a larva.

Single-dotted Wave
Idaea dimidiata (Hufnagel) Category one.
Widespread. Recorded in all counties, although never as frequently encountered as Small Fan-footed Wave. It is most commonly found in the south-east. Rather local in the north and largely absent from inland parts of Antrim, Londonderry and Tyrone, but present on Rathlin Island (Antrim). Its apparent scarceness in these areas, especially in Londonderry, may well be due to the low level of recording at the appropriate time of year. Adults are generally cream with a series of small dark dots on the upper surface of the forewings. There is also a diagnostic dark patch near the outer margin on each forewing and a dotted line that is continuous around the outer margins of all wings. This species is found in damp woodland, hedges, heaths and bogs. It is a common visitor to some garden traps. Adults are attracted to light and appear in small numbers. The recorded flight period is 18 June to 31 August. The larvae can be found from September to May on Cow Parsley, Hedge Bedstraw and Burnet-saxifrage. It overwinters as a larva.

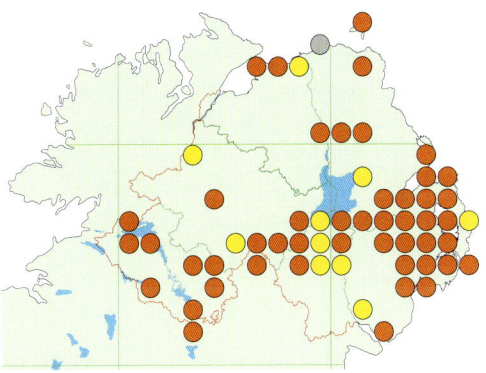

Riband Wave
Idaea aversata (Linnaeus) Category one.
Widespread. This is a common species in southern counties, but has a more localised distribution in Antrim and Londonderry. It was found on Rathlin Island (Antrim) in 2001 and it also occurs at Copeland Bird Observatory (Down). Adults are either red or grey, with three distinctive cross-lines on the forewings and two on the hindwings. There is also a banded form which has a well-defined, broad dark band across all wings. The unbanded form is very like Plain Wave which is generally smaller with a smooth silky appearance and a slightly curved rather than indented postmedian cross-line. This is a generalist species present in most habitats including woodland, heathland, fens, and rural gardens. Adults appear frequently at light in moderate numbers. The recorded flight period is 10 June until 5 September. The larval foodplants are not known in Northern Ireland but, in Britain, the larva is reported to feed on a variety of low-growing herbaceous plants which include bedstraw, Primrose, dock and Knotgrass. It overwinters as a small larva.

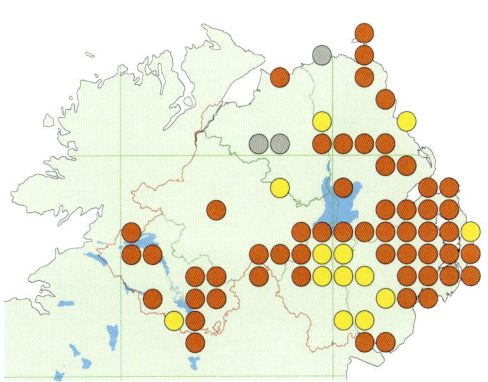

Plain Wave
Idaea straminata (Borkhausen) Category three.
Scarce. This species was described by Baynes as being local and rare, but recent recording puts its rarity in question. Since 1990, it has been recorded in every county except Londonderry, although there is an historical record from the county by D.C. Campbell. There have been many records from north Armagh at the Argory, Brackagh Moss, Clare Glen and Peatlands Park. The only Antrim sites are Ballinderry in the south of the county and, more recently, Breen Wood in the north. It is known from six widely scattered sites in Down

at Belvoir Park, Bohill Forest, Helen's Bay, Killard Point, Lackan Bog and Maghery Bog. In the west, there is an old record for Tamnamore and, more recently, Carnagat Forest and Rehaghy Mountain (Tyrone) and Correl Glen and Garvary Wood (Fermanagh). Adults are similar to the unbanded form of Riband Wave, but have a silky sheen to the wings. The subterminal line is distinct and curved towards the costa rather than indented as in Riband Wave. The Northern Ireland records have been from broadleaved woodland, wooded bogs and fens and coastal grassland. The recorded flight period is 1 July to 9 August. The larval foodplants in the wild are unknown. It overwinters as a larva.

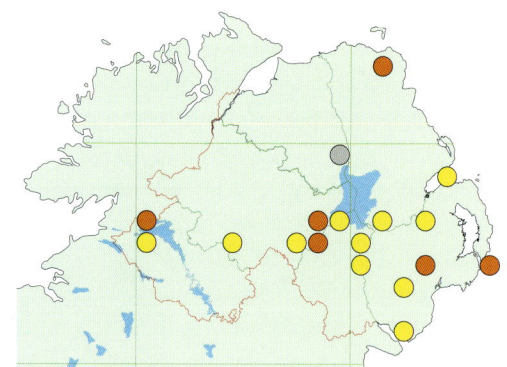

Vestal

Rhodometra sacraria (Linnaeus) Category four.
Notable. A southern European and North African species that migrates northwards into central and western Europe, in some years reaching Northern Ireland. It was first recorded here in 1945 at Lisnabreeny (Down). As one would expect with a migrant, records are concentrated in the south and east. Prior to 1990 it had been recorded just four times, but since then, the frequency of records has increased markedly. There have been records in 1992, 1995 (four), 1996 (one), 1998 (one), 1999 (one), 2000 (two), 2001 (one) and 2003 (two). Apart from one instance when two adults were trapped together at Loughgall, the records have all involved singletons. Many of the records have come from the coast of Down including Helen's Bay (twice), Murlough NNR and Quintin

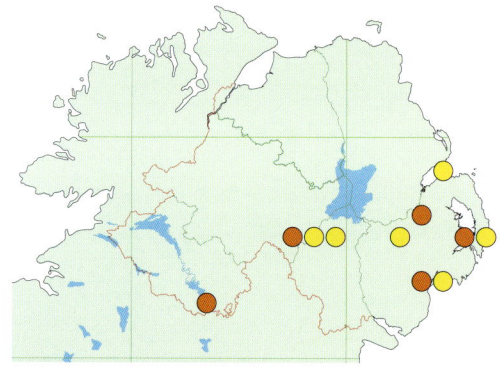

Bay. There are two records from gardens in the Antrim part of Belfast. In Armagh it has appeared at the Argory, Aughinlig and Loughgall. It has been seen twice at Stillago in Tyrone in 1998 and 2003 and on a single occasion in Fermanagh at Aghalane in 2001. Adults can show some variation in colour but most are generally pale ochreous. There is a distinctive oblique stripe that runs from the apex of the forewing to the inner margin; this is mainly brown but can vary in colour. Waring and Townsend mention specimens with a pink stripe, but to our knowledge this form has not been recorded in Northern Ireland. The wings are held close to the body in a high arched position. As is typical for a migrant, Vestal shows no strong habitat association. It has been seen in gardens on at least five occasions. Adults have been recorded from 4 August to 27 October and once in June. The larvae are continually brooded and have been found in the wild in Britain on Knotgrass. It is unlikely to breed here since it requires constant warmth.

Oblique Carpet

Orthonama vittata (Borkhausen) Category one.
Widespread. Generally distributed in southern counties, but much more local and sporadic in the north where it is probably under-recorded. North of Lough Neagh it has been found at Ahoghill, Ballymena and Rathlin Island (Antrim). There are also records for the Londonderry shores of Lough Neagh and Lough Beg as well as at Magilligan on the north coast. It is frequently encountered at the well-trapped sites in north Armagh and in south Fermanagh. Adults are generally straw-coloured with a distinctive oblique stripe and a small dark discal spot on each forewing. This is a wetland species, chiefly found in damp woodland, fens, bogs, wet heaths and dune slacks. Adults are attracted

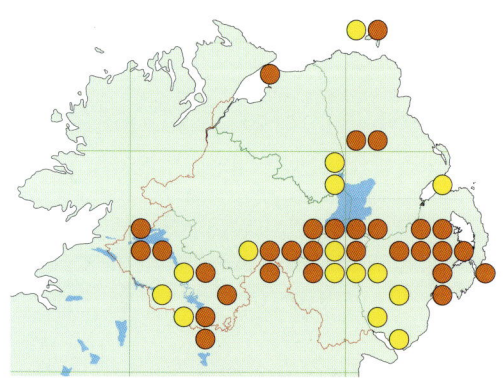

Vestal.

Belfast, County Antrim.

An attractive migrant that appears in small numbers in Northern Ireland in some years.

to light but appear only in small numbers. The recorded flight period is 18 May to 28 September. The data indicates there are two annual generations, the first from May to early July and the second from late July to the end of September. The larvae overwinter and can be found from September until April and again in midsummer on bedstraws and related species.

Gem
Orthonama obstipata (Fabricius) Category four.
Rare. This immigrant moth, originating from the warmer regions of southern Europe, is of almost annual occurrence in the southern half of both Britain and Ireland. The numbers vary

considerably from year to year, but the frequency of sightings has been increasing. However, this is not apparent in Northern Ireland where it has appeared only four times, the last occasion being 28 July 1989 at Castle Caldwell (Fermanagh). There are just three previous records, all over 100 years ago, from Ballycastle and Cushendall (Antrim) and Derry City (Londonderry). Adult males are pale brown with a dark, often irregular, central band. The females are darker, usually deep reddish-brown, with a small, white-ringed spot in the dark band. Adults are most likely to appear during the summer months particularly in coastal regions. The foodplants are unknown in the wild. It is continually brooded in southern Europe but the adults cannot survive the winter months as far north as Ireland.

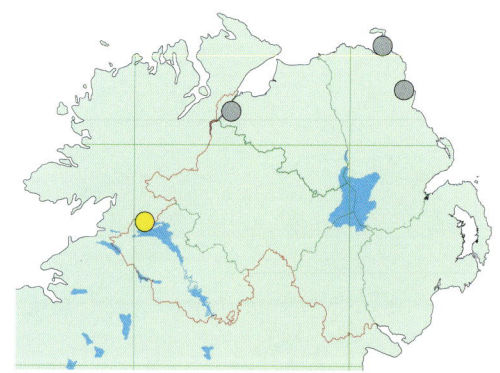

Flame Carpet
Xanthorhoe designata (Hufnagel) Category one.
Ubiquitous. This is a very common and widespread species especially in the south of Northern Ireland. It is known from many northern sites including a number on Rathlin Island (Antrim) and it has also been taken on Copeland Bird Observatory (Down). Adults (which are often variable in colour) are mainly pale grey with a distinctive reddish-brown central band, which has a double projection close to the outer area of the forewing. Flame Carpet is a generalist species that can be found in most habitats, but particularly damp woodland, meadows, hedgerows and suburban gardens. Adults come frequently to light in moderate numbers and have been disturbed from low vegetation by day. The flight season is prolonged, from 4 April to 29 September. The pattern of records indicates that two generations are produced at many Northern Ireland sites. There is no clear distinction between the two broods, suggesting more northerly populations are single-brooded. The larval foodplants in the wild are unknown. It overwinters as a pupa.

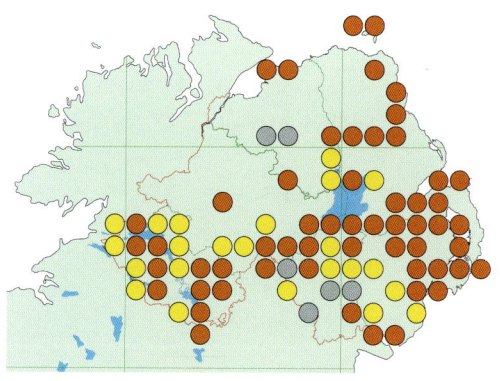

Red Carpet
Xanthorhoe decoloraria (Esper) Category four.
Rare. Recorded from all counties, but since 2000, this rare species has been recorded only in Antrim at Divis Mountain, Slievenacloy and on Rathlin Island. A 1994 record from Corbet Fen is its only recent occurrence in Down. In Armagh it was taken in 1977 and 1979 at Aughinlig, but not subsequently. Older, often vague, records for Antrim, Armagh, Down, Londonderry and Tyrone are given in Donovan and Kane. The lack of any recent records from some of its former haunts may suggest a decline, but the low level of recording in its preferred upland habitat is an equally plausible explanation. Adults have pale greyish forewings with a dark central band that appears much darker towards the edge. There is a single projection on the rear margin of the central band. The similar Flame Carpet has two clearly-defined projections on the outer edge of the central band. Red Carpet is a moth of upland heaths and grassland with rocky outcrops, although found at lower altitude in the north. Adults reportedly often rest on rocks and old stone walls. The recorded flight period is 6 July to 5 August, but this is based on few records. In Britain, the main flight period is late June to mid-August. The larval

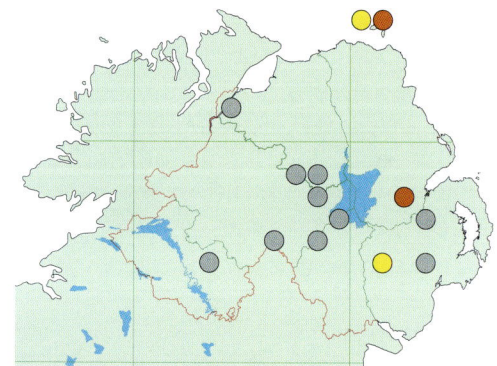

foodplant is Lady's-mantle which occurs at all recently recorded sites, but larvae have never been found in Northern Ireland. It overwinters as a larva.

Red Twin-spot Carpet
Xanthorhoe spadicearia (Denis & Schiffermüller) Category three.
Notable. Current records indicate this species is restricted to Fermanagh, south Down and Rathlin Island (Antrim). Apart from a single late nineteenth-century record from Armagh, it is unknown in other counties. Baynes described the species as common, an assessment that is unjustified today. It is found at seven sites in Fermanagh – Carrick Lough, Crom, Eshywulligan, Garvary Wood, Gorteen, Legatillida and Monawilkin. On Rathlin Island it was found at Kinramer at the western end of the island in 2001. The only recent record for Down is from Rostrevor Wood in 2001. Red Twin-spot Carpet closely resembles the red-banded form of the commoner Dark-barred Twin-spot Carpet and close examination of specimens is advisable. 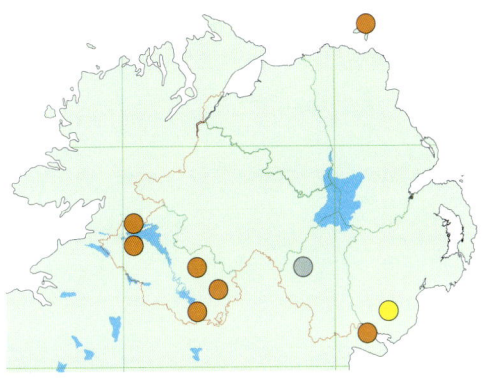 The Northern Ireland records have come from heath, broadleaved woodland, wooded bogs and calcareous grassland. It has two generations each year in southern England, but one in the north. The Northern Ireland data suggests that the double-brooded pattern is followed here; however, this is based on fewer than 20 records. The recorded flight period is 7 May to 25 June and 28 July to 18 August. The larval foodplants are not known in Northern Ireland but various bedstraws and Ground-ivy have been recorded in Britain. It overwinters as a pupa.

Dark-barred Twin-spot Carpet
Xanthorhoe ferrugata (Clerck) Category one.
Widespread. This species has been recorded in all counties but it has mainly a southern distribution. It is apparently much scarcer further north, and largely confined to north coast sites such as the Umbra and Magilligan (Londonderry). It is also present on Rathlin Island (Antrim). The adults are usually pale grey with a dark central band across the forewings. It is dimorphic, the forms differing in the colour of the central band. The black-banded form f. *unidentaria* is the dominant and most widespread form. The less frequent red-banded form is very similar to Red Twin-spot Carpet and some individuals may be difficult to distinguish. The best diagnostic feature is the presence of a notch on the rear edge of the central band towards the costa in Dark- 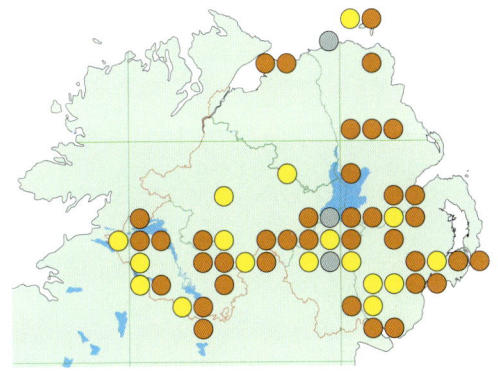 barred Twin-spot Carpet. This is absent in the other species. Dark-barred Twin-spot Carpet can be found in many lowland habitats including bogs, fens, and some suburban gardens. Adults come to light frequently and in moderate numbers. The recorded flight period is 25 April to 2 September in two generations – late April to mid-June and late June to September. The larvae feed on herbaceous plants including docks, bedstraws and Ground-ivy. It overwinters as a pupa.

Silver-ground Carpet
Xanthorhoe montanata (Denis & Schiffermüller) Category one.
Ubiquitous. This is one of the commonest of the carpets. It is found throughout Northern Ireland including Rathlin Island (Antrim) and Copeland Bird Observatory (Down). Adults are generally pale grey, although this is prone to slight variation. There is a prominent band across the centre of each forewing which varies in colour from light to dark greyish-brown. The

Silver-ground Carpet.

Lackan Bog, County Down.

margins of the band are wavy and the band itself varies in width. Forms exist in which the band is split or reduced to a patch near the costa. Each segment on the upper surface of the abdomen has a distinctive pair of small black dots. This is a generalist species that can be encountered in most lowland habitats where there are damp patches of tall herbaceous vegetation. It can be a common moth in suitable gardens and it is often disturbed by day from ground vegetation. Adults are attracted to light but appear only in small numbers. There is a single annual generation. The recorded flight period is 2 May to 24 August. The larval foodplants include Primrose, bedstraws and possibly other herbaceous plants. It overwinters as a larva.

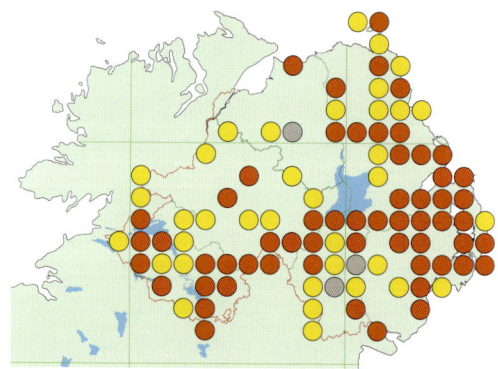

Garden Carpet

Xanthorhoe fluctuata (Linnaeus) Category one.

Widespread. Recorded in all counties and generally distributed in south Antrim, Armagh, Down and Fermanagh. It occurs at many sites along the north coast and is also found on Rathlin Island (Antrim). Adults appear pale grey, although the colour can vary between different individuals. The central band is usually incomplete, appearing as a dark patch at the costal edge of each forewing. There is also a dark basal patch and a broken line around the periphery of the forewing. As its name suggests, this can be a common species in gardens and urban habitats, but it can also be encountered in most natural habitats. Adults are attracted to light in moderate numbers and are also found occasionally at rest on stone walls and fences. This species has a long flight period. The phenology suggests that there are two generations but there is much overlap. There may be just a single generation at more northerly and upland

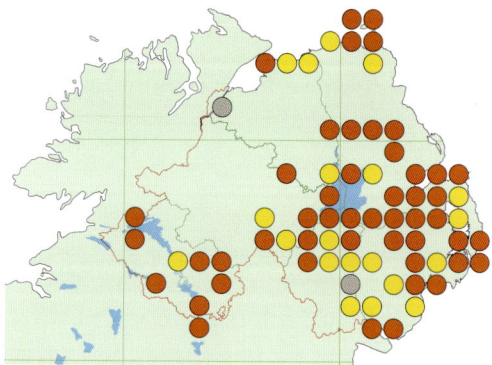

Lead Belle.

Rehaghy Mountain, County Tyrone.

localities. The recorded flight period is 22 April to 23 October. The larvae can be found from June until October on Garlic Mustard and other crucifers. It overwinters as a pupa.

Shaded Broad-bar
Scotopteryx chenopodiata (Linnaeus) Category one.

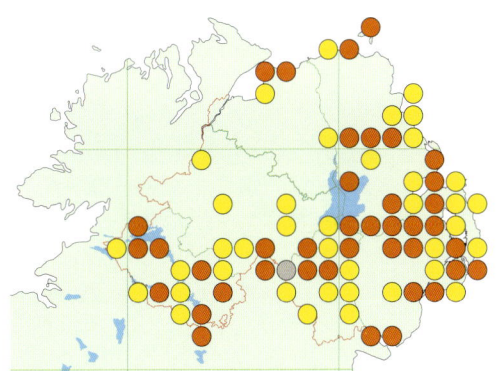

Widespread. This is a common species present in all counties throughout Northern Ireland. It has also been found on Rathlin Island (Antrim). The forewings of most individuals are ochreous-brown with a prominent darker central band; in contrast, the hindwings are greyish-white. Variants are common, especially darker, more reddish-brown individuals, but the general pattern of markings is consistent. Shaded Broad-bar occurs in grassy areas including dunes, heaths, semi-improved and unimproved grassland, woodland rides and glades and many gardens. Adults come sparingly to light but visit flowers, especially Common Knapweed and Common Ragwort, after dusk. Individuals are often disturbed during the day from low-growing vegetation. There is a single annual generation and the recorded flight period is 14 July to 8 September. The larval foodplants are clovers and vetches. It overwinters as a larva.

Lead Belle
Scotopteryx mucronata (Scopoli) Category three.
Scarce. This species has been recorded in all counties but it has a southern and western distribution. Most of the recent records are from Fermanagh and Tyrone including Garvary Wood, Legatillida, Monawilkin and Rehaghy Mountain. It has also been recorded on Rathlin Island (Antrim). In Down it has a predominantly coastal distribution with records from Cranfield, Helen's Bay and Killard Point. It also occurs inland from Newcastle at Tullybrannigan. This species closely resembles the July Belle, and where both species occur together, a voucher

specimen is required. Adults are usually darkish-grey with two median cross-lines – the central area between these lines is normally darker. There is also a distinctive apical streak and black discal spot, which in the case of this species is normally situated in the centre between the antemedian and postmedian lines, although this is not a reliable method for identification. The later flight period of July Belle can be a useful guide but examination of the genitalia is the most reliable method, especially when recording at a site for the first time. Lead Belle occurs in coastal and inland heaths and mixed woodland. Adults appear at light in small numbers. The recorded flight period is from 1 May to 24 July. The larval foodplants are Gorse and Broom. It overwinters as a larva.

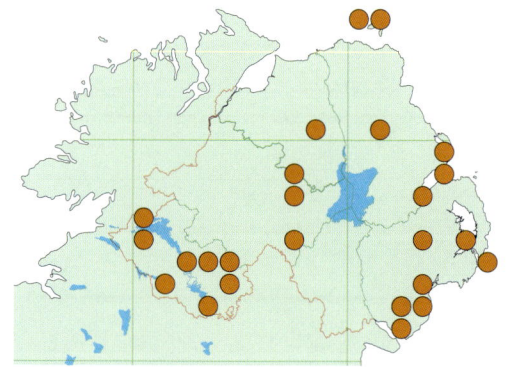

July Belle

Scotopteryx luridata (Hufnagel) Category three.
Scarce. There have been claimed records of this species from all counties except Antrim. However many of these records are difficult to verify since in most cases there are no voucher specimens and identifications have been made based on visual characteristics rather than examination of the genitalia. Many of these localities need to be visited again and specimens verified to give credence to earlier records. The most recent records emanate from coastal localities in Down, such as Bloody Bridge and Glasdrumman, where there are claims of an established population. Elsewhere there are inland records from Camlough (Armagh) in the early 1990s and Florencecourt and Marble Arch (Fermanagh); however, there have been no records from Fermanagh since 1992. North of Lough Neagh it was reported from Castlerock (Londonderry) in 1980. In Tyrone there are records for Aughnacloy and Ravella. Adults are very similar to the Lead Belle and the most reliable distinction between the two species is the position of the discal spot on the forewing. In this species this lies, in most cases, closer to the antemedian line than it does in Lead Belle. However, this character is not infallible. The two species can occur at the same site but the evidence suggests that July Belle is much the scarcer of the two in Northern Ireland. Flight period is also an indicator of species identity as this species flies later in summer than Lead Belle, although there is some overlap. Adults have only been recorded in Northern Ireland during July. In Britain, the main flight period is mid-June to early August. The larvae share the same characteristics and foodplants as Lead Belle. It overwinters as a larva.

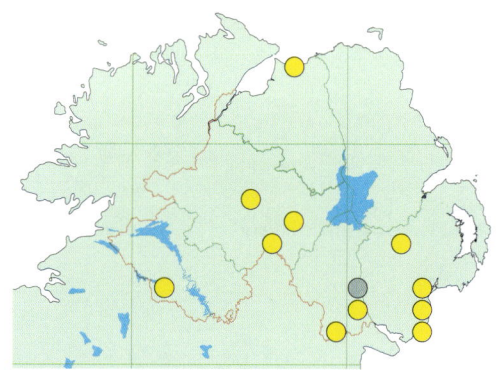

Small Argent and Sable

Epirrhoe tristata (Linnaeus) Category four.
Rare. This species has been recorded in Antrim, Armagh, Down and Fermanagh, but there have been only two post-1990 records. There have been several reports from the high ground on the north shore of Belfast Lough at Carrickfergus, Cave Hill and Knockagh (Antrim). The last record from this area was in 1943. Watts found it in Down in the 'Mournes' and at Tollymore Forest, where it was last seen in 1938 by the Rankins. It was recorded from Enniskillen (Fermanagh) in 1937 by Allen although this is almost certainly not the precise locality. There have been four reported occurrences in Armagh at Drumman More Lough in 1966, Armagh City in 1969, Slieve Gullion on 26 June 1993 and Altnaveigh in 2000. The last record (of two adults) is the most recent from the region. This species appears to have declined, and is deserving of specific effort to locate extant populations. It is a small but distinctive geometrid. The basal colour of the

forewings is white with a light or dark brown central band. The outer margins of both wings are suffused with black with an irregular, fine white line; the edges of the wings have a chequered appearance. It bears some resemblance to the larger Argent and Sable, but this has a greater area of white on the forewings and more contrasting chequered patterning. There is little information on the habitat occupied in Northern Ireland, but it would appear to be mainly a species of upland heaths and grassland as it is in Britain. It is also found on some lowland heaths and bogs. The few dated Northern Ireland records have all been in June. Elsewhere, its main flight period is May to early July with a partial second brood in August reported from southern England and parts of Ireland. Individuals will often fly in afternoon sunshine as well as at dusk. The larval habits are not known in Northern Ireland. In Britain the recorded foodplant is Heath Bedstraw which is common throughout most of Northern Ireland. It overwinters as a pupa.

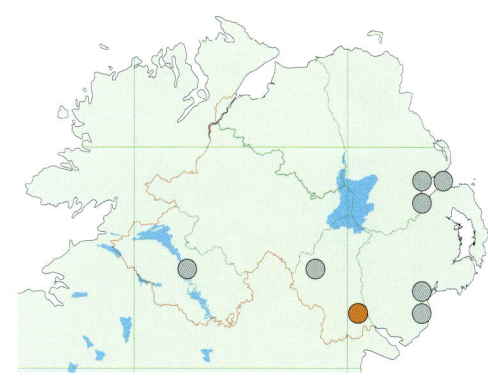

Common Carpet
Epirrhoe alternata (Müller) Category one.
Ubiquitous. This is the commonest of the *Epirrhoe* species in the region as well as one of the most widespread moths. There are significant gaps in the distribution that cannot be explained by under-recording. It is recorded from Rathlin Island (Antrim) and also Copeland Bird Observatory (Down). Adults are richly marked but quite variable in patterning. The ground colour is generally white, although most of this is obscured by the greyish markings. Most individuals show a dark central band, but this may be incomplete. Behind this band there is a thinner white bar enclosing a pale grey line. Common Carpet is a generalist species that can be found in most lowland habitats including rides and clearings in woodland, bogs, heaths, fens and suburban gardens. Adults become active from dusk onwards and are attracted to light in moderate numbers. They are frequently encountered (often in moderate numbers) during the daytime and are easily flushed from low-growing vegetation. In Britain, Common Carpet is double- or even triple-brooded in the south, but single-brooded in the north. The phenology of this species here is unclear as there is no noticeable separation between broods. The recorded flight period is 29 April to 12 October. The larva feeds on bedstraws and the overwintering stage is the pupa.

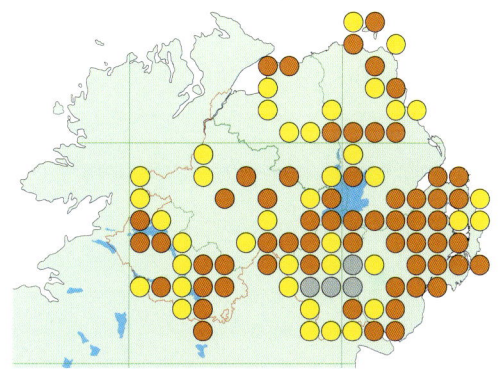

Galium Carpet
Epirrhoe galiata (Denis & Schiffermüller) Category three.
Notable. A very local species recorded from all counties. The most recent records originate from coastal localities including Copeland Bird Observatory and Murlough NNR (Down), the Umbra (Londonderry) and Rathlin Island (Antrim). Inland, there are records from Crom and Gortmaconnell Rock (Fermanagh). In the 1990s it was found at Ballykinler Dunes, Castlereagh and Killard Point (Down). There are older records from a scattering of sites on the Down coast, inland around Lough Neagh and in the Sperrins. Adults have pale forewings with a broad, darker central band. The costa is slightly concave centrally which provides a diagnostic feature. Galium Carpet is recorded from coastal heaths, dune grassland, limestone grassland and woodland.

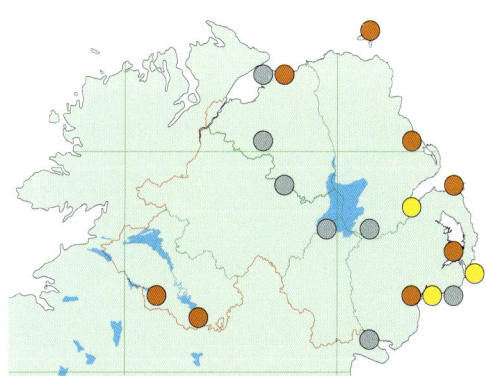

Yellow Shell.

Aughinlig, County Armagh.

Adults are active from dusk onwards and appear at light in small numbers. They are occasionally flushed from grassy vegetation during the day. The recorded flight period is 29 June to 3 September. In southern Britain and parts of Ireland it has two generations, but this appears not to be the case here. However, this is based on just a few records and more data is needed to ascertain if this is the case. The larval foodplants are bedstraws. It overwinters as a pupa.

Yellow Shell

Camptogramma bilineata (Linnaeus) Category one.

Widespread. Generally distributed in southern counties but more local in the north where it is mainly confined to coastal localities at Magilligan, Portstewart Dunes and the Umbra (Londonderry). It is also found at several sites on Rathlin Island (Antrim). Recorded locations in the south include Brackagh Moss (Armagh); Belvoir Park, Helen's Bay and Murlough NNR (Down) and Crom and Monawilkin (Fermanagh). Adults of this species vary in colour and the intensity of markings on the wings. Most individuals are deep yellow, heavily marked with dark wavy cross-lines. Some, however, show a central band but are still readily identifiable by the colouration. Adults trapped in coastal localities are often much paler and less heavily marked than

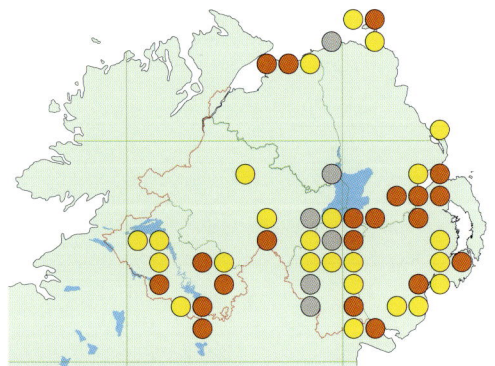

typical individuals. Yellow Shell is present in most lowland habitats. It can be found in rank grassy vegetation along hedges and other field margins, at the edges of woods and scrubby patches, on mature dunes, the drier margins of fens and many gardens. Adults appear at light, but never abundantly. They become active around dusk and are attracted to various flowers such as Common Ragwort and Red Valerian. Individuals have occasionally been found during the day when disturbed from the vegetation. The recorded flight period is 8 June to 5

September. The larvae can be found from September to May on herbaceous plants including docks, bedstraws and chickweeds. It overwinters as a larva.

Yellow-ringed Carpet
Entephria flavicinctata (Hübner) SOCC (P) Category four.
Rare. This species was first found in Ireland at Murlough Bay in northeast Antrim by Kane in 1897. The only other Irish record is from Legacurragh on Benaughlin in west Fermanagh where it was found in 1914. Recent attempts to rediscover it here have so far proved unsuccessful. The Murlough Bay population is, however, still extant with adults or larvae having been found in 1995, 1997, 1999, 2000 and 2004. Adults are generally pale grey and cryptically patterned to resemble the weathered surface of the rocks in their habitat. The basal, central and outer marginal cross-lines are suffused with yellow. This rare species occurs in flushed, unimproved grassland in limestone and basalt uplands and coastal scarps. Adults have only been recorded 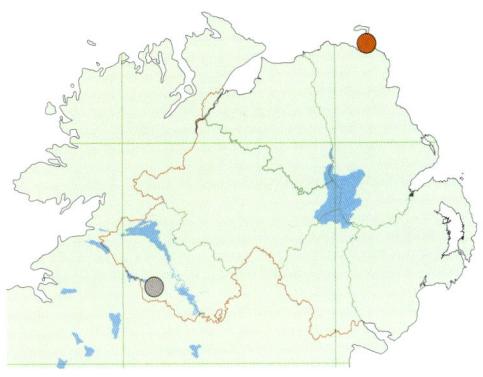 on one occasion at light in Northern Ireland, all other records have been either of individuals disturbed from their resting places during the day or of larvae on their foodplant. In Northern Ireland adults have only been recorded in August. In Britain the species is seen in May and August where bivoltine and in July and August in univoltine populations. The larvae, which overwinter, can be found from September to June on Yellow Mountain Saxifrage or Mossy Saxifrage. In Northern Ireland larvae have been found only on Mossy Saxifrage although Yellow Mountain Saxifrage does occur at both its known localities. Mossy Saxifrage occurs widely on the chalk and limestone in Antrim and Fermanagh uplands and there may be undetected populations of Yellow-ringed Carpet in these areas. It overwinters as a larva.

Grey Mountain Carpet
Entephria caesiata (Denis & Schiffermüller) Category two.
Notable. This moth is known from every county except Armagh. Since 1980 it has been found at Rathlin Island, Slieveanorra and Slievenacloy (Antrim); Benaughlin, Cuilcagh, and Marlbank area of west Fermanagh and at Murley Mountain (Tyrone). There are three historical records from Down, all from 1938. These were from Bloody Bridge and Slieve Bernagh in the Mourne Mountains and Drumkeeragh on Slieve Croob. It has not been seen in these areas since, but there is no reason to assume that it does not still occur. Magurran recorded it at Banagher Glen in 1978, the first, and to date, only record from Londonderry. Adults are greyish-brown with a series of dark median cross-lines. The hindwings are pale cream. There is a 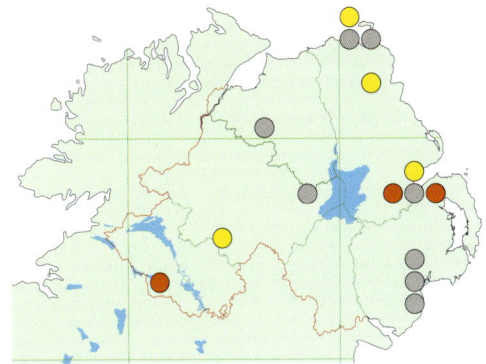 certain amount of variation in the appearance of the adults which seems to reflect the dominant colour of the rocks in the vicinity of each colony. So adults in Antrim, where the rock is basalt, are darker than those found on limestone in Fermanagh. Grey Mountain Carpet is found on rocky heathland chiefly in upland areas, but down almost to sea level in the Mourne Mountains and on Rathlin Island. Adults appear at light sparingly but are best looked for during the day (most of the recent specimens have been found in daytime) when they can be flushed from Heather and other low vegetation. They also rest on outcrops of rocks, stone walls, fence posts and tree trunks. The recorded flight period is 27 June to 13 September. The

The Butterflies and Moths of Northern Ireland

Grey Mountain Carpet.

Benaughlin, County Fermanagh.

larval habits and foodplants are not known in Northern Ireland. In Britain, the recorded foodplants are Heather and Bilberry. It overwinters as a larva.

Shoulder Stripe
Anticlea badiata (Denis & Schiffermüller) Category three.
Scarce. This species has been recorded frequently in Down, north Armagh, south Antrim, Tyrone and south Fermanagh. The only records outside this area are from Garvary Wood in north Fermanagh and the Umbra and Magilligan Dunes on the north coast of Londonderry. It is a regular species at the most consistently trapped localities including Aughinlig and Loughgall (Armagh); Helen's Bay (Down); Crom and Monmurry (Fermanagh) and Rehaghy Mountain (Tyrone). The adults are quite distinctive although they vary in colour and size. The majority of individuals are pale ochreous-brown with a broad buff or, in some individuals, a brownish-yellow central band. The Shoulder Stripe is a woodland species but it is also found along hedgerows, in scrubland, on wooded bogs and in mature gardens. Adults appear frequently at light in small numbers. This is an early spring species with a single annual generation. The recorded flight period is 4 March to 23 May. Its absence in much of the north and west is likely to be due to the low level of trapping in the early spring. The larvae can be found from May to July on wild roses. It overwinters as a pupa.

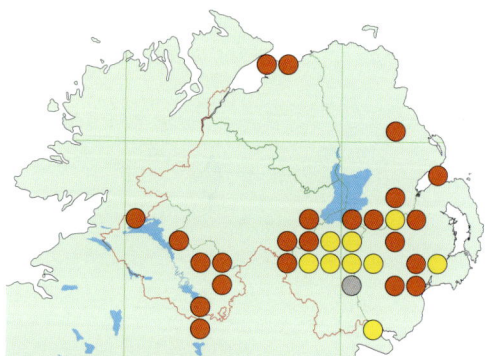

Shoulder Stripe.

Lackan Bog, County Down.

Streamer
Anticlea derivata (Denis & Schiffermüller) Category three.
Widespread. The Streamer is found in all counties but appears to be absent from the northern half of Northern Ireland. Across the southern part of the region it is a generally distributed moth. It is especially common in north Armagh and the well-wooded districts of south Fermanagh. The most northerly record is from Killybegs Bog in mid-Antrim. Reliable sites in southern counties include the Argory and Gosford Forest (Armagh); Belvoir Park and Helen's

Bay (Down); Crom (Fermanagh) and Rehaghy Mountain (Tyrone). Like the previous species, the lack of northern records may be due to poor coverage during its main flight period. This is a well-marked and easily recognised species. The adults are generally pale greenish-grey, although some individuals show a violet tinge when fresh. It derives its name from the wavy black line near the tip of the forewing that runs obliquely down from the costa. The Streamer is found on the edges of woods, in patches of scrub and along hedgerows. Adults are attracted to light in moderate numbers. The recorded flight period is 12 March to 25 May. The larvae can be found in June and July on wild roses. It spends the winter period as a pupa.

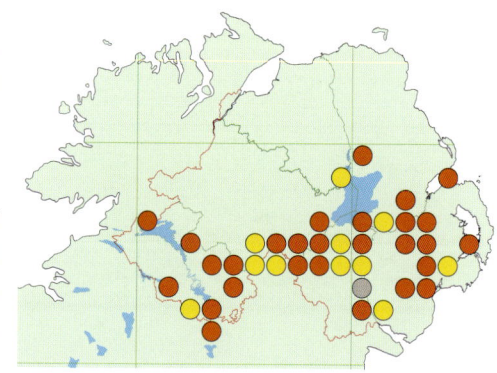

Beautiful Carpet
Mesoleuca albicillata (Linnaeus) Category two.
Widespread. Recorded in all counties, but more frequently encountered in north Armagh, Down and south Fermanagh. It has been recorded as far north as Magherafelt and Banagher Glen in mid-Londonderry. There is also an old record from Rathlin Island (Antrim) in 1937, but it has not been rediscovered here during recent periods of trapping. Other records from Antrim include Montiaghs Moss, Portglenone Wood, Tardree Forest and several suburban gardens, mainly in the south of the county. Brackagh Moss and Peatlands Park (Armagh); Belvoir Park, Bohill Forest and Helen's Bay (Down) and Crom and Garvary Wood (Fermanagh) are amongst the most reliable sites in southern counties for this species. This strikingly pale moth with sharply contrasting dark patches on the forewings is unmistakable. There is a broad white central band and purplish blotches near the apex and at the base of the forewings. The Beautiful Carpet is found in glades and along the edges of woodland, in scrub-covered fens, mature gardens and Hazel woodland on limestone. Adults are attracted to light but they only appear occasionally and always in small numbers. There is a single annual generation and the recorded flight period is 26 May to 8 August. The larvae can be found from July to September on Bramble, Hazel and Raspberry. It overwinters as a pupa.

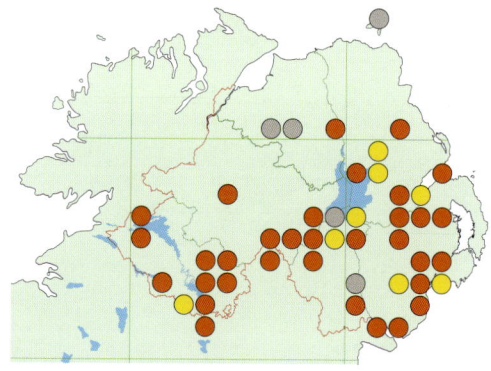

Dark Spinach
Pelurga comitata (Linnaeus) Category three.
Scarce. This species appears to have declined, but it may also be under-recorded as it occurs in habitats that are seldom trapped by recorders. Since 2000, there have been only two records (both from Down) at Clandeboye and Murlough NNR. In the 1990s it was found at Larne and Magheramorne (Antrim), Killard Point (Down) and Crom (Fermanagh). The other records are less recent and include eight from several sites in north Armagh up to 1980, including Derryadd Lough, Loughgall and Portadown. The Rankins found it on three occasions at Portballintrae (Antrim) in July 1937 and once, in July 1939, on Rathlin Island (Antrim). It has not been found anywhere on the north coast since and it also appears to have been lost from north Armagh. There are also old records from Rostrevor Wood and Strangford (Down) in Foster's list. The monitoring data from the Rothamsted trap network indicates a decline in abundance of this species in Britain of 95 per cent between 1968 and 2002. Adults are straw-coloured with a

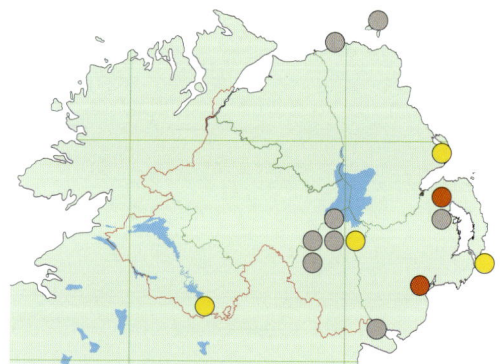

Streamer.

Derryleckagh Wood, County Down.

Beautiful Carpet.

Brackagh Moss National Nature Reserve, County Armagh.

darker, irregular, wavy median band. Adults normally rest with the tip of the abdomen curled up. Dark Spinach is found on disturbed ground on the coast and inland. The recorded flight period is 26 June to 19 August. The larvae have not been found in Northern Ireland. In Britain it feeds on goosefoot and oraches. It overwinters as a pupa.

Water Carpet
Lampropteryx suffumata (Denis & Schiffermüller) Category one.
Widespread. Recorded in all counties, this species has mainly a south-eastern distribution where it is frequently encountered at many sites including the Argory, Aughinlig and Gosford Forest (Armagh) and Belvoir Park and Helen's Bay (Down). It is especially common in south Fermanagh and regularly recorded from sites such as Crom, Garvary Wood, Legatillida and Monmurry Bog. There is a scattering of records from Antrim north to Glenariff Forest and on Rathlin Island. It also occurs at Banagher Glen and Traad Point (Londonderry), but the last record from either site was in 1990. Most of the records from Tyrone are all post-1996. Adults are variable in both markings and colour. The forewings are generally light brown with a darker brown central band 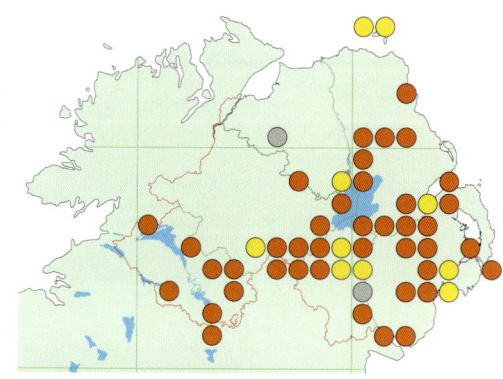 edged thinly in white. The rear margin is jagged with a distinctive double projection. Despite its name this species is not found in very wet places, but rather in damp woodland, heaths, bogs, fens and occasionally, mature suburban gardens. Adults are frequently found at light in moderate numbers. The recorded flight period is 11 March to 8 June. The larval foodplants are bedstraws. It overwinters as a pupa.

Purple Bar
Cosmorhoe ocellata (Linnaeus) Category one.
Widespread. There are records from all counties of this species. It is found at numerous sites around the coast of Northern Ireland, and it is also commonly encountered inland in parts of Tyrone and throughout Fermanagh. There are also records from Rathlin Island (Antrim) in 1988 and 1999. It is especially frequent on the coast of Down at Helen's Bay, Killard Point and Murlough NNR. Adults are normally pale grey with a silky appearance. There is a broad central band and a basal patch of dark bluish-grey, both edged by a thin ochreous-brown line. This species appears to favour sites with well-drained soil and the largest populations are found on sandy coastal sites. Inland, unimproved grassland and open woodland, often at altitude are its preferred habitats. Adults are active 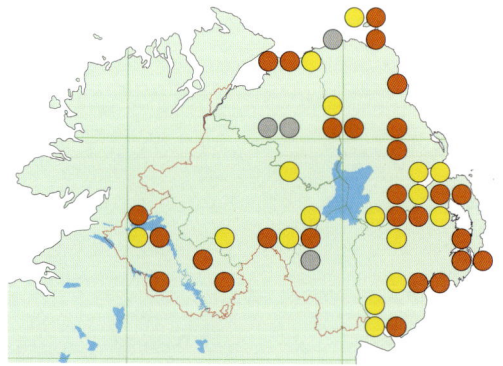 from dusk onwards and come frequently to light, often in good numbers at some of its coastal localities such as Killard Point and Murlough NNR. It is occasionally disturbed by day from bushes and other vegetation. The recorded flight period is from 6 May to 29 September with the majority of records in July and August. The pattern of records suggests there is just a single generation here rather than the bivoltine pattern prevalent in southern Britain. The larva feeds on bedstraws. It overwinters as a fully-grown larva and pupates in the spring.

Striped Twin-spot Carpet
Nebula salicata (Denis & Schiffermüller) Category three.
Scarce. The majority of recent records of this species originate from the Marlbank area of west Fermanagh including Legalough. Other Fermanagh localities include Garvary Wood,

Legatillida and Monawilkin. It has been frequently recorded in the extensive dunes on the north Londonderry coast at Benone, Magilligan and the Umbra. In other counties it has been taken at just a few sites in each including Fair Head (Antrim); Goragh Wood and Slieve Gullion (Armagh); Bloody Bridge, Murlough NNR and Rostrevor Wood (Down) and Davagh Forest (Tyrone). The adults appear grey with heavy mottling and fine dark and light cross-lines. There is variation in the intensity of the colour within populations. Striped Twin-spot Carpet is chiefly found inland on rocky hill slopes. Coastal populations exist on large sand dune systems and heaths. A few populations occur close to, or in, woodland. It is probable that there are undetected populations of this inconspicuous moth in upland areas throughout Northern Ireland. There are two annual generations in Northern Ireland. The first brood is recorded from 29 April to 20 June and the second from 29 July to 9 September. Adults are attracted to light in small numbers and may occasionally be disturbed by day from vegetation. The larvae can be found from summer through to autumn on bedstraws. It overwinters as a full-grown larva encased in its cocoon.

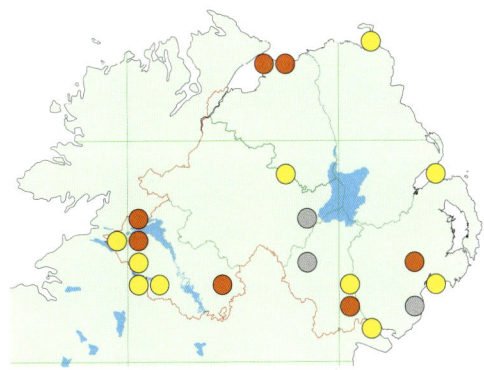

Phoenix

Eulithis prunata (Linnaeus) Category one.

Scarce. The greatest proportion of records of this scarce species is from gardens in the Greater Belfast area and wooded estates such as Belvoir Park, Castleward, Mount Stewart and Seaforde in Down. It is recorded from all the other counties, but the number of identified sites in each is small. Localities outside Belfast include Crom (Fermanagh), Favour Royal (Tyrone) and the Umbra (Londonderry). Adults have richly marked reddish-brown forewings. The wavy central band, which broadens towards the costa, is dark brown and edged with white. The adults resemble Small Phoenix, but are normally larger and much darker in appearance. The male normally rests with the abdomen curled up at the tip. The Phoenix is found in mature gardens and woodland especially on old estates where its foodplants are cultivated or are long naturalised. The recorded flight period is 5 July to 27 August. The larvae can be found from April to June on Black Currant, Red Currant and Gooseberry. It overwinters as an egg attached to the foodplant.

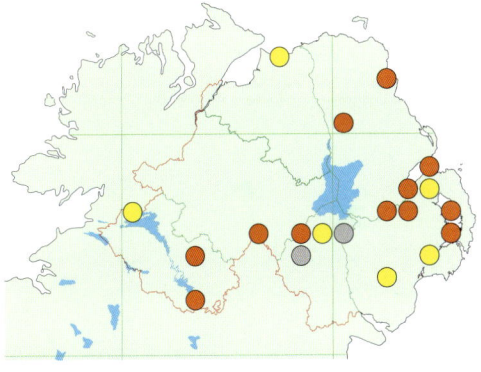

Chevron

Eulithis testata (Linnaeus) Category one.

Widespread. This species has a wide but rather patchy distribution across southern counties. It is more local across the north and west although still present in many areas. Recently recorded sites here include Breen Wood, Clare Wood and Glenariff Forest (Antrim) and Ballynagilly and Rehaghy Mountain (Tyrone). In Down, it is commonly met with at Ballykinler and Murlough NNR. There are also recent records from Rathlin Island (Antrim). The adults have distinctive markings and are readily identifiable. The most obvious feature is the chevron-shaped cross-lines on the forewings. There is also a reddish blotch on the hind edge and towards the tip of each forewing. Males are larger than females and yellow rather than brown. Chevron occurs on coastal heaths, wooded lowland bogs and fens. Adults come frequently to light in

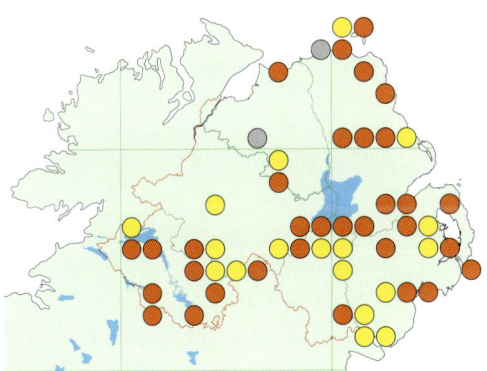

Chevron.

Ballykinler Dunes, County Down.

moderate numbers especially in coastal localities. Individuals have occasionally been found by day at rest in low vegetation. The recorded flight period is 15 July to 2 October. The larval foodplants are not known in Northern Ireland but it has been recorded on Creeping Willow, Aspen, birch, Hazel and Heather in Britain. It overwinters as an egg.

Northern Spinach

Eulithis populata (Linnaeus) Category two.

Scarce. This species is recorded from all counties, but it is infrequent in most. Many of the recent records are from Fermanagh and Tyrone. In both counties it has been discovered since 2000 at a number of sites including Altaclanabryan, Ballyreagh Bog, Carmacullagh, Garvary Wood and Randalshough in Fermanagh and Altadavan, Crockacleaven Lough, Davagh Forest and Muntober Bog in Tyrone. Away from these two counties the species occurs at several sites in north Antrim including Breen Wood, Glenariff Forest and Kebble on Rathlin Island and in south-east Down at Bohill Forest, Castlewellan Forest, Killard Point and Rostrevor Wood. Adults vary in colour from pale yellow to dark brown. The central band is darker with a double tooth-like projection on the outer margin. Adults normally rest with the abdomen curled up. Confusion is possible with Chevron, but adults of that species are normally darker and have a later flight period. Northern Spinach is found in upland and lowland bogs and heaths and acid woodland. Adults appear at light usually in very small numbers. The recorded flight period is 20 June to 4 September. The larvae have not been found in Northern Ireland, but in Britain they feed on Bilberry during April to June. It overwinters as an egg.

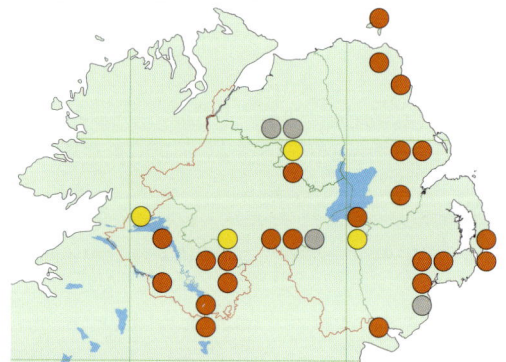

Barred Straw

Eulithis pyraliata (Denis & Schiffermüller) Category one.

Widespread. Generally distributed in Armagh, Down and south Fermanagh, but much more local elsewhere. It is found at Glenariff Forest in north Antrim and on the Londonderry coast between Magilligan Point and the Umbra. It has been seen at a scattering of sites inland at Banagher Glen (Londonderry); Ballynasollus, Rehaghy Mountain and Tattanellen Bog (Tyrone) and Aghalane, Brookeborough, Crom and Monmurry (Fermanagh). This is a distinctive species and unmistakable due to its peculiar resting posture. The narrow forewings are held well away from the body and the hindwings are hidden beneath them. When the moth is at rest, the edge of the hindwing and the tip of the abdomen are both held in a raised 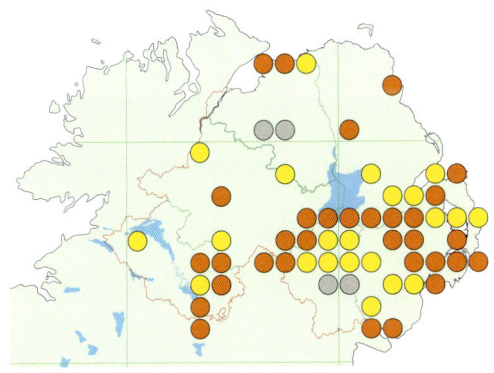 position. The forewings are generally lemon-yellow with brown cross-lines. Barred Straw is found in woodland, coastal grassland, unimproved grassland, bogs and gardens. Adults appear at light in good numbers. The recorded flight period is 2 July to 4 September. The larva feeds on bedstraws and can be found from mid-April until June. It overwinters as an egg.

Small Phoenix

Ecliptopera silaceata (Denis & Schiffermüller) Category two.

Widespread. This species has been recorded in every county except Londonderry. It is currently unknown north of a line between Larne (Antrim) and Omagh (Tyrone) although this is likely due to under-recording. It is a frequently recorded species at many of the regularly trapped sites in the south including Belvoir Park, Helen's Bay and Hillsborough Forest (Down) and Crom (Fermanagh). Its absence from the northern half of Northern Ireland is puzzling as there would appear to be plenty of suitable habitat. Adults are generally brown with a broad, dark brown central band which is often fragmented and enclosed with fine white lines. Beyond the central band is a series of small dark wedge-shaped spots. It is similar to Phoenix, but smaller and generally lighter.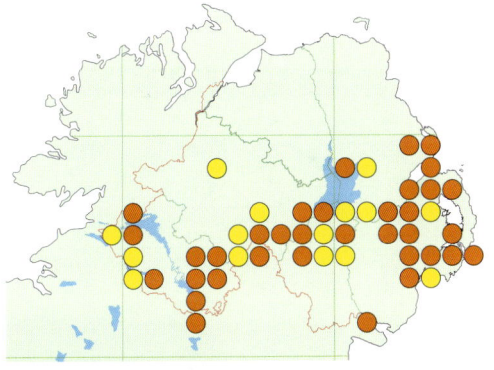
The tip of its abdomen is raised in the resting adult. Small Phoenix is found particularly in clearings and the ungrazed margins of woods, bogs and gardens with stands of tall willowherbs. The recorded flight period is 24 April to 10 September. There are two peaks in the pattern of records, indicating at least some of the Northern Ireland population may be double-brooded. The larva feeds on tall willowherbs, especially Rosebay Willowherb. It overwinters as a pupa.

Red-green Carpet

Chloroclysta siterata (Hufnagel) Category two.

Widespread. Generally distributed across southern counties, but very much scarcer north of Lough Neagh. The increased level of recording since the mid-1990s has produced many records and new sites, particularly in Fermanagh and parts of south Down. The northernmost sites are Breen Wood and Skerry East in north Antrim. Adults are rather variable, with some individuals being more strongly marked than others. The forewings are mottled green, often with a reddish tinge, and a dark central band. In some specimens the band is replaced by dark wavy cross-lines. Distinguishing them from Autumn Green Carpet can be difficult. They are best told by the colour of the hindwings, dark grey in this species, pale grey in Autumn Green Carpet. The forewings are more pointed in Red-green Carpet, giving the moth a more

Red-green Carpet.

Castlewellan Forest, County Down.

This is a widespread species which is similar in appearance to Autumn Green Carpet. Adults hibernate during the winter months and reappear in early spring.

streamlined appearance than Autumn Green Carpet. Red-green Carpet is a species of broadleaved woodland and mature hedgerows. Adults also appear occasionally in mature gardens and come frequently to light in moderate numbers. It visits Ivy blossom during the autumn and, after hibernation, willow catkins. Adults appear in late summer and early autumn and females hibernate, reappearing in early spring when they lay their eggs. Occasional specimens may be tempted to fly during mild periods in winter. The recorded flight period is from 24 August to 28 May of the following year. The larvae feed on the leaves of broadleaved trees and shrubs including oak, birch, Blackthorn, Rowan, Ash and Dog Rose.

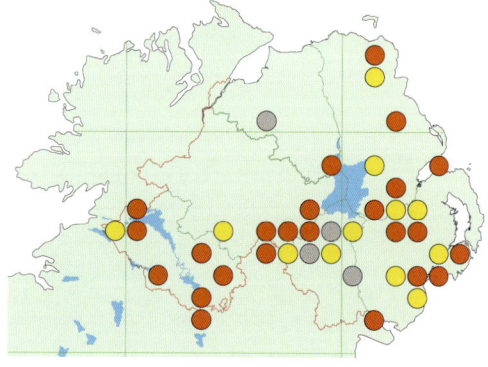

Autumn Green Carpet

Chloroclysta miata (Linnaeus) Category three.

Notable. This species is recorded from a small number of sites in south Down and north Armagh. The Down localities are Castlewellan Forest, Lackan Bog, Murlough NNR, Rostrevor Wood and Seaforde. In Armagh, there is an old record from Drumman More Lough in 1969. This species has been claimed from some additional sites but examination of vouchers has shown these to have been misidentified Red-green Carpets. There are historical records from Down, Fermanagh, Londonderry and Tyrone. Adults are similar in appearance to Red-green Carpet and specimens require careful examination where both species coexist. The forewings lack the reddish tinge often present

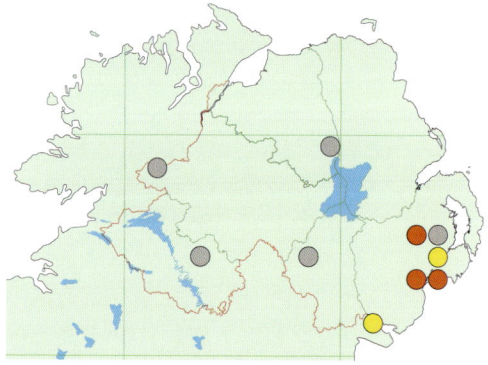

Moth accounts

Autumn Green Carpet.

Castlewellan Forest, County Down.

A much scarcer species than Red-green Carpet and often confused with it.

in Red-green Carpet and are more rounded, making the moth appear larger when at rest. The hindwings are also paler than in Red-green Carpet. Autumn Green Carpet is associated mainly with mature, broadleaved woodland and occasionally, bogs. The moth is less often seen than Red-green Carpet, which generally appears a few weeks earlier. The females hibernate during the winter months and reappear in early spring. Adults come to light sparingly but also visit Ivy blossom and willow catkins. The recorded flight period is 22 September to 11 October. There have been no confirmed records of hibernated females in the spring in Northern Ireland. In Britain these are recorded from March to May. The larva feeds on birch, willow, Alder, and Rowan.

Dark Marbled Carpet

Chloroclysta citrata (Linnaeus) Category one.

Widespread. Common and generally distributed in all counties and present on Rathlin Island (Antrim). This is an extremely variable species with many different colour forms. This species is similar to Common Marbled Carpet and specimens of both species should be carefully examined for correct identification. The most common variety has dark grey forewings with a pale brown central band. Paler specimens with pale grey forewings and a greatly reduced band are commonly seen. The most reliable feature is the pointed projection on the outer edge of the central band on the forewing. Dark Marbled Carpet is recorded from a wide range of habitats including woodland margins and clearings, bogs and heaths. Adults come to light frequently and in moderate numbers. They are occasionally disturbed by day from their resting position on tree trunks and stone walls. The recorded flight period is 5 June to 9 October. The larva feeds on birch, willow, Bilberry and Heather. It overwinters as an egg.

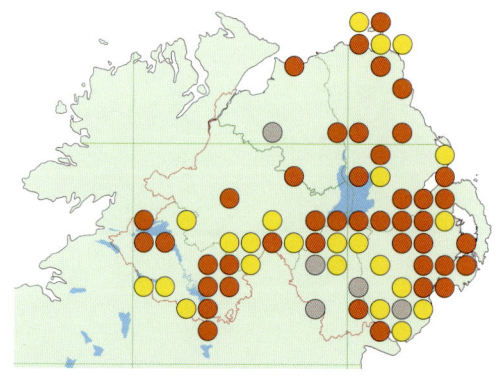

Common Marbled Carpet

Chloroclysta truncata (Hufnagel) Category one.

Ubiquitous. Common Marbled Carpet is one of the most frequently recorded of the small geometrid moths in Northern Ireland. It is present at many sites in all counties although, like many species, it appears to be absent from much of central Londonderry and west Tyrone. It was recorded on Rathlin Island (Antrim) for the first time in 2001. It is, like the related and similar Dark Marbled Carpet, a very variable moth and worn examples of both species may be difficult to distinguish without careful examination. The commonest form has dark grey forewings with a darker median band. The hind edge of the band has a rounded projection. A very distinctive form, unique to this species, has a large rusty patch in the centre of the wings. Common Marbled Carpet is a generalist species and a frequent visitor to garden moth traps. It is probably present in most habitats that have woody plants. Adults appear frequently at light in moderate numbers. There are two annual generations, the first peaking in late May, the second in mid-August. The recorded flight period is 29 April to 5 November. The larval foodplants are willow, birch, Bramble, Bilberry, Hawthorn, Wild Privet and Heather. It overwinters as a larva.

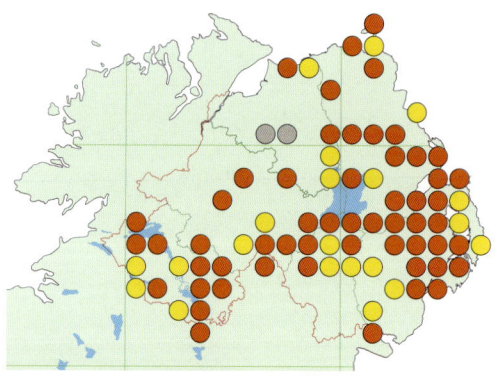

Barred Yellow

Cidaria fulvata (Forster) Category one.

Widespread. Recorded from all counties and generally distributed in north Armagh, Down, south Fermanagh and south Tyrone. It is found in scattered sites in south Antrim and south Londonderry and along the north coast between Magilligan (Londonderry) and White Park Bay (Antrim). It is also known from several localities on Rathlin Island (Antrim). Recently recorded sites include Derryola Bridge (Antrim);, Brackagh Moss (Armagh); Bohill Forest, Castleward and Mill Bay (Down); Aghalane and Gortmaconnell Rock (Fermanagh) and Rehaghy Mountain (Tyrone). This is a small, instantly recognisable species. The adults have orange-yellow forewings with a darker reddish-

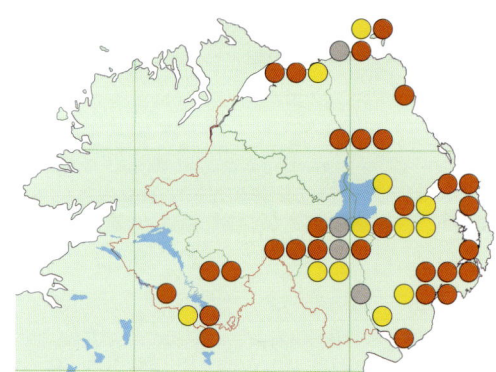

Moth accounts

Dark Marbled Carpet.

Rostrevor Wood, County Down.

Variable both in colour and markings throughout its range.

Common Marbled Carpet.

Castlewellan Forest, County Down.

Like the Dark Marbled Carpet, this species exhibits variation in colour and patterning.

113

brown central band. Barred Yellow has been recorded from woodland, cutover bogs, maritime heath, scrub-invaded rough grassland and sand dunes. It has rarely been reported from gardens. Adults appear at light in small numbers. The recorded flight period is 19 June to 3 September with a pronounced peak in mid-July. The larva feeds on Dog Rose and Burnet Rose. It overwinters as an egg attached to the foodplant.

Blue-bordered Carpet

Plemyria rubiginata (Denis & Schiffermüller) Category two.

Scarce. This is a widely distributed but rather infrequently recorded species, although it is known from all counties. At most of its localities records have been sporadic, often with a gap of several years between successive occurrences. It appears most frequently in north Armagh where it is known from Brackagh Moss, Oxford Island, Peatlands Park and Raughlan. In Down it has been recorded at Belvoir Park, Bohill Forest, Castle Espie, Castleward, Murlough NNR and Seaforde (Down). It also occurs at Florencecourt and Reilly Wood in Fermanagh, Duncrun and Traad Point in Londonderry and Favour Royal and Rehaghy Mountain, Tyrone. In Antrim it has been found on Rathlin Island and at Ballymena, Montiaghs Moss and Portglenone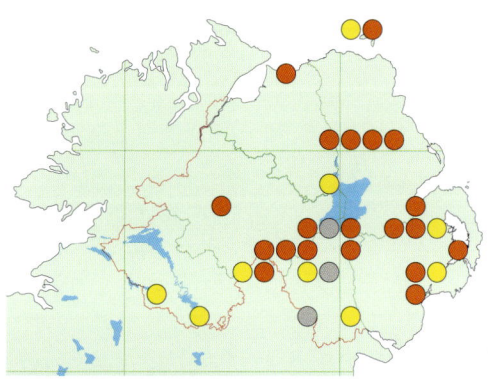
Wood. It may be more widespread than current records indicate especially where Alder is common. Adults are white with a bluish-grey outer border to the forewings. The central band is normally reduced to a brown blotch on the costa. Blue-bordered Carpet inhabits fens, bogs, damp woodland and other marshy areas. It has also been taken in garden traps. Adults appear not to be overly attracted to light and have been seen flying past actinic traps, which perhaps explains the paucity of records. They become active around dusk and have been observed in good numbers flying along woodland rides and clearings where they are easily taken with a net. They have also been disturbed during the day from vegetation. The recorded flight period is 22 June to 5 September. The larval foodplants are Alder, Blackthorn, Hawthorn and birch. It overwinters as an egg attached to the branches of the foodplant.

Pine Carpet

Thera firmata (Hübner) Category two.

Scarce. Since 1990, this species has been recorded from the eastern half of Down, north Armagh, south Tyrone, and south Fermanagh. It was recorded at Banagher Glen in Londonderry in the 1970s, but this remains the only record from the county. It has been regularly recorded from just a few sites including the Argory, Aughinlig and Loughgall (Armagh); Helen's Bay, Murlough NNR and Seaforde (Down) and Aghalane, Crom and Monmurry (Fermanagh). This is a late-flying species and this undoubtedly contributes to a degree of under-recording. Unlike other conifer-feeding moths, there does not appear to have been a significant increase in the range and abundance of this species. Adults are quite variable. The forewings are generally reddish-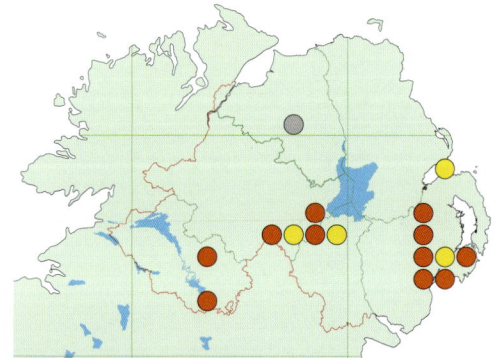
brown with a darker central band. This band is not always well defined, but is strongly indented on the inner edge unlike the closely-related Grey Pine Carpet, where the line is more uneven rather than indented. As its common name suggests, this species is associated with pine and it is found in habitats with mature Scots Pine trees including heathland, cutover bogs, estate woodland and gardens. It is generally absent from extensive conifer plantations. Adults show

Blue-bordered Carpet.

Brackagh Moss National Nature Reserve, County Armagh.

some attraction to light but are never seen frequently or abundantly. They can be disturbed by day from the branches of the foodplant. The recorded flight period is 8 September to 8 November. The larva feeds on Scots Pine. The larval habits are not known in Northern Ireland.

Grey Pine Carpet
Thera obeliscata (Hübner) Category one.
Widespread. This is a generally distributed species known from many sites across the southern counties. It is much more local to the north of Lough Neagh, but it has been found at Glenariff Forest and Breen Wood in north Antrim and at the West Lighthouse on Rathlin Island. The only known sites in Londonderry are Banagher Glen and the Umbra. This is an extremely variable species throughout its range. Adult colouration varies from grey to reddish-brown, with a well-defined darker central band which widens towards the costa. In most specimens the band is not edged with white as seen in Spruce Carpet. Grey Pine Carpet is found in mature conifer woods and mixed woodlands. It also occurs on cutover bogs and heathland with conifers, gardens and parks.

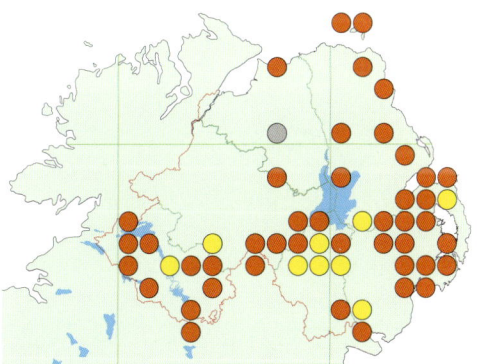

Adults come frequently to light in moderate numbers and can often be found during the day by beating the branches of the foodplant. There are two annual generations, the first from the end of April until late June and the second from late July until early November. The recorded flight period is 28 April to 28 November. The larva feeds on Scots Pine, Norway Spruce and probably other species of conifer. It spends the winter months as a larva.

Spruce Carpet
Thera britannica (Turner) Category one.

Widespread. The Spruce Carpet was first recorded in Ireland in 1958 at Glenageary (Dublin). Since then it has clearly increased its range and it has colonised much of Northern Ireland. It had certainly appeared by 1977 as Magurran recorded it at Banagher Glen (Londonderry) in that year, but it may well have been present but undetected before that. It occurs at many sites in southern counties, but it is rather less frequently recorded in the north of the region. The only recent records are from the Umbra. However, even here it is widespread as the records from Rathlin and the north coast testify. Adults are similar in size and appearance to Pine Carpet and Grey Pine Carpet, but this is a more richly marked species. The adults are grey or brown with a prominent 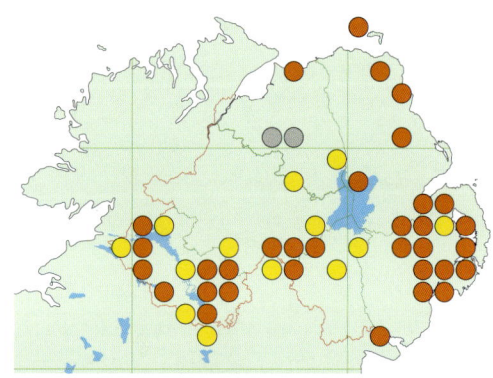 darker central band. The adults of the spring brood are browner than those of the second brood, which have a distinctly greyer tint. The central band is edged in white. Spruce Carpet is closely associated with conifer plantations and the greatly increased extent of these has clearly allowed it to colonise and spread throughout Ireland. It is also found, although much less frequently, in mixed woodland, gardens and other habitats with suitable stands of mature conifers. Adults come to light frequently in small numbers but are never as numerous as Grey Pine Carpet. Individuals rest during the day on the branches and trunks of the foodplant from where they are easily disturbed. The flight period lasts from 22 April to 21 November, in two clearly separated generations. The first is from the end of April to early August, the second from late August to late November. The larva feeds on Norway Spruce and Douglas Fir. It overwinters as a larva.

Chestnut-coloured Carpet
Thera cognata (Thunberg) Category four.

Rare. Chestnut-coloured Carpet is confined to Rathlin Island (Antrim), where it was discovered new to Northern Ireland in July 2001. Single adults were trapped at four different locations on the island at Cleggan, Cleggan South, Cooraghy Bay and Kinramer South on 28 July. According to Baynes this species was scarce and local in Ireland and found only in western Ireland between Clare and Donegal (where it was recently recorded again in 2003). The forewings are chestnut with a broad, darker, central band. On well-marked adults the central band is finely and conspicuously outlined in white. Adults resemble the similarly sized Juniper Carpet which has a black apical streak, absent in this species, and a much later flight period. Chestnut- 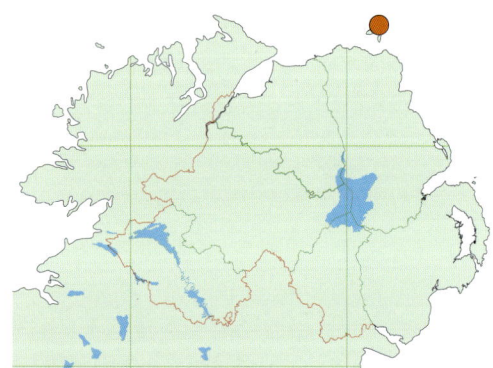 coloured Carpet is found on rocky heathland, moorland and sandy coastal heaths with established stands of its foodplant, Juniper. Oddly, Juniper appears to be rare on Rathlin Island although it does occur at Fair Head. There may be undetected colonies elsewhere within the native range of Juniper, especially in north-east Antrim and west Fermanagh. There is

Spruce Carpet.

Castlewellan Forest, County Down.

A conifer-feeding species which has clearly increased its range in recent years.

insufficient phenological data to determine the flight period in Northern Ireland. The main flight period in Britain is July and August and it overwinters as a small larva. The larval habits are unknown in Northern Ireland.

Juniper Carpet
Thera juniperata (Linnaeus) Category four.
Rare. This species was recorded for the first time in Northern Ireland in a suburban garden in Finaghy, south Belfast (Antrim), on 25 October and 4 November 2004. These are currently the only adult records for Ireland. Donovan considered the records of this species in Ireland to be doubtful, but Kane was of the opinion that it may have been introduced with imported, ornamental varieties of Juniper. Baynes gives records of larvae found near Lough Creag Dubh, Connemara (Galway) in 1956. There are also doubtful larval records from the Burren in 1963, which were apparently reared to adults. This is a rather small, plainly-marked species. Adults are brownish-grey with a darker median cross-band which widens significantly at the costa.

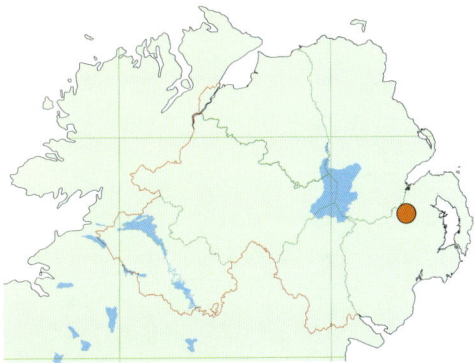

The distal edge of the band facing the outer end of the forewing is quite wavy. There is a dark apical streak which is conspicuous against the paler ground colour of the forewing. It seems highly likely that this species has been imported on cultivated plants rather than coming from an established population. At present there is no evidence to suggest that there are any resident populations even though ornamental junipers are common in gardens throughout Northern Ireland. Further research is needed before any firm conclusion can be reached. In Britain,

The Butterflies and Moths of Northern Ireland

Broken-barred Carpet.

Brackagh Moss National Nature Reserve, County Armagh.

Juniper Carpet is widespread on chalk downland and limestone grassland where its foodplant is found and also in many additional sites including parkland, estates and gardens. There is insufficient data to give an accurate flight period in Northern Ireland. In Britain the flight period is late September to early November. The larval foodplants are wild and cultivated junipers. It overwinters as an egg.

Broken-barred Carpet
Electrophaes corylata (Thunberg) Category one.
Widespread. This species seems almost entirely confined to the area south of a line from Belfast Lough to Pettigo (Fermanagh). It is most frequently recorded in the east, especially parts of north Down and Greater Belfast. The northernmost site is Shane's Castle on the north coast of Lough Neagh (Antrim), where it was last recorded in 1989. Other recently recorded sites include Portmore Lough (Antrim); the Argory, Carganamuck and Drumask (Armagh); Copeland Bird Observatory and Mount Stewart (Down); Aghalane, Altaclanabryan, Brookeborough, Cleen, Garvary Wood and Monmurry (Fermanagh) and Altadavan (Tyrone). Adults are easily distinguished from similar species by the distinctive, irregular cross-band on the forewing which appears broken towards the centre. The band is edged with a well-defined white line. Broken-barred Carpet occurs in broadleaved woodland, on wooded cutover bogs and heaths. It has rarely been reported in the large, well-established woodlands, seemingly being more typical of open, secondary woodland and scrub. It is a very regular species in gardens in the east of

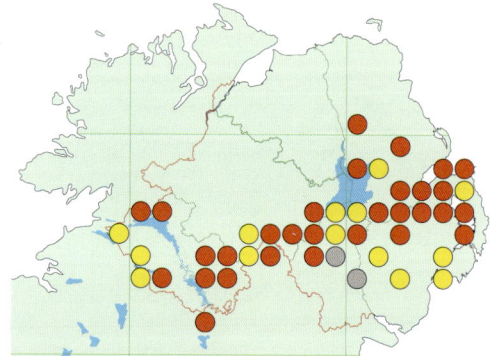

Northern Ireland. Adults come frequently to light, but never abundantly. They reportedly rest during the day on tree trunks and bushes. The recorded flight period is 11 May to 18 July. The larva feeds on birch, Hawthorn and oak. It overwinters as a pupa.

Beech-green Carpet
Colostygia olivata (Denis & Schiffermüller) Category four.
Unknown. The current status of this species here is unknown as there appear to have been no confirmed records for at least 60 years. There have been only three fully acceptable records. It was first recorded at Glenariff Forest (Antrim) by C.W. Watts in 1894 (who described it as numerous). The other records are from Tempo (Fermanagh) in 1894 by Langham and Murlough Bay (Antrim) in 1942. There have been some recent reports, but they have not been substantiated or supported by voucher specimens. It has not been rediscovered at any of the former sites. Fresh adults are green with a darker median band which is finely edged in white. The cross-lines are wavy and dark. However, like many other green species the colour fades to yellowish-brown as the moth ages.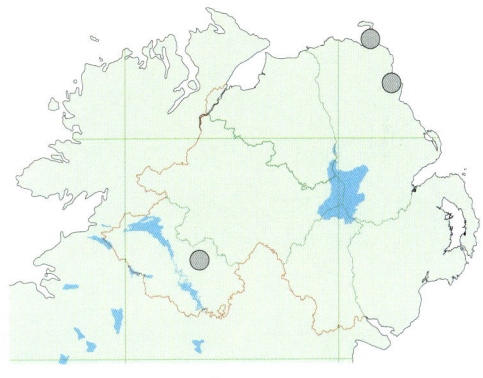
The hindwings are brownish-grey with a chequered fringe. There is no information on the habitat of the species in Northern Ireland. In Britain, Beech-green Carpet is found in woodland and moorland and, in southern Britain at least, it is largely restricted to calcareous soils. It reportedly rests during the day on rocks, stone walls and the trunks of trees. The main flight period in Britain is July to August. The larval habits and ecology of this species are not known in Northern Ireland. In Britain the larva, which overwinters in a young instar, feeds on bedstraws.

Mottled Grey
Colostygia multistrigaria (Haworth) Category two.
Scarce. Recorded from all counties including Rathlin Island (Antrim), but known from relatively few sites. It was regarded by Donovan as common throughout Ireland, but Baynes knew of only a few records. This species has probably been overlooked to some extent due to its early flight period and possible confusion with Early Tooth-striped. There are old records from Belfast Castle and Carrickfergus (Antrim) and Cookstown and Parkmount (Tyrone). Recent records have come from Breen Wood, Cleggan South, Glenmakeeran and Taylorstown (Antrim); Hillsborough Forest, Lackan Bog, Rostrevor Wood and Seaforde (Down); Aghalane, Crom and Monmurry (Fermanagh), and Curran Bog, Magilligan, Roe Valley and the Umbra (Londonderry).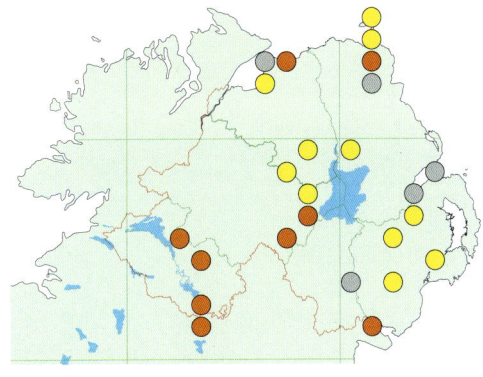
Surprisingly, it is only known from Peatlands Park in Armagh. Davagh Forest and Rehaghy Mountain are its only recent sites in Tyrone. Adults have mottled greyish-brown forewings with an indistinct darker central band and pale grey hindwings. Mottled Grey is an early spring species favouring woodland clearings, cutover bogs and heaths. Adults come to light in moderate numbers. The recorded flight period is 17 February to 5 May. The larval foodplants are bedstraws. It overwinters as a pupa.

Green Carpet
Colostygia pectinataria (Knoch) Category one.
Widespread. Generally distributed across the southern part of the region, but more scattered towards the north. It has also been recorded on Rathlin Island (Antrim). Recently recorded

Green Carpet.

Lackan Bog, County Down.

sites include Ballymena, Divis Mountain, Glenariff Forest, Lagan Meadows and Portglenone Wood (Antrim); Carganamuck (Armagh); Bohill Forest, Castleward, Castle Espie, Mill Bay and Rostrevor Wood (Down); Aghalane, Altaclanabryan, Brookeborough, Castle Caldwell and Garvary Wood (Fermanagh) and Tattanellen Bog (Tyrone). When freshly emerged, the forewings are generally a deep green with a darker central band which is edged with white and narrows towards the centre. The green pigment begins to weaken soon after emergence and the moth turns much paler. There are two black patches on the costa of each forewing. Green Carpet is a common moth, present in many open habitats including dunes, unimproved grassland, heaths and bogs. It can also be seen in open woodland, but it rarely ventures into gardens. Adults come frequently to light in small numbers. On occasions, adults can be flushed from the vegetation during the day. The recorded flight period is 29 May to 14 October with a peak in June and July. The data indicates that a partial second brood is developing in Northern Ireland in late August to mid-October. Two generations are now the normal pattern across southern Britain. The larval habits and foodplants are not known in Northern Ireland. In Britain, it feeds on bedstraws and overwinters as a larva.

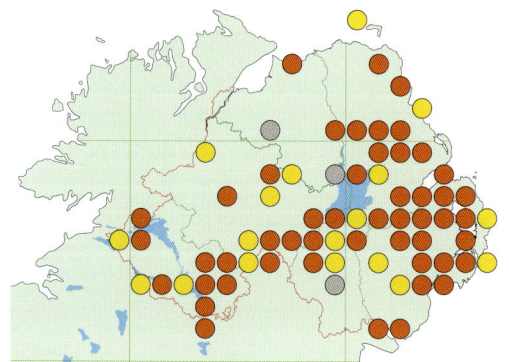

July Highflyer

Hydriomena furcata (Thunberg) Category one.

Ubiquitous. This is easily the most common of the three highflyer species found in the region. It has been recorded throughout Northern Ireland and on both Rathlin Island (Antrim) and Copeland Bird Observatory (Down). It is an extremely variable species, exhibiting many different colour forms ranging from green to brown. Individuals from upland sites, or where Bilberry is the main foodplant, are said to be smaller and darker than those found in lowland habitats. There is some evidence this applies in Northern Ireland, but it is by no means conclusive. The forewings have darkish bands and cross-lines which vary in intensity and pattern. Adults bear a close resemblance and share similar habitats to both May and Ruddy Highflyer; however, both these species have earlier flight periods, overlapping only slightly with this species. July Highflyer is a generalist species and present in a wide range of upland and lowland habitats. It is a frequent inhabitant of gardens. Adults are attracted to light, often in large numbers. They are occasionally disturbed from branches and foliage during the day. The recorded flight period of the single generation is prolonged, from 17 June to 14 October. The larval foodplants are Bilberry, Heather, Hazel and willow. It overwinters as an egg.

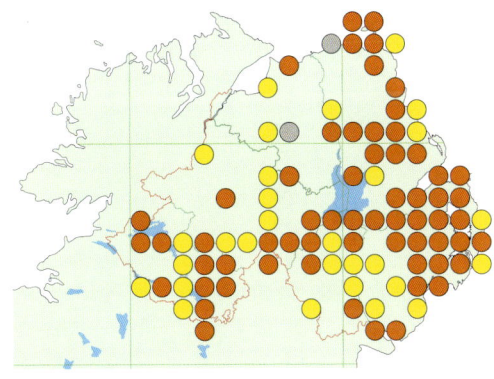

May Highflyer

Hydriomena impluviata (Denis & Schiffermüller) Category two.

Scarce. This species appears to have a restricted range in the southern counties. There have been no records from Londonderry and west and north Tyrone although it has been recorded from Rathlin Island in 1988. In the south it has been recorded from a range of sites including the Argory, Brackagh Moss, Clare Glen, Gosford Forest, Loughgall and Peatlands Park (Armagh) and Belvoir Park, Hillsborough Forest and Seaforde (Down). Most of the recent records, post-2000, originate from Fermanagh including Aghalane, Altaclanabryan, Crom, Eshywulligan, Garvary Wood, Lough Corry and Monmurry Bog. Other sites with recent records include Colin Glen, Montiaghs Moss and Portmore Lough (Antrim) and Annaloughan Bog, Crilly and Rehaghy Mountain (Tyrone). It has been trapped regularly in two Belfast gardens, one in the north of the city and one at Finaghy. The moth is probably more widespread than the map indicates, especially in parts of Armagh and the west where its foodplant occurs commonly. Adults are variable in colour. Most are greyish-brown with a pale central band but some are dark brown with pale cross-lines and an apical streak. The May Highflyer is found in wet woods, wooded lakeshores and fens containing stands of its sole foodplant, Alder. Adults appear at light in small numbers. The recorded flight period is 6 May to 1 July. The larvae can be found from August to October. It overwinters as a pupa.

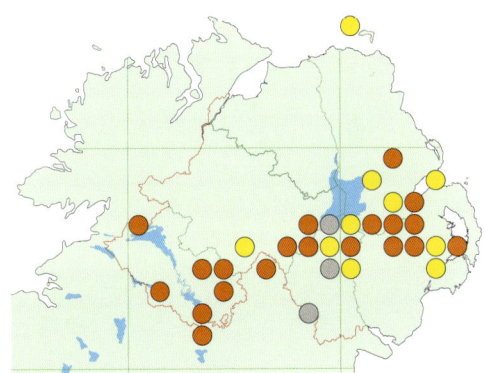

Ruddy Highflyer

Hydriomena ruberata (Freyer) Category two.

Scarce. This has a similar range to the May Highflyer but occurs slightly further north, extending to Ballymena in mid-Antrim. It has been recorded once on Rathlin Island (Antrim) in 1988 and more recently at Portmore Lough (Antrim). There appear to be no recent records from Londonderry. The majority of records are from Fermanagh especially in the south-east of the county at Fox's Hill Bog, Gorteen, Knocknalosset and Lough Corry. It was rediscovered in Tyrone at Annaloughan Bog in 1996, which was the first county record since the early part of the twentieth century, when it was recorded from the Cookstown area and Favour Royal. There have been few records from Down at Belvoir Park, Lackan Bog and Murlough NNR. Adults are mainly reddish-brown but like other 'highflyers' are prone to some variation in Northern Ireland. The forewings are narrow and have dark cross-lines and a short, oblique dash at the apex of the forewing. The Ruddy Highflyer is associated with heaths, damp woodland, and wet marshy areas. The recorded flight period is 28 April to 19 June. The larval habits are unknown in Northern Ireland. In Britain it can be found from July to September on willows. It overwinters as a pupa.

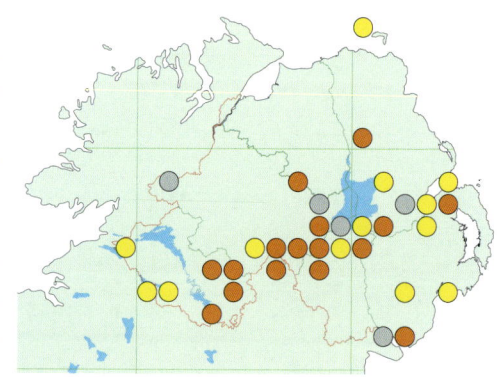

Slender-striped Rufous
Coenocalpe lapidata (Hübner) SOCC (P) Category four.
Unknown. There have been just two records of Slender-striped Rufous in Northern Ireland; both are more than 90 years old and it is feared the species may now be extinct. It was first recorded on Divis Mountain (Antrim) in 1893. The second record came from Benaughlin (Fermanagh) on 20 September 1914, where it was seen by J.E.R. Allen. Recent searches for the moth in the vicinity of Benaughlin and Cuilcagh in west Fermanagh have been unsuccessful; however, populations may remain undetected because of the remoteness of the habitat and the late flight period. Elsewhere in Ireland it was known from Donegal, Sligo, Mayo and Galway but there are no recent records from any of these counties. Adults have pointed, pale brown forewings which are marked with fine reddish-brown cross-lines. Research in Scotland has clarified its habitat preferences. There it occurs principally on flushed rough pasture which is influenced with lime. Similar habitat appears to exist in Fermanagh and possibly the Belfast Hills. The adults fly during the day in sunshine, but also come to light. The main flight period in Britain is September to early October. The larval foodplant is still unconfirmed, but a species of buttercup is considered the most likely candidate. It overwinters as an egg.

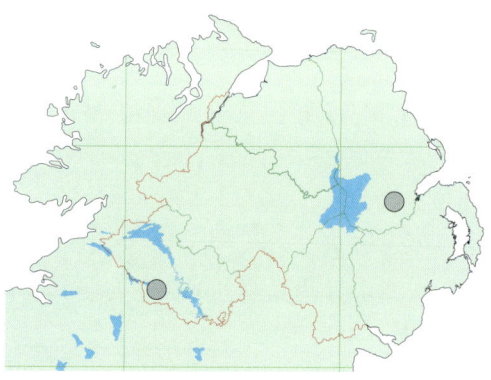

Argent and Sable
Rheumaptera hastata (Linnaeus) SOCC (P) Category three.
Rare. This species has always appeared to have been a rarity in Northern Ireland. Most records, including the only recent ones, have come from Fermanagh. Historically, it has also been reported from Down (Tollymore Forest), Londonderry (unknown locality) and Tyrone (Altadavan), but the last records from these counties were in 1941, 1875 and 1970, respectively. The first reference to Argent and Sable in Fermanagh was made by Partridge in 1893. Langham recorded it in 1902, thereafter it went unrecorded until 1997 when adults were seen at Lough Namanfin. Over the next two years, 1998 and 1999, the moth was seen at four different localities in west Fermanagh at Glennasheevar, Killybeg, Lough Alaban and Mullanawinna, but it has gone unreported since. This attractive and conspicuous moth has black and white wings and a striped abdomen. The outer margin of both wings is chequered. Argent and Sable is a day-

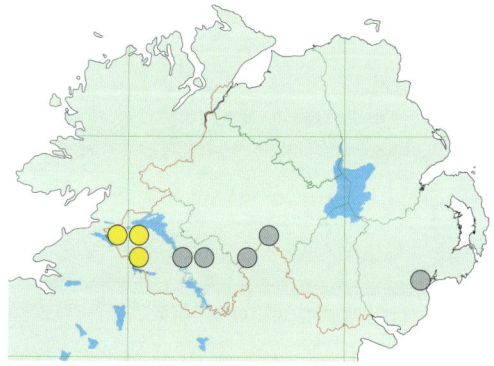

Scallop Shell.

Brackagh Moss National Nature Reserve, County Armagh.

flying species, most active in sunshine, when it flies in the manner of a butterfly. The sightings have been from wet heathland and flushed bog with stands of Bog Myrtle, open birch woodland and along rides and tracks in conifer plantations. The recorded flight period is 27 May to 20 June. The larval foodplant is not known in Northern Ireland, but Bog Myrtle, the foodplant used in Scotland, is common at all the recent Fermanagh localities, whereas birch (the usual foodplant in southern Britain) is sometimes absent. It overwinters as a pupa.

Scallop Shell
Rheumaptera undulata (Linnaeus) Category three.
Scarce. Recorded recently in 2003 and 2004 from a suburban garden in south Belfast (Antrim) and from several sites in Armagh, Down and Fermanagh. Many of these sites have been discovered relatively recently. The Armagh localities include Brackagh Moss, Carganamuck, Peatlands Park and Selshion Moss. In Down, it has been taken at Belvoir Park, Castle Espie, Helen's Bay, Hillsborough Forest and Lackan Bog. There are also recent records from Ardunshin, Crom, Florencecourt, Garvary Wood, Legatillida and Reilly Wood in Fermanagh. There is one old record from Lissan Estate in Tyrone and from Banagher Glen and Traad Point (Londonderry). The only recent record for Antrim away from Belfast, is from Breen Wood in 2004.

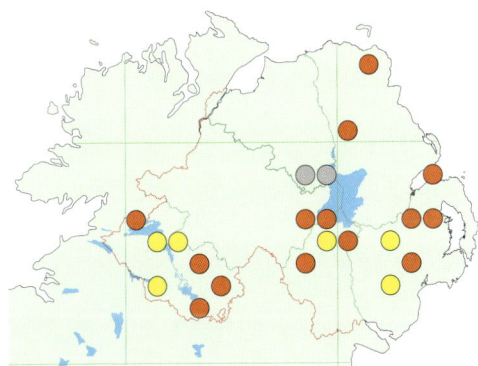

This is a distinctive moth and unlikely to be confused with any other species. The forewings are creamy white and heavily contoured with dark wavy cross-lines. The outer margin is pale sandy brown with a scalloped fringe. The hindwings are similar to the forewings. Scallop Shell is associated with fens, woodland and cutover bogs where willows are common. Adults appear

Tissue.

Correl Glen, County Fermanagh.

An elusive species that is seldom seen at light.

sparingly at light, usually as single individuals. They become active at dusk and have been seen flying along woodland rides where they have been observed feeding at Bramble flowers. The recorded flight period is 28 June to 10 August. The larvae can be found from August to October on willow and Bilberry. Scallop Shell overwinters as a pupa.

Tissue

Triphosa dubitata (Linnaeus) Category four.

Rare. There have been five records since 1990 from separate sites in Armagh, Down and Fermanagh, all involving single adults. It was also recorded from Cave Hill (Antrim) by the Rankins on 28 October 1938. The only site in Armagh is Clare Glen where a single adult was taken on 11 May 1991. The Fermanagh records come from Correl Glen, the walled garden at Crom and the hills around Legalough. The respective dates for these were 4 September 1999, 3 August 1997 and 18 August 1995. There are also historical records from Rostrevor Wood and Strangford (Down) both in 1933, and Stewartstown and Drumquin (Tyrone) in 1922 and 1923 respectively. Adults are predominantly greyish-brown with a rather broad forewing which is

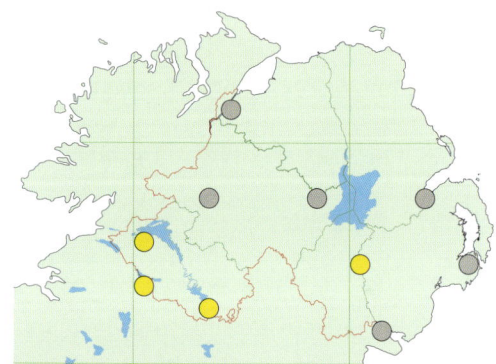

scalloped at the edges and has darker cross-lines. There is a darker central band (although on some individuals this is indistinct) which has a thin white line along the outer margin. Three of the recent records have come from within or near broadleaved woodland and one from unimproved calcareous grassland with scrub. All of the records have been at light. There is a single generation annually which appears in late summer and after hibernation in spring. In

Britain, adults are most often reported when hibernating in caves, disused buildings and unheated outhouses. The larval habits and foodplants are not known in Northern Ireland. In Britain, larvae have been recorded on Buckthorn and Alder Buckthorn. With the exception of Crom, these shrubs are absent from the sites where the Tissue has been found, suggesting it is using another foodplant or that the adults are prone to wander considerable distances.

Brown Scallop
Philereme vetulata (Denis & Schiffermüller) Category four.

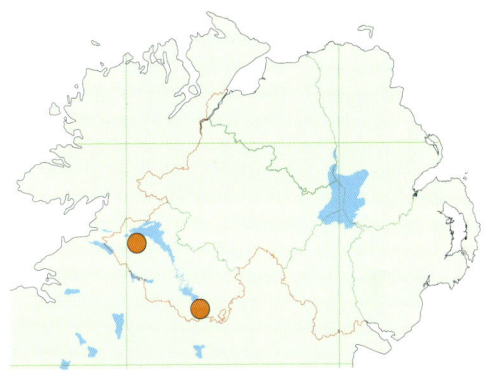

Rare. This is a rarely-reported species in Ireland. Baynes listed it only from Kenmare Demesne (Kerry) in 1939, but there are more recent records from Clare at Lough Bunny (1995), Coolorta (2002) and Dromore Lough (2002) and from Galway at Lough Cutra (1996) and Coole Lough (2002). In 1996, it was discovered in Northern Ireland at Crom (Fermanagh) from where it has been reported frequently since. It was discovered at a second Fermanagh locality, Castle Caldwell at the northern end of Lower Lough Erne, on 9 July 2003. There is a high likelihood that there are other populations elsewhere on Upper and Lower Lough Erne where its foodplant, Buckthorn, grows. Adults are greyish-brown with fine cross-lines and scalloped fringes. Buckthorn grows on the shores of Lough Erne on the landward side of fens and the scrubby edge of broadleaved woodland, particularly where the shoreline is rocky. Since adults seldom come to light, Brown Scallop is easily overlooked, exaggerating its rarity. The recorded flight period in Northern Ireland is 1 to 23 July. In Britain it appears in late June. The larvae are nocturnal and spend the day between spun leaves of the foodplant. It overwinters as an egg.

Dark Umber
Philereme transversata (Hufnagel) Category four.

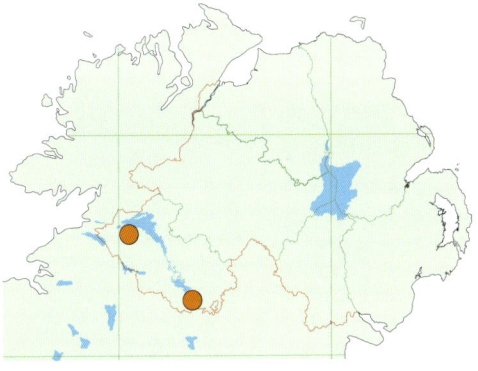

Rare. Like the closely-related Brown Scallop, this moth is confined to Castle Caldwell and Crom in Fermanagh. The first population was discovered in 2000 at Crom and adults have been seen here each year since. Most of the Crom records have been from the eastern shore of Upper Lough Erne, but there has also been a record from Derryvore on the western shore. It was found at Castle Caldwell on 9 July 2003. Elsewhere in Ireland, it occurs only in the Burren where it was discovered in 1952. Adults are normally brown with a darker central cross-band. The leading edge is angled sharply towards the base just before it meets the costa. Both wings are scalloped, but particularly the hindwings. Although similar in appearance to Brown Scallop which shares the same foodplant and habitat, Dark Umber is larger and darker in appearance. It occurs in lakeshore scrub and the landward edge of fens with bushes of Buckthorn. Adults are attracted to light but appear sparingly. The recorded flight period is 6 July to 2 September. The larval foodplant is Buckthorn. It overwinters as an egg.

Sharp-angled Carpet
Euphyia unangulata (Haworth) Category two.
Widespread. Generally distributed and frequently recorded in the southern counties, especially north Armagh and most of Down. It is known from Shane's Castle (Antrim) and Traad Point (Londonderry) on the north shore of Lough Neagh and further north at Breen Wood (Antrim)

and the Umbra (Londonderry). In Tyrone there are recent records for Altadavan, Crockacleaven Lough, Favour Royal and Rehaghy Mountain. Sharp-angled Carpet is a distinctive species. Adults have a well-defined, brownish central band on the forewing. Roughly halfway along the rear edge of this band is a triangular pointed projection, the feature that gives the species its vernacular name. There are two dark discal spots in the central area of the band towards the costa. Sharp-angled Carpet is found in broadleaved woodland, wooded fens and cutover bogs and, rarely, gardens. Adults become active from dusk onwards and are attracted to light but always in small numbers. The recorded flight period is 7 June to 25 August, in a single generation peaking in July. The larval habits and foodplants are unknown in Northern Ireland. In Britain, it has been reared in captivity on chickweeds and stitchworts. It overwinters as a pupa.

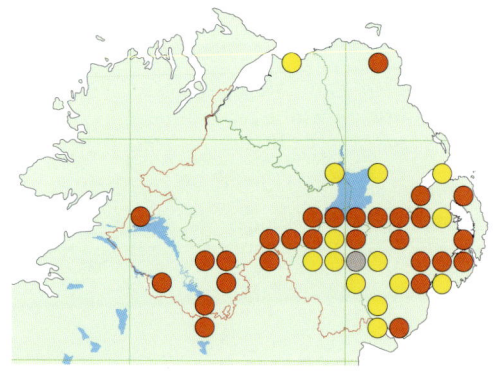

Notes on the four *Epirrita* species in Northern Ireland

The four species of Epirrita *are grey, or occasionally brown, moths with a series of light and dark transverse bands on the forewings. Identification of this genus poses the greatest challenge of any of the Northern Ireland species as they overlap in size, appearance and flight period. Only males can be identified with certainty by examination of the genitalia. It is recommended that a voucher specimen is retained when an* Epirrita *is encountered in a previously unrecorded site so that the species can be correctly determined. The voucher specimens can be used to assess other specimens from the area without having to collect large numbers of adults. Very few* Epirrita *records in the database were determined to species, so a composite map is given rather than maps for individual species.*

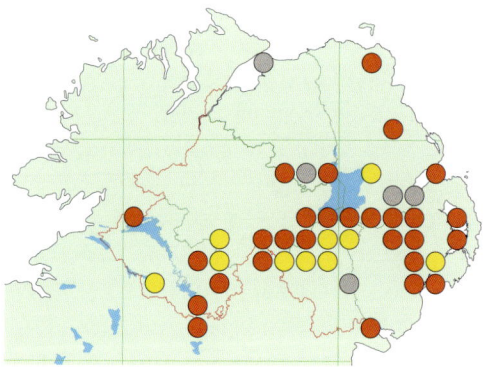

November Moth
Epirrita dilutata (Denis & Schiffermüller) Category three G.
Scarce. This is the most commonly encountered species of the genus. It has been reliably recorded from many of the regular trapping sites in Armagh, Down and Fermanagh. It is unrecorded north of Lough Neagh. Adults are quite variable in colour ranging from a uniform grey-brown to light grey-brown with darker banding. The cross-lines are often not as well defined. Examination with a hand lens of on the underside of the abdomen of the males is necessary for a firm diagnosis. The eighth abdominal segment has two tooth-like projections which are larger and more widely spaced than in Pale November Moth. The presence of a curved spur on the clasper also distinguishes it from Autumnal Moth. November Moth is found in upland and lowland broadleaved woodland, fens, heaths and bogs. Adults rest among foliage, on the trunks of trees and occasionally, fence posts. They are attracted to light in moderate numbers. The recorded flight period is 23 September to 26 November. The larval foodplants are oak, birch, Blackthorn, Hawthorn, willows and possibly other broadleaved trees and shrubs. It overwinters as an egg.

Pale November Moth
Epirrita christyi (Allen) Category three G.
Notable. This species has only been reliably recorded from Castlewellan Forest and Murlough NNR (Down). Adults are quite variable in appearance. The forewings are often darker than in

November Moth.

Rostrevor Wood, County Down.

Pale November Moth.

Castlewellan Forest, County Down.

Autumnal Moth.

Ballynagilly, County Tyrone.

November Moth, but the wavy cross-lines are usually well defined in fresh specimens. The discal spot when present is usually isolated from the cross-band, unlike in November Moth where the inner margin of the outer band passes through the discal spot. Males can be distinguished from those of the November Moth by the size of the backward-projecting teeth on the ventral side of the eighth abdominal segment; these are small and close together in this species. Pale November Moth occupies similar habitat to November Moth and occurs mainly in mature woodland and heaths. Adults are attracted to light in reasonable numbers. The larval foodplants are Beech, birch, Alder, Hazel, willows and Hawthorn. The recorded flight period is early October to mid-November. It overwinters as an egg.

Autumnal Moth

Epirrita autumnata (Borkhausen) Category three G.

Scarce. Widely distributed but local and recorded mainly in southern counties, particularly parts of north Armagh and south Fermanagh. It is scarce and sporadic further north with a few scattered records from Traad Point (Londonderry) and Muntober Bog (Tyrone). It has been reliably recorded from Gosford Forest (Armagh); Belvoir Park, Castlewellan Forest, Hillsborough Forest, Rostrevor Wood and Murlough NNR (Down); Aghalane, Brookeborough, Crom, Garvary Wood, Legatillida, and Monmurry (Fermanagh) and Ballynasollus, Caledon Estate, Crilly and Rehaghy Mountain (Tyrone). Adults are brownish-grey with two median bands. The postmedian line is normally at a right angle to the discal spot which helps to distinguish it from Small Autumnal Moth when both are present at the one site. Autumnal Moth is an autumn-flying species associated with heathland and woodland rides and clearings. Adults appear frequently at light in moderate numbers. The recorded flight period is 28 September to 11 November. The larval foodplants are birch, Heather and Alder. It overwinters as an egg.

Small Autumnal Moth.

Ballynasollus, County Tyrone.

This is essentially an upland species. The largest populations appear to be in Tyrone.

Small Autumnal Moth
Epirrita filigrammaria (Herrich-Schäffer) Category three.
Rare. This is a scarce upland species, more localised in its distribution than Autumnal Moth with only a few recent records from scattered localities. It has been confirmed from Ballynagilly, Ballynasollus and Davagh Forest (Tyrone), where it has been taken in reasonable numbers; the Argory and Aughinlig (Armagh) and Seaforde (Down) where it was found in the late 1980s and early 1990s. Old records exist for Magilligan (Londonderry) and Benaughlin, Enniskillen and Florencecourt (Fermanagh). Adults are greyish-brown with small darkish bands across the forewings; usually well defined when fresh but often faint or absent in some individuals. Its smaller size, habitat preferences and earlier flight period are usually sufficient

The Butterflies and Moths of Northern Ireland

Winter Moth.

Rehaghy Mountain, County Tyrone.

This small plain-looking moth is without doubt more widespread than current records suggest.

to confirm its identity. Small Autumnal Moth inhabits upland areas and moorland. Adults are attracted to light in small numbers and can be found after dark, resting on vegetation and, occasionally, rocks. The recorded flight period is 3 August to 20 October. The larval foodplants are Heather and Bilberry. It overwinters as an egg.

Winter Moth
Operophtera brumata (Linnaeus) Category one G.
Scarce. Recorded from all counties and most frequently in north Armagh, north Down and the Greater Belfast area. It is undoubtedly an under-recorded species elsewhere particularly in the west although recent trapping has produced records from Aghalane, Correl Glen, Crom, Legatillida and Monmurry Bog (Fermanagh) and Crilly, Rehaghy Mountain and Stillago

Moth accounts

Northern Winter Moth.

Monmurry, County Fermanagh.

Many early lepidopterists were of the opinion that this species was absent from Ireland. Its recent discovery in Fermanagh and Tyrone has confirmed its status in Northern Ireland.

(Tyrone). It has been recorded regularly from sites such as Belvoir Park and Helen's Bay (Down) which are trapped or monitored throughout the winter. Surprisingly, there has been only a single record from north of Lough Neagh from Portstewart Dunes (Londonderry). There are also no records from south Down and much of Fermanagh where there appears to be much suitable habitat. Adults are variable in colour but are mainly darkish-brown with a darker, but often vague, central band. Like many of the winter-active moths it is only the male that has wings. The wings of the female are vestigial, reduced to small black-banded stumps that barely reach the abdomen. The males resemble those of Northern Winter Moth but are smaller and darker. The females of the two species are simply told by the length of the wing stumps. Winter Moth can be found wherever there are trees including many gardens, parks and woods of many types. Most of the recent records are of adults found after dark by torchlight at rest on low vegetation. However, males do come to light and are often caught in the beams of car headlights along tree-lined roads. The flightless females rest on the trunks and the finer branches of trees. The recorded flight period is 2 November to 13 February. The larvae can be found from April to June on many species of broadleaved tree. In Britain it is sometimes considered a pest, especially in orchards. It overwinters as an egg.

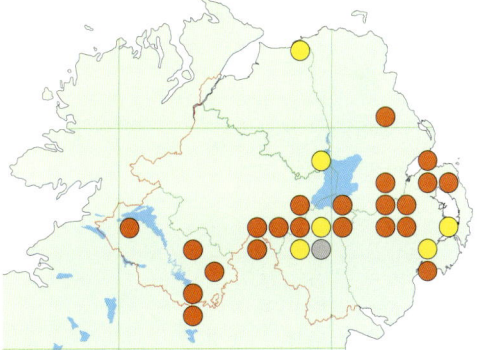

Northern Winter Moth
Operophtera fagata (Scharfenberg) Category three G.
Rare. The status of this species in Ireland was contested by earlier authors. Donovan considered it scarce and listed it from just five Irish counties including Fermanagh and Down. Baynes considered it doubtful having 'never seen an Irish specimen' and the general opinion has been that the species was unconfirmed. However, the discovery of a colony at Monmurry Bog (Fermanagh)

on 4 November 2000 suggests that the earlier records were probably correct. In the same year it was found at nearby White Park (Fermanagh). Adults have been regularly recorded at Monmurry since the initial discovery. In Tyrone, it was recorded at Rehaghy Mountain in 2001 and Crilly in 2002. No other sites are known, here or elsewhere in Ireland, but more colonies may well await detection in western birch woodlands. The adult males are very similar to males of the Winter Moth, but they are larger and paler with a glossy sheen. The females are flightless with vestigial wings. The wing stumps are long in this species, reaching at least halfway down the abdomen. The Northern Ireland records come from birch woodland on heathland and cutover bogs. Males are reported to 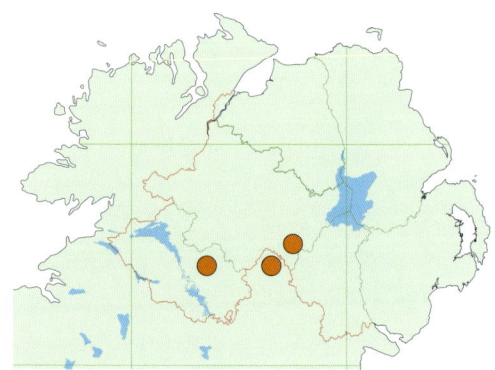 come to light, often in large numbers, but this has not been the experience of recorders here. The majority of records in Northern Ireland have been of males (in large numbers) observed by torchlight sitting on the branches and trunks of birch trees. The recorded flight period is 2 to 28 November. In Britain the main flight period is October to December, somewhat shorter than for Winter Moth. The larval foodplants have not been confirmed in Northern Ireland. In Britain, larvae can be found from April to June on birch and Alder.

Barred Carpet

Perizoma taeniata (Stephens) Category two.
Notable. According to Baynes, this species was 'very local but abundant where it occurs.' He listed it from Antrim and Down. There are also old records from Altadavan and Favour Royal (Tyrone), Newtownbreda (Antrim) and Newcastle (Down). It went unrecorded between 1932 and 1998 when a population was discovered at Monawilkin in west Fermanagh. There have been several subsequent Fermanagh records from Castle Caldwell, Correl Glen, Garvary Wood and Monawilkin. It was rediscovered at Altadavan in 2001. Adults have ochreous forewings with a distinctive dark central band, finely outlined in white. Barred Carpet is mainly associated with damp woodland and hedgerows. Adults show little attraction to light and are best found by beating trees and 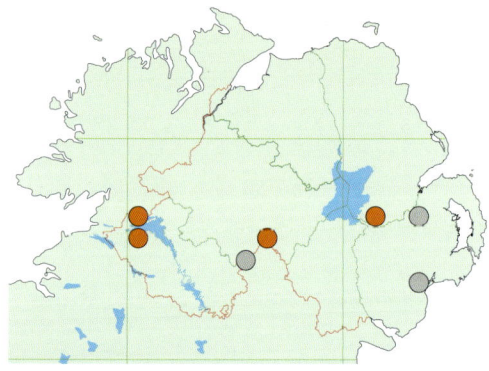 shrubs during the day in likely habitats. The Monawilkin population was discovered in this way. The recorded flight period is 11 to 30 July, but this is based on just a few records. In Britain, the main flight period is from late June to mid-August. The larval foodplant in the wild is unknown, but it has been suggested that it may feed on mosses. It apparently overwinters as a larva.

Rivulet

Perizoma affinitata (Stephens) Category two.
Scarce. This is present mainly in north Down where it has been taken annually in small numbers in the Rothamsted trap located at Hillsborough Forest. Other sites in Down are Banbridge, Belvoir Park and Seaforde. Further west, it occurs at Altadavan in Tyrone and the Argory and Peatlands Park in Armagh. There have been no recent reports from Fermanagh. It is rare in Londonderry where it was last recorded in 1978 at Banagher Glen. The only Antrim localities are Aghalee, Lagan Meadows and Portrush. Adults have brownish forewings with a distinctive white, wavy cross-band containing a fine dark line. There is a single indentation on the inner margin of the band. It closely resembles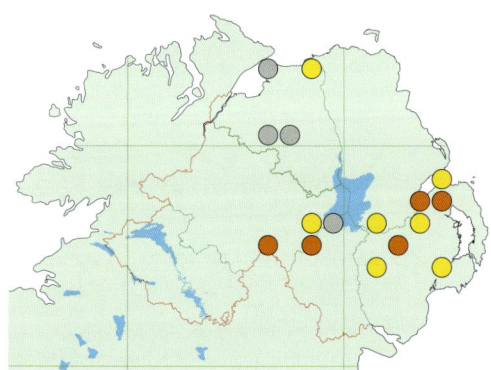

Small Rivulet.

Rehaghy Mountain, County Tyrone.

Small Rivulet, but that species is smaller and has a double indentation on the band. Rivulet is a woodland species that is much less frequently encountered than the closely related Small Rivulet. The lack of records in the west is probably due to the scarcity of its foodplant. Adults are attracted to light in very small numbers. The recorded flight period is 6 June to 1 August. There appears to be just a single generation here. The larval habits are not known in Northern Ireland. In Britain, it can be found from July to September in the seed capsules of Red Campion. It overwinters as a pupa.

Small Rivulet
Perizoma alchemillata (Linnaeus) Category two.
Widespread. Generally distributed in Armagh and Down, but much less frequent in Antrim, Fermanagh and Tyrone. There appear to be no recent records from Londonderry, although it was listed from this county by Baynes. In Antrim, it occurs on Rathlin Island and has recently been recorded from the Greater Belfast area and Rea's Wood on the northern end of Lough Neagh. Frequently recorded sites include the Argory and Peatlands Park (Armagh) and Belvoir Park, Helen's Bay, Hillsborough Forest, Lackan Bog, Murlough NNR and Seaforde (Down). Fermanagh sites include Crom, Garvary Wood, Gorteen and Monmurry. Adults resemble Rivulet in colour and markings, but are obviously smaller and have a double indentation on the inner margin of the central band. Small Rivulet is typically associated with woodlands, bogs, and marshy areas. Adults come frequently to light in small numbers and are more commonly encountered than the Rivulet. The recorded flight period is 26 June to 23 August. The larval foodplant is Common Hemp-nettle. It overwinters as a pupa.

Barred Rivulet

Perizoma bifaciata (Haworth) Category three.
Rare. This small species has been recorded on Rathlin Island (Antrim), at several sites on the coast of Down and inland in Antrim and Armagh. There are old records (in the 1970s) from Banagher Glen (Londonderry) and the Cookstown area (Tyrone). The sites in Down are Castle Espie (1995), Killard Point (1996 and 2003), Ballydorn (1998) and Ballyquintin Point (2001). It was discovered at Cleggan South and Kinramer on Rathlin Island (Antrim) in 2001 and Aughinlig (Armagh) in 2004 but has not yet been recorded from Fermanagh. The forewing colour is generally pale grey-brown with a darker central crossband. The inner and outer margins of the band are much lighter. Adults are seldom encountered at light. The recorded flight period is 26 July to 4 September. Larvae, to our knowledge, have not been reported in Northern Ireland. In Britain they can be found in September and October inside the seed heads of Red Bartsia and eyebrights (the latter being common at most of its recent localities). Its elusive nature makes it difficult to ascertain its true distribution. Most recent records are from coastal grassland; however, small populations may well exist in other localities on the south Down coastline. It apparently overwinters as a pupa in the seed cases of the foodplant for more than one year.

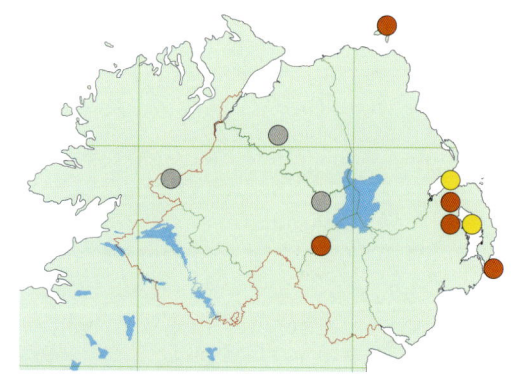

Pretty Pinion

Perizoma blandiata (Denis & Schiffermüller) SOCC (P) Category three.
Rare. This species has been recorded on just four occasions since 1980 at three widely separated sites. It was reported from Rathlin Island (Antrim) in the late 1980s, but has not been recorded since. On 30 May 1992 it was found at Lisblake Bog (Fermanagh). A population was reportedly discovered at the Umbra (Londonderry) in the early 1990s. This was confirmed in August 1998, but adults were not seen at this coastal site again until 2004. There are historical records from Glenarm (Antrim), Newcastle and Rostrevor Wood (Down) and Lough Fea (Tyrone), but it has not been seen recently at any of these areas. Adults have very pale forewings with a dark grey central band and a dark round discal spot near the costa. Pretty Pinion is associated with coastal grassland and moorland. In Britain it is found where its foodplant, eyebright species, grows amongst taller vegetation. Adults may occasionally be seen in late afternoon but are usually more active from dusk onwards. They appear at light but usually in very small numbers. The flight period here is uncertain. The main flight period in Britain is late May to early August. The larval habits are not known in Northern Ireland. In Britain, the larvae have been reported feeding on the flowers and seed heads of eyebrights from early August to mid-September. It overwinters as a pupa.

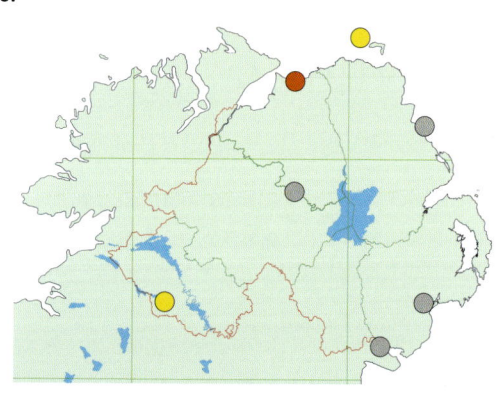

Grass Rivulet

Perizoma albulata (Denis & Schiffermüller) Category three.
Scarce. This small moth is known from all counties. The majority of records originate from Armagh and most date from the 1970s; sites include Drumman More Lough, Loughgall, Milford and Peatlands Park. The only reported occurrence from Armagh since 1990 was at Oxford Island on 19 June 1999. It has been seen only once in Fermanagh at Crom in 2001. The Londonderry records are all from the Umbra in 1998, 1999 and 2004. The Antrim sites are Aghalee, Glengad, Montiaghs Moss and Rathlin Island. In Down there have been records

from Belvoir Park, Hillsborough Forest, Seaforde and Strand Lough. There are historical records from east Tyrone by Greer who described it as 'abundant in meadows' in that area. The monitoring data from the Rothamsted trap network indicates a decline in abundance of this species in Britain of 97 per cent between 1968 and 2002. Adult forewings are predominantly pale grey with a series of greyish-brown cross-lines which can vary in intensity. There is little information on habitat usage in Northern Ireland, but it has been recorded from unimproved dry and damp grassland and dunes. In Britain, it is especially associated with grassland on calcareous sites. Adults rest among grassy tussocks and fly usually from late afternoon when the weather is favourable. The recorded flight period is 2 June to 23 August. The larva feeds inside the seed heads of Yellow-rattle, its sole foodplant. It overwinters as a pupa.

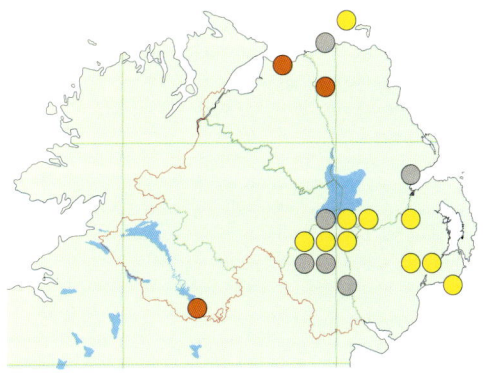

Sandy Carpet
Perizoma flavofasciata (Thunberg) Category two.
Rare. According to Baynes, this is mainly a northern species in Ireland. Historically, it has been recorded from scattered sites in every county of Northern Ireland except Armagh. It has not, however, been reported for many years at a number of these sites including Ballygalley Head (Antrim), Moneymore (Londonderry) and Killymoon (Tyrone). Virtually all of the recent records emanate from Down. In 1993, an adult was taken at Copeland Bird Observatory and large numbers were seen there in June 2003. It was also found on the neighbouring Copeland Island in 1999. Around Belfast there have been records from Belvoir Park and Castlereagh (Down) and the Lagan Meadows (Antrim). It has also been reported once at Hillsborough Forest (Down). In the north-west there has been one recent record from Duncrun (Londonderry). Adults are generally white with sandy cross-lines and cross-bands. The few Northern Ireland records have come from woodland or low, rocky coasts. Adults become active at dusk but are not strongly attracted to light which may account for its absence at many apparently suitable localities. Individuals have been seen in daylight on both of the Copeland Islands. New colonies might be detected by searching patches of the foodplants during the day for resting adults. The recorded flight period is 29 May to 4 August. Larvae have not been reported in Northern Ireland. In Britain, they can be found from July until September inside the seed heads of Red Campion. It overwinters as a pupa.

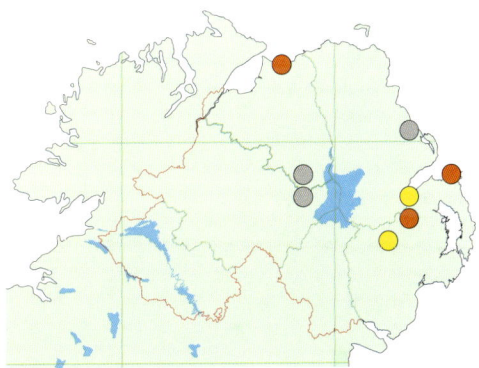

Twin-spot Carpet
Perizoma didymata (Linnaeus) Category one.
Widespread. This species is present at many sites in Armagh, Down, south Fermanagh and the extreme south of Tyrone. It is also found widely in the Glens of Antrim and on the north coast including Rathlin Island. It is apparently absent from most of north Fermanagh, Tyrone and Londonderry apart from recent records at Derrygonnelly (Fermanagh), Davagh Forest (Tyrone) and Duncrun, Kilrea and the Umbra (Londonderry). Adult forewings are generally grey or brownish-grey. A diagnostic feature of this species is the presence of two dark spots near the apex and costa of the forewing. Twin-spot Carpet is chiefly associated with lowland bogs, heaths and moorland.

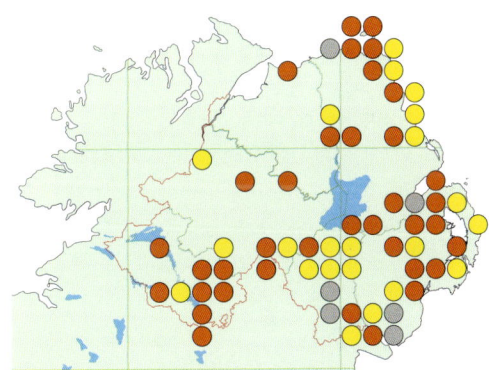

Adults come frequently to light in moderate numbers and have occasionally been disturbed by day from rocks and the trunks and branches of trees. The recorded flight period is 13 June to 4 September. The larval foodplants are Bilberry, Heather, willows, Sea Campion and willowherbs. It overwinters as an egg.

Notes on the *Eupithecia* (Pugs)

There are 52 species of these small moths on the British and Irish list and 38 have been recorded in Northern Ireland. As a group the pugs are easy to distinguish from other macromoths on account of their small size, elongated wing shape and characteristic resting posture, where the forewings are spread in a straight line away from the abdomen. The lower third of the abdomen is often elevated in many species. Species identification is difficult and undoubtedly provides a great challenge to recorders. Consequently, the pugs are the least known group of the larger moths in Northern Ireland. Historical information on the species in Northern Ireland is patchy. Many old records are impossible to assess, mainly due to taxonomic changes, and these have only been included when supported by correctly named voucher specimens. The information on the species in Northern Ireland presented here is therefore very incomplete and the maps should be interpreted with care.

A recently published monograph (Riley and Prior 2003) on the British and Irish fauna should greatly aid study and identification of the pugs. This comprehensive book provides much information on identification, biology and ecology and readers are advised to consult it. The taxonomy of Riley and Prior is adopted in the following species accounts.

Slender Pug
Eupithecia tenuiata (Hübner) Category four.
Rare. This species has been reported from several sites in north Armagh and a single site in Down. These are Ardress, the Argory, Brackagh Moss, Derryhubbert and Loughgall in Armagh and Hillsborough Forest in Down; however the validity of some records is questionable since there are no voucher specimens to substantiate them. The occurrences at the Argory on 26 July 2001 and Hillsborough Forest on 4 August 1997 are the only fully authenticated records. Adults are normally greyish-brown and have well-rounded forewings and a black discal spot. Slender Pug is found in damp woodland, fens, wet meadows and other types of wetlands with willows. The main flight period in Britain is June to July, but records here have been between 12 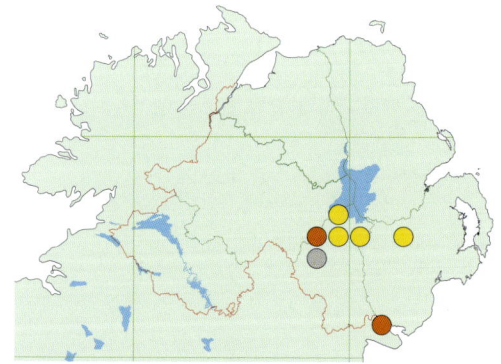 July and 10 August. The larva feeds inside the catkins of large species of willow in March and April. It overwinters as an egg.

Haworth's Pug
Eupithecia haworthiata Doubleday Category four.
Rare. There are old records of this moth from Favour Royal and Stewartstown in Tyrone. Since 1999, it has been reported from Portrush in north Antrim, Castle Espie and Murlough NNR in Down and Aughinlig in north Armagh where it has been recorded almost every year. Adults have well-rounded forewings and are normally greyish-brown with a series of fine dark

cross-lines and broader pale bands, which are obvious on fresh specimens. The base of the abdomen has a reddish band which helps distinguish it from closely-related species. In Northern Ireland, Haworth's Pug has been recorded regularly from a mature garden and unspecified habitats at three coastal sites. It may be more widespread than the current distribution map indicates, especially in gardens. Adults are attracted to light in small numbers. The recorded flight period is 19 June to 3 August. The larvae feed during July and August on the flower buds of cultivated species and varieties of clematis and Traveller's-joy, which is widely naturalised here. It overwinters as a pupa.

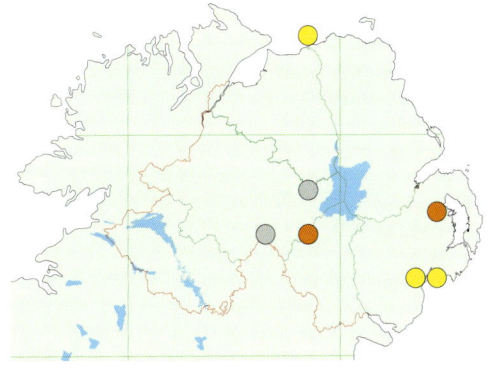

Lead-coloured Pug

Eupithecia plumbeolata (Haworth) Category four.

Rare. There were four confirmed records in 2001 and 2002, the first since 1921. A single adult was taken in Peatlands Park (Armagh) on 28 June 2001. There have been two further records from the same site on 7 and 14 July 2002. It has been found at two sites in Fermanagh — Aughalun (3 July 2001) and Glennasheevar (2004). There have been a few historical records from Derry in 1875, Favour Royal in 1902, Slaghtfreeden and Lissan Estate in 1921. Adults are similar in appearance to Haworth's Pug, but lack the reddish band on the abdomen. The forewings are pale grey, occasionally with a yellowish tinge. The wavy cross-lines are darker in colour and usually distinct on fresh specimens. The Lead-coloured Pug is associated mainly with open woodland, meadows and sandy areas in Britain. Here it has been found on wooded bogs. The main flight period is June and July in Britain. The larvae are said to feed from late June to mid-August on the flowers of Common Cow-wheat and possibly Yellow-rattle. It overwinters as a pupa.

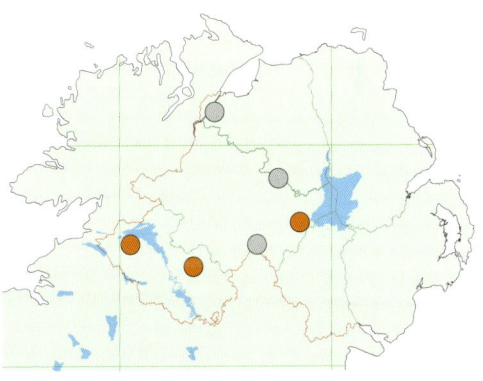

Cloaked Pug

Eupithecia abietaria (Goeze) SOCC (P) Category three.

Rare. This large and distinctively marked pug was listed by Donovan and Baynes from every county except Londonderry. It appears to have been sufficiently common in the early twentieth century for Donovan to describe it as 'widely spread among spruce.' It then appears to have declined dramatically both here and in Britain. There were no reports of the species in Northern Ireland for many years until 1999 when a single specimen was taken at Bohill Forest (Down). The identification of the voucher specimen was confirmed by Bernard Skinner. Since then other adults have been recorded at the same locality in 2000, 2001, 2003 and 2004. All have come from a small area within this wood. It was found at Castlewellan Forest (Down) on 23 July 2000 and Rostrevor Wood (Down) on the 4 August 2004. There is an historical record from Luke's Mountain in the Mourne Mountains (Down) in the early 1930s. Adults are large by comparison with other pugs and not easily confused with other species. The forewings are pale to darkish-grey with broad, dark cross-lines. There is a large black discal spot near the median area of the forewing close to the costa, and a fine, broken black line along the outer margin.

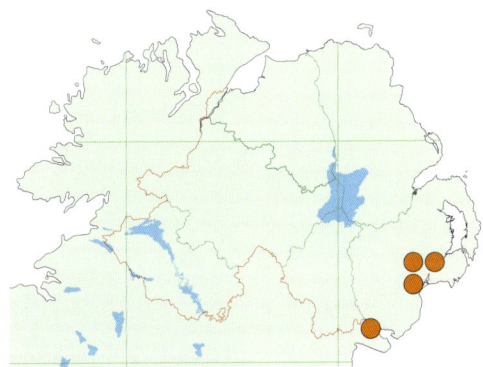

Cloaked Pug.

Bohill Forest, County Down.

Foxglove Pug.

Rehaghy Mountain, County Tyrone.

The Cloaked Pug is found in woods and plantations containing mature cone-producing spruce and fir trees; the larva feeds inside the cones of these conifers. Adults come to light in moderate numbers and in Northern Ireland they have also been beaten from foliage during the day. The records have been between 18 June and 4 August. Norway Spruce, Sitka Spruce, Noble Fir and Silver Fir are the most frequently reported foodplants, some of which are present at its known sites. It overwinters as a pupa.

Toadflax Pug
Eupithecia linariata (Denis & Schiffermüller) Category four.
Unknown. This species is included on the Northern Ireland checklist as there is a single specimen in Kane's collection in the National Museum of Ireland that is labelled Favour Royal (Tyrone). The date of capture is not given, but is possibly around 1900. Common Toadflax (an introduced species mainly found on roadsides) seems to be scarce in Tyrone. There appear to be no recent records of this plant from the area. Adults of this species are similar to Foxglove Pug, but smaller. The distal edge of the dark central cross-band is curved rather than kinked as in Foxglove Pug. Nothing is known about the habits or ecology of the adult and larva in Northern Ireland. In Britain, it occurs on limestone grassland and roadside verges. It has also been taken in
gardens where snapdragons are grown. The flight period in Britain is July and August, slightly later than the peak of the flight period of Foxglove Pug. It is possible that this species has been overlooked due to its resemblance to Foxglove Pug. In Britain, the larva feeds on the flowers and seed capsules of Common Toadflax and snapdragons. It overwinters as a pupa.

Foxglove Pug
Eupithecia pulchellata Stephens Category one.
Scarce. This is generally distributed in Armagh, Down and Fermanagh. It has also been found at a few sites in Antrim and Tyrone as far north as Newtowncrommelin in mid-Antrim and Davagh Forest and Lough Fea on the border of Londonderry and Tyrone. The Argory (Armagh); Belvoir Park and Hillsborough Forest (Down); Garvary Wood and Legatillida (Fermanagh) and Rehaghy Mountain (Tyrone) are considered reliable sites for the species. This is a distinctive pug with pale ochreous-brown forewings and a grey central band which is kinked near the costa. The ground colour is, however, variable with paler forms often predominating at coastal sites. Foxglove Pug is found in woodland, rough grassy hillsides, hedge banks, moorland and heathland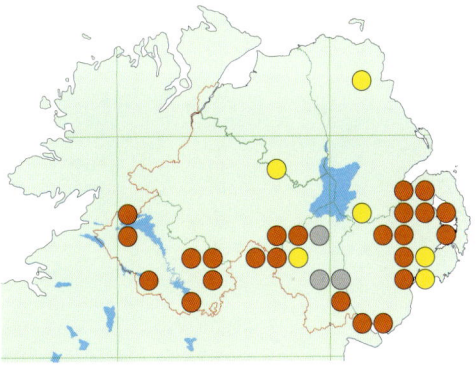
and other situations where its foodplant, Foxglove, is present in abundance. Adults come frequently to light in small numbers. The recorded flight period is 13 May to 1 August. The larvae feed during July and August on the flowers of the foodplant. It overwinters as a pupa.

Mottled Pug
Eupithecia exiguata (Hübner) Category two.
Scarce. This species is entirely confined to the south of the region especially Fermanagh, south Tyrone and north Armagh. It is also present at a few sites in Down, but there have been no records from Londonderry. It has frequently been recorded from a garden in Finaghy (Antrim) and at Aughinlig and Loughgall (Armagh). Other recently recorded sites include the Argory,

Brackagh Moss, Carganamuck and Peatlands Park (Armagh); Belvoir Park and the Quoile Pondage (Down); Brookeborough, Cleen, Crom, Hanging Rock, Legatillida, Marlbank and Monmurry Bog (Fermanagh) and Altadavan (Tyrone). Adults have rather pointed forewings and are usually light greyish-brown, with a pale cross-band. There is a small black discal spot near the costa and a series of small, dark, wedge-shaped marks below the lower edge of the band. This woodland species is found in broadleaved woods, especially those on old estates. There are also records from mature hedgerows, wooded bogs and rural gardens. Adults come quite often to light traps, but usually in small numbers. The recorded flight period is 14 May to 9 July. The larval foodplants are Hawthorn, Blackthorn and Sycamore. It overwinters as a pupa.

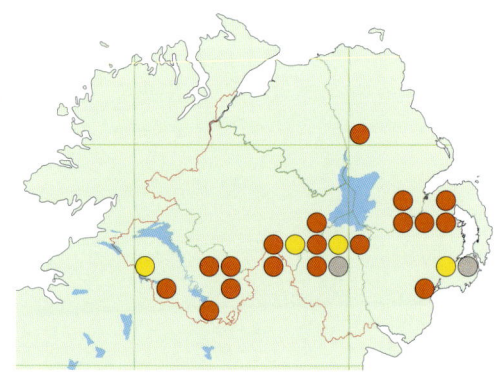

Valerian Pug
Eupithecia valerianata (Hübner) Category four.
Rare. This pug was recorded for the first time in Northern Ireland at Brackagh Moss (Armagh) in 1995. Adults have been seen at the same site annually from 2001 to 2004, but this remains the only known location. It appears to be uncommon throughout Ireland as it has been recorded in just a few counties. Baynes listed it from Cavan, Clare, Cork and Donegal. Adults are brownish-grey with a fine, wavy, white subterminal line which is not always evident. There is also a small, but distinct, white tornal spot. This is a wetland species occurring only where there are large stands of its sole foodplant, Common Valerian. Brackagh Moss is an enriched fen so appears to be typical habitat. This species seems to be elusive at this site despite being one of the most intensively trapped localities in Northern Ireland. It is likely that populations exist in suitable habitat elsewhere, but so far none have been discovered. Most of the records are of adults attracted to light traps, but several individuals have been netted at dusk, flying among stands of the foodplant. It has also been reported that individuals will fly in the late afternoon. The recorded flight period is 11 June to 6 July. The larva feeds during July and August on the flowers of Common Valerian. It overwinters as a pupa.

Marsh Pug
Eupithecia pygmaeata (Hübner) Category four.
Unknown. The present status of this wetland species in Northern Ireland is currently unknown. The most recent record, from Loughgall (Armagh) on 23 August 1973, is over 30 years old. It was described by Donovan as rare and local in Ireland and Baynes listed it from Fermanagh and Tyrone, amongst other counties. Greer recorded it at three sites — Lissan Estate (1921), Stewartstown (1920) and Tullyhogue (1921) — in east Tyrone. The forewings are pointed and generally dark brown with a small white tornal spot and brown and white chequered fringes. As its name suggests, this is a wetland species in Britain associated chiefly with fens and wet grasslands, but also found in drier sandy grasslands and dunes. The larval habits and the habitat occupied in Northern Ireland are not known. In Britain, adults are occasional visitors to light

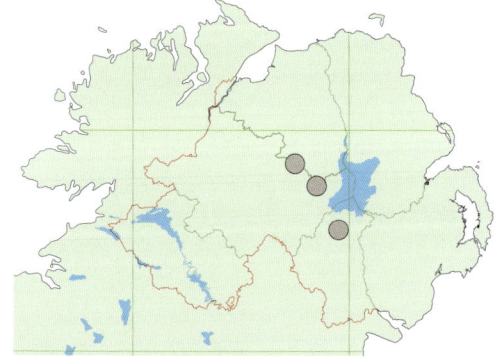

Netted Pug.

Copeland Bird Observatory, County Down.

One of the rarest pugs in Northern Ireland. It was rediscovered in 2003 after an absence of 28 years.

but may also be seen flying in sunshine around the foodplant. May to June is the main flight period in Britain. The larva feeds on the flowers and seed capsules of Field Mouse-ear and possibly other mouse-ears. Field mouse-ear is rare in Northern Ireland with only a few recent records and none from the areas where this moth has been reported. It overwinters as a pupa.

Netted Pug
Eupithecia venosata (Fabricius) Category four.
Rare. The only site currently known for this species is Copeland Bird Observatory off the north coast of Down. It was recorded here on 13 and 14 June 2003, the first records from this island since 1975. Historical records from Crawfordsburn (Down) and Magilligan (Londonderry) in 1875, Strangford Narrows (Down) in 1933 and Rathlin Island (Antrim) in 1971 indicate a wider distribution around the Northern Ireland coast, but there have been no recent reports from any of these sites. This is perhaps one of the most distinctive and easily recognised pugs. The forewings are rounded at the apex and variable in colour, ranging from light to dark brownish-grey with the distinctive net-like pattern of cross-lines. Specimens from the west of Ireland (especially the Burren) are generally much darker in appearance then northern populations. Adults are cryptic when at rest on lichen-covered rocks in coastal sites. Netted Pug occurs on low rocky coasts where its foodplant Sea Campion grows. There would appear to be suitable habitat around much of the Northern Ireland coastline especially on Rathlin Island, north Down and the Ards Peninsula, but surprisingly, no specimens have been found. Adults are attracted to light in small numbers. The recorded flight period in Britain is May and June. The larva feeds during June and July on the seed pods of the foodplant and it overwinters as a pupa.

Lime-speck Pug

Eupithecia centaureata (Denis & Schiffermüller) Category two.
Scarce. Since 1980, Lime-speck Pug has been recorded from Magilligan (Londonderry) and Rathlin Island (Antrim). However, most of the recent records have been from the Greater Belfast area and several coastal localities in Down, including Ballykinler Dunes, Copeland Bird Observatory, Killard Point, Mill Bay and Murlough NNR. It appears to be rare inland and is known from only a few sites in north Armagh at Ballyrea, Loughgall, Navan Quarry and Richhill and at Ahoghill in mid-Antrim. This is a very distinctive pug as the resting adults resemble a bird-dropping. The forewings are pale grey with an obvious dark blotch that merges with the black discal spot; the size of the blotch can vary between individuals. Lime-speck Pug has been recorded from coastal heathland, dunes, unimproved grassland, gardens and saltmarsh. Adults come frequently to light and are occasionally found by day at rest on flat surfaces. The data indicates it has just a single annual generation in Northern Ireland with a recorded flight period of 6 June to 4 September. The larva feeds from June to October on a wide variety of herbaceous plants including ragworts, knapweeds, Field Scabious, Wild Angelica. It overwinters as a pupa.

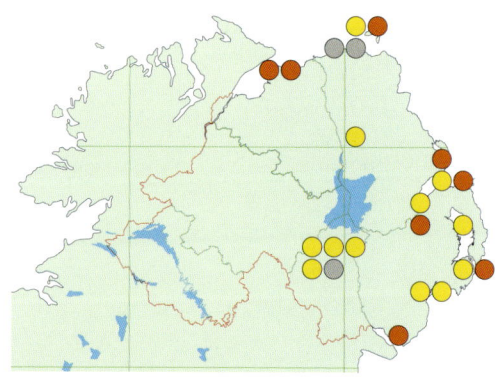

Triple-spotted Pug

Eupithecia trisignaria Herrich-Schäffer Category four.
Notable. The recent confirmed records of this pug have come from Finaghy (Antrim), the Argory (Armagh), Hillsborough Forest (Down), Aghalane and Crom (Fermanagh) and Crockacleaven Lough and Favour Royal (Tyrone). There have been no records from

Lime-speck Pug.

Killard Point National Nature Reserve, County Down.

A pale, slender-winged pug, which bears a resemblance to a bird-dropping when at rest.

Londonderry. Historically, it was regarded as very rare in Ireland. The first Northern Ireland record came from the Enniskillen area of Fermanagh in 1912 and was of larvae found feeding on Wild Angelica. Adults have uniformly brown forewings and these appear glossy. The most useful diagnostic feature is the group of three dark spots on each forewing (two on the costa and one towards the centre of the wing). Triple-spotted Pug has been found in fens, wet woodland margins and other marshy habitats. Adults are not very attracted to light which might account for its rarity and apparent absence from many suitable localities. The recorded flight period is 7 July to 5 August. The larva feeds on the fruits of the foodplant. Hogweed is cited as a foodplant in Britain in addition to Wild Angelica. It overwinters as a pupa.

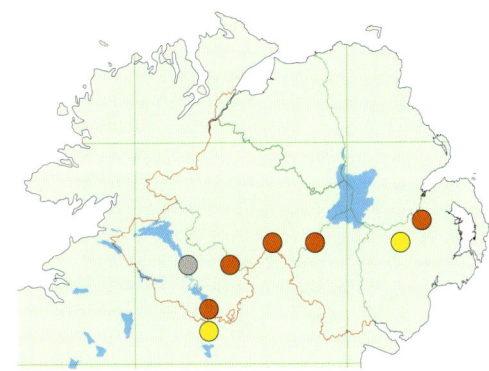

Satyr Pug
Eupithecia satyrata (Hübner) Category four.
Notable. All the recent records of Satyr Pug are from the southern half of Northern Ireland, in Armagh, Down and west Fermanagh. There are old records from Londonderry and Tyrone. The most recent records are from Dartry Lodge, Drumnahavil and Peatlands Park (Armagh); Castlewellan Forest (Down) and Gortmaconnell Rock and Monawilkin (Fermanagh). Adults are pale greyish-brown with poorly-defined cross-lines and appear mottled. There is a small dark discal spot and a white tornal spot with darker blotches along the costa of the forewing that are easily seen in fresh individuals. It is similar in appearance to the Common Pug, but this is more orange-brown and has a larger discal spot. The Satyr Pug has been recorded from wet heath and lowland cutover bogs and, in Fermanagh, from calcareous grassland. These last may have been strays from nearby upland heaths. The recorded flight period is the 9 June to 17 August. The larval foodplants are not known in Northern Ireland. In Britain larvae have been reported on Devil's-bit scabious, Common Knapweed, Common Ragwort, Meadowsweet, Heather and Cross-leaved Heath. It overwinters as a pupa.

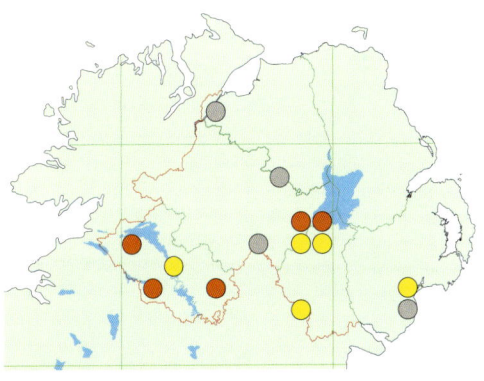

Wormwood and Ling Pug
Eupithecia absinthiata (Clerck) Category one.
Widespread. This is one of the most commonly reported pugs in Northern Ireland and is generally distributed across southern counties, especially parts of north Down and the fens of north Armagh. It has been found recently at several of the well-recorded sites in south Fermanagh. In the northern half of Northern Ireland it is found at Duncrun, Magilligan, Portstewart and the Umbra (Londonderry) and on Rathlin Island (Antrim). Ling Pug is now considered as a form (f. *goossensiata* Mabille) of Wormwood Pug, rather than a distinct species. This form has been reported from a few localities in north Armagh and east Tyrone including Altadavan, the Argory, Peatlands Park and Rehaghy Mountain. It is also reported from several sites in Down and one in south Fermanagh. Adults of Wormwood Pug have reddish-brown forewings with prominent black costal blotches and a central discal spot. There is a faint broken white subterminal cross-line and a white tornal spot; this is often poorly defined. Specimens of Ling

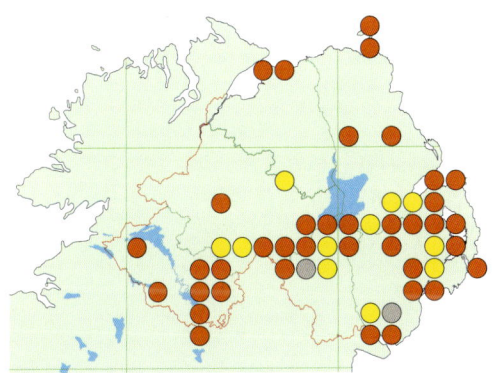

Currant Pug.

Aughinlig, County Armagh.

Pug are smaller and greyer, but otherwise similar to typical examples of Wormwood Pug. This species occurs in a wide variety of habitats, but especially woodland, fens, waste ground, coastal heaths and dunes and occasionally mature suburban gardens. The Ling Pug form is associated with moorland and heathland and is not generally seen outside these habitats. Adults appear frequently at light in moderate numbers. The recorded flight period is 8 May to 4 September. The larva feeds from August to October on various composite species including Yarrow, Sea Aster, Common Ragwort and Michaelmas-daisy and Heather. It overwinters as a pupa.

Currant Pug

Eupithecia assimilata Doubleday Category two.

Scarce. There have been recent records from every county except Londonderry. The majority of these have come from gardens and woodland in the Greater Belfast area and north Down. Further west it has been regularly recorded from Aughinlig in north Armagh. Other sites include Hillsborough Forest and Seaforde (Down), Ardess, Brookeborough and Monmurry Bog (Fermanagh) and Rehaghy Mountain (Tyrone). It is likely to be overlooked in many areas and therefore the map probably does not truly reflect its distribution in Northern Ireland. Adults are similar in shape and markings to Wormwood Pug, but the wings are shorter and more rounded and the white tornal spot is more obvious. The wings are typically reddish-brown with prominent black costal blotches and a central discal spot. Currant Pug is recorded from rural and urban gardens, parkland and mature woodland where its foodplants are grown or have become naturalised. The recorded flight period is 15 May to 16 August which may

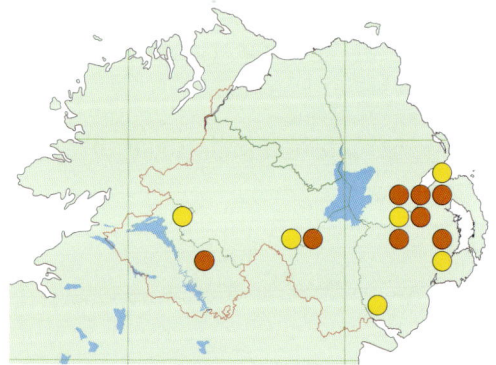

indicate that there are two generations. Further evidence is needed to determine whether this is the case. The larva feeds on Black Currant and Red Currant. It overwinters as a pupa.

Bleached Pug

Eupithecia expallidata Doubleday Category four.
Notable. This species has only been reliably recorded from Down and historically, from Londonderry, although there is a claimed record from Gorteen in Fermanagh. This record has not been confirmed, although a voucher specimen apparently exists. Donovan stated that it occurred at Magilligan (Londonderry) and Portaferry (Down), but it has not been seen at either locality for many years. The only recent record is from Knockbarragh on 1 August 1999. It appears to be rare elsewhere in Ireland, with records from just a few eastern counties. Adults are pale in colour and usually light brown with broad, well-rounded forewings. There is a series of small but distinct black marks on the costa and a large black elongated discal spot; the abdomen has a broad black band. The 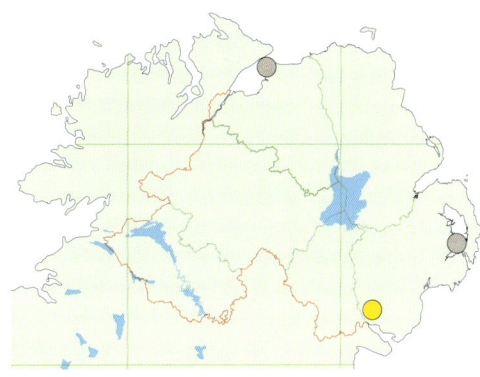 Bleached Pug has been recorded from woodland rides and clearings. Adults are not commonly attracted to light but reportedly individuals can be disturbed from its foodplant by day. In Britain, the main flight period is late June to August. Goldenrod is stated to be the principal larval foodplant, although other composites are accepted in captivity. It overwinters as a pupa.

Common Pug

Eupithecia vulgata (Haworth) Category one.
Widespread. A frequently recorded pug in southern counties especially at regularly trapped sites in north Down, Greater Belfast, north Armagh and Fermanagh. It is recorded from scattered localities in mid- and north Antrim including Magheramorne, Portglenone Wood and Rathlin Island. In the west, recent sites include Ballyreagh Bog, Brookeborough, Cleen, Crom, Eshywulligan and Mountgibbon Bog (Fermanagh) and Rehaghy Mountain (Tyrone). There have been no records from Londonderry since 1978 when it was found at Banagher Glen. Adults are variable in colour, ranging from pale brown to dark reddish-brown. The wings have cross-lines which are kinked towards the costa. The cross-lines vary in strength and are scarcely visible on 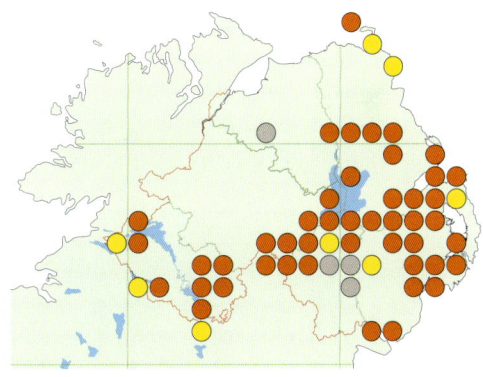 some individuals. There is a small discal spot and a fine, pale, broken subterminal line ending in a small but distinctive white tornal spot. Some individuals from coastal localities are more richly marked than their inland counterparts. Common Pug is a generalist species that has been recorded from most lowland habitats in the region, but is especially found in woodland, heaths and suburban gardens. The flight period is extensive, from 8 April to 9 September with a peak in late May and June. Two generations are the normal pattern in the populations in southern Britain, but there appears to be only a single generation here. The larva feeds throughout the summer and autumn on a wide range of plants and shrubs including willows, Hawthorn, Bilberry, Bramble and Common Ragwort. It overwinters as a pupa.

White-spotted Pug

Eupithecia tripunctaria Herrich-Schäffer Category three.
Scarce. This species is confined to the south and east and has rarely been recorded outside Armagh and Down. It is known from just four sites in south Fermanagh (Aghalane, Black

The Butterflies and Moths of Northern Ireland

White-spotted Pug.

Peatlands Park, County Armagh.

Lough, Crom and Legatillida) and two sites in Tyrone (Altadavan and Favour Royal). The northernmost records are from Ballymena and Larne in Antrim. In Down, it was frequently recorded from Seaforde in the late 1980s and more recently has been found at Hillsborough Forest and on Copeland Bird Observatory. In Armagh, it was regularly reported from the Loughgall area during the 1980s and in recent years at the Argory and Peatlands Park. Adults are greyish-brown, with conspicuous white spots on the thorax and tornus and a pale subterminal line. The hindwings are a similar colour to the forewings with a small white dot at the anal angle. This species has been taken in woodland, fens and other damp habitats. Adults are not overly attracted to light, which might account for its absence at some apparently suitable sites. It is probably more widespread in the west than current records indicate. The recorded flight period is 14 May to 27 August, in two generations, peaking in early June and August. The larval habits and foodplants are not known in Northern Ireland. In Britain, the larva feeds on the flowers of Elder in the first generation and the seed heads of umbellifers, especially Wild Angelica, in the second generation. It overwinters as a pupa.

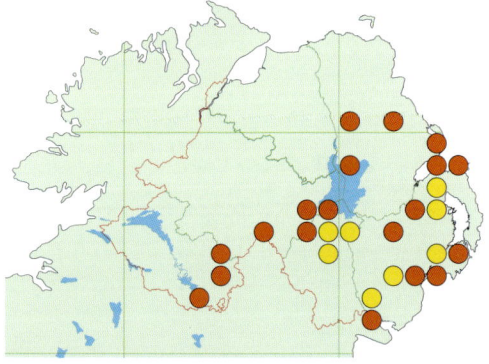

Grey Pug

Eupithecia subfuscata (Haworth) Category two.

Widespread. This is a frequently recorded pug present in all counties, also found on Rathlin Island (Antrim) and at Copeland Bird Observatory (Down). It is common and widespread across southern counties, but much more local further north and almost certainly under-recorded. It is regularly encountered at the well-trapped sites throughout Northern Ireland. Adults are pale grey and appear mottled. Some individuals show an ochreous tinge. The forewings have several cross-lines and a conspicuous dark discal spot. It closely resembles Golden-rod Pug which has a larger, more rounded discal spot; also the crosslines are less obvious. Examination of the eighth abdominal segment is a useful aid to identification (see account of Golden-rod Pug). Grey Pug is a widespread species that has been recorded from woodland, bogs, coastal dunes and mature suburban gardens, Adults are attracted to light in small numbers. The flight period is 7 May to 16 August. There are two peaks apparent in the records, one in May and June and a second in July and August, indicating two generations. The larva feeds on a wide variety of trees and shrubs such as Blackthorn and Hawthorn and also herbaceous plants including Common Ragwort and knapweeds. It overwinters as a pupa.

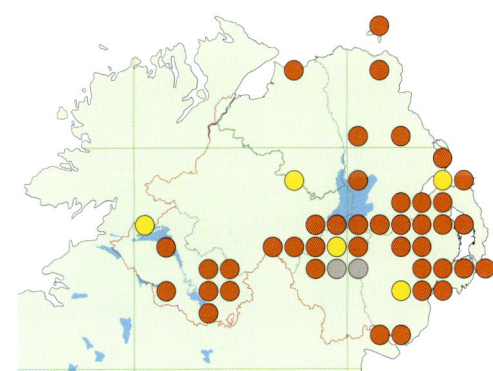

Tawny-speckled Pug
Eupithecia icterata (Villers) Category three.
Notable. This species has been recorded from several sites in Armagh, including Ballynewry Lough, Brackagh Moss and Drumherriff, and more recently from the Argory, Aughinlig and Dartry Lodge. Killard Point, Lackan Bog and Murlough NNR are the only known sites in Down. It was recorded from Portmore Lough (Antrim) in 2003. Greer, in the early part of the last century, described it as being local but widespread in Tyrone, but it has not been seen recently in this county despite regular trapping at a number of sites. The known Fermanagh sites are Garvary Wood and White Park. This is a fairly large pug which has two colour forms. The most common form is predominantly grey with a small discal spot and a reduced amount of red in the central area of the forewing. The other colour form is less often seen but has a strong reddish patch in the middle of the forewing. The hindwings are greyish-brown with indistinct markings. The Tawny-speckled Pug has been recorded from woodland, coastal grassland, fens and cutover bogs and rural gardens. It is probably more widespread than the distribution map indicates since one of its foodplants (Sneezewort) is commonly encountered throughout all counties. Adults are attracted to light and the flowers of Common Ragwort. The recorded flight period is 15 May to 5 September. The larva feeds during September and October on Yarrow and Sneezewort. It overwinters as a pupa.

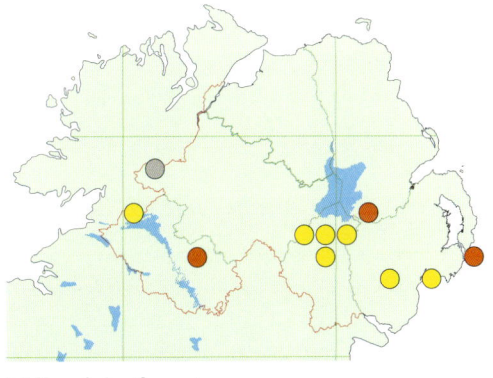

Bordered Pug
Eupithecia succenturiata (Linnaeus) Category three.
Notable. This is a large distinctive pug recorded mainly in north Down and the Greater Belfast area, west along the Lagan Valley to Dunmurry and Lisburn in south Antrim. It has also been recorded in Armagh and Fermanagh, but there have been no recent records from either Londonderry or Tyrone. The majority of records come from regularly trapped gardens in Belfast and Helen's Bay. Other sites in Down are Belvoir Park and Castle Espie. In Armagh, it has been reported from Derryadd Lough in 1978 and twice in Portadown, the last occasion being in 1998. It has been seen once in Fermanagh at Crom in 2000. One of the most distinctive pugs, the adults have white forewings entirely bordered with a greyish-brown suffusion. There is a small dark discal spot in the median area of the forewing. The thorax is white, contrasting sharply with

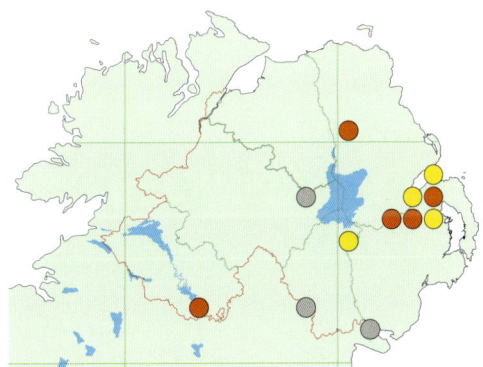

The Butterflies and Moths of Northern Ireland

Bordered Pug.

Belfast, County Antrim.

A large distinctive species with a conspicuous black spot on the forewings. Adults are seldom seen at light in more than ones and twos.

the dark brown abdomen. The species has been recorded in urban and suburban gardens, woodland and rough grassland. Adults appear sparingly at light, usually as single individuals. The recorded flight period is 2 July to 30 August. The larvae to our knowledge have not been reported in Northern Ireland. Mugwort and Yarrow are reported to be the main foodplants in Britain. It overwinters as a pupa.

Shaded Pug

Eupithecia subumbrata (Denis & Schiffermüller) Category four.
Notable. Baynes described this species as not uncommon although he listed it from just nine Irish counties, including Fermanagh. The most recent records are from Killard Point (Down) in 2004. Wright reported it from Tollymore Forest (Down) and from an unspecified site in Antrim. It has also been recorded in Armagh at Lough Gullion in 1974, Drumman More Lough in 1977 and, most recently, Peatlands Park in 2000. The only known Tyrone site is Davagh Forest where it was found in 2002. The remaining sites are all in Fermanagh at Aghalane, Colebrooke, Crom, Gortmaconnell Rock and Tullyreagh Bridge. Adults are whitish with a darker broad band running around the outer margin of the forewing. The cross-lines are often faint and

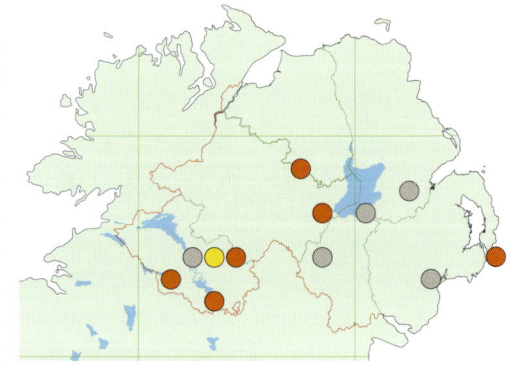

difficult to see in all but fresh specimens. There is a small discal spot near the costal margin and a fine broken black line along the outer margin of the forewing. The Shaded Pug is found in rank grassland on the margins of woods, road verges and the edge of wetlands. Adults are attracted to light and may also be disturbed from grassy vegetation by day. The recorded flight period is 4 June to 11 August. The larval foodplants are Field Scabious, hawk's-beards, St. John's-worts and Common Ragwort. It overwinters as a pupa.

Plain Pug
Eupithecia simpliciata (Haworth) Category four.

Rare. There have been four records of this pug in Northern Ireland; it would seem to be rare throughout Ireland. Donovan listed it from just two counties, Armagh and Dublin, and Baynes added Antrim, Cork and Wicklow to this list. It was recorded at Shane's Castle (Antrim) in 1875 and at Lurgan (Armagh) in 1936. There were no other records of this species until 2004 when an adult was recorded at light at Rostrevor Wood (Down) on 4 August. Adult colouration is light sandy brown, but occasionally grey, with a distinct white cross-band and a well-defined white subterminal line. There is a small black discal spot near the costa. In Britain, Plain Pug is found most commonly in coastal localities and along rivers. Habitats occupied include the landward margins of saltmarshes, coastal cliffs, disturbed ground in urban areas and weedy arable land. The precise habitat of the Fermanagh specimen is unknown. Since the majority of its foodplants have a mainly coastal distribution in Northern Ireland it may be worth targeting suitable localities along the Antrim and Down coastline where its foodplants are well established. In Britain, adults are active from dusk onwards and apparently visit the flowers of Common Ragwort and Marram. The flight period in Britain is mid-June to early August. The larval habits and foodplants are unknown in Northern Ireland. In Britain the larva feeds during August and September on the flowers and seed heads of goosefoot and orache species. It spends the winter months as a pupa.

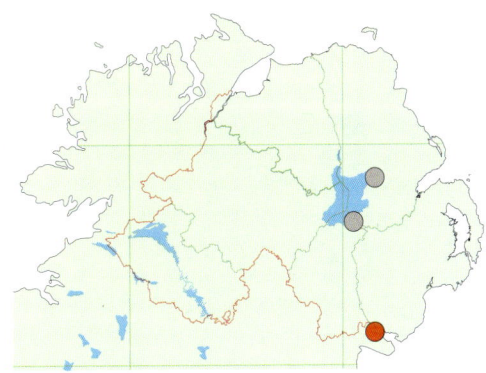

Thyme Pug
Eupithecia distinctaria Herrich-Schäffer Category three.

Rare. This species is known from all counties, but only from a few localities in each. The only post-1980 records are from Fermanagh and Londonderry. Historically, it has been recorded from Armagh City, Kilkeel and Tollymore Forest in Down, Favour Royal in Tyrone and Belfast Castle and the Gobbins in Antrim. Campbell listed it from Londonderry and it was found again in this county at the Umbra in 1994. It was discovered in Fermanagh in 1988 at Blackhill Bog and subsequently recorded from several sites (Gortmaconnell Rock, Killykeegan and Monawilkin) in the limestone hills in the west of the county. It also occurs at Brookeborough in east Fermanagh. Adults are small and grey with dark costal patches and a black discal spot. The colour darkens towards the outer margin and there are fine dark cross-lines which are visible on fresh specimens. Ochreous Pug is similar, but the forewings are narrower and more pointed. Thyme Pug is found on rocky coasts and sand dunes and rocky inland localities with sparse vegetation, particularly in Northern Ireland on limestone. There may be undetected populations within the range of Wild Thyme, its principal foodplant, especially along the north and east coasts. Adults become active from dusk onwards and appear at light sparingly. They may also be found by examining patches of the foodplant during the day. The recorded flight period is 15 May to 21 July. The larva feeds on the flowers of the foodplant. It overwinters as a pupa.

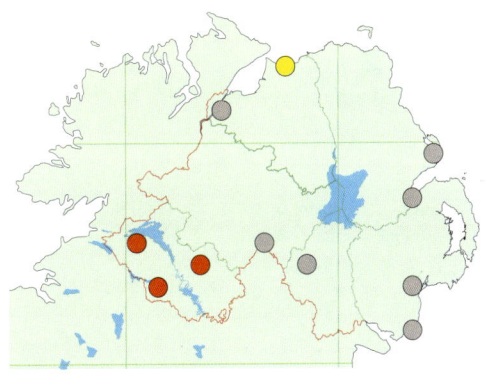

Ochreous Pug
Eupithecia indigata (Hübner) Category three.

Rare. Ochreous Pug has a western distribution in Armagh, Fermanagh, Londonderry and Tyrone and has never been reliably reported from Antrim or Down. It has been recorded just

The Butterflies and Moths of Northern Ireland

once in Londonderry by Campbell in 1902. There are two old records from Tyrone, from Altadavan and Lissan Estate, both in 1902 by Kane and Greer, respectively. Donovan listed it from Fermanagh (at Enniskillen) and it was rediscovered here in 2000 at Carnmore and Crom. It was recorded again in Fermanagh at Crom in 2001, Slisgarrow Quarry in 2002 and, most recently, Monmurry in 2004. In Armagh, the species was reported from Loughgall in 1978 and the Argory in 1998. Adults are sandy or greyish-brown and even freshly-emerged specimens appear washed out. The cross-lines are fine and indistinct. There is a small black discal spot near the costa; this is the only reliable character. Specimens thought to belong to this species should ideally be confirmed by genital examination especially when recording a site for the first time. Ochreous Pug is found in pine woodland and possibly other habitats where pine is well established. It is likely to be more widespread than current records suggest but the difficulty in distinguishing it from closely-related species may account for the paucity of records. Adults become active from dusk onwards and come to light in very small numbers. The recorded flight period is 12 May to 2 July. In Britain, the main flight period is late April to early June. The larvae have not been reported in Northern Ireland. In Britain, it feeds from June to mid-September on the young buds of Scots Pine and possibly European Larch. It overwinters as a pupa.

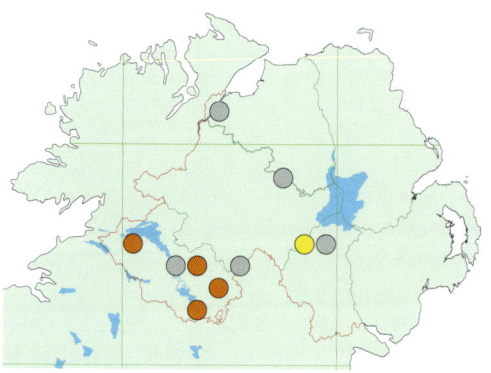

Pimpinel Pug
Eupithecia pimpinellata (Hübner) Category four.
Rare. This has been recorded on just three occasions in Northern Ireland. Campbell listed it from Londonderry in 1902. It went unrecorded for almost 100 years until 1999 when adults were taken at Killard Point (Down) on 10 June. The only other recent records were in 2000 from a suburban garden in Belfast and 2004 from a mature garden in Lisburn, both in Antrim. Pimpinel Pug is considered a rare species throughout Ireland and the only other counties with confirmed records are Cork, Dublin, Kerry and Limerick. Adults are pale greyish-brown with a black elongated discal mark. There are two additional black marks on the costa. The subterminal line is wavy and quite pale. In Britain, Pimpinel Pug inhabits grassland on light, usually calcareous, soils in rocky places inland and on the coast. Killard Point is a rocky coastal headland overlaid with sand, covered in herb-rich unimproved grassland and appears to be typical habitat. Adults are not particularly attracted to light, so the species may have been overlooked at suitable sites. The flight period in Northern Ireland is unknown but in Britain, it has a single generation in June and July. The larvae feed on the seed capsules of Burnet-saxifrage. It spends the winter months as a pupa.

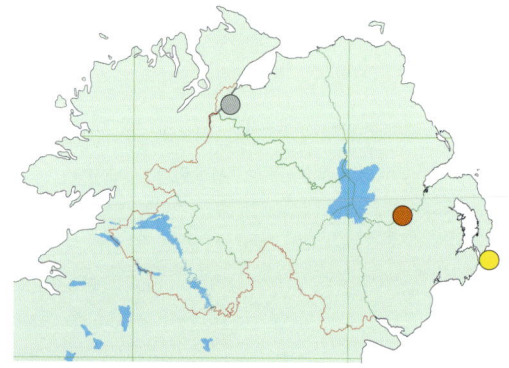

Narrow-winged Pug
Eupithecia nanata (Hübner) Category one.
Widespread. This is one of the more frequently recorded pugs in Northern Ireland. It is present in all counties and is generally distributed across the southern half of the region. To the north of Lough Neagh it is known only from the dunes at Benone, Magilligan and Umbra on the Londonderry coast, several sites on Rathlin Island and Breen Wood and Glenariff Forest in north Antrim. It has long been known from Loughgall and Peatlands Park (Armagh)

150

Moth accounts

Pimpinel Pug.

Killard Point National Nature Reserve, County Down.

Narrow-winged Pug.

Annagarriff National Nature Reserve, Peatlands Park, County Armagh.

A strongly-marked species associated with heathland and moorland.

151

and other reliable sites are Garvary Wood (Fermanagh) and Rehaghy Mountain (Tyrone). This is a variable species ranging in colour from light to dark grey with a marble-like appearance. The wings are narrow and pointed and traversed by strongly angled cross-lines. There is also a small white spot that borders the black discal mark. Narrow-winged Pug is characteristic of heaths, bogs, and moorland. Adults are attracted to light in small numbers and may occasionally be flushed from the foodplant by day. The recorded flight period is 7 May to 28 August. It is bivoltine in Britain, but the data does not indicate that pattern here. The larva feeds from July until September on Heather. It overwinters as a pupa.

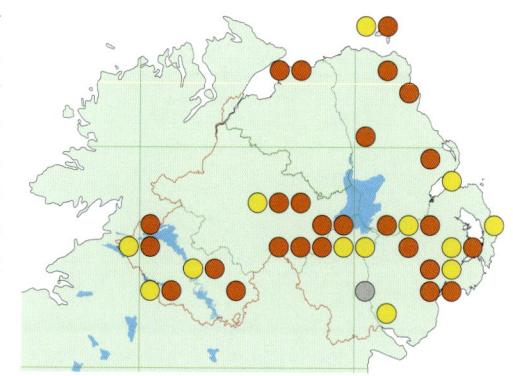

Angle-barred and Ash Pug
Eupithecia innotata (Hufnagel) Category three.

Notable. In recent checklists, the Ash Pug has been treated as a form (f. *fraxinata* Crewe) of the Angle-barred Pug rather than as a separate species. The species is known from Antrim, Armagh, Fermanagh and Tyrone. It was recorded for the first time in Northern Ireland in 1974 at Drumman More Lough (Armagh). Subsequent records came from Loughgall, also in Armagh, in 1990 and Ballinderry (Antrim) in 1991. It has not been reported from Antrim since, but additional Armagh records have come from the Argory, Aughinlig, Lislasly and Loughgall. It was first found in Tyrone in 2001 at Favour Royal and seen again in 2004 at Crilly. The known Fermanagh sites are Brookeborough, Crom and Monmurry. This is a large pug and the distinctive adults are unlikely to be confused with any other species. The forewings are narrow and appear elongated and pointed at the apex. The colour is uniform but variable, ranging from dark to light greyish-brown. Individuals that emerge early in May and June tend to be much darker that those that appear in late July and early August. There is a small black discal spot towards the costa. Like many pugs it often rests with its abdomen raised at the tip and curved laterally to the right or left. Angle-barred Pug is chiefly a woodland species but it has also been taken in mature, rural gardens. Adults appear very sparingly at light which is likely to account for its absence from many suitable sites. The recorded flight period is 19 May to 25 August. There have been too few records to determine the phenology here, but in Britain it is bivoltine. The main larval foodplant is Ash but it has also been recorded in Britain on Blackthorn, Hawthorn and Elder. It overwinters as a pupa.

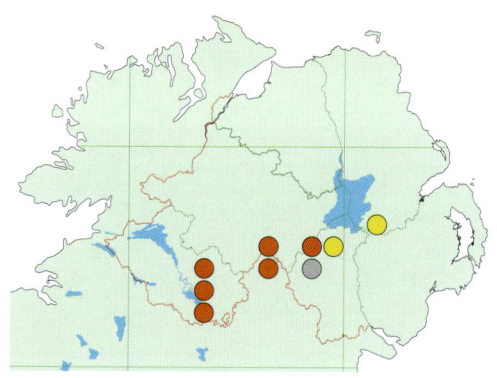

Golden-rod Pug
Eupithecia virgaureata Doubleday Category three.

Scarce. Recorded from all counties but the majority of records have come from north Armagh, east Tyrone and east Down. It has been recorded just twice in Antrim from a suburban garden (Belfast) in 1999 and Portmore Lough in 2002. The only record for Londonderry is by Campbell in 1875. In the west there are recent records from Brookeborough, Crom and Monmurry (Fermanagh) and Ballynasollus, Favour Royal and Rehaghy Mountain (Tyrone). Recently recorded sites in Armagh include Aughinlig, Carganamuck, Lislasly and Peatlands Park and and Castle Espie, Castlewellan Forest and Rostrevor Wood (Down). Adults are greyish-brown with

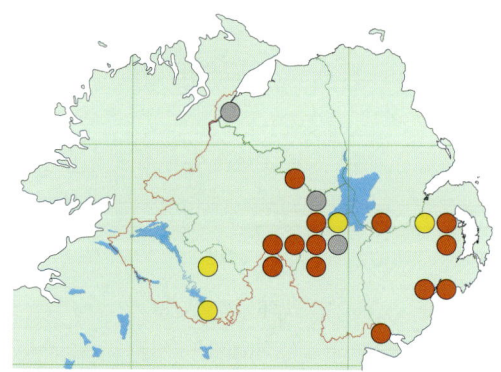

Brindled Pug.

Rostrevor Wood, County Down.

chequered veining and a small tuft of white scales is usually visible on the thorax of fresh individuals. There is a black discal mark on each forewing and a pale subterminal line which ends in a white spot. The hindwings are similar in colour. It closely resembles Grey Pug and the two species may be indistinguishable without close inspection of the discal spot, which is larger and more rounded in Golden-rod Pug. Also the underside of the eighth abdominal segment is pale and indented rather than pincer-like in Grey Pug. Golden-rod Pug is primarily a woodland species preferring open areas and clearings. It has also been found in suburban gardens and heathland. Adults appear occasionally at light in very small numbers. The recorded flight period is 4 May to 29 August. There has been only one July record (on 31 July) suggesting there are two annual generations, but more records are needed to confirm this. The larval habits are unknown in Northern Ireland. In Britain it is bivoltine in the south. The larval foodplants are Goldenrod and ragworts. It overwinters as a pupa.

Brindled Pug
Eupithecia abbreviata Stephens Category two.
Scarce. Generally distributed across the southern counties but there have been no recent records north of Lough Neagh, although Wright did list it from Londonderry but without giving a precise locality. It was also described by Donovan as common and widespread in oakwoods throughout Ireland. Its early flight period and the lack of fieldwork at the appropriate season may account for its apparent absence from the north as suitable habitat would seem to be present. Recently recorded sites include Portmore Lough and a suburban garden in Belfast (Antrim); the Argory, Clare Glen, Gosford Forest and Loughgall (Armagh); Belvoir Park, Derryleckagh, Helen's Bay, Hillsborough Forest and Rostrevor Wood (Down);

Crom, Garvary Wood, Knocknalosset, Legatillida and Monmurry (Fermanagh) and Rehaghy Mountain (Tyrone). Adults have a mottled brown appearance becoming lighter in the median area, with darker angled cross-lines. There is a small dark discal mark near the costa but this is not always clearly defined. The Brindled Pug is a woodland species found in mature, broadleaved woodland with an abundance of oak. Adults are active from dusk onwards and come frequently to light in small numbers. They rest among the branches and on the tree trunks of the foodplant during the day. The recorded flight period is 9 March to 21 June. The larva feeds during June and July on oak and possibly Hawthorn. It spends the winter months as a pupa.

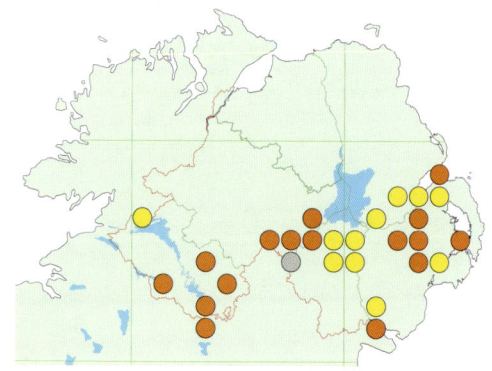

Oak-tree Pug
Eupithecia dodoneata Guenée Category three.
Notable. The distribution of this pug is restricted to a few localities in north Armagh, Down and south Tyrone. The Armagh sites are the Argory, Kinnegoe, Loughgall and Portadown. In Down it has been reliably recorded only from Hillsborough Forest. The recent Tyrone records have come from Crilly, Favour Royal and Rehaghy Mountain. There have been no recent records from Antrim, Londonderry and Fermanagh. Its absence from the latter is particularly strange considering the abundance of habitat and the amount of trapping undertaken in Fermanagh throughout its flight period. Adults bear a close resemblance to Brindled Pug, but tend to be smaller in size and have a large, distinct, elongated discal spot. The forewings are pale brown with darker flecks along the veining giving it a mottled appearance. It emerges somewhat later than Brindled Pug, but there is much overlap in flight period. This is, as its name indicates, associated principally with oak and found in mature broadleaved woodland and large hedgerows. Adults become active from dusk onwards and are attracted to light, but never in large numbers. The recorded flight period is 5 April to 14 June. The larva feeds from late June until August on the flowers and leaves of oak and Hawthorn. It overwinters as a pupa.

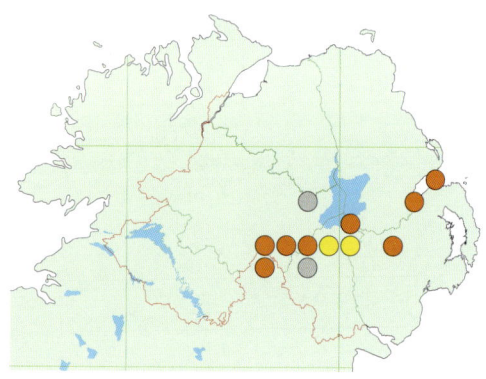

Juniper Pug
Eupithecia pusillata (Denis & Schiffermüller) Category four.
Rare. Donovan listed Juniper Pug from Antrim, Armagh and Down. These records are from Divis (Antrim), Armagh City and Slieve Donard (Down); all are over 100 years old and there have been no recent records from these sites. Since 1995, there have been records from several sites in Antrim, Down and Fermanagh. It has twice been reported from the Rothamsted trap at Hillsborough Forest in Down in 1995 and 2002. In 2001 specimens were taken at Kebble and the West Light on Rathlin Island (Antrim). It was discovered in Fermanagh at two sites, Altaclanabryan and Legatillida, in the east of the county in 2002. The most recent record is of a single adult which was taken in a suburban garden at Finaghy (Antrim) on 6 August 2004. Adults vary from light to dark brown. There is a black central spot and two other dark marks between the central spot and the apex of the forewing. The cross-lines are angled but not always clearly defined. The Northern Ireland records have come from an estate with mature conifers, coastal

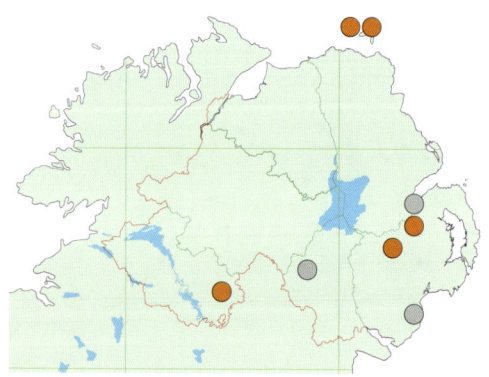

and inland heathland and a suburban garden. With the exception of the recent Rathlin records none of the sites are within the natural range of its principal foodplant, Juniper. It may be more widespread than records suggest, especially in west Fermanagh and north Antrim where native populations of this conifer exist, but clearly at other sites it seems likely that the species utilises ornamental junipers. Individuals have been recorded from 2 to 25 August in Northern Ireland, but the main flight period in Britain is July to September. Larvae have not been found in Northern Ireland. In Britain adults become active from dusk onwards, although there have been observations of individuals flying in sunshine around Juniper bushes. The larva feeds on Juniper and probably other related introduced varieties. It passes the winter months as an egg.

Larch Pug
Eupithecia lariciata (Freyer) Category three.
Scarce. There have been recent records of this species from every county except Londonderry. The majority of records are from Down and Fermanagh. Sites in Down include Castle Espie, Castlewellan Forest, Murlough NNR, Rostrevor Wood and Tollymore Forest. In Fermanagh, there have been records from Colebrooke, Correl Glen, Crom, Eshywulligan, Legatillida, Monmurry and Stranfeley. It has also been recorded from Gosford Forest and Slieve Gullion (Armagh) and Altadavan and Rehaghy Mountain (Tyrone). In Antrim, it was found at Breen Wood in 2001 and there is an historical record from Colin Glen in 1895. This species is predominantly greyish-brown with a sharply-angled median line around the small black discal spot. There is

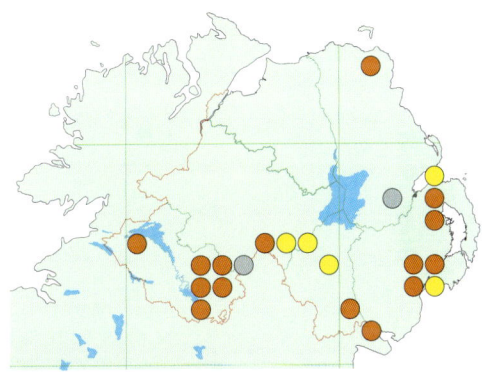

also a white crest or spot on the thorax. Larch Pug is a woodland species recorded mainly from old estates and the older conifer plantations, but also mixed woodland where larches have been planted or have become established. Adults are attracted to light and appear in small numbers. Individuals can occasionally be found during the day by beating the lower branches of the foodplant. The recorded flight period is 2 May to 27 August. The larva feeds from the end of June until the beginning of August on larch and possibly related species. It spends the winter months as a pupa.

Dwarf Pug
Eupithecia tantillaria Boisduval Category three.
Scarce. There have been records since 1984 from Armagh, Down, Fermanagh and Tyrone. The first confirmed Northern Ireland record came from Loughgall (Armagh) in 1984 and it was taken again in the same county in 1997 at Gosford Forest. In Fermanagh, it was first recorded at Correl Glen and subsequently found at Brookeborough, Carmacullagh Bog, Cooneen, Crom, Glen Wood, Knocknalosset and Monmurry. The first record from Tyrone was in 1998 at Ballynasollus. Since then it has been found at Altadavan and Favour Royal. There have been three records in Down from Bohill Forest in 2000, Rostrevor Wood in 2001 and, most recently, Hillsborough Forest in 2004. Baynes knew of just one Irish record of this species from Kerry, but it

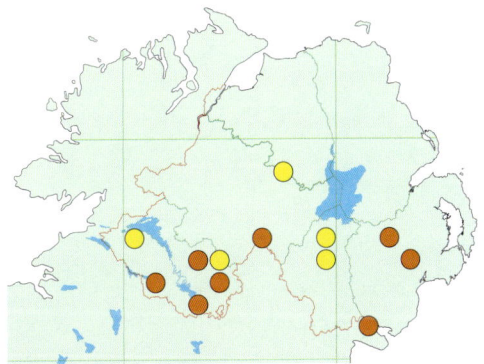

appears to have spread as its habitat and foodplant, woodland with spruce, has become common and reached maturity. Adults are brownish-grey with a conspicuous dark discal spot. There are darker patches along the costa of the forewing and banding on the abdomen. The hindwings are paler with brown cross-lines and a small but distinctive black discal spot. Dwarf

Pug is a spruce-feeding species that favours conifer plantations and old estates where the foodplants are found. Adults are active from dusk onwards and come to light in small numbers. The recorded flight period is 15 May to 18 August. The larva feeds throughout July and August on spruce and Douglas Fir. It overwinters as a pupa.

V-Pug

Chloroclystis v-ata (Haworth) Category one.

Widespread. Recorded from all counties and generally distributed across the southern half of the region where it is frequently met with at many well-trapped sites. In the north it has recently been recorded at several sites in mid- and north Antrim, on Rathlin Island and at Duncrun and the Umbra in Londonderry. Adults are green when freshly emerged, with a characteristic black V-shaped mark near the costa. There is also a small black mark on the top of the thorax and another along the costa. The abdomen has parallel rows of small black dots. As with many other similarly coloured species, the green colouration begins to fade fairly quickly after emergence. V-Pug occurs in many habitats, particularly woodland margins and coastal grasslands and occasionally, fens and bogs. It has been trapped regularly in gardens in Belfast, Helen's Bay and Lisburn. Adults come frequently to light in moderate numbers. There are two generations, the first from mid-April to late May and the second from early July. The recorded flight period is 6 April to 5 September. The larva feeds throughout summer and autumn on a wide variety of plants including Elder, Bramble, Traveller's-joy, Goldenrod and Wild Angelica. It overwinters as a pupa.

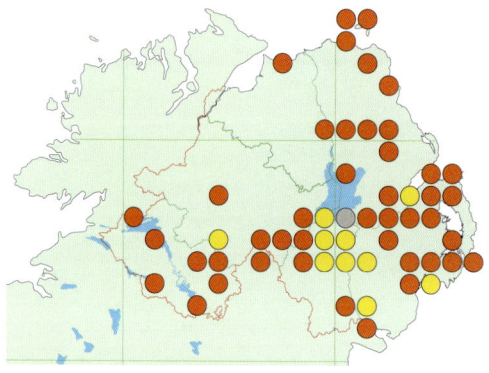

V-Pug.

The Argory, County Armagh.

Green Pug
Pasiphila rectangulata (Linnaeus) Category one.
Widespread. Recorded from many sites in Armagh, Down and south Fermanagh. The increased level of recording has produced many new records, especially in the Greater Belfast area. It is very local in the north in Antrim, Londonderry and Tyrone. It was reported in 2001 at the East Lighthouse on Rathlin Island (Antrim) and at Duncrun (Londonderry). Adults of this species are distinctive and not easily confused with any other species except V-Pug, which has a conspicuous heavy black V-shaped mark on the forewing. Most freshly-emerged individuals are green and heavily marked with darker cross-lines. There is also a dark abdominal band. The green fades quickly and many individuals look light brown when viewed in the trap. Green Pug occurs in many lowland habitats with 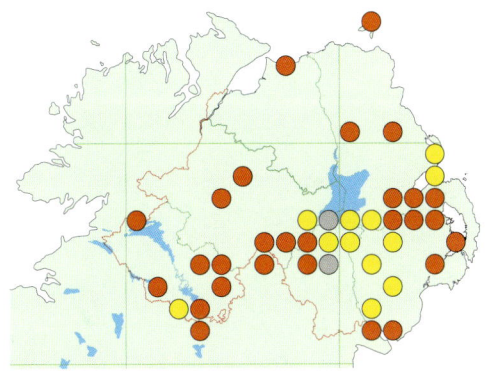 trees and shrubs including woodland margins, orchards, mature suburban gardens and hedgerows. Adults appear frequently at light in small numbers. The recorded flight period is from 26 May to 2 September. The larva feeds from April to May on the flowers of fruit trees including cultivated pear, apple, and cherry, but also Blackthorn and Hawthorn. It overwinters as an egg.

Bilberry Pug
Pasiphila debiliata (Hübner) Category four.
Rare. Recorded for the first time in Northern Ireland at Bohill Forest (Down) on 30 June and subsequently on 6 and 19 July 2004. Baynes describes it as local and not uncommon but only gave records for counties Cork, Galway, Kerry, Sligo, Waterford and Wicklow. This species is easily overlooked and may occur in low density, especially in the west where its foodplant is particularly common. Adults, when fresh, are pale green with a dark discal spot and cross-lines which are composed of black dots. Like many other green species the colour weakens and most specimens, when captured, appear pale brown. Bilberry Pug favours open woodland, heathland and moorland with an undergrowth of Bilberry. Adults are attracted to light from dusk onwards and are best looked for around the 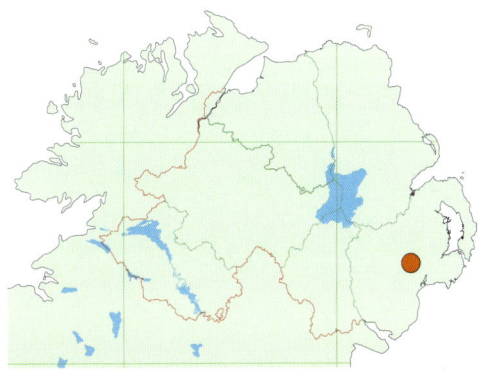 foodplant. There is insufficient data to determine its phenology in Northern Ireland. In Britain, it is normally single-brooded, flying in June and July. The larva feeds at night, remaining concealed among spun leaves of Bilberry when inactive. It overwinters as an egg.

Double-striped Pug
Gymnoscelis rufifasciata (Haworth) Category one.
Widespread. Recorded in all counties and generally distributed south of Lough Neagh. It has been recorded frequently at many of the regularly trapped localities including, Finaghy (Antrim) and Belvoir Park, Helen's Bay, Hillsborough Forest and Murlough NNR (Down). To the north of Lough Neagh, it has been found at a scattering of sites in mid- and north Antrim and on the north coast of Londonderry on the Magilligan dune system. It was found at six separate sites on Rathlin Island (Antrim) in 2001. It is also recorded from Copeland Island off the north coast of Down. Adults are quite variable in colour, ranging from light to dark reddish-brown. The two dark cross-lines are quite distinct, but often fade away towards the centre of the forewing.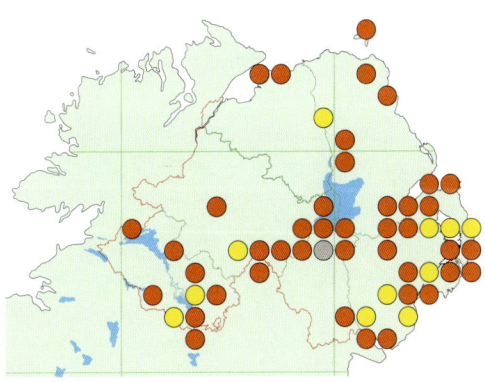

The Butterflies and Moths of Northern Ireland

Streak.

Corbet Lough, County Down.

The long narrow wings and characteristic tent-like resting posture are distinctive features of this autumn-flying moth.

There is also a darker band across the upper segments of the abdomen. Second generation adults tend to be smaller and more richly marked than those of the first generation. This pug can be seen in most habitats but is particularly frequent in mature hedges, woodland, parkland, bogs, heaths, moors and gardens. Adults become active around dusk and come to light frequently in small numbers. The recorded flight period is 16 March to 15 September. There are two annual generations, one peaking in early May and a second in early July. The larva feeds from May until late autumn on a variety of trees, shrubs and herbaceous plants including Holly, Gorse, Broom, Common Ragwort and Heather. It overwinters as a pupa.

Streak

Chesias legatella (Denis & Schiffermüller) Category two.

Notable. This is predominantly a south-eastern species and, apart from two records – Tempo (Fermanagh) and Traad Point (Londonderry) – seems confined to Armagh and Down. The most recent records are from the Argory and Peatlands Park in Armagh. Other older records from Armagh include Aughinlig, Drumman More Lough, Loughgall and Richhill. In Down it has been reported from Belvoir Park, Derryleckagh, Lackan Bog and most recently Corbet Lough in 2001. Adults are generally brown with a distinctive long, white streak running the entire length of the forewings. There is an oval-shaped mark in the central area of the forewing and a short white cross-line at the outer margin. Unusually for a geometrid, the adults rest with their wings folded in a tented position over the body. The Streak has been taken in disturbed and abandoned ground at the edges of bogs, woods and amenity sites where its foodplant, Broom, has been planted or has naturally colonised. Adults are active from dusk onwards but come to light only sparingly. Some individuals in Northern Ireland have been found resting on the stems of Broom after dark. This is an autumnal moth recorded between 18 September and 13 November. The lateness of its flight season and the patchy, ephemeral nature of the stands of its foodplant probably exaggerate its rarity. The larva can be found in May and June on Broom. It overwinters as an egg.

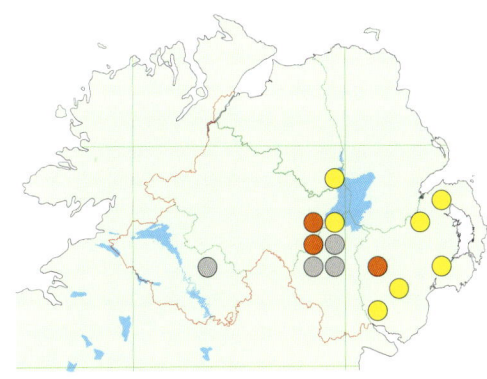

Treble-bar

Aplocera plagiata (Linnaeus) Category two.

Scarce. Recorded in all counties, but local in all and mainly coastal in the east and north. It is unrecorded from much of Tyrone and inland parts of both Antrim and Londonderry. Survey work in 2001 showed that it is widespread on Rathlin Island (Antrim). Other sites with recent records are Breesha Plantation, Fair Head, Murlough Bay and Slievenacloy (Antrim); Murlough NNR (Down); Garvary Wood, Gortmaconnell Rock and Legatillida (Fermanagh) and the Umbra (Londonderry). Adults have pale grey forewings marked with three dark prominent bands. The bands vary in width and intensity on individuals. There is a short brown apical streak and a chequered fringe on the forewings. The hindwings are unmarked and pale grey. Treble-bar has been recorded from sand dunes, upland grassland and open woodland. It prefers well-drained sites, so inland, it is only found on the basalt of Antrim and limestone in Fermanagh. Adults appear frequently at light traps and are occasionally disturbed from vegetation during the day, especially in coastal sites where the largest populations exist. The recorded flight period is 25 June to 25 September. The larva, which is the overwintering stage, can be found throughout the autumn and again in spring on St. John's-wort.

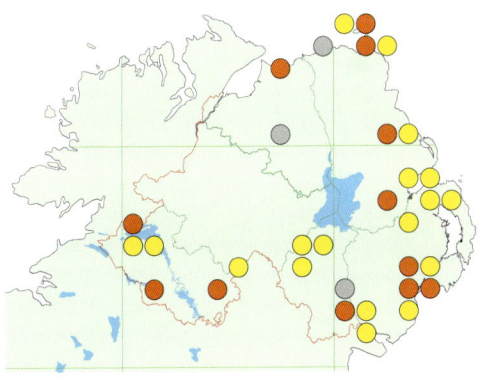

Chimney Sweeper

Odezia atrata (Linnaeus) Category three.

Scarce. This easily recognised species has an unusual distribution, being confined almost entirely to Tyrone. It was discovered at many of its sites in the county in the 1990s, having undoubtedly been overlooked previously. Sites for it include Altdrumman, Coneyglen, Craigs, Davagh Bridge, Killeter, Semples Bridge and Wellbrook. Outside Tyrone it is very rare and has been found since 1990 at just three sites – Murlough Bay (Antrim) in 1994, Tullynacor (Fermanagh) in 1997 and Bealnaslaght Bridge (Londonderry) in 1996. Three old records exist from near Derry City by Campbell in 1875, Tempo (Fermanagh) by Langham in the early twentieth century, and a 1974 record from Markethill (Armagh). It has never been reported in Down. The forewings of fresh specimens are a uniform dark brown or black with a distinctive white tip at the apex, but older individuals often look paler. Chimney Sweeper is found in damp, unimproved grassland usually close to streams and rivers or woodland and is probably in decline as the extent of suitable habitat dwindles. Adults fly in sunshine, but can be disturbed from vegetation in duller weather. The recorded flight period is short, from 15 June to 10 July. The larva can be found from April to June on the seeds and flowers of Pignut. It overwinters as an egg.

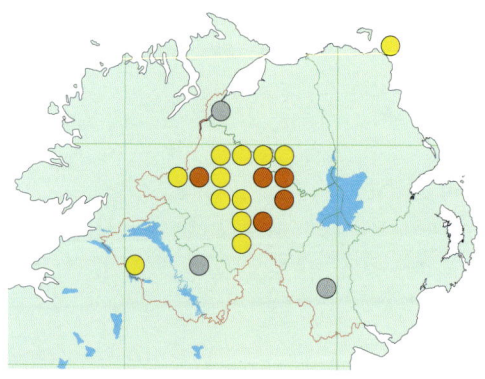

Welsh Wave
Venusia cambrica Curtis Category three.
Scarce. This woodland moth has been recorded from several sites in Antrim, Down, Fermanagh and Tyrone since 1990. It was recorded by Campbell near Derry City and at sites in Tyrone by both Greer and Kane. Watts reported it from Newcastle and Saintfield (Down). It has not yet been recorded from Armagh. The known sites are Correl Glen, Glennasheevar and Gortmaconnell Rock (Fermanagh); Altaveedan, Mountjoy and Rehaghy Mountain (Tyrone); Altnaveigh, Ballymacashen Bog, Bohill Forest, Castlewellan Forest, and Clandeboye (Down) and Drumadion, Glenariff Forest and Glenarm (Antrim). Adults are grey and marked with blackish cross-lines. The central line is the darkest and has a pair of spikes projecting from the rear edge. There is also a dark dashed line around the outer area of the forewing. Welsh Wave has been recorded from open, semi-natural woodland where it adjoins moor or heathland. Adults come to light in small numbers and are apparently found by day on rocks and the trunks of trees. The recorded flight period is 16 June to 10 August. Rowan is the main larval foodplant, but it also occasionally uses birch. It overwinters as a pupa.

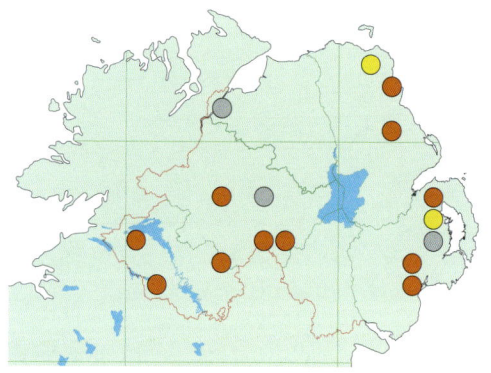

Small White Wave
Asthena albulata (Hufnagel) Category three.
Rare. Apart from a record from Bellaghy (Londonderry) in 1976, this species seems to be confined to Down. It was first recorded in the county by Foster in 1905 at Rostrevor Wood. It still occurs here and at nearby Derryleckagh where it was discovered in 1990. It was recorded here again in 1999, when adults were beaten by day from the foliage of the foodplant. Similarly, it was found at Bohill Forest in central Down in 1990 and seen again in 1999 and 2000. Finally, there is a single record from Crawfordsburn in 1994. There have been no other records from other counties, although suitable habitat would appear to be present in most, especially Fermanagh and Antrim. Adults are small and richly marked. The wings are silky white with fine brown cross-lines and there is a fine black broken line around the outer margin of all wings. Adults are

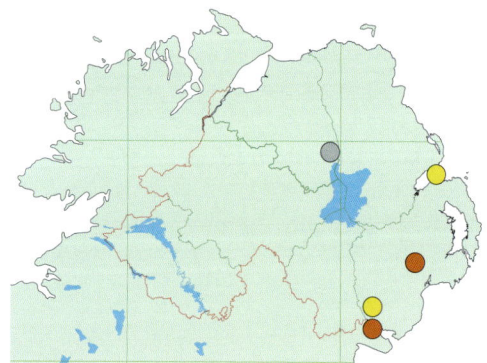

seldom seen at light which may account for its absence at some western localities where Hazel is common. Most of the recent records are of adults disturbed by day – beating Hazel branches has proved to be the most productive method for recording adults. This small moth is associated primarily with semi-natural broadleaved woodland with stands of Hazel. The recorded flight period is 9 May to 30 June. The larval foodplants are Hazel and possibly birch. It overwinters as a pupa.

Small Yellow Wave
Hydrelia flammeolaria (Hufnagel) Category four.
Rare. This small attractive moth was discovered for the first time in Northern Ireland in 2000, when a single adult was taken at light in a regularly trapped suburban garden on the outskirts of Belfast (Antrim). It was recorded again from the same garden in 2003 and 2004. Since its discovery in Belfast, adults have been recorded at Mount Stewart and Bangor (Down), in 2001 and 2003, respectively. Elsewhere in Ireland the species is known from Killarney (Kerry) and Corofin (Clare). Its current status in Northern Ireland is unclear – it is either a recent colonist or an established, but overlooked, resident. Adults are straw yellow with ochreous cross-lines on both wings, which are variable in thickness, and a small black discal spot. In Britain, it is 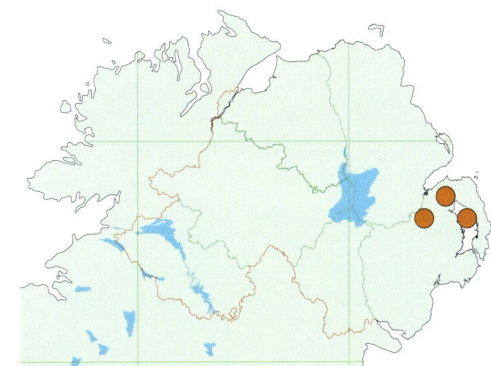 associated with damp woodland and stands of trees along rivers containing Alder, its main foodplant. The Belfast records appear at first atypical, but suitable habitat exists along the nearby Lagan Valley. The Mount Stewart record seems to be from a more normal habitat. Adults are attracted to light and can apparently be flushed from the foliage of Alder during the day. The northern Irish specimens were caught between 2 and 6 July. The main flight period in Britain is late May to July in a single generation. The larva can be found during August and September on Alder and possibly Sycamore. It overwinters as a pupa.

Seraphim
Lobophora halterata (Hufnagel) Category four.
Rare. This rarely-recorded species is known from all counties with the exception of Londonderry. It almost certainly has a wider distribution than current records suggest, especially in Fermanagh where plenty of suitable habitat exists. Many of the records from the west are over sixty years old. It is not known if the species has declined or simply been under-recorded; the latter seems the most plausible explanation. Since 1990 it has been recorded from Hillsborough Forest (Down) in 1997 and Carn (Armagh) in 2000. In Fermanagh there are records from Crom and Derrygonnelly in 2003 and Legatillida in 2004. It was also recorded from both Belfast (Antrim) and Crilly (Tyrone) in 2004. There are historical records from Tempo Manor (Fermanagh) and 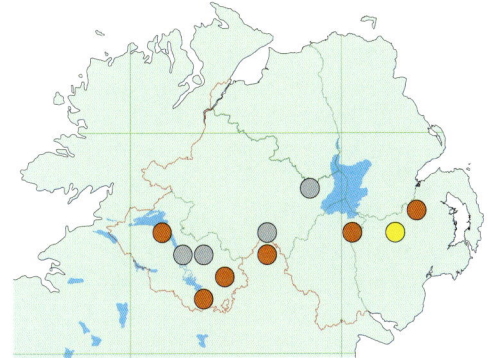 Cookstown and Favour Royal (Tyrone). Adults have mottled grey wings with darker grey cross-bands. The band at the outer margin of the forewings is rather broad and a reliable diagnostic feature. The area between the cross-bands often has conspicuous fine buff cross-lines. Seraphim is associated with woodland and other habitats where Aspen and other poplars grow. Adults do not have a strong attraction for light, which might explain the lack of records from seemingly suitable sites. It rests reportedly during the day on the trunks of trees which may be worth investigating for resting adults. The recent records were taken between 13 May

The Butterflies and Moths of Northern Ireland

Seraphim.

Peatlands Park, County Armagh.

This woodland species has always been considered rare in Northern Ireland, although this may be largely attributed to under-recording rather than genuine absence.

and 18 June which corresponds with the main flight period in Britain. The larva can be found in July on Aspen and other poplar species. It overwinters as a pupa.

Early Tooth-striped
Trichopteryx carpinata (Borkhausen) Category one.
Scarce. Recorded from all counties, but frequent only in parts of north Armagh, south Fermanagh and east Tyrone. Most of the sites have been discovered since 1991, undoubtedly because of regular trapping in the early part of the year. Reliable sites in this core area include the Argory and Peatlands Park (Armagh); Brookeborough, Crom and Monmurry (Fermanagh) and Crilly and Rehaghy Mountain (Tyrone). It is very locally distributed in Antrim and Londonderry. Curran Bog is the only known site in Londonderry and the only Antrim sites are Montiaghs Moss and Sharvogues Bog. In Down, it occurs at Bohill Forest, Lackan Bog and Rostrevor Wood. Adults are brownish-grey with dark brown cross-lines on the wings, which can vary in intensity. In some

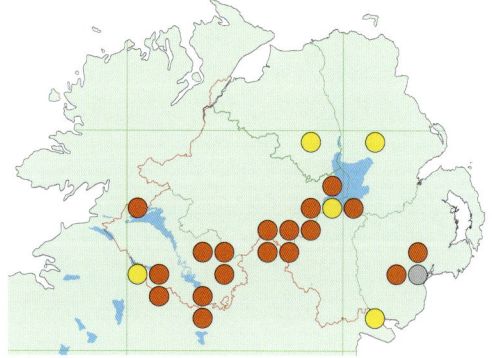

individuals they are barely evident. Slightly worn specimens could be mistaken for Mottled Grey which has an overall mottled appearance with weakly-defined cross-lines and a more pointed forewing. It also overlaps in flight period. Early Tooth-striped is found mainly on heaths and bogs, especially those with adjoining birch scrub. Adults are attracted to light in reasonable numbers and rest by day on the trunks of trees and other similar supports. The recorded flight period is 15 March to 1 June. The larval foodplants are willows, birch, Honeysuckle and Alder. It overwinters as a pupa.

Early Tooth-striped.

Lackan Bog, County Down.

An early spring species associated mainly with heaths and bogs.

Small Seraphim
Pterapherapteryx sexalata (Retzius) Category four.
Scarce. Recorded recently from Fermanagh and Tyrone since 2000. There are old records for Antrim and Londonderry, but it never has been recorded reliably from Armagh or Down. It was listed by Campbell from Londonderry in 1875 and Wright recorded it from Ballymacilrany in south-east Antrim in 1951. It was discovered in Fermanagh at Florencecourt in 1988 and, since 2000, at many more sites in the county including Cuilcagh Mountain, Derrygonnelly, Glen Wood, Gortmaconnell Rock, Legatillida and Monmurry. Greer was the first to report it from Tyrone at Stewartstown. It was rediscovered in Tyrone in 1997 at Rehaghy Mountain and reported subsequently at Altadavan in 2001 and 2002; also another site close to Fivemiletown in

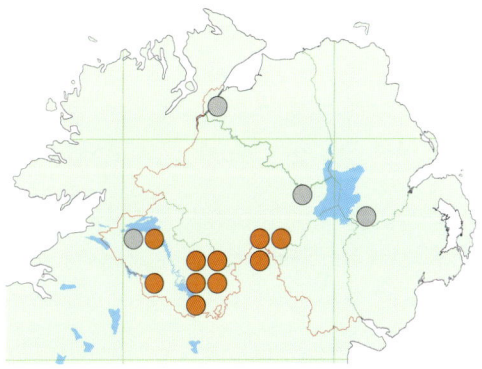

The Butterflies and Moths of Northern Ireland

Small Seraphim.

Monmurry, County Fermanagh.

2003. In view of the many recent records it seems this species has simply escaped detection. It probably awaits discovery at other suitable sites especially in Tyrone and Fermanagh. The adults have pale grey forewings with a darker grey, central cross-band and a small dark central spot. There are fine buff cross-lines between the banding and a fine, white, wavy submarginal line which is quite conspicuous. This species has been found in woods and wetlands including bogs. Despite the existence of suitable habitat in many areas, it seems elusive and is seldom encountered at light. The recorded flight period is 1 June to 7 August. The larval foodplant is willow. It overwinters as a pupa.

Yellow-barred Brindle

Acasis viretata (Hübner) Category two.

Scarce. Widely distributed (but never commonly encountered) across southern counties, particularly Down and Fermanagh. It is found also in south Tyrone and north Armagh and from a suburban garden in Finaghy (Antrim). Surprisingly it has not been recorded from Londonderry although Holly, its major foodplant, is widespread. The majority of records have come from the Argory, Aughinlig and Peatlands Park (Armagh); Belvoir Park and Helen's Bay (Down); Crom and Garvary Wood (Fermanagh) and Rehaghy Mountain (Tyrone). Adults are olive green when freshly emerged, but the colour quickly fades to yellowish-green. This change in appearance often leads to misidentification. Most individuals have a darker, broad central cross-

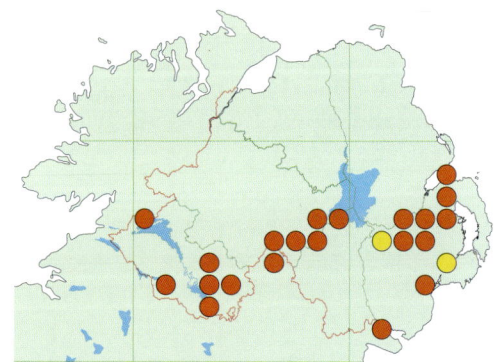

band, but this is occasionally reduced to a series of fine, wavy cross-lines. This species occurs in broadleaved woodland with Holly, especially oak-birch woodland on acid soils. It has also been taken in gardens and estate woodland. Adults are attracted to light in very small numbers and can be found during the day on the trunks of trees. The pattern of records in Northern Ireland suggests there are two generations, as is the normal pattern in southern Britain. Two peaks are apparent in the phenology, the first in mid-May and the second, a smaller one, in August. The recorded flight period is 5 April to 18 September. The larval foodplants are Holly, Ivy and Wild Privet. It overwinters as a pupa.

Magpie

Abraxas grossulariata (Linnaeus) Category one.

Widespread. Recorded in all counties and on offshore islands. Like most species, it seems to be more scarce in the north and west, but this probably reflects the level of recording. In recent times, however, it seems to have become less common even at some of its reliable sites. These include Loughgall, Oxford Island and Peatlands Park (Armagh); Castle Espie and Killard Point (Down) and Aghalane, Crom and Garvary Wood (Fermanagh). It is also regular at Magilligan and the Umbra (Londonderry) on the north coast. Adults are very variable in appearance, but remain distinctive and instantly recognisable. The numerous named forms and aberrations which appear, especially in captive-bred individuals, are seldom encountered in the wild. The 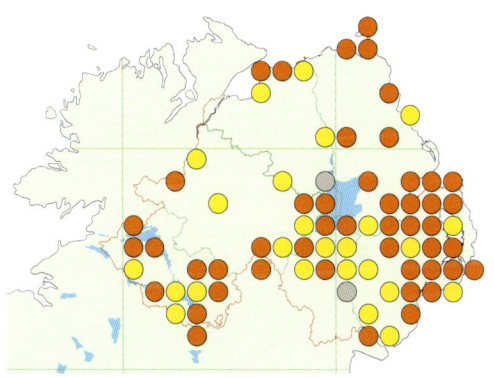 wings are predominantly white, marked with heavy black spotting. There is a pale orange cross-line running across the forewing near the outer margin. The abdomen is marked with orange and black stripes. The Magpie is found in many habitats particularly woodland, scrub and suburban gardens. Adults are attracted to light, but appear sparingly in traps. Individuals are occasionally found or disturbed by day from vegetation. The flight period is prolonged from 3 June to 4 September, peaking in early July. The larvae have been reported in Northern Ireland on Spindle and Hawthorn. In Britain other recorded foodplants include Blackthorn, cultivated and wild currants and Hazel. In northern Scotland, Heather is the principal foodplant supporting large moorland populations. It overwinters as a larva.

Clouded Magpie

Abraxas sylvata (Scopoli) Category three.

Scarce. Recorded from several sites in Down, two sites in Tyrone and single sites in Antrim, Armagh and Fermanagh. There are no confirmed records from Londonderry. The most reliable sites are in Down including Belvoir Park, Castle Espie, Castleward, Hollymount, Lackan Bog, Murlough NNR and Rostrevor Wood. The colonies at Murlough NNR and Rostrevor Wood are considered two of the largest in the region. Its localities in other counties are Glenarm (Antrim), the Argory (Armagh), Carricknabrattoge (Fermanagh), Favour Royal and Strabane Glen (Tyrone). At all these sites the species has been recorded on just one occasion. Resting adults resemble a bird-dropping. The ground colour is predominantly white with irregular patches of black, 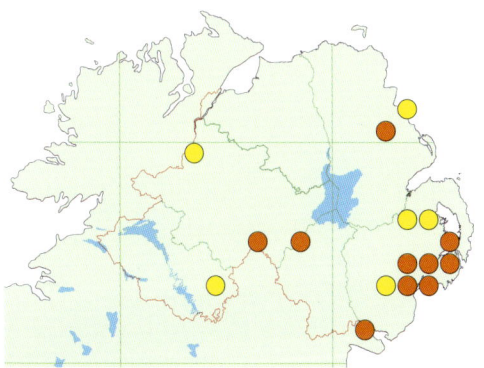 brown and grey scattered across all four wings. The size and density of these markings varies between individuals. Clouded Magpie is an elm-dependent species. Colonies are known only from areas with a long history of woodland cover. Some colonies may have been lost following the epidemic of Dutch elm disease in the 1970s and 1980s, although this may have been less serious here than in Britain, as the common elm in Northern Ireland is the native Wych Elm, which was less affected by this disease. Adults are attracted to light in moderate numbers where locally abundant; elsewhere it is found usually as a single individual. Their resemblance to a bird-dropping means they will often rest on the upper surface of leaves during the day. The recorded flight period is 20 May to 24 August. Larvae have not been recorded in Northern Ireland, but in Britain they can be found from August to October on Wych Elm and English Elm. It overwinters as a pupa.

Clouded Border

Lomaspilis marginata (Linnaeus) Category one.

Ubiquitous. A common and widespread species occurring at many sites in all counties. The gaps in the distribution are likely to be due to under-recording. It is also known from Copeland Bird Observatory (Down). This is a small, distinctive moth with black-bordered white wings. The width and shape of the black margins is subject to much variation. There is a rare form (f. *diluta*) in which the black is replaced by gold. Baynes mentions two records of this form being taken at Dingle (Kerry) in 1962 and 1963. This form has also been recorded in Killarney National Park and more recently in Northern Ireland at Aghalane (Fermanagh) on 15 July 2001 and the Argory (Armagh) on 11 June 2003. Clouded Border occurs in many habitats but is especially typical of wet woodland, wooded heath, fens and cutover bogs. Adults are often seen in gardens. They come frequently to light and are occasionally seen or disturbed from foliage during the day. The flight period is prolonged, extending from 16 April to 7 October. The data indicates that there is only a single generation peaking in midsummer. The larval foodplants are willows, Aspen and other poplars and possibly Hazel. It overwinters as a pupa.

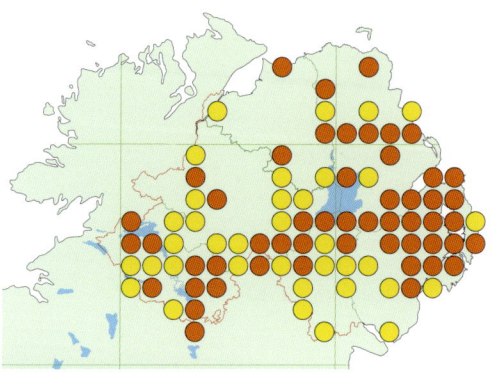

Scorched Carpet

Ligdia adustata (Denis & Schiffermüller) Category four.

Rare. Recorded from all counties except Londonderry; however, many of the records are pre-1980. It appears on a nineteenth-century list of moths seen at Glenarm (Antrim). In the same county, Wright recorded it from Ballymacilrany in 1951 and in 1991, it was trapped at nearby Ballinderry. There are historical records for Maloon, Stewartstown and Tullyhogue (Tyrone) in the early 1920s, but it was not recorded in this county again until 2001, when a single adult was taken at Crilly. There have been two records in Down, the first from Gilford in 1976 and a second, more recent one, from Rostrevor Wood in 2001. It has been recorded twice in Armagh at Drumman More Lough in 1967 and more recently from Loughgall in 2001. Finally, in Fermanagh, there is a vague record from Enniskillen in 1937 and two more recent records, both from Crom, in 2001 and 2003. Adults are distinctive and unlikely to be confused with any other species. The forewings are creamy white with a wavy blue-grey cross-band near the outer margin. The basal areas of the forewing, head and thorax are also a dark blue-grey. This is a species of wet woodland and scrubby fens where its foodplant, Spindle, is present. Adults become active at dusk and appear sparingly at light. In Britain, this is single-brooded in the north and double-brooded in the south. The precise phenology of the Northern Ireland populations is unclear. The few records with full data have been in the period 17 May to 23 July. It overwinters as a pupa.

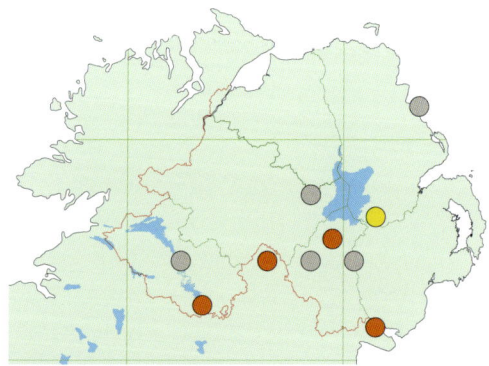

Moth accounts

Clouded Border.

The Argory, County Armagh.

This is a small, commonly recorded species found in a wide range of habitats throughout Northern Ireland.

Clouded Border.

Killarney National Park, County Kerry.

This rare form (f. diluta) was thought to be confined to County Kerry, but adults have been found at two sites in Northern Ireland.

Scorched Carpet.

Rostrevor Wood, County Down.

Tawny-barred Angle
Macaria liturata (Clerck) Category three.

Scarce. The main distribution of Tawny-barred Angle is in the eastern half of Down and, further west, in a band running south-west from the south-western corner of Lough Neagh through south Tyrone to south Fermanagh. It is also found north of Lough Neagh in a few scattered sites in north Antrim and Londonderry including Glenariff Forest and Portglenone (Antrim) and the Umbra (Londonderry). It is probably more widespread than current records indicate, especially in the extensive conifer plantations in the north. Frequently recorded sites include Loughgall and Peatlands Park (Armagh); Belvoir Park, Bohill Forest and Rostrevor Wood (Down) and Legatillida (Fermanagh). There are two main colour forms in this species which coexist in most populations. The 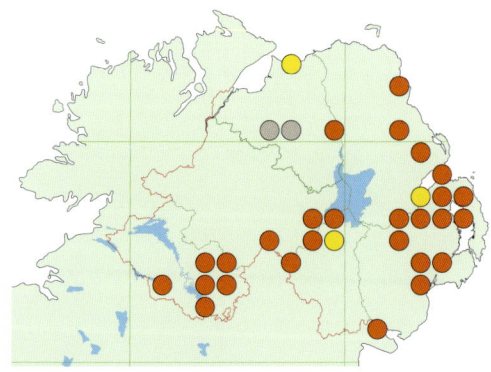 most frequently encountered form is speckled greyish-brown with an orange-brown band across the forewings. The other is a melanic form (f. *nigrofulvata*) in which the entire wings, apart from the tawny band, are dark brown. The hindwings are sharply pointed. This moth is found in woodland with mature conifers including recent conifer plantations (large and small) and older, estate woodland. It can also be found in other habitats with mature conifers. Adults appear frequently at light in small numbers. By day the adults often rest in the foliage of conifer trees. The recorded flight period is 23 May until 8 August. This species is double-brooded in southern Britain, but here the data indicates that there is just a single annual generation. The larval foodplants are Scots Pine, Norway Spruce and Sitka Spruce. It overwinters as a pupa.

V-Moth
Macaria wauaria (Linnaeus) Category four.

Unknown. There have been no records of this moth since the early 1930s and its current status here is not known. It has been recorded four times in Northern Ireland: in 1893 from Derry City; at Strangford Narrows (Down) in 1931, where it was described by Foster as 'increasingly numerous' and Favour Royal and Stewartstown (Tyrone) in respectively 1902 and 1921. The lack of any recent records suggests that it has become extinct, although suitable habitat probably still exists throughout most counties. A search of some of its former haunts may reveal undetected populations. The monitoring data from the Rothamsted trap network in Britain indicates a decline in abundance of this species of 97 per cent between 1968 and 2002. Adults are pale grey 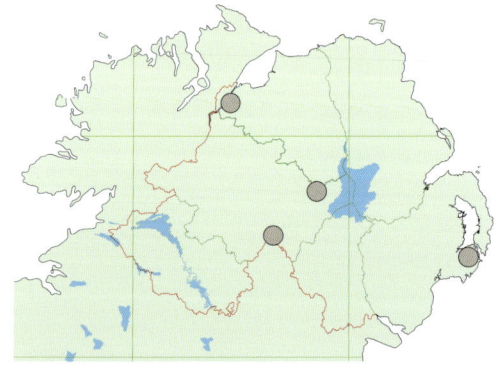 with a conspicuous black V-shaped mark on the forewing and dark marks along the costa. V-Moth is associated particularly with mature gardens and orchards with established bushes of currants and related species. Recent garden trends may have had an effect on its survival here since its foodplants are rarely seen in modern gardens and allotments. The flight period is unknown in Northern Ireland. In Britain, it is given as July and August in a single generation. The larval foodplants are Red Currant, Black Currant and Gooseberry. It overwinters as an egg.

Latticed Heath
Chiasmia clathrata (Linnaeus) Category one.

Widespread. Recorded from all counties and generally distributed in the south-west in Fermanagh, south Tyrone and Armagh. More patchy in the east and north and unrecorded from much of east Down and the inland parts of Antrim and Londonderry. Adults are

Moth accounts

Latticed Heath.

Brackagh Moss National Nature Reserve, County Armagh.

generally brown with white or cream speckled markings on both wings giving a latticed appearance from which it takes its name. It rests in a similar manner to a butterfly with its wings slightly raised. It is found on heathland but also in unimproved grassland, sand dunes, cutover bogs and woodland clearings. Adults are frequently seen by day, resting among Heather and other low vegetation from where it is easily disturbed. It normally flies in sunshine, but is also active at night and is attracted to light in very small numbers. The recorded flight period is 15 May to 11 September with a peak in June and July. The larvae can be found in summer and early autumn, feeding on clovers and trefoils. It overwinters as a pupa.

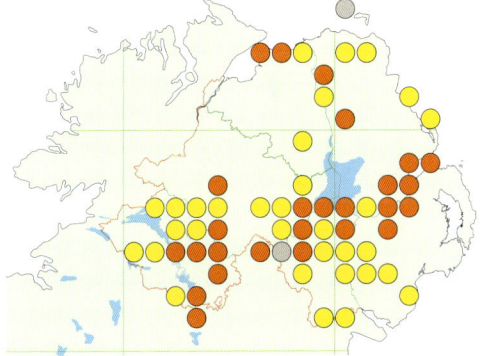

Brown Silver-line
Petrophora chlorosata (Scopoli) Category one.
Widespread. Present throughout the region and generally distributed across southern counties, especially Down. More local in north Tyrone and much of Antrim, although it has been recorded from Rathlin Island. It has been regularly recorded from the Argory and Peatlands Park (Armagh); Lackan Bog and Murlough NNR (Down); Crom and Garvary Wood (Fermanagh) and Rehaghy Mountain (Tyrone). There has been just a single record in Londonderry from Traad Point in the 1980s, although it is probably under-recorded in the north. Adults have reddish-brown forewings with two white cross-lines edged with brown and a small central discal spot. It rests like a noctuid with the hindwings covered. Brown Silver-line is found in the presence of

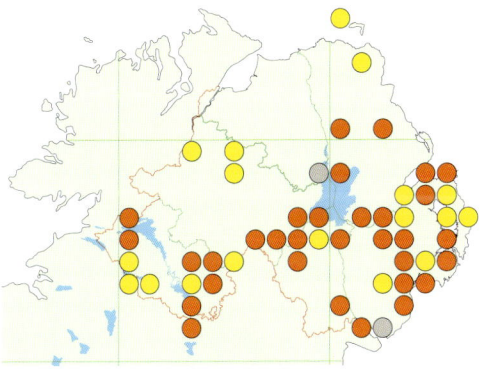

Bracken in grassland, woodland, bogs and heaths. Adults are most often seen at light in moderate numbers, but are occasionally disturbed during the day from stands of Bracken. The recorded flight period is 28 April to 8 July. The larvae can be found from July to September on Bracken. It overwinters as a pupa.

Barred Umber

Plagodis pulveraria (Linnaeus) Category one.
Scarce. The distribution of this species is mainly south-western. The majority of records originate from Fermanagh and south Tyrone. It is known from just three localities in Down (Belvoir Park, Bohill Forest and Rostrevor Wood) and one locality in Antrim (Glenarm). The only Londonderry record was from Banagher Glen in 1978. It was found at Carganamuck in Armagh in 2001, the first county record for 25 years. Sites in Fermanagh and Tyrone included Crom, Derrylin, Gorteen, Marlbank and Tullyreagh Bridge (Fermanagh) and Altadavan, Crilly, Cullinane and Rehaghy Mountain (Tyrone). Adults are readily identifiable by the brown forewings with a wide, reddish-brown crossband, which has jagged projections on the outer edge. Barred Umber is found in semi-natural and old plantation woodlands. Some of the records from west Fermanagh and Armagh are from scrub on limestone. It is likely to be present in other old woodland sites, especially in the east. Adults appear at light, but usually only in ones or twos. The recorded flight period is 30 April to 13 June. The larvae can be found from June until August on a wide variety of trees including Hazel, willow, birch, and Hawthorn. It overwinters as a pupa.

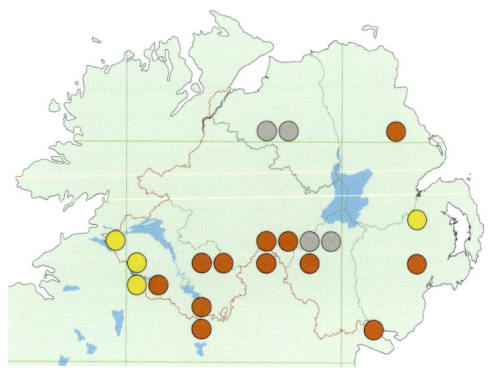

Scorched Wing

Plagodis dolabraria (Linnaeus) Category four.
Scarce. A southern species known from only a scattering of sites in Fermanagh, north Armagh and Down. Greer reported it from several sites in Tyrone including Cookstown, Killymoon, Lissan Estate and Stewartstown and Wright apparently found it at Aghalee in Antrim. It is not currently known from Londonderry. In Fermanagh, Scorched Wing is recorded from many of the well-trapped localities including Crom, Garvary Wood, Gortmaconnell Rock and Monmurry. In Armagh, there is a long series of records from Loughgall up to 1991 and more recent records from the Argory and Brackagh Moss. The Down localities are Castlewellan Forest and Rostrevor Wood. This unmistakable and striking moth has uniquely-shaped wings, with markings and colouring which suggest they have been scorched. It is recorded from mature woodland and scrub and also secondary woodland on bogs and fens. Adults are attracted to light, but are seldom seen in more than ones and twos. The recorded flight period is 13 May to 28 June. The larvae can be found from July to September on oak, Beech, birch and willow. It overwinters as a pupa.

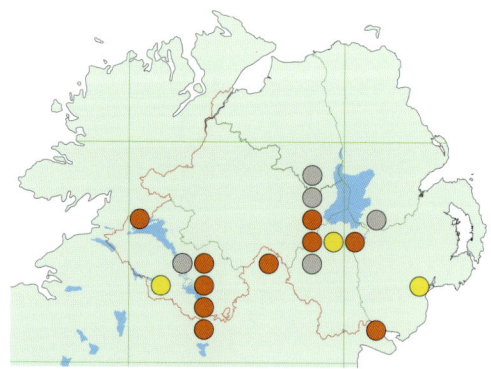

Brimstone Moth

Opisthograptis luteolata (Linnaeus) Category one.
Ubiquitous. One of the commonest moths in the region, found in all counties and on Rathlin Island (Antrim) and the Copeland Islands (Down). The bright yellow adults are unlikely to be confused with any other species. The forewings are yellow with pale brown patches at the tip and

Scorched Wing.

Castlewellan Forest, County Down.

An unmistakable moth which appears at light in small numbers.

at the costa; the hindwings are similar in colour. There is a small white dash surrounded with brown near the costa. Brimstone Moth is a common species in woodland, gardens and wherever there are patches of trees and scrub in other habitats. The phenology of this species in the region is unclear. Adults have been recorded continuously between 2 May and 3 November. There are two peaks in the records, one in early June and the second in late July. These presumably represent different broods, but there is much overlap between the two. It may also reflect that some populations have a single generation, as is the normal pattern in northern Britain. The larval foodplants are Blackthorn, Hawthorn, and possibly Rowan. Its overwintering habits in Northern Ireland are unknown. In Britain, some individuals apparently overwinter as partially-grown larvae, some as pupae.

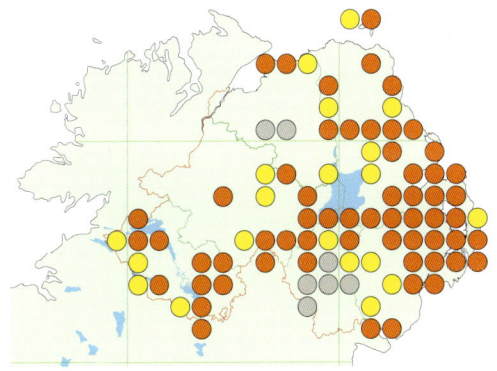

Bordered Beauty
Epione repandaria (Hufnagel) Category two.
Scarce. Widespread in the southern counties, especially in north Armagh and Fermanagh. Sites in these counties include the Argory, Aughinlig, Brackagh Moss, Clare Glen, Loughgall and Peatlands Park (Armagh) and Aghalane, Cliffs of Magho, Crom, Garvary Wood and Monmurry (Fermanagh). In Down and across the northern half of the region it is known from just a few localities, principally Castle Espie, Hillsborough Forest, Hollymount and Lackan Bog. Clare Wood (Antrim) is currently the only known site north of Lough Neagh where it has been seen since 2000. Wright listed it from Aghalee (Antrim) but the date of his record is

unknown. This species in probably under-recorded, especially in the north where recording has been less intense. Adults are normally orange-brown with prominent veining and a darker brown band along the outer edge of both wings. This band is wide with a sinuous inner margin and it tapers to a point at the apex of the forewing. There is a small discal spot near the costa of the forewing. The males have comb-like antennae. Bordered Beauty is found in wet woodland and wooded cutover bogs and fens. Adults are attracted to light, but seldom in more than ones or twos. The recorded flight period is 28 July to 8 October. The larva can be found from May to July on willows, Alder and Hazel. It overwinters as an egg.

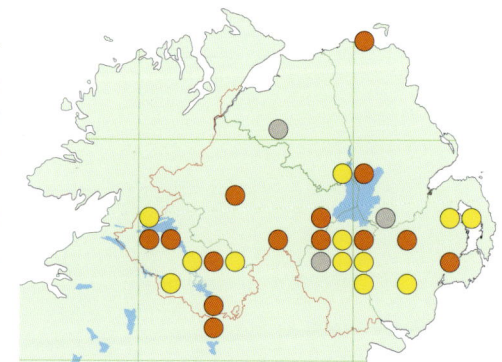

Lilac Beauty

Apeira syringaria (Linnaeus) Category two.
Scarce. This very local species occurs in north Armagh, parts of central Down and Greater Belfast and a few localities in Tyrone, south Fermanagh and Antrim. The majority of records are from north Armagh including the Argory, Brackagh Moss, Navan Fort and Peatlands Park. Other sites include Bohill Forest, Hillsborough Forest, Lackan Bog, Hollymount (Down); Aghalane and Crom (Fermanagh) and Rehaghy Mountain and Strabane Glen (Tyrone). The only records for Antrim are from the early 1990s from Aghalee and Derryola Bridge. This is an easily recognised species that has a distinctive resting posture with the forewings folded in a crease parallel to the costa, so it resembles a withered leaf. The ground colour of the male is reddish-

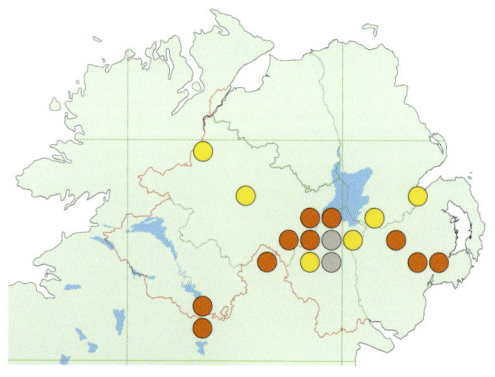

Left
Bordered Beauty.

Lackan Bog, County Down.

Right
Lilac Beauty.

Peatlands Park, County Armagh.

Adults of this species are highly cryptic when at rest among withered leaves.

Moth accounts

173

brown, whereas the larger female is a duller greenish-grey. The markings on the forewings include a dark red cross-line, an orange patch near the apex and a series of lilac patches along the costa. This is a woodland species, encountered here most frequently on wooded heaths, bogs and fens, and occasionally, gardens. Adults are active from dusk onwards and show some attraction to light, but are never seen commonly. The recorded flight period is 25 June and 18 August. The larvae have been found in Northern Ireland on Honeysuckle. Other recorded foodplants in Britain are Wild Privet and Ash. It overwinters as a larva.

August Thorn
Ennomos quercinaria (Hufnagel) Category one.
Widespread. August Thorn has been recorded from most of the regularly trapped localities in Down, Armagh, south Tyrone and south Fermanagh. It is apparently absent from north Fermanagh and most of Tyrone, Londonderry and Antrim. It was found for the first time in north Antrim in 2002 at Breen Wood which is the northernmost known locality. It is likely that other populations await discovery in the north. Like all the 'thorns', it has scalloped margins to the forewings. The males are orange-yellow; the females are lighter, normally pale lemon, although there is individual variation in colour. The forewings have scalloped rear margins and are marked with double cross-lines. The antemedian line curves sharply upwards near the costa, whilst the 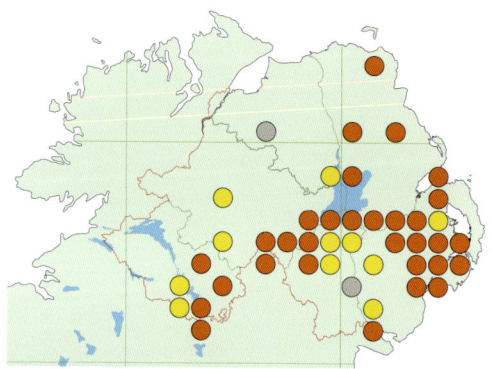 longer postmedian line has a slight kink where it meets the costa. The shape of these two lines distinguishes it from the very similar September Thorn, which has recently been found in the region, although only as a larva. Like all thorns, this is a woodland species also present in wooded heaths and mature gardens. Adults are sometimes found by day resting in vegetation or by beating the branches of trees and shrubs. Adults are attracted to light and frequently appear in traps, but usually in small numbers. As its name suggests, the August Thorn is a late summer species. Adults fly in a single generation and have been recorded from 22 July to 20 October. The larval foodplants are oak, Beech, birch, Hawthorn and possibly Blackthorn. It overwinters as an egg attached to the twigs of the foodplant.

Canary-shouldered Thorn
Ennomos alniaria (Linnaeus) Category one.
Widespread. Recorded from many sites across the south, but rarely reported north of a line from Belfast Lough to Lower Lough Erne. It has occurred at Shane's Castle (Antrim) and at Lough Beg, Traad Point and the Umbra (Londonderry), but nowhere else across the northern half of Northern Ireland. Regularly recorded sites in southern counties include Aughinlig, Brackagh Moss, Loughgall, Oxford Island and Peatlands Park (Armagh); Belvoir Park (Down) and Crom, Garvary Wood, Gorteen and Monmurry Bog (Fermanagh). Adults are similar to both August Thorn and September Thorn, but most specimens of this species can be recognised by the hairy and canary yellow thorax which is noticeably brighter than the rest of the body. The forewings 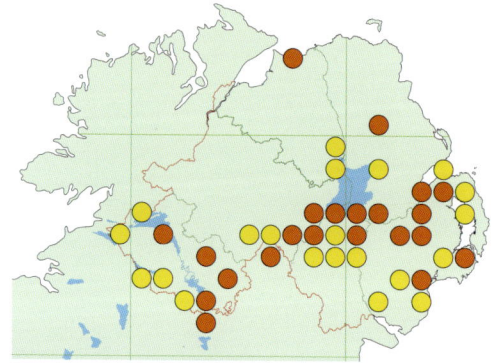 are scalloped along the outer edges and in some individuals are heavily speckled. Like the other *Ennomos* species there are two cross-lines on the forewings, but these curve gently to meet the costa in this species. In the others, these lines are sharply-angled or kinked. It is found in many habitats with trees including broadleaved woods, wooded heaths, bogs and lake shores. This is

Left
Canary-shouldered Thorn.

Aughinlig, County Armagh.

The most conspicuous of the thorns, easily recognised by the bright yellow thorax.

Overleaf
Early Thorn.

The Argory, County Armagh.

a late-summer species recorded between 15 July and 22 October. The larval foodplants include birch, willow, Alder and lime. It overwinters as an egg.

Early Thorn
Selenia dentaria (Fabricius) Category one.
Ubiquitous. Recorded from all counties and present on Rathlin Island (Antrim) and the Copeland Islands (Down). Early Thorn is easily distinguished from the other 'thorns' by its characteristic resting posture in which the wings are held tightly together, butterfly-like, vertically above the body. Adult colouring in this species is subject to variation both between sites and the two generations. Individuals that emerge in the spring are larger and more richly marked than second generation adults. The wings are scalloped and there are three cross-lines on the forewings. There is also a dark patch at the apex of the forewing. This common moth

is a generalist species found in broadleaved woodland and wherever there are trees including heaths, wooded bogs and urban and suburban sites. Adults appear frequently at light usually in good numbers, especially in early spring. They are occasionally found by day resting among leaves on branches or disturbed from among the leaf litter. Adults have been recorded in virtually every week from 20 February to 8 September. The pattern of records suggests that whilst many Northern Ireland populations are double-brooded, some are single-brooded. There are three peaks in the records, in late March, early to late May and late July to early August. The first and third are tentatively ascribed to the spring and summer generations of double-brooded populations; the May peak to the appearance of single-generation populations. This pattern needs to be further investigated. The larvae can be found during May and June and again in September on a variety of trees including Blackthorn, Hawthorn, birch, alder, willow, Hazel and probably other broadleaved species. It passes the winter months as a pupa.

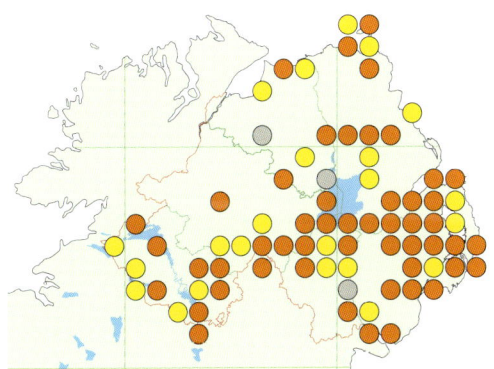

Lunar Thorn
Selenia lunularia (Hübner) Category two.

Widespread. Considered by Baynes as scarce in Ireland, although recent records show it to be widespread across southern counties. It has been recorded from most of the regularly trapped sites in Armagh, Down and Fermanagh. Further north there have been few records. It is known from Davagh Forest in Tyrone, Binevenagh and Magherafelt in Londonderry and Ballymena, Glenarm and Rathlin Island in Antrim. Adults are similar in appearance to the Early Thorn but, unlike it, they normally rest with wings held slightly apart. There is also a small silver crescent on both the upperside and underside of each forewing. This is a reliable diagnostic feature and distinguishes it from related species. Adult colouration is subject to much individual variation. The wings have scalloped edges and there are dark patches on the apex of the forewings. Lunar Thorn is a woodland moth, but it has also been recorded from wooded bogs and heaths and suburban gardens. Adults are attracted to light in small numbers. They can occasionally be beaten from the branches of trees during the day and have also been disturbed from leaf litter. The recorded flight period is 14 April to 1 August in a single generation. The peak period for records is early May to mid-June. The larval foodplants are oak, Ash, birch and possibly other trees. It overwinters in the pupal stage.

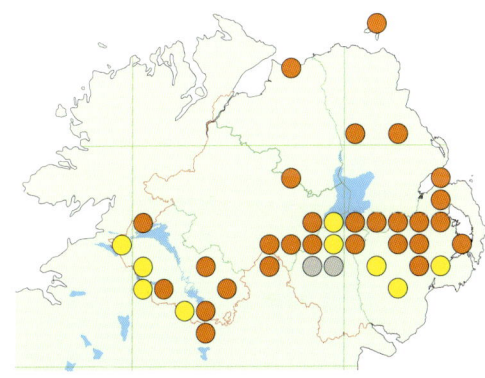

Scalloped Hazel
Odontopera bidentata (Clerck) Category one.

Widespread. The distribution pattern of this species is typical of many moths in the region, being generally recorded south of Lough Neagh, but much more local to the north and north-west. It is known from a handful of sites in mid-Antrim and south Londonderry and also on Rathlin Island (Antrim). Apparently absent from much of Tyrone away from the south-east, but this area is under-recorded. Adults are variable with different colour forms, ranging from light to dark brown, which occur throughout most of its range. The forewings, which are heavily scalloped, have two cross-lines and a small discal spot. The area between the cross-lines can often appear darker, forming a central band. It is similar in appearance to Scalloped Oak, but the outer edges of its wings

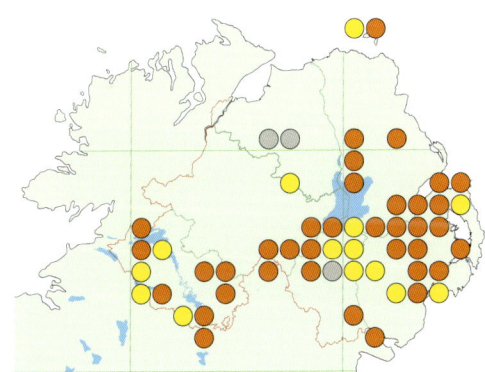

The Butterflies and Moths of Northern Ireland

are only very lightly scalloped. Although its name would suggest that it is associated with Hazel, the larva feeds on a wide variety of trees and shrubs. It is a woodland moth which can be found in many habitats providing there are some mature trees, including heaths, bogs and, frequently, gardens. Adults are attracted to light in moderate numbers and may occasionally be found during the day by beating the foliage of its foodplants. The recorded flight period is 16 April to 27 July with a short, pronounced peak from mid-May to mid-June. The larvae can be found in summer and early autumn, feeding on a wide variety of trees and shrubs such as oak, birch, Hawthorn, Blackthorn, pine and willow. It overwinters as a pupa.

Scalloped Oak

Crocallis elinguaria (Linnaeus) Category one.

Widespread. A common and widespread species found throughout Northern Ireland especially in Down, north Armagh and Fermanagh. It appears scarcer north of Lough Neagh, particularly inland, but there are sufficient records to suggest it is likely to be present wherever there is suitable habitat. It has also been recorded from Rathlin Island (Antrim). Although prone to some variation in colour, most individuals of this distinctive moth should be easily recognised. Adults are generally yellow with a brown central band and a distinct black discal spot. The wings are only slightly scalloped. Scalloped Hazel is similar in appearance but can be quickly distinguished by the deeply scalloped edges on its wings. Scalloped Oak can be found in broadleaved woodland, wooded heaths and bogs, scrub on dunes and limestone and gardens in both urban and rural settings. Adults come frequently to light in small numbers. There is a single annual generation and adults have been recorded from 15 June to 24 September. The larvae have been found on Hawthorn and oak in Northern Ireland. Other recorded foodplants in Britain are Blackthorn, oak, Bilberry, Heather and birch. It overwinters in the egg stage.

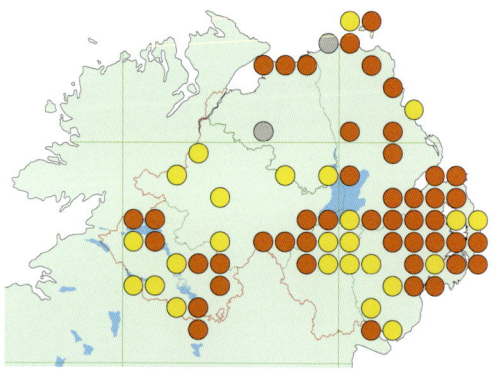

Swallow-tailed Moth

Ourapteryx sambucaria (Linnaeus) Category one.

Widespread. Recorded in all counties but very local north of Lough Neagh. It is known from Rathlin Island (Antrim). Regular sites for it include Brackagh Moss, Loughgall and Peatlands Park (Armagh); Belvoir Park, Castleward and Helen's Bay (Down) and Aghalane, Crom and Garvary Wood (Fermanagh). There are also recent records from several urban gardens and the Lagan Meadows in the Greater Belfast area. Northern localities include Glenariff Forest, Portglenone and Rea's Wood (Antrim) and Duncrun (Londonderry). This moth derives its name from the small projections on the apex of the hindwings, similar to those present on a swallowtail butterfly. The adults are large and butterfly-like in appearance. The lemon-yellow wings have two widely spaced cross-lines. The bright lemon-yellow colour, so distinctive of fresh specimens, fades quickly and the vast majority of adults trapped at light are pale. Many also show signs of damage to the wings. This is a woodland species which also occurs in gardens, wooded bogs and fens. Adults appear at light, but never abundantly. They have a skittish nature and fly at the least sign of disturbance. The recorded flight period is 15 June to 19 August. The larvae have been recorded on Hawthorn in Northern Ireland. Other recorded foodplants in Britain are Ivy and Blackthorn. It overwinters as a larva.

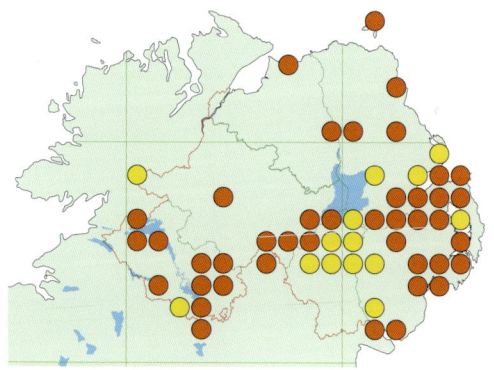

Feathered Thorn
Colotois pennaria (Linnaeus) Category one.

Widespread. Feathered Thorn has been recorded from all counties and it is generally distributed in Down, Armagh and adjacent areas of Tyrone and around Lough Neagh. It is much scarcer in north Antrim, Fermanagh, and west Tyrone. Known localities in these areas include Breen Wood (Antrim); Aghalane, Crom, Garvary Wood and Monmurry (Fermanagh) and Altadavan, Crilly, Rehaghy Mountain and Strabane (Tyrone). Traad Point is the only known site in Londonderry. It probably occurs more widely in the north and west of Northern Ireland, but has been overlooked due to the low level of recording late in the year. Adults are reddish-brown with two dark cross-lines and a small discal spot near the costa of the forewing. This 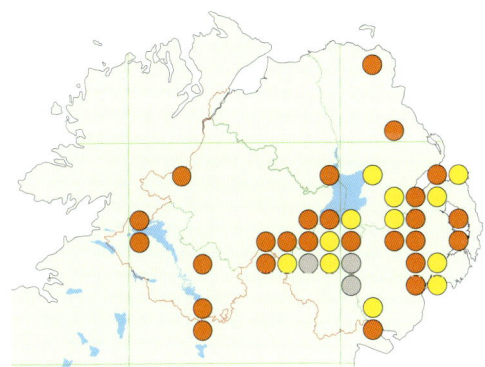 is a variable species throughout its range in Northern Ireland with some individuals appearing much darker and redder than others. The males have feathery antennae. Feathered Thorn is found in broadleaved woods, well-wooded heaths and bogs and suburban gardens. Adults come frequently to light in reasonable numbers, but have also been disturbed from leaf litter by day. They also reportedly rest low down on the trunks of trees. The recorded flight period is 17 September to 28 November. Larvae have been found in Northern Ireland on birch and, in Britain, on oak, willow, Hawthorn, Hazel, poplar, and Blackthorn. It overwinters as an egg.

Feathered Thorn.

Rehaghy Mountain, County Tyrone.

Pale Brindled Beauty
Phigalia pilosaria (Denis & Schiffermüller) Category one.

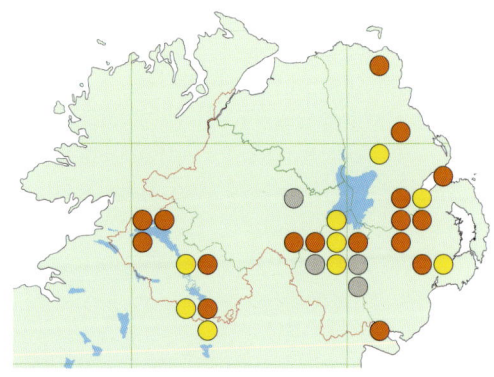

Scarce. This late winter species is recorded from all counties but the distribution is rather irregular in most. The largest populations appear to be in sites in and around the Lough Neagh basin, including Randalstown Forest and Shane's Castle (Antrim) and the Argory, Brackagh Moss, Loughgall and Peatlands Park (Armagh). Many of the other sites have only recently been discovered. Some of these include Breen Wood (Antrim); Rostrevor Wood (Down) and Aghalane, Correl Glen, Crom, Garvary Wood and Monmurry (Fermanagh). Belvoir Park is perhaps the most reliable known site for it in Down. Adult males are generally greyish-green with faint irregular cross-lines. They have thick hairy bodies with large comb-like antennae. It closely resembles Brindled Beauty, but it flies earlier in the year and lacks the conspicuous heavy black cross-lines present in that species. The female is wingless, lacking even the stumps of the wings. Pale Brindled Beauty occurs in broadleaved woodland and also wooded bogs and heaths. Adult males are attracted to light in moderate numbers in early spring. The moth may occasionally be found by day at rest low down on the trunks of trees. The recorded flight period is 15 January to 22 April. The larva feeds throughout late spring and early summer on a variety of trees including oak, birch, Hawthorn, willow, Hazel, Ash, Alder and Blackthorn. It overwinters in the pupal stage.

Brindled Beauty
Lycia hirtaria (Clerck) Category four.

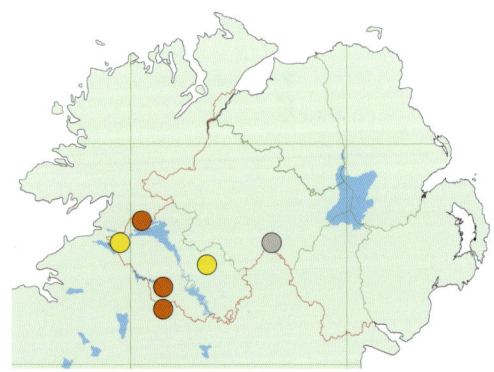

Rare. Apparently restricted in recent times to Fermanagh. It has also been recorded once in Tyrone, at Favour Royal, in the nineteenth century. The Fermanagh sites have all been discovered since 1992. The majority of records originate from Castle Caldwell and Garvary Wood in the north west of the county. It occurs also at several sites in the Marlbank area, including Florencecourt, Gortmaconnell Rock, Marble Arch and Monastir Gorge, the site where it was first found. The males of this species are similar in appearance to the Pale Brindled Beauty, but they are more robust with a hairier thorax and broader body. The forewings have a mottled appearance and are a mixture of grey and dark yellow and there are conspicuous black cross-lines and banding. The colour and strength of the markings is, however, prone to variation. Females, unlike those of Pale Brindled Beauty, are fully winged and told from the males by the lack of feathering on the antennae. Brindled Beauty is associated with broadleaved woodland and mature scrub on limestone and, once, a wooded bog. Adults are attracted to light in moderate numbers and are known to rest during the day low down on the trunks of trees and, occasionally, stone walls. This elusive spring species may be overlooked in other areas in the west or confused with Pale Brindled Beauty. The recorded flight period is 23 April to 30 May. There has also been one March record of an adult, but the precise date is not known. The larva feeds from late spring to early summer on a variety of trees including birch, oak, Alder, Hawthorn, and Beech. It overwinters in the pupal stage.

Moth accounts

Brindled Beauty.

Marble Arch, County Fermanagh.

This species is mainly confined to County Fermanagh where it was discovered during the early 1990s. This moth is stockier than the Pale Brindled Beauty.

Belted Beauty
Lycia zonaria (Denis & Schiffermüller) SOCC (P) Category four.
Unknown. This species is presumed extinct at its single Northern Ireland haunt at Ballycastle (Antrim). It was first reported here in 1889 by Bristow and recorded several times by a number of naturalists until at least 1900. Bristow found five adults at Ballycastle on 29 March. Larvae were found at the site by a Mr. Milne in 1893. The last record was sometime around 1900 and since then it has gone unreported. Recent searches have proved fruitless. The appearance of adults from Ballycastle was described by Kane and considered different from the form in Connemara. Males of this species are fully winged, but the wings of the female are vestigial. The abdomen of both sexes is black with orange rings. The male has silvery-white forewings with prominent dark veining and a series of brown cross-lines. The two outermost of these cross-lines are the broadest and most pronounced. Males have long feathery antennae and are quite hairy. Apart from the Antrim site, the Belted Beauty is confined to Galway and Mayo, where it occurs on machair habitat, similar to that occupied in the Scottish islands. The males can be found by day resting conspicuously on low vegetation, stones and fence posts. The main flight period in Britain is March and April and on the west coast of Ireland it has been recorded in large numbers during April. The larvae have been reported feeding on Bird's-foot Trefoil (specifically the flowers) in Antrim and Mayo. In Britain other published foodplants include plantain, clover, Creeping Willow and Burnet Rose. It overwinters in the pupal stage.

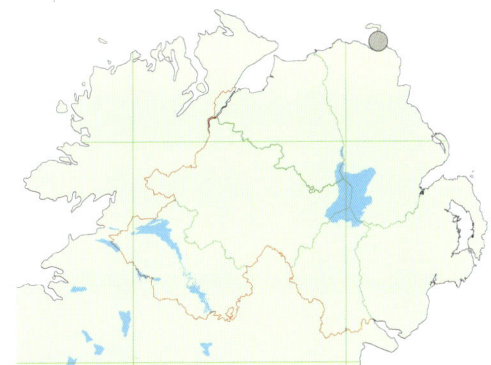

Oak Beauty

Biston strataria (Hufnagel) Category three.

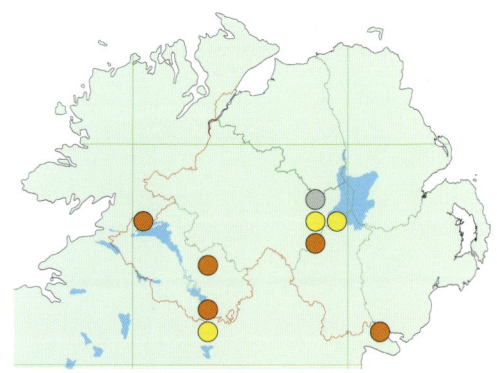

Notable. This attractive species is known from a handful of sites in Armagh, Down and Fermanagh and, historically, from two localities in Tyrone (Loughry and Stewartstown). Greer mentions finding a female on a tree trunk near Loughry and a larva on Wild Cherry in 1921. In Fermanagh, it occurs at Aghalane, Castle Caldwell, Crom, Garvary Wood and Monmurry. The Argory and Peatlands Park are the only Armagh sites and Rostrevor Wood the only known locality in Down. It was taken at the latter site in 2001 for the first time since the early part of the twentieth century and again in 2004. The Armagh sites were discovered in the late 1990s. This is a late winter and early spring species and it is highly likely that it remains undetected at many sites, particularly in Antrim and Londonderry, where little early season trapping has been done. The adults are white with pale grey mottling and with two brown cross-bands edged with black. Their mottled appearance provides camouflage for the moth when it is resting on tree trunks. The males have long feathery antennae and the body and thorax are both quite hairy. Oak Beauty is found only in mature, lowland broadleaved woodland, usually containing some mature oak trees. Adult males are frequently seen at light, but the females are seldom encountered. The recorded flight period is 21 February to 14 April. The larvae have been recorded on Wild Cherry in Northern Ireland. Other recorded foodplants include oak, elm, Hazel, Aspen and Alder. It overwinters in the pupal stage.

Peppered Moth

Biston betularia (Linnaeus) Category one.

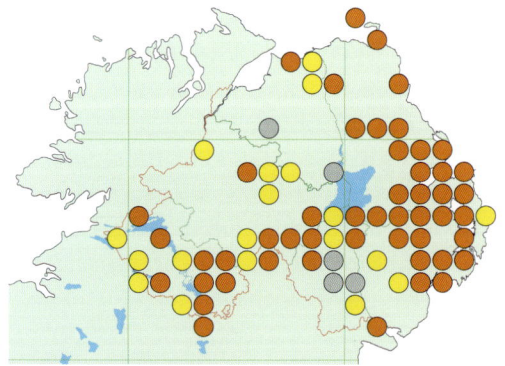

Widespread. This moth has been recorded from all counties. It also occurs on Rathlin Island (Antrim). Frequently recorded sites include the Argory, Loughgall and Peatlands Park (Armagh); Belvoir Park, Helen's Bay and Lackan Bog (Down) and Crom, Garvary Wood, Gorteen and Legatillida (Fermanagh). The adults are distinctive and easily recognised by the black and grey speckling on the white wings and abdomen. The amount of speckling varies between individuals. The thorax is quite hairy and heavily marked. Males can be distinguished from females by the large combed antennae. Peppered Moth famously has a melanic form f. *carbonaria* in which the body and wings are a uniform sooty black. This was once common in areas heavily affected by smoke and soot pollution. With cleaner air the incidence of *carbonaria* has declined significantly, but the occasional specimen is still reported. One was seen at Newcastle (Down) in 1991 and another at Jordanstown (Antrim) in 1996. Specimens of *carbonaria* have also been recorded from Aghalee and Cushendun (Antrim); Conlig and Rostrevor Wood (Down) and Belfast. This is a woodland moth which can also be seen in parkland, along hedgerows and in gardens with mature trees. Adults come regularly to light in small numbers, but they are seldom seen by day. The recorded flight period is 2 May to 5 September. The larvae have been found on birch and Hawthorn and oak in Northern Ireland. Other recorded foodplants include willow, poplar, Beech and Bramble. It overwinters in the pupal stage.

Spring Usher
Agriopis leucophaearia (Denis & Schiffermüller) Category four.
Rare. Until recently it was generally believed that this species was extinct in Northern Ireland, as there had been no records since 1893 when it was reported from Enniskillen (Fermanagh). This belief was confounded when a single adult was found on the door of a building at Belvoir Park (Down) on 7 February 2000. Despite regular moth recording at this site, no other individuals have been reported. The only other Irish record of Spring Usher was from Bray Head (Wicklow) in 1914. This record has been generally accepted, although it was viewed with some suspicion by Donovan and Baynes who suggested that it was 'possibly a vagrant from England'. The ground colour of the forewing in the males is pale brown with large areas 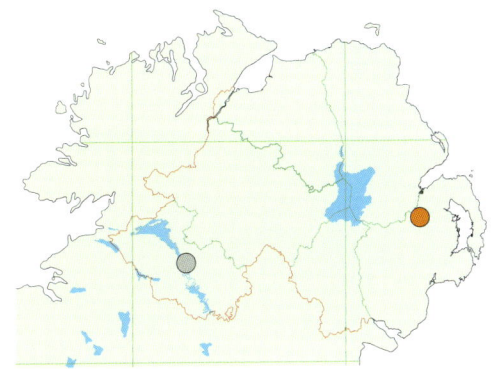 of darker brown along the base, although there is individual variation in both colour and pattern. The cross-lines are normally dark and conspicuous. The antemedian line is strongly curved while the postmedian line is wavy. The hindwings are light brown and speckled. The females have vestigial wings. In Britain, this species is found most abundantly in broadleaved woodland, but also occurs in more open areas provided there are some mature trees. Males are attracted to light and may occasionally be found during the day at rest on fences and tree trunks. The flightless females are best looked for after dark by examining the trunks of trees. The main flight period in Britain is early February to mid-March, but adults can be seen in January, if the winter is mild. The larval habits are not known in Northern Ireland. In Britain larvae can be found from April to June, mainly on oak.

Scarce Umber
Agriopis aurantiaria (Hübner) Category three.
Scarce. Recorded from all counties, but the records are well scattered. Most of the recent records originate from the west in Fermanagh and Tyrone where it occurs at Castle Caldwell, Crom and Monmurry (Fermanagh) and Caledon Estate and Rehaghy Mountain (Tyrone). Old records exist from Enniskillen and Tempo Manor (Fermanagh) as well as Favour Royal, Killymoon and Lissan Estate (Tyrone). In Down it is known from Belvoir Park, Helen's Bay and Rostrevor Wood, although it has not been seen at the last site since 1933. In Armagh it has been found at the Argory, Gosford Forest and Loughgall. Sites in Londonderry include Traad Point and an unspecified area in Coleraine dating back to 1950. This species 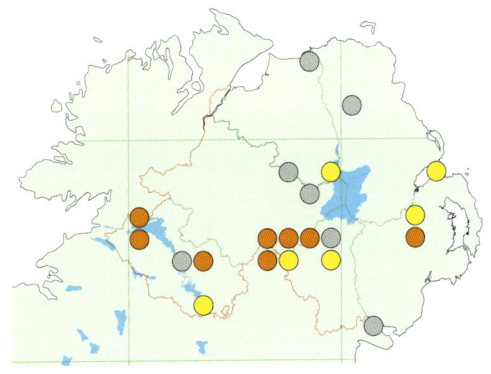 probably occurs more widely than current records suggest and is under-recorded due to its late flight period. This is a northern species in Ireland, not recorded south of Monaghan with the exception of Kerry. Adult males are variable in colour ranging from orange-yellow to a darker brown. The forewings appear mottled with two brown cross-lines. The females are flightless with vestigial, stumpy wings. It is similar in appearance to Dotted Border which normally flies earlier than this species; however, in Northern Ireland, both have been recorded in January. Scarce Umber is recorded from semi-natural, broadleaved woodland, heaths and old estate plantations. It is seen frequently at light. The females reportedly can be found on the trunks of trees after dark or early in the morning. This has a late flight period from 31 October to 24 January, somewhat later than the main flight period in Britain.

Scarce Umber.

Rehaghy Mountain, County Tyrone.

The larval foodplants are oak, birch, Hazel and possibly Blackthorn and Hawthorn. It overwinters as an egg.

Dotted Border
Agriopis marginaria (Fabricius) Category two.
Widespread. Recorded in all counties, but particularly common in sites around the Lough Neagh basin and parts of east Down and Fermanagh. Many more records have been produced in recent years as more fieldwork has been undertaken in the early spring. However, it still remains under-recorded in many areas. It is well known at many of the regularly trapped localities across Northern Ireland. Other sites include Breen Wood, Gawley's Gate and Sharvogues Bog (Antrim); Castle Archdale and Gorteen (Fermanagh) and Curran Bog (Londonderry). Adult males are generally reddish-brown, although the colour varies between individuals. The forewings have a darker band and a series of small dots towards the outer margin. There is a melanic

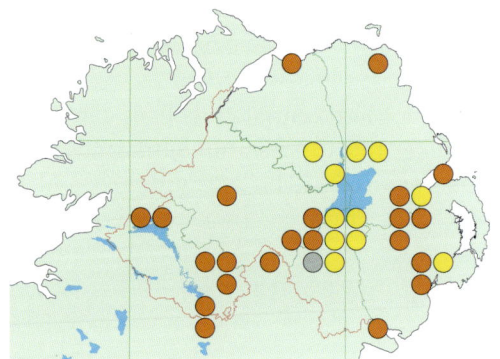

form called f. *fuscata* which is present in small numbers throughout its range in Northern Ireland. The females are flightless, but have fairly large wing stumps. The Dotted Border is found in broadleaved woodland, heaths, bogs and, occasionally, urban gardens close to woodland. Males are found commonly at light and occasionally by day. The females can be found resting on the trunks of trees. The recorded flight period is 7 January to 5 May. The larval foodplants are oak, birch, willow, Hawthorn, Blackthorn and Hazel. It overwinters as a pupa.

Mottled Umber
Erannis defoliaria (Clerck) Category one.

Scarce. Present in a wide band across south-western and central parts of Northern Ireland. Apparently absent from south Armagh and south Down and most of Tyrone, north Londonderry and north Antrim. It has frequently been reported from many of the regularly trapped woodland sites within its recorded range. Trapping at the appropriate season has also produced records from Belshaw's Quarry and Leathemstown Reservoir (Antrim); Ballygowan and Castleward (Down); Castle Caldwell (Fermanagh) and Stillago (Tyrone). Adult males are generally reddish-brown, although the ground colour and the pattern can vary considerably between individuals. The cross-bands (when present) are dark brown and irregular in shape and there is 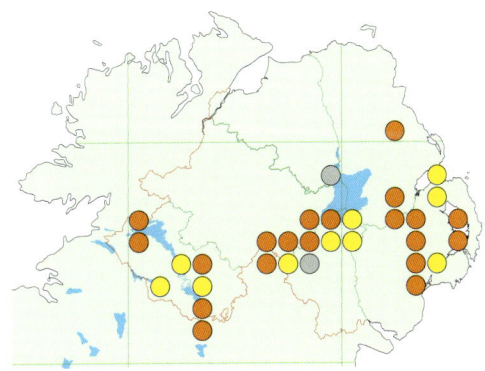 normally a dark central spot. The males are similar in appearance to Dotted Border, but lack the dots along the outer margin of the hindwing. The females are flightless and the wing stumps are barely visible. Mottled Umber inhabits woodland and bogs and other habitats with mature trees. Males come frequently to light in moderate numbers and the females are best found by searching tree trunks after dark. The recorded flight period is 29 September to 20 February. The larvae can be found from April to June on a variety of trees, particularly oak, birch, willow, Hawthorn, Blackthorn and Hazel.

Willow Beauty
Peribatodes rhomboidaria (Denis & Schiffermüller) Category one.

Widespread. Frequently encountered throughout most of Down, north Armagh and south Fermanagh, but only found in scattered sites elsewhere. It does occur in north Antrim at Glenariff Forest, Portglenone Wood and on Rathlin Island. The gaps probably reflect a low level of recording effort in the north-west. Adults closely resemble Mottled Beauty and some individuals can be difficult to tell apart, especially when slightly worn. The ground colour is generally smoky-grey or a pale brown, although there is some variation between individuals. The cross-lines are not always clearly visible; however, the line distal to the centre of the forewing is strongly angled or kinked at the costa. Both central cross-lines converge at the dorsal aspect of the 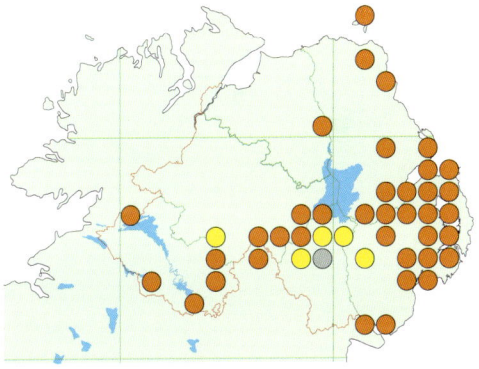 forewing; this can be a useful diagnostic feature. Willow Beauty is found in many habitats wherever there are broadleaved trees and shrubs including parkland, mature hedgerows and suburban gardens. Adults come frequently to light in moderate numbers. The recorded flight period is 8 June to 7 September. The larvae, which overwinter, can be found from late summer until late spring of the following year on Hawthorn, birch, Ivy, Honeysuckle, Alder and other cultivated and wild woody plants.

Bordered Grey
Selidosema brunnearia (Villers) SOCC Category four.

Rare. Bordered Grey in recent times has only been found at two sites, Peatlands Park (Armagh) and Kebble on Rathlin Island (Antrim). The Peatlands Park records all come from Mullenakill Bog, the only intact piece of raised bog remaining in the park. It has not been seen beyond the confines of this bog despite suitable habitat existing throughout the site. There are historical records for this area dating back to the early twentieth century and it was described as common

Bordered Grey.

Mullenakill National Nature Reserve, Peatlands Park, County Armagh.

This was the only known site for this rare species in Northern Ireland, until its discovery on Rathlin Island in 2001.

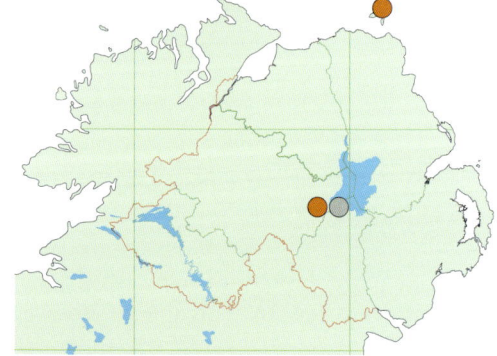

in this part of Armagh by Johnson. The Kebble population was discovered in 2001. Greer makes reference in one of his publications to presence of the moth on bogs near Washing Bay in the south-west corner of Lough Neagh (Tyrone) and the occurrence of melanic adults. Adults from this area were considered to be smaller than average with a narrower marginal band on the forewing and were described as a distinct subspecies *tyronensis*; however, it is debatable whether these subtle differences ever justified this status. The surviving areas of bog in the Washing Bay area are very degraded and it is highly unlikely that they still support this species. Adults are usually greyish-brown with a darker broad band along the outer edges of the wings and black markings across the forewing. The males have large comb-like antennae. This species occurs on intact raised bog and heathland in Northern Ireland. It is found on bogs elsewhere in Ireland and also on calcareous grassland in the Burren. Adults are attracted to light in small numbers. The males have been seen flying during late afternoon and have occasionally been disturbed from vegetation during the day, when they are easily netted. The few records in Northern Ireland fall between 26 July and 11 August. In Britain, the main flight season extends over the whole of July and August. The larvae are best looked for between early autumn and late spring of the following year on its foodplant, Heather. It overwinters as a larva.

Ringed Carpet

Cleora cinctaria (Denis & Schiffermüller) Category four.

Unknown. Reported once in Northern Ireland at Portaferry (Down) by Langham in 1914. This remains the only record and its status here is uncertain. Very little is known about its habits elsewhere in Ireland, where it is known to occur mainly in western counties from Donegal to Kerry, but also in Meath and Offaly. Its continued existence in Northern Ireland must be considered doubtful, but it is possible there are small undetected populations, most likely in the western counties. It was recently recorded from Glenveagh National Park (Donegal) in 2004, confirming its continued existence at least in this region. This is a mottled grey and brown moth. The forewings are triangular in shape and have a small, whitish central crescent or dash on both the fore- and hindwings. There is also a well-defined basal cross-band. The habits and ecology of this species are not known in Northern Ireland. In Britain, Ringed Carpet is found on  heathland with scattered trees in the south and boggy moorland in the north. Adults are attracted to light in small numbers and may occasionally be found by day, resting on the trunks of trees and fence posts. The flight period in Britain is late April and May and the larvae can be found from June to August. Southern populations in Britain feed on birch, Bilberry and Bell Heather, but the Scottish populations appear to use Bog Myrtle. It overwinters in the pupal stage.

Satin Beauty
Deileptenia ribeata (Clerck) Category two.
Scarce. Since its discovery in 1999 at Rostrevor Wood (Down), populations of this species have been found in every county except Londonderry, although its discovery here seems highly probable. It would seem likely that the species has spread into the region as its habitat (conifer woodland) has matured. Most of the records have come from Down, Fermanagh and east Tyrone. Sites include Belvoir Park, Bohill Forest, Castlewellan Forest and Hillsborough Forest (Down); Aughalun, Castle Caldwell and Glen Wood (Fermanagh) and Altadavan, Favour Royal and Rehaghy Mountain (Tyrone). In Antrim it is known only from Breen Wood. The Armagh localities are the Argory and Peatlands Park. This species was unknown to Donovan and 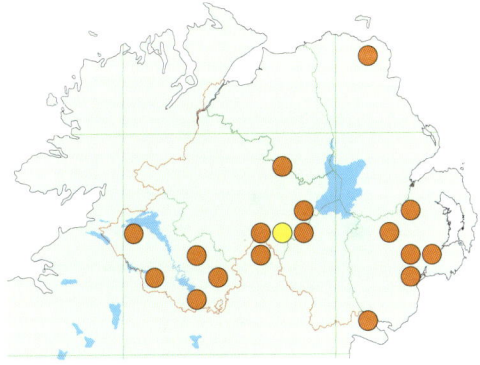 Baynes listed it from just three Irish counties, Kildare, Laois and Wicklow. Adults are usually a darkish-brown with a well-rounded forewing. The males are smaller than the females and have feathery antennae. This species closely resembles Mottled Beauty and individuals should be examined carefully to confirm identification. Most of the Northern Ireland records have come from sites with long-established conifer woodland and mature trees, particularly old estates, but also wooded heaths and bogs. Adults are attracted to light in small numbers and may occasionally be found during the day by beating the foliage of the foodplants. Adults have been recorded from 30 June to 13 August. The larvae, which overwinter, are found from late August to the May of the following year on Yew, Scots Pine, spruce and Douglas Fir.

Mottled Beauty
Alcis repandata (Linnaeus) Category one.
Widespread. This common species has been recorded from many localities in all counties including Rathlin Island (Antrim). Adults of this moth are large and robust. Males can be told from the females by the feathered antennae. This is an extremely variable species with many different colour forms, ranging from pale to dark brown. A banded form f. *conversaria* is found chiefly in coastal localities but also, in smaller numbers, at some inland sites. The wings are

mottled, but the degree and strength of mottling varies considerably between individuals. The most reliable diagnostic feature is the crossline distal to the centre of the forewing which is wavy and continuous, with a distinct kink near the costa. It is often confused with the Willow Beauty, but that has a slightly later flight period. The cross-lines in this species, however, are more irregular and converge at the dorsum of the forewing. The Mottled Beauty is found in a wide range of habitats but is most common in woodland, heathland and bogs. It is also a frequently encountered species in many gardens. Adults come freely to light in good numbers and may be disturbed by day from tree trunks. The recorded flight period is 2 June to 5 September. The larvae, which overwinter, can be found from the end of August to May of the following year on a variety of trees and shrubs including Bilberry, Hawthorn, Blackthorn, birch, Bramble, oak and Honeysuckle. It overwinters as a larva.

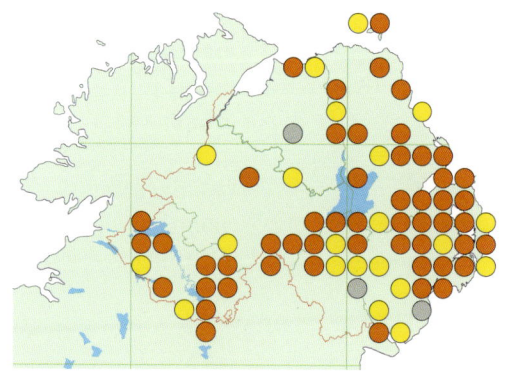

Dotted Carpet
Alcis jubata (Thunberg) Category four.
Rare. This species is only known from Castlewellan Forest (Down) and Breen Wood (Antrim). There has been just one other Irish record from Kerry in 1941. It was discovered close to the Arboretum at Castlewellan Forest on 17 July 1999. Despite frequent trapping at this locality no other individuals have been found. The Breen Wood population was discovered in August 2001 and others have been found here in August 2002 and 2004. Adults are generally white with light grey dusting giving a mottled appearance. There are small, dark grey patches along the costa, a

Dotted Carpet.

Breen Wood, County Antrim.

central dark spot and an area of heavy grey shading at the apex of the forewing. The hindwings have a small discal spot and a well-defined cross-line. Dotted Carpet occurs in mature woodland with abundant growth of beard lichens. In Britain, it is associated with both broadleaved and conifer woodland and is mainly found in the west where lichen growth is most common. Adults are attracted to light and have been taken in moderate numbers at Breen Wood. The larval habits are unknown in Northern Ireland. In Britain, larvae have been beaten from lichen-covered branches. The main flight period in Britain is July and August. The larva feeds on lichens, especially beard lichen, from August until early summer of the following year. It overwinters as a larva.

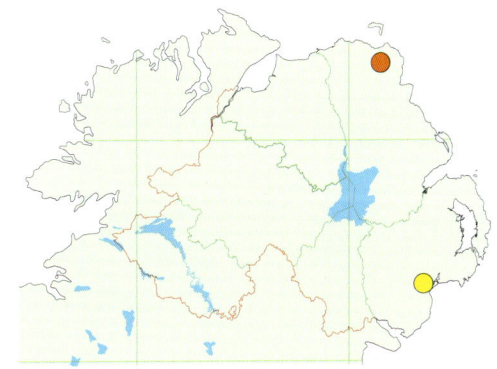

Brussels Lace
Cleorodes lichenaria (Hufnagel) Category one.
Widespread. Recorded from all counties, but the distribution is almost entirely southern and it is known from just a few sites in Antrim and Londonderry. It has been found widely across Fermanagh, north Armagh and east Tyrone and throughout Down. The only Antrim records are from Aghalee and Ballinderry in the extreme south of the county and around Bushmills in the north. It was last recorded at the latter site in 1938. The Umbra is the only known site in Londonderry. In the south of the region, the species is present at most of the regularly trapped sites. Adults of this moth are fairly small and its colour is a mixture of black and olive green mottling on a light background. The postmedian cross-line has a jagged appearance which is diagnostic. Like

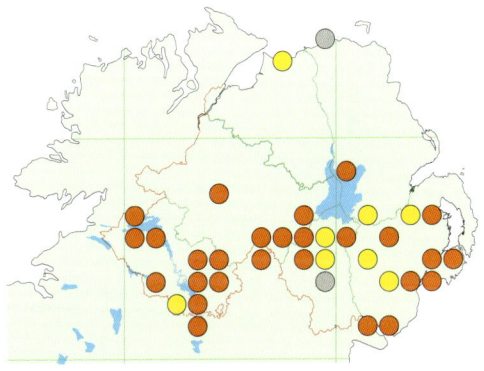

other green species, the colour tends to fade rather quickly after emergence. Brussels Lace occurs where there are healthy lichen populations in damp woodland, mature hedgerows and scrub. It can also be found on open, rocky ground and heathland, especially on the coast. Adults come frequently to light in small numbers, but are seldom encountered during the day. The recorded flight period is 11 June to 11 August. The larvae can be found from September to May of the following year on various species of lichens found on trees, and occasionally old fence posts and rocks in coastal localities. It overwinters as a larva.

Engrailed
Ectropis bistortata (Goeze) Category four.
Rare. The taxonomy of the Engrailed and Small Engrailed is debated, with some European experts considering them to be a single species; however, in Britain, they have been traditionally recognised as separate species on the basis of their phenology. Separate accounts are given, but the records are mapped together. No consistent and reliable differences between these two species have been reported in either external morphology or in the form of the genitalia. The ground colour is greyish-brown, often heavily freckled with darker wavy cross-lines. The inner cross-line has a double-pointed projection which is conspicuous. The degree of freckling varies considerably between individuals. Specimens of the Engrailed are stated to be darker with less clearly defined cross-lines than Small Engrailed. The main criterion distinguishing the species is the flight period, but this is confounded by the variability in the number of generations across

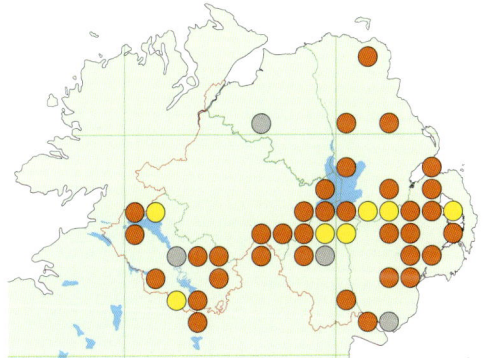

Britain and Ireland. The Engrailed is considered to have more than one generation in southern Britain, but just one in the north including Northern Ireland. The first generation flies in March to early May and from April to mid-June in single-brooded populations. Small Engrailed is considered to fly later. Specimens ascribed to the Engrailed have been reported from Peatlands Park (Armagh); Castlewellan Forest, Hillsborough Forest, Lackan Bog, Murlough NNR and Rostrevor Wood (Down) and Crom, Marlbank and Monmurry Bog (Fermanagh). The records come mainly from damp woodlands and lowland bogs. Adults come frequently to light and may occasionally be found at rest on the trunks of trees. The recorded flight period is early March to mid-May. The larval habits are unknown in Northern Ireland. In Britain, recorded foodplants include birch, oak, willow, Hawthorn, Hazel, Broom. It overwinters in the pupal stage.

Small Engrailed

Ectropis crepuscularia (Denis & Schiffermüller) Category three.
Notable. The taxonomic status of this species is uncertain and distinguishing it from the Engailed is problematic. There are many more records ascribed to this species than to the Engrailed. These are largely confined to southern counties and the northernmost localites are Capanagh Wood and Shane's Castle (Antrim). It has frequently been reported at Peatlands Park (Armagh); Belvoir Park, Bohill Forest and Helen's Bay (Down); Crom Estate, Garvary Wood, Legatillida and Monmurry (Fermanagh) and Rehaghy Mountain (Tyrone). Small Engrailed is stated always to be single-brooded. Adults cannot be reliably told from the Engrailed on appearance, although it is stated they are paler and more strongly marked. Small Engrailed is associated with woodland, bogs, limestone uplands and other scrubby areas. Adults are attracted to light, but seldom appear in any numbers at most localities with the exception of Peatlands Park (Armagh) and Crilly (Tyrone) where it has been frequently recorded in large numbers. Freshly-emerged individuals have been found at rest by day on limestone rocks in the Fermanagh uplands. The recorded flight period is the beginning of May to the end of June, occasionally early July. The larval habits and foodplants are not known in Northern Ireland. In Britain, the larva has been found on birch and willow in July and August. It overwinters in the pupal stage.

Grey Birch

Aethalura punctulata (Denis & Schiffermüller) Category four.
Unknown. This species has not been seen in Northern Ireland for many years and its status here is uncertain. It has recently been found in Donegal, raising hopes that there may be undetected populations in the north and west. There are just two accepted records from Rostrevor Wood (Down) in the early 1930s by Johnson and Tempo (Fermanagh) in 1936 by Langham. There have been a few recent claims of this species but none have been substantiated. Baynes stated it was scarce and of sporadic distribution in Ireland and, apart from the two Northern Ireland records, found only in the southern half of the island. Adults have a mottled appearance and are predominantly grey with four conspicuous black blotches along the costa. As its name indicates, 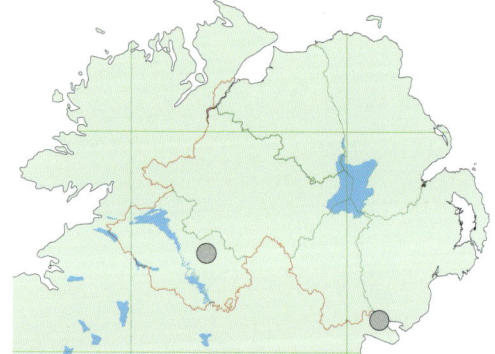 this moth is associated with birch, and in Britain is found in birch woodland and scrub, especially where this is long established. Adults are reportedly attracted to light and can occasionally be found by day resting on the trunks of trees and fence posts. The main flight period in Britain is May and June. The larval habits are not known in Northern Ireland. In

Moth accounts

Engrailed.

Lackan Bog, County Down.

This is a photograph of the first adult recorded in Northern Ireland. It was confirmed by a leading British expert.

Small Engrailed.

Monawilkin, County Fermanagh.

The freshly-emerged adult was found resting on this rock in early morning. This is a much paler insect than the Engrailed.

Britain larvae occur in July and August and feed mainly on birch, but occasionally on Alder. It overwinters in the pupal stage.

Common Heath

Ematurga atomaria (Linnaeus) Category one.

Ubiqutous. This is one of the most widespread species in the region and is known from many sites in all counties and on Rathlin Island. It is a bogland species and is more prevalent in the north and west than in the south and east, but even here it persists on small relict areas of habitat. Adults show considerable variation in colour and the males also differ from females. The general pattern is of a white, yellow or brown moth with brown speckling and banding. Females, which lack the feathering on the antennae present in the males, are generally white with brown bands, but many are brown with just a dusting of white. Males range from yellow to dark brown. Common Heath is almost exclusively found on bogs and heathland in Northern Ireland, especially in the wetter west and rarely in grassland and open woodland. It is a day-flying species, encountered in sunshine. Adults sit on plants with their wings held fully open. During dull and inclement weather they settle low down among the heather and other vegetation, making them difficult to detect. The recorded flight period is 16 April to 9 August. The larvae occur from July to September on heathers. It overwinters in the pupal stage.

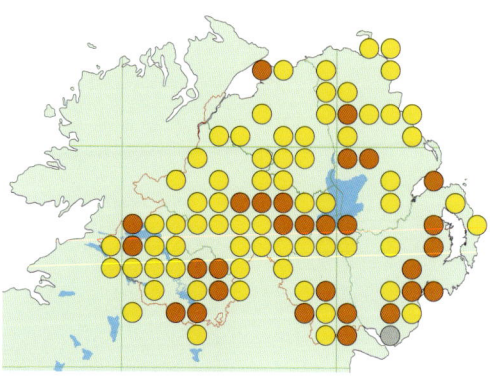

Bordered White

Bupalus piniaria (Linnaeus) Category two.

Scarce. Recorded from all counties, but it is uncommon in the north. It is widespread in north Armagh, east Down and Fermanagh. Sites with recent records include Peatlands Park (Armagh); Bohill Forest, Donard Wood and Leitrim Lodge (Down) and Eshywulligan, Legatillida, Lough Alaban, and Tullyhona (Fermanagh). In northern counties its distribution is more sporadic, with few records, but including Garry Bog (Antrim) and the Umbra (Londonderry). Males have feathery antennae and dark brown wings with whitish centres, but some individuals have yellow patches. The females are lighter brown with yellow centres to the wings. Specimens from Northern Ireland tend to be paler and more richly marked than their English counterparts. Adults rest with their wings held tightly together in a vertical position showing the underwing patterning, which is brown with a white streak. Bordered White is found on heaths, bogs and open woodland with mature pine trees. The adults are attracted to light, but are also said to fly during the day around the crowns of pine trees. They may be disturbed from their resting places during the day by tapping the lower branches of the foodplant. The recorded flight period is 23 May to 14 July. The larval foodplants are Scots Pine, possibly spruce and larch. It overwinters in the pupal stage.

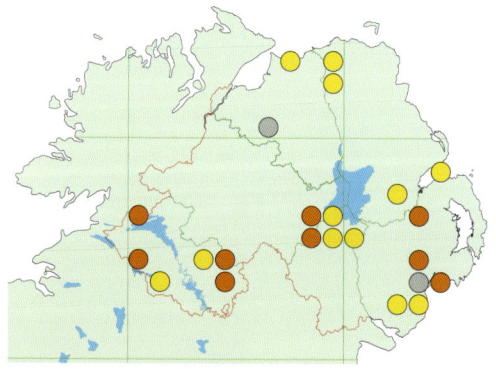

Common White Wave

Cabera pusaria (Linnaeus) Category one.

Widespread. Generally distributed across the southern counties, but records are more patchy over much of the north and west. Frequently recorded sites include Montiaghs Moss (Antrim); the Argory, Brackagh Moss and Peatlands Park (Armagh); Belvoir Park, Bohill Forest and

Bordered White.

Peatlands Park, County Armagh.

Helen's Bay (Down); Brookeborough, Crom, Legatillida and Monmurry (Fermanagh) and Rehaghy Mountain (Tyrone). It has also been recorded from Rathlin Island (Antrim) and at Copeland Bird Observatory (Down). Adults have white wings which appear speckled where they are heavily dusted with fine grey scales. There are three distinct cross-lines across the forewings and two on the hindwings. The cross-lines are light pinkish-grey, and those on the forewings are straight. Common Wave is similar, but greyer in appearance and the cross-lines on the forewings are browner and curved. Common White Wave is found in woodland, wooded bogs and heaths and suburban gardens. Adults come frequently to light in moderate numbers. It can also be disturbed from the foliage of trees and bushes during the day. Adults have a long flight period from 8 May to 25 October, which may indicate two generations, although this is not obvious from the records. The peak period is July. The larvae can be found from June to October on a variety of trees including birch, willow and Alder. It overwinters in the pupal stage.

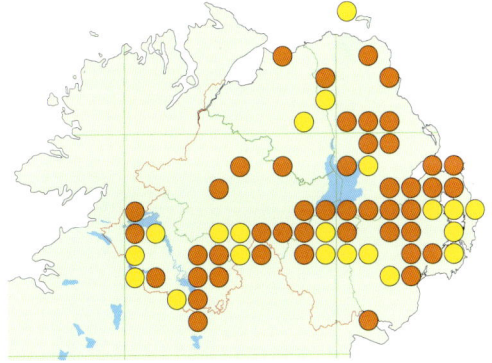

Common Wave

Cabera exanthemata (Scopoli) Category one.

Widespread. This species has been found in all counties and recorded from many sites across the south. It is less common north of Lough Neagh where the records are more scattered. It was first recorded at Copeland Bird Observatory (Down) in 2003, but has not been seen on Rathlin Island (Antrim) since 1937. Adults are similar in appearance to the Common White

Wave and it may be difficult to distinguish worn individuals of the two species. Common Wave is a more ochreous-grey colour on the forewings than Common White Wave. The shape of the outer cross-line is the best distinguishing character. In Common Wave, the line is curved and approximately parallel to the hind margins of the wings while in Common White Wave, the line runs straight across the forewing. Common Wave is found chiefly in damp woodland, fens and bogs where willows are common, but adults can appear in other habitats, including gardens. Adults become active from dusk and appear frequently, but not abundantly, at light. Like many other 'waves' it is easily disturbed during the day from the foliage of willows and other bushes. Like the Common White Wave, adults have a long flight period from 6 May to 4 September which may indicate two generations, although this has not been proven. The larvae can be found from June to October feeding on willow and possibly Aspen, especially in the west. It overwinters as a pupa.

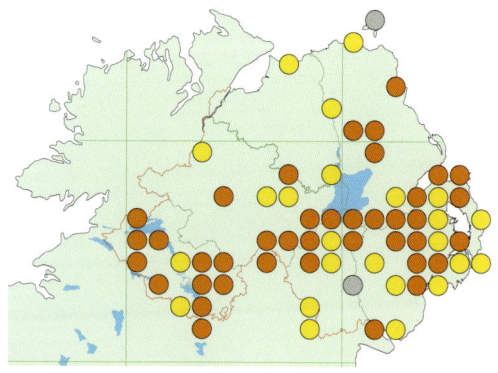

Clouded Silver

Lomographa temerata (Denis & Schiffermüller) Category one.
Widespread. This has been recorded in all counties, but is essentially confined to the south of Northern Ireland. The lack of northern records is probably due to under-recording. The only recent records from north of Lough Neagh are from Ballymena and Killybegs Bog (Antrim) and Binevenagh Forest (Londonderry). It has recently been seen at Copeland Bird Observatory (Down). The forewings are white with a silky sheen and a series of fine dark flecks. There is a dark suffusion of scales near the apex and a small but distinctive discal spot towards the costal edge of the forewing. Clouded Silver occurs in broadleaved woodland, wooded bogs, hedgerows and, occasionally, gardens. Adults are attracted to light and are often seen in small numbers in garden traps. Individuals have been disturbed from vegetation by day. There is a single annual generation and the recorded flight period is 11 April to 28 July. The larval foodplants are Hawthorn, Blackthorn, plum and cherry. It overwinters as a pupa.

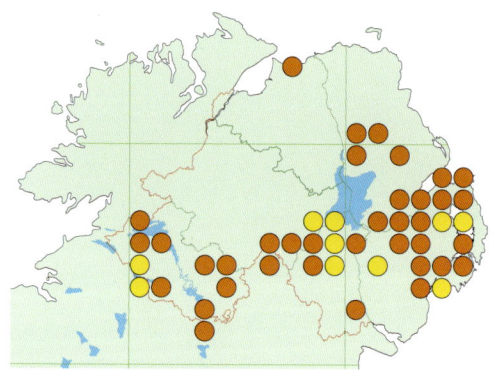

Early Moth

Theria primaria (Haworth) Category three.
Scarce. Widely distributed and present in all counties but only at scattered localities. It is undoubtedly under-recorded in many areas as it is one of the few moths that fly only during the winter months. Recent records have come from the Argory and Peatlands Park (Armagh), Helen's Bay (Down), Castle Archdale and Legatillida (Fermanagh) and Sixmilecross and Stillago (Tyrone). The males have reddish-brown forewings with a darker curved cross-line and central spot. The females, like many of the winter-flying species, have vestigial wings. Early Moth occurs in woodland rides and clearings, hedgerows and suburban gardens. Males are attracted to light and become active from dusk onwards. The recorded flight period is 31 December to 26 March. The larvae can be found in April and May on Hawthorn and Blackthorn.

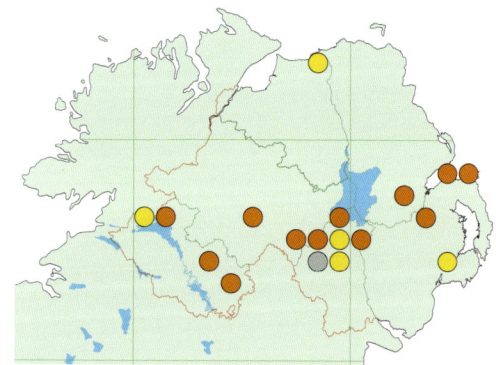

Clouded Silver.

Aughinlig, County Armagh.

Light Emerald

Campaea margaritata (Linnaeus) Category one.

Widespread. This large, distinctive species is found throughout Northern Ireland. It occurs on Rathlin Island (Antrim) and also at Copeland Bird Observatory (Down). Many of the northern sites have been discovered since 2000, suggesting it is under-recorded in this area. These sites include Breen Wood, Ecos Park, Glenariff Forest and Portglenone (Antrim) and Curran Bog and Duncrun (Londonderry). Freshly-emerged adults are an unmistakable greenish-white with two white cross-lines edged with a darker greenish-brown line on each forewing. The green colouration quickly fades to ochreous-white. The tips of the forewings are pointed with a small reddish patch. The hindwings are similar in appearance, but have only a single cross-line. This woodland moth is found in broadleaved woods, wooded bogs and fens and, occasionally, gardens. Adults appear frequently at light in small numbers. In southern Britain, there are two annual generations but in Northern Ireland the species appears to have just a single generation. Adults have been recorded from 12 June to 13 August. The larvae can be found in September to May of the following year on oak, birch, Beech, willow, Hawthorn, Blackthorn and Hazel. It overwinters as a larva.

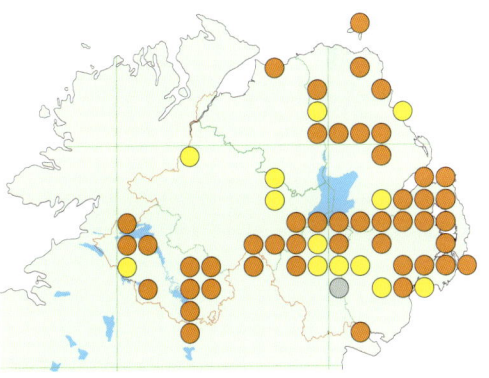

Barred Red

Hylaea fasciaria (Linnaeus) Category one.

Widespread. Recorded in all counties but most common in the south, especially Down, which has the greatest number of known sites. Regular localities in Down include Belvoir Park, Bohill Forest, Helen's Bay, Hillsborough Forest and Rostrevor Wood. In recent years it has been recorded more frequently, which may reflect a genuine increase as well as the greater level of recording. It

Barred Red.
Argory, County Armagh.

seems likely that it remains under-recorded in Antrim, Londonderry and Tyrone where there are extensive coniferous plantations. Adults are usually reddish-grey with two cross-lines, which are normally edged with white, across the forewings. The area between the cross-lines is often slightly darker giving the impression of a broad central band. Barred Red is associated with conifer plantations, mixed woodland, parkland and estates with mature conifers and other habitats that have established conifers, including many suburban habitats. Adults are attracted to light and appear in small numbers in favoured localities. The recorded flight period is 18 June to 27 August. The larval foodplants are Scots Pine, Norway Spruce and probably Larch. It overwinters as a larva.

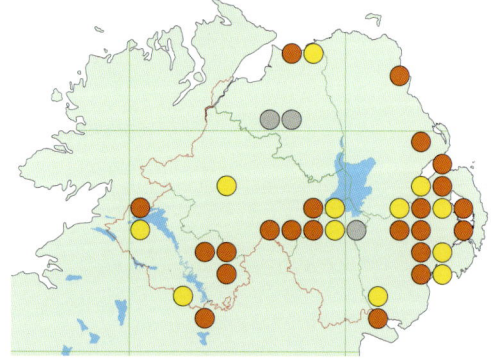

Annulet

Charissa obscurata (Denis & Schiffermüller) Category three.
Notable. This species has an entirely coastal distribution in Northern Ireland, but in the south of Ireland it occurs frequently inland on heathland and rocky habitats. In Down it is regularly recorded from Murlough NNR, where it appears common on the heathland and dune slacks. It has also been taken recently at Killard Point. There have also been records on the north Down coast at Grey Point and Helen's Bay. However, it has not been seen at either site since 1993. In Antrim it is known from Magheramorne and several sites on Rathlin Island. There are historical records from the Giant's Causeway, Portballintrae and Portrush in the late nineteenth and early twentieth century. It was rediscovered on the north coast in 2003 at the Umbra (Londonderry). Adults are brown or grey with

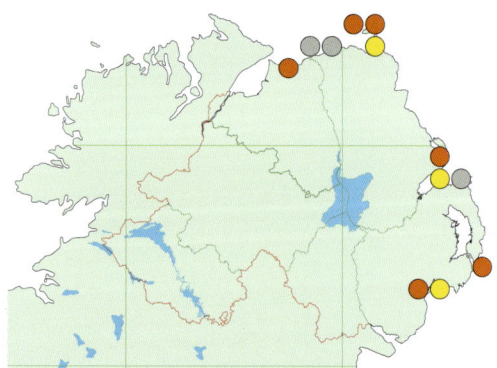

jagged cross-lines and a circular spot on all wings. However, the ground colour is variable, usually reflecting the geology of the habitat. Populations on heaths and on dark rock types tend to produce dark individuals, but at many sites a mixture of dark and light forms can be seen. Annulet is found on coastal heaths, rocky coasts and sand dune systems. Adults come to light in small numbers. In Northern Ireland it has seldom been encountered by day, but in Britain, it is reported that adults can be found resting on bare sand and rocky ledges. The recorded flight period is 16 July to 25 August. The larval habits and foodplants are not known in Northern Ireland. In Britain, the larvae can be found in September to May of the following year on Sea Campion, Thrift, Bird's-foot Trefoil, Kidney Vetch, Wild Thyme and Heather. It overwinters as a larva.

Grey Scalloped Bar
Dyscia fagaria (Thunberg) Category one.
Scarce. This moth is found widely in Fermanagh and north Antrim, but is very local elsewhere and present in just a few scattered localities. There have been no recent records from Down, where it was last seen in 1903 at Newcastle. It occurs throughout Fermanagh at Carnmore, Cavans Bog, Garvary Wood, Lisblake Bog and Monawilkin. In north Antrim it has been seen on Rathlin Island and on the mainland at Breen Wood, Martinstown and Tamnabrack. In Armagh, the few sites are all in the north-west of the county at the Argory, Aughinlig and Peatlands Park. It has been seen many times at the latter site. Adults of this species are variable in size and colour. The typical form tends to be light greyish-brown with two dark, curved cross-lines and a 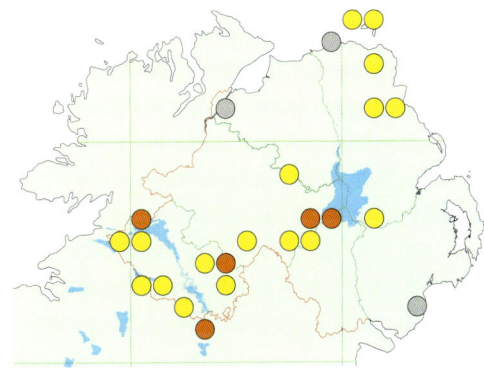 conspicuous discal spot near the costa. Paler forms are also common among populations. Grey Scalloped Bar is found in upland and lowland heaths and bogs. Adults are occasionally seen at light but usually in small numbers. They have also been observed after dark, resting on Heather. The recorded flight period is 15 May to 16 August. The larval foodplants are Heather and Cross-leaved Heath. It overwinters as a larva.

Grass Wave
Perconia strigillaria (Hübner) Category three.
Rare. Apparently recorded only from Antrim, Armagh and Fermanagh. It is known from the Argory and Peatlands Park in north Armagh. The majority of the records come from the latter site where it was first found by Johnson in 1893. Within Peatlands Park it has been seen at Annagarriff Wood, Derryhubbert and Mullenakill. The Fermanagh sites are in the north-west of the county at Braade and Correl Glen. Several old records exist including one from Ballymacilrany (Antrim) by Wright in 1953. It is currently unknown in Down, Londonderry and Tyrone. The adults show some variation in size, colour and markings on the wings. The typical form is light, greyish-brown, with darker brown cross-lines and an overall freckled appearance. Darker, 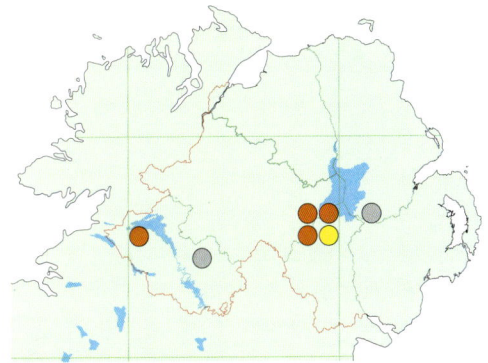 more heavily marked individuals are common in some populations. This is strictly a species of wet heaths and bogs, but why it is so restricted is unclear. Adults can be flushed from the vegetation during the day, but its flight period is from dusk onwards. The recorded flight period is short, from 5 June to 12 July. It flies from May to July in Britain. The larval habits are not known in Northern Ireland. In Britain, it can be found from August to May of the following year on Heather, Broom, Gorse and possibly Blackthorn. It overwinters as a larva.

SPHINGIDAE

Hawkmoths are medium-sized to very large moths and are undeniably the most eye-catching insects found in Northern Ireland. There are twelve species on the Irish list and all apart from one, the Bedstraw Hawkmoth Hyles galii, *have been reported in Northern Ireland. Five of the Northern Ireland species are resident and the other six are immigrants originating from southern Europe and the Mediterranean. Some of these species have been recorded almost annually in Northern Ireland, others are irregular or very rare visitors. Most hawkmoths are nocturnal and the adults show some attraction to light, but two are day-flying. Some species have long tongues and hover while feeding from flowers, but others have underdeveloped mouthparts and do not feed in the adult stage.*

The larvae of many of our hawkmoth species are, like the adults, strikingly coloured and patterned. Those of the larger species grow to an impressive size in their last instar. As a group, the larvae are characterised by the presence in some shape or form of a horn-like projection from the anal plate on the eighth segment of the body. Many of the larvae adopt a characteristic sphinx-like posture when at rest on the foodplant, or when threatened, have the ability to retract the head into the larger thoracic segments making them appear larger and more intimidating towards a potential predator. The pupae are all subterranean, spending the winter below ground in a constructed chamber. Most of the resident species occur as adults from May to late July, while the immigrants are most often seen between early summer and late autumn.

Convolvulus Hawkmoth

Agrius convolvuli (Linnaeus) Category three.
Scarce. This large moth, which originates from the Mediterranean, is an immigrant to northern Europe, including Ireland. It is generally a late summer visitor but there is no evidence that it has ever bred here. Prior to 1990 there were few records (although 1987 was an exceptional year when thirteen individuals were recorded in Ireland), but since then it has been reported more frequently with records almost every year from 1991 to 2000 with the exception of 1992 and 1995. The most recent records are from Ballycastle and Rathlin Island (Antrim) in 2003. This increase reflects the higher level of moth recording and also increased northward vagrancy by the species as the climate has warmed. The majority of sightings have been in the 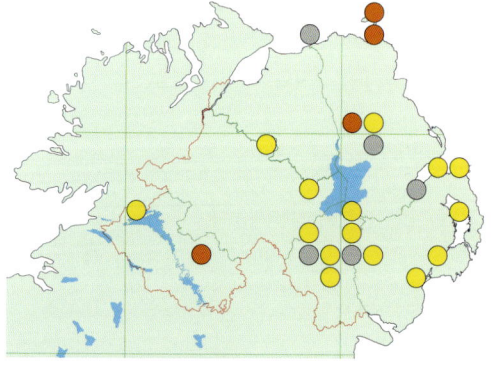 south and east, although individuals can appear virtually anywhere and there have been occurrences in every county. The forewings, thorax and head of the adult moth are mottled grey and brown. The abdomen is marked dorsally with a broad, grey longitudinal stripe and pink and black transverse bands edged with white on the sides. This species has an exceptionally long proboscis. At rest, the wings are held along its body. Adults are attracted to light but individuals can also be found during the day, resting on tree trunks, fence posts and walls. It is occasionally seen in gardens, particularly around dusk, visiting plants such as Petunia which have long tubular flowers. The Northern Ireland records have been between 18 August and 30 October, the majority in September. The larvae feed on bindweeds, from which it takes its name. Larvae have occasionally been found in Britain (although not in Ireland), but it is highly unlikely that pupae could overwinter successfully.

Death's Head Hawkmoth.

Adult captive bred from a larva found in Crete.

A migrant moth that appears occasionally in Northern Ireland.

Death's Head Hawkmoth
Acherontia atropos (Linnaeus) Category three.
Notable. This very large moth, with its famous marking on the thorax resembling a human skull, is an occasional vagrant to Northern Ireland. Most specimens have been seen in Down, reflecting its southerly origins around the Mediterranean. Since 1990, individuals have been seen in 1991, 1994, 2002 and 2003. In September 1956 a total of seven adults was recorded — two in both Belfast and Newtownards (Down), and one each in Armagh City (Armagh), Downpatrick (Down) and Strabane (Tyrone). This is the largest number reported in a single year in Northern Ireland. Most of the records emanate from the general public who find the moth by day at rest on walls or fences. The adult moth is unmistakable. The forewings are mainly black suffused with

brown with some patches of lighter scales. The abdomen has a wide dorsal stripe ribbed with yellow lateral patches. Unlike other species in the family, Death's Head Hawkmoth has a short proboscis and so cannot obtain nectar from deep, tubular flowers. The moth is known to raid beehives for honey and it was often referred to as the 'Bee Robber' by early entomologists. When in the hive it emanates a squeaking sound that apparently has a calming effect on the bees safeguarding it from attack. The adults also have a liking for sap exuded by trees. Adults have been recorded in Northern Ireland in May, June, August, September and October. In southern Europe, this is a continually-brooded species producing at least two generations a year. The larvae feed on the leaves of Potato, Deadly Nightshade and related species. Larval records were formerly common in Britain (they have declined due to increased usage of pesticides), but were less often reported here. It is highly unlikely that the larvae could complete development in Ireland.

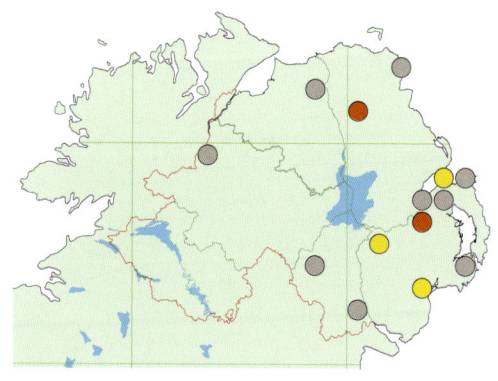

Eyed Hawkmoth
Smerinthus ocellata (Linnaeus) Category one.
Scarce. The Eyed Hawkmoth is present in all counties. Its stronghold appears to be the wetlands around the south shore of Lough Neagh, particularly Peatlands Park (Armagh) where individuals are recorded in most years. Frequently recorded sites in other counties include Derryola Bridge, Montiaghs Moss and Portmore Lough (Antrim) and Crom and Garvary Wood (Fermanagh). It is local in Londonderry with no recent records since 1992 when it was recorded from Traad Point. The only recent record for Tyrone was from Killeter Bog in 1989. It is very rare in Antrim away from the extreme south and the Lough Neagh shore. It is very rare in Down; the most recent record in Down was in 2002 at Keel Point. Eyed Hawkmoth occurs in low density so is likely to have been overlooked at suitable sites. Adults are pale brown with pinkish-brown forewings which are slightly scalloped. The hindwings are ochreous-brown, flushed with a rosy pink hue and large bluish-grey eyespots. The moth adopts a curious resting posture, in which the forewings are held over the hindwings and the tip of the abdomen is curled upwards. When threatened, it exposes its hindwings to reveal the large eyespots. This species seems to prefer acidic sites where willows are common, including cutover bogs, fens and damp woodland. Adults are attracted to light but only in small numbers. The recorded flight period is 1 May to 27 July. The larvae have been found in Northern Ireland on willow. They seem to prefer small, wiry, and isolated trees growing in sheltered spots. Other foodplants which may be used include apple. It overwinters as a pupa below ground.

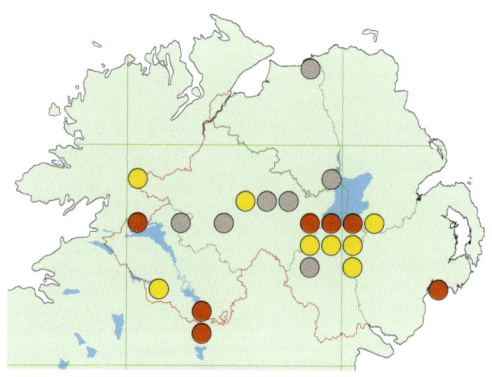

Poplar Hawkmoth
Laothoe populi (Linnaeus) Category one.
Ubiquitous. This is by far the most frequently encountered hawkmoth in Northern Ireland. It is common and widespread in all counties and present on Rathlin Island (Antrim) and Copeland Bird Observatory (Down). The idiosyncratic adults are unmistakable with their ash grey or pinkish-brown forewings with a broad darker central band. The forewings are finely scalloped and have a small white spot. The hindwings are similarly coloured, but also have a distinctive large reddish-brown basal patch. The moth adopts an unusual resting position, where the hindwings are clearly visible and held in front of the forewings. There is a scarcer buff-coloured form which has been recorded in Northern Ireland. Poplar Hawkmoth occurs in

Narrow-bordered Bee Hawkmoth.

Monawilkin, County Fermanagh.

Adults are mimetic, resembling bumblebees. They are most active in sunshine, especially in late morning, visiting flowers.

many habitats including fens, bogs, heaths, offshore islands, many coastal localities and urbanised areas. Adults appear frequently at light in moderate numbers. The general public frequently finds them during the day resting on walls and fence posts. The adults have an underdeveloped proboscis and do not feed. The recorded flight period is 30 April to 13 September. The larvae have been found on willow in Northern Ireland. They are more frequently seen during August and September when almost fully grown. They feed on the leaves of willows and poplars. It overwinters as a pupa below ground.

Narrow-bordered Bee Hawkmoth
Hemaris tityus (Linnaeus) SOCC (P) Category three.
Scarce. This day-flying bumblebee mimic has been recorded in all counties. It was described by early entomologists as common and widespread, but it appears to have suffered a marked

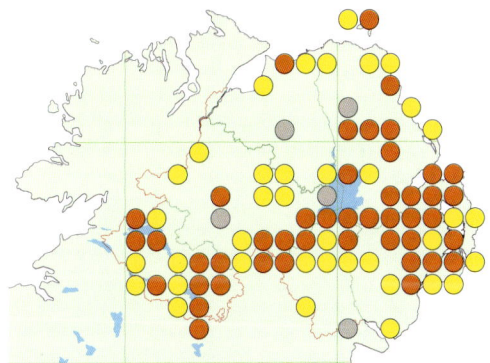

decline, although the extent of this is difficult to quantify. Recent records indicate a definite retreat westwards especially to Fermanagh, its present stronghold. It has been recorded at twelve sites in Fermanagh, most regularly at Braade, Monawilkin and Eshywulligan, the latter site having one of the largest populations. Its appearance in other counties has been more sporadic and only exceptionally has the species been seen at an individual site on more than on one occasion. The most recent records from other counties have been from Antrim in 1970 at White Park Bay, Armagh in 1998 at Drumnahavil, Londonderry in 1970 at the Umbra and Tyrone in 1998 at Keeran Moss. There have been no reliable records from Down since 1932. The

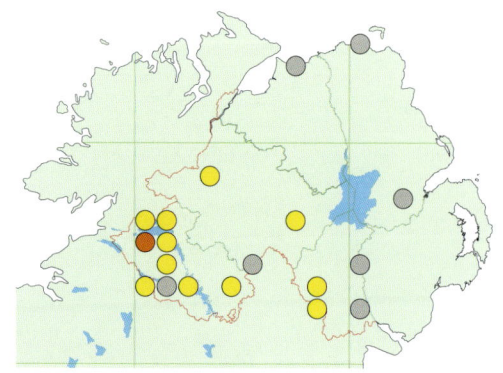

adults are densely hairy with an ochreous-yellow abdomen with two black central bands and a dark anal tuft. The wings of the freshly-emerged insect are covered with scales, but these fall off after the maiden flight leaving a transparent window with well-defined black veins. The antennae are long and clubbed. The Narrow-bordered Bee Hawkmoth has been found on damp moorland, unimproved calcareous grassland, sand dunes, cutover bogs and fens wherever its foodplant occurs. Adults are usually seen in late morning and early afternoon, flying swiftly in sunshine. They will visit flowers to feed especially Bugle, Lousewort and Bird's-foot Trefoil. During dull and inclement weather the moths will remain concealed in vegetation. The recorded flight period is 21 May to 25 June. The larvae can be found on the undersides of Devil's-bit Scabious leaves from early July to late August. It overwinters as a pupa.

Hummingbird Hawkmoth
Macroglossum stellatarum (Linnaeus) Category two.
Widespread. This, the commonest migrant hawkmoth in Northern Ireland, has been recorded from all counties and has occurred on both Rathlin Island (Antrim) and Copeland Bird Observatory (Down). It is most regularly seen along the east coast in Down and Antrim. There has been a marked increase in the numbers reported annually since 1980 and it is now appearing virtually every year. Prior to 1980, the sightings were very sporadic with none in most years. The number reported each year fluctuates enormously. For instance, there were large influxes in both 2000 and 2003 involving many individuals, whereas in 2002 only one was reported. Adults are unmistakable in appearance and behaviour. They are generally brown with

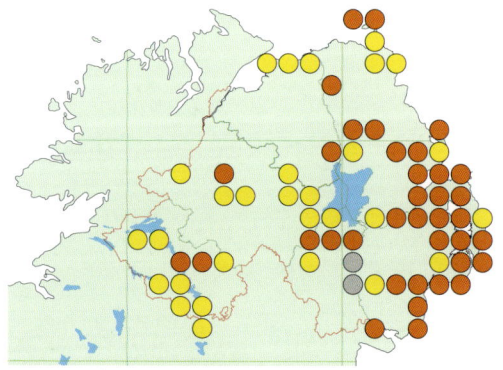

conspicuous white patches on the sides of the abdomen and a black anal tuft. The forewings are a dull brown with darker cross-lines, whereas the hindwings are mainly orange which shows conspicuously in flight. Adults can be seen anywhere, but they especially frequent sites with a supply of nectar-rich flowers. Tubular flowers are particularly favoured by feeding individuals, for example Buddleia, Lilac, Petunia and Red Valerian. Many are seen in gardens where the behaviour of the moth of hovering in front of a flower and inserting its proboscis to obtain nectar, convinces some owners that their garden has been visited by a hummingbird. There is no reliable evidence that this species has successfully bred here, even in the exceptional years such as 1984 and 1992 when large numbers of individuals were observed. Adults have been seen in Northern Ireland between 3 April and 20 October. The peak period for sightings is mid-June to late July. The larvae feed on bedstraws.

Oleander Hawkmoth

Daphnis nerii (Linnaeus) Category four.

Rare. This large migrant hawkmoth from southern Europe has been recorded on three occasions in Northern Ireland – at Ballywalter (Down) on 28 October 1938; Belfast on 8 September 1954 and Carrickfergus Marina (Antrim) in August 1995. It has rarely been reported elsewhere in Ireland, the only records being from Carlow in 1953 and Dublin in 1944, 1953 and, most recently, 1997. Adults of this beautiful species have intricately patterned forewings in several shades of green, with pink blotches and cream lines. There is also a pale apical streak on each forewing. It rests in a similar posture to Eyed Hawkmoth. It is resident in southern Europe around the Mediterranean where it produces more than one generation a year; however, it does not breed in northern Europe. The larvae feed on Oleander and Periwinkle.

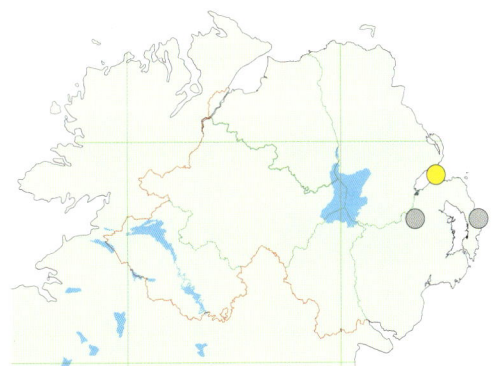

Oleander Hawkmoth.

Adult captive bred from a larva found in Crete.

Striped Hawkmoth

Hyles livornica (Esper) Category four.

Rare. This migrant has appeared nine times in Northern Ireland, the last occasion in 1992 when a single adult was found at Bangor (Down). It has been recorded in all counties except Fermanagh. The earlier records have been from Belfast in 1888, 1943 and 1966; Larne (Antrim) 1932; Jerrettspass (Armagh) 1940; Helen's Bay (Down) 1943; Derry City (Londonderry) 1923 and Ballymagorry (Tyrone) 1985. Adults are greenish-brown with two lateral white stripes on the thorax. The abdomen has black and white bands with a series of small white marks along the dorsal surface. The forewings are greenish-brown, with a pale ochreous stripe running from the base to the apex of the wing. This is overlain with a series of fine white lines. The hindwings are 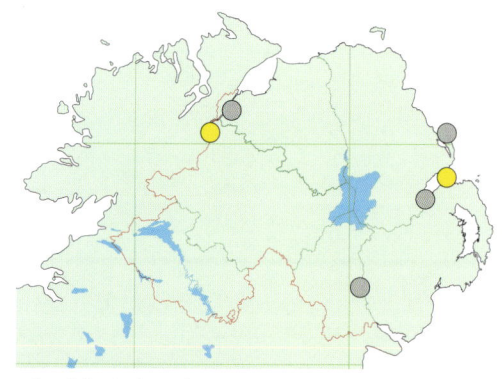 pinkish-red with a small white patch near the basal area of each wing and a black band near the outer margin. Adults have a long proboscis which they insert into tubular flowers whilst hovering to feed. Feeding usually happens at dusk on plants such as Red Valerian and Petunia. In Northern Ireland, adults have been recorded in every month between February and June and in October and November. The larvae are known to feed on bedstraws, willowherbs and Knotgrass. They are unlikely to survive in Northern Ireland.

Elephant Hawkmoth

Deilephila elpenor (Linnaeus) Category one.

Widespread. Recorded throughout Northern Ireland but most frequent in the southern counties and Antrim (including Rathlin Island). It is apparently absent from much of Tyrone and Londonderry. The plump, pink and brown adults are unmistakable. Reliable sites include the Argory, Brackagh Moss and Oxford Island (Armagh); Helen's Bay and Rostrevor Wood (Down); Aghalane, Crom, Garvary Wood, Legatillida and Monmurry (Fermanagh) and Rehaghy Mountain (Tyrone). Elephant Hawkmoth is found in large woodland clearings that have been colonised with Rosebay Willowherb, damp grassland, sand dunes, hedgerows and gardens. Adults become active around dusk and visit flowers, especially Honeysuckle, to feed. They 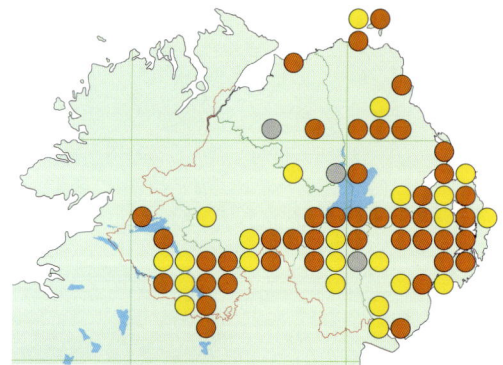 appear frequently at light in small numbers and are occasionally found in early morning. The recorded flight period is 17 May to 28 July, peaking in mid-June. The larvae (which have two colour forms) are generally nocturnal but are often conspicuous during the day and have been found in Northern Ireland on Rosebay Willowherb, Greater Willowherb, Marsh Willowherb and Fuchsia. When fully grown, the larvae leave the foodplant to pupate and they are frequently found crawling on the ground by curious members of the public. The enormous size of the caterpillar, its striking markings and behaviour when threatened, can engender fear in some people. To defend itself, a caterpillar dilates its thoracic segments, retracting its head and displays its large eyespots while it sways from side to side. It overwinters as a pupa.

Small Elephant Hawkmoth

Deilephila porcellus (Linnaeus) Category two.

Notable. This is the rarest of the resident species of hawkmoth in the region. It is currently known from only three sites on the coast of south Down, a single site in Londonderry, and two inland sites in Armagh. It has never been found in Antrim, Fermanagh or Tyrone. The Down sites are Killard Point, Murlough NNR and the Quoile Pondage. The first is its most regular site

Elephant Hawkmoth.

Brackagh Moss National Nature Reserve, County Armagh.

in Northern Ireland and where it was most recently reported in 2002 and 2004. A population was discovered at Magilligan (Londonderry) in 1971, and there have been sporadic records from here up to 1994. In Armagh, it has been recorded once from a mature garden at Aughinlig, in 1992 and, once, in 1922 at Poyntzpass. It is likely that small undetected populations may exist in low density possibly on the coast and, inland, on the limestone in Fermanagh. Adults are an unmistakable mixture of pinkish-red, yellow and brown. The dorsal surface of the abdomen is ochreous-yellow. The centres of the forewings are deep ochreous-yellow with brownish cross-lines. This species occurs mainly on unimproved coastal grassland in Northern Ireland. Adults are attracted to light from dusk onwards and visit flowers such as Viper's bugloss, campions, Honeysuckle and Red Valerian. The recorded flight period is 8 June to 20 July. The larvae feed mainly at night on bedstraws, but are occasionally found by day at rest among vegetation. It overwinters as a pupa.

Silver-striped Hawkmoth
Hippotion celerio (Linnaeus) Category four.
Unknown. An immigrant species recorded on one occasion in Northern Ireland on 28 November 1938 in Belfast. The voucher specimen is in the Ulster Museum. There have been

only two other Irish records from Sligo and Dublin, both in the late nineteenth century. The adults are quite distinctive having two pale lateral stripes at the side of the thorax. The forewings are ochreous-brown, with a smooth silver stripe that runs the length of the wing, curving towards the apex. The hindwings are pinkish-red with dark brown veining. The vast majority of records in Britain are adults taken at light from late summer until early autumn. The larvae have been reported in Britain, but are unlikely ever to reach maturity in the wild. The larval foodplants include Grapevine, Virginia Creeper, bedstraws, willowherbs and Fuchsia.

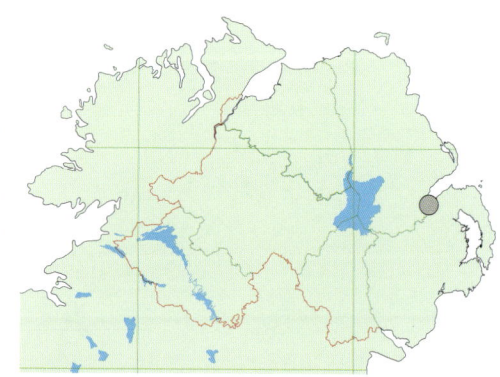

NOTODONTIDAE

The prominents are thick-bodied medium-sized moths, which take their name from the tuft-like scales that are present in many of the species and project from the inner margin of the forewings. When the moth is at rest these tufts come together, forming a raised projection. There are fourteen resident species in Northern Ireland. The adults are rarely found by day, but all species are attracted to light and appear in moderate numbers with the exception of the Small Chocolate-tip (adults of this species have seldom been recorded at light, most of the records have been of larvae). In common with other moth families, the males are most commonly encountered; the females of many species are seldom found in traps. The majority of adults fly during the summer months, but a few appear in the spring or early autumn.

Many species are distinctively marked making identification in most cases a simple process. Some species are superbly camouflaged, for example, the Buff-tip. Others such as the Puss Moth and the 'kittens' are remarkably hairy.

The larvae of this family are equally attractive and diverse. They are characterised by the presence of dorsal humps or anal appendages. Many sit with the head and anal segments raised. The anal appendages have thread-like structures called flagellae which are protruded when the larva is threatened. All species feed on the foliage of trees and shrubs. Most of the species pupate among leaf litter or underground. The kittens construct a hard cocoon on the trunk of a tree which is cryptic and difficult to detect.

Puss Moth

Cerura vinula (Linnaeus) Category one.
Widespread. Generally distributed in the southern counties, but very local in the north, although it has been recorded on Rathlin Island (Antrim) and at Magilligan (Londonderry) on the north coast. Its strongholds appear to be in the bogs and fens of north Armagh such as Brackagh Moss and Selshion Moss and similar sites in Fermanagh including Crom, Eshywulligan and Garvary Wood. Many of the records refer to the striking and distinctive bright green larvae, which are easier to find than the secretive adults. This is a fairly large species with conspicuous hairy legs and body and pale grey forewings with prominent yellow veining. The abdomen is a similar colour, rather hairy and marked with a series of black spots. The Puss Moth favours bogs, marshes and damp woodland where willows thrive. Adults come sparingly to light, but seldom more than one or two will be found in a single trap. The recorded flight period is 30 April to 3 July. The larvae have been recorded on willows in Northern Ireland. They can be

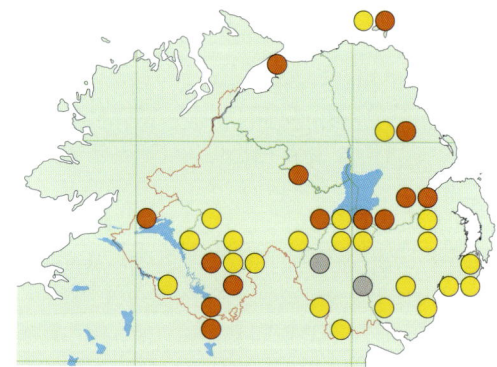

Puss Moth.

Lackan Bog, County Down.

found by searching the lower branches of the foodplant between July and September. It overwinters as a pupa inside a tough cocoon attached to the trunk or branch of a tree.

Alder Kitten
Furcula bicuspis (Borkhausen) Category four.
Rare. The only confirmed Northern Ireland (and Irish) record is from Legatillida (Fermanagh), involving two individuals trapped on 26 June 2001. There have been earlier, unverified claims of Alder Kitten and in view of the Legatillida record, some of these now appear more plausible. It seems likely that populations of this species may exist in other parts of Fermanagh, especially as suitable habitat is plentiful in this county. Adults have white forewings with a blackish central band. This dark band is irregularly shaped and narrowed in the centre of the forewing. This

The Butterflies and Moths of Northern Ireland

Alder Kitten.

Southern England.

distinguishes it from the two other kitten species. Alder Kitten is mainly associated in Britain with damp woodland and marshy areas with Alder and also drier sites with birch, including gardens. Both sexes are attracted to light. The main flight period in Britain is mid-May to early July. Late June to early September is the most productive period for finding the larvae. These are similar in appearance to the other kittens and should be examined carefully to confirm identification. The recorded foodplants are birch and Alder. It overwinters inside a tough cocoon attached to the trunk or branch of the foodplant.

Sallow Kitten
Furcula furcula (Clerck) Category two.
Scarce. This is the most commonly recorded of the three species of kitten in Northern Ireland. It has been found in all counties but apparently exists at only a small number of localities in

Iron Prominent.

Lackan Bog, County Down.

each. The majority of records have come from sites in Armagh, Down and Fermanagh including Brackagh Moss (Armagh); Belvoir Park, Seaforde and Willy Wood Island (Down) and Crom and Legatillida (Fermanagh). Favour Royal and Traad Point are respectively the only current localities in Tyrone and Londonderry. It occurs only in the extreme south of Antrim at Dunmurry and Portmore Lough. Adults are white with a broad, pale grey central band on the forewing. There are also dark apical patches and a series of black dots above the band and along the outer edge of the forewing. The thorax is dark grey-brown and quite hairy. The hind margin of the central band is irregular, whereas in the very similar Poplar Kitten it is smoothly curved. Adults and larvae need close examination to distinguish between these two species. Sallow Kitten is found in open, usually damp, woodland, fens and cutover bogs where willows are common. Adults appear sparingly at light and may occasionally be found resting on the trunks and stems of the foodplant. The data indicates there is just a single generation here with adults recorded from 10 May to 21 July. The larvae have been found on willows in Northern Ireland. It overwinters in the pupal stage inside a tough cocoon attached to the trunk or branch of a tree.

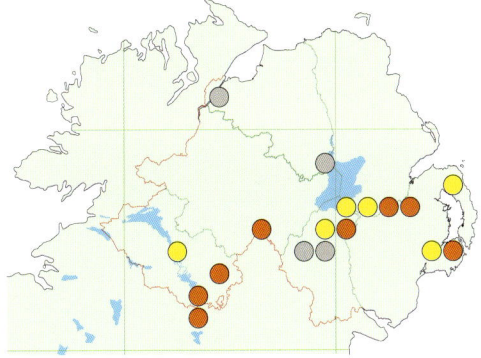

Iron Prominent
Notodonta dromedarius (Linnaeus) Category one.
Widespread. This species is known from many sites in Armagh, Down and Fermanagh but is apparently much scarcer and more sporadic in Antrim, Londonderry and Tyrone, although this may be due to under-recording in some areas. It occurs along the north coast at Magilligan

and the Umbra (Londonderry) and also on Rathlin Island (Antrim). Regularly recorded sites in the south include Aughinlig and Peatlands Park (Armagh); Helen's Bay (Down); Crom, Garvary Wood, Legatillida and Monmurry (Fermanagh) and Rehaghy Mountain (Tyrone). Adults are distinctive and easily recognised due to their characteristic resting posture. The forewings are generally dark reddish-brown with a small discal spot. The postmedial cross-line is often broken and the outer margins have rust-coloured streaks. The colouring is prone to variation with locality and very dark specimens are not uncommon. Iron Prominent is found in damp woodland, heathland, wooded bogs and fens, where the foodplant is common. Adults come frequently to light in small numbers. The recorded flight period is 1 May to 28 August. The larvae have been found in Northern Ireland on small, often isolated, birch trees. Other reported foodplants in Britain include Alder, oak and Hazel. It overwinters as a pupa.

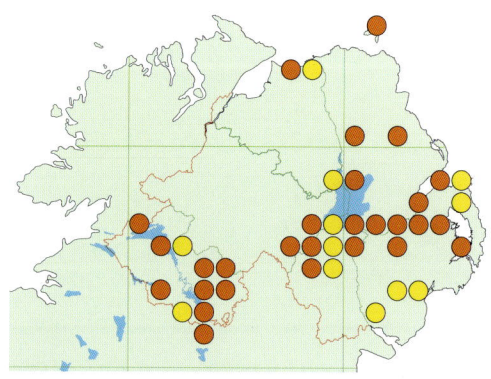

Pebble Prominent

Notodonta ziczac (Linnaeus) Category one.

Widespread. This species is found in much of Northern Ireland including Rathlin Island. Gaps exist in the north and west, but these would probably be filled with more intensive recording. Recently recorded sites include Breen Wood and Derryola Bridge (Antrim); Gosford Forest and Oxford Island (Armagh); Bohill Forest and Castlewellan Forest (Down); Aghalane, Altaclanabryan and Legatillida (Fermanagh); Duncrun, Magheramorne and the Umbra (Londonderry) and Crilly (Tyrone). Adults are very distinctive and easily recognised. There is a dark, pebble-like marking at the tip of the forewing which contrasts sharply with the pale central and basal areas. Like other prominents, they rest with their wings held tightly over the body. There is little variation in the markings or colouration. Pebble Prominent occurs chiefly in wooded bogs and fens and damp woodland, but can be encountered in other habitats. Adults come frequently to light in small numbers and are regularly seen in garden traps. The flight period is 3 April to 8 September. Two generations are normal in southern Britain and the data suggests that some populations in Northern Ireland also have two broods; however, more research is needed on this aspect. The larvae are often found on the lower branches of small willows growing on the margins of woodland and on isolated trees. Other foodplants include Aspen and other poplars. July and August appear to be the best months to find the larvae, when they are large and conspicuous on the foodplant. It overwinters as a pupa.

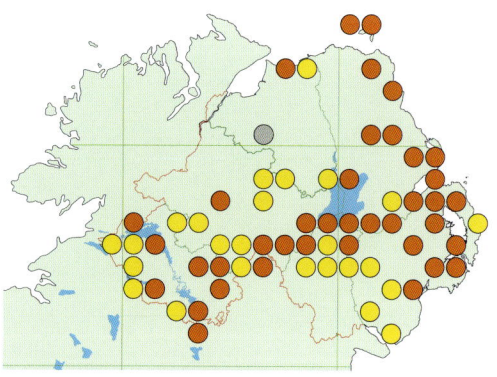

Lesser Swallow Prominent

Pheosia gnoma (Fabricius) Category one.

Widespread. This is a commonly recorded species in Armagh, Down and south Fermanagh. It is much more localised in the other counties, but likely to be under-recorded in some areas. Adults show little variation and appear whitish in colour. The forewings are heavily tapered with brown and black markings along the costa and a distinctive wedge-shaped, white tornal streak. The hindwings have a small dark anal patch. The closely-related Swallow Prominent is very similar but lacks the white tornal streak on the forewing. This species is found in wooded bogs and fens, heaths and woodland and,

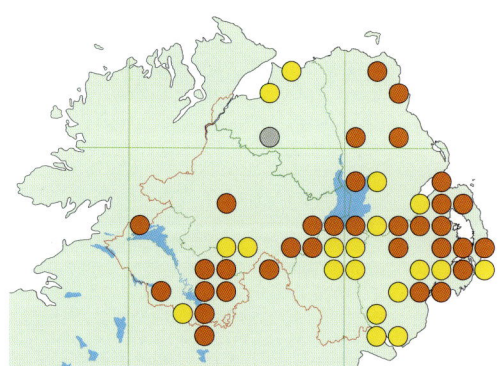

Pebble Prominent.

Annagarriff Wood National Nature Reserve, Peatlands Park, County Armagh.

Overleaf
Scarce Prominent.

Crom, County Fermanagh.

occasionally, gardens in urban areas. Adults come frequently to light in small numbers. There are two broods of adults, one peaking in May, the second in early August, but there is considerable overlap between the two generations. The recorded flight period is 24 April to 16 September. In Northern Ireland, larvae have been recorded feeding on birch, but they are elusive and have only been seen on a few occasions. Late July to early September is perhaps the best time to find them. It overwinters as a pupa.

Swallow Prominent
Pheosia tremula (Clerck) Category three.
Widespread. Scarcer than the previous species and mainly restricted to east Down, north Armagh and south Fermanagh. Records are very sparse in Tyrone and north of Lough Neagh. In Londonderry, it occurs at Banagher Glen and Traad Point. Breen Wood, Capanagh Wood and Portmore Lough are currently the only Antrim sites. Adults are very similar to Lesser Swallow Prominent and inexperienced recorders especially may confuse the two species, particularly where they occupy the same habitat. The differences between the two are rather subtle. The Swallow Prominent is usually larger and has a long, thick white line rather than a distinct wedge-shaped streak in the tornal area of the forewing. The hindwings have a dark anal patch which has a thin wavy white line passing through it. Swallow Prominent is found mainly in damp woodland, fens and wooded bogs and, occasionally, gardens. Adults appear commonly

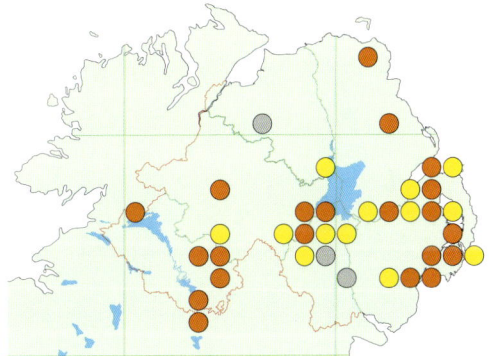

at light but always in small numbers. The recorded flight period is 1 May to 5 September. The peak period for records is July and early August and there are indications of a second, smaller peak in early June, suggesting some of our populations have two annual generations as in southern Britain. The larval foodplants are willows and poplars. It overwinters as a pupa.

Coxcomb Prominent
Ptilodon capucina (Linnaeus) Category one.
Widespread. Recorded throughout Northern Ireland but generally distributed in southern counties. It is more local in the north and west where it has been taken recently at Capanagh Wood, Glenariff Forest and Portglenone Wood (Antrim) and Duncrun and the Umbra (Londonderry). It has also been recorded from Copeland Bird Observatory (Down). The adults are reddish-brown and camouflaged to resemble a withered leaf when resting amongst vegetation. The forewing is scalloped and there are two conspicuous tufts, one on the thorax which is cream and a second on the dorsal edge. This is a broadleaved woodland species that also occurs in other well-wooded places including parks, birch-covered relict bogs and suburban 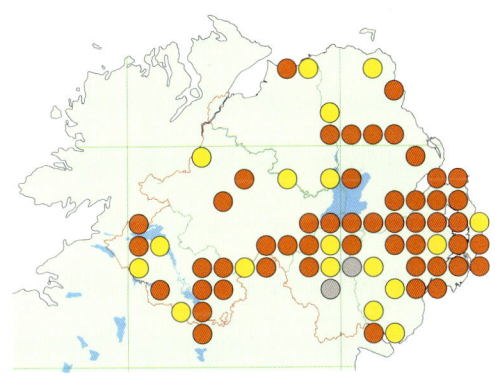 gardens. Adults are seldom, if ever, encountered by day, but come frequently to light in small numbers. The recorded flight period is 6 May until 25 August. Two annual generations are normal in Britain except in the far north, but the data suggests that there is just a single generation in Northern Ireland. The larvae have been found in Northern Ireland on birch. Other recorded foodplants in Britain include Hazel, Alder, oak, Beech and Rowan. It overwinters as a pupa.

Scarce Prominent
Odontosia carmelita (Esper) Category four.
Notable. This species is confined to six sites in Fermanagh and three in south Tyrone. It was not listed from Northern Ireland by Baynes and was first found here at Crom in south Fermanagh on 20 April 1992. This remained the only known site until 1997 when it was taken at Rehaghy Mountain (Tyrone). It has since been found at other sites in Tyrone at Altadavan, Crilly and Stillago, the latter being a mature garden well away from the other sites. In Fermanagh it has been recorded from Ardunshin, Knocknalosset, Legatillida, Marlbank and Monmurry. These discoveries were largely the result of regular trapping early in the year. Surprisingly, it has not been recorded from Peatlands Park (Armagh) which would appear to have suitable habitat. It was 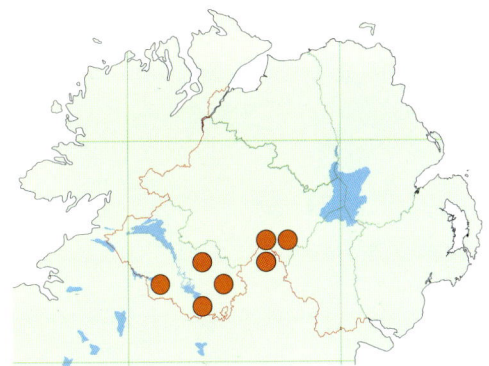 discovered in Ireland by Donovan near Kenmare in Kerry in July 1932 when larvae were beaten from birch. Elsewhere in Ireland it is known from Cork, Tipperary and Wicklow. The adults rest in a similar posture to Coxcomb Prominent with wings held tightly closed. The forewings are scalloped and rufous, suffused with grey scales, which give the moth a purplish-grey appearance. Scarce Prominent is found in woodland with mature stands of birch. Adults are attracted to light but are never seen commonly. This is a spring species recorded here between 23 March and 27 May. The larvae feed on the leaves of mature birch, but none to our knowledge have been found in Northern Ireland. It spends the winter months in the pupal stage.

Pale Prominent.

Lackan Bog, County Down.

An obvious moth, easily recognised by the inverted palps and scalloped forewings.

Pale Prominent

Pterostoma palpina (Clerck) Category one.

Widespread. Recorded from all counties, but mainly a southern species. It is frequently encountered in north Armagh, parts of Down and Fermanagh but is apparently absent from much of Tyrone and Londonderry, with the exception of Banagher Glen and Traad Point. Capanagh Wood is the most northerly site in Antrim. Regularly recorded sites in southern counties include Brackagh Moss, Loughgall and Peatlands Park (Armagh); Lackan Bog (Down); Crom, Garvary Wood, Legatillida and Monmurry (Fermanagh) and Rehaghy Mountain (Tyrone). The adults are pale brownish-grey with fine blackish streaks and heavily scalloped forewings. The diagnostic feature is the prominent inverted palps. Adults rest on the trunks of trees or among ground vegetation with the wings firmly wrapped around the body, resembling a piece of seasoned wood. Pale Prominent is found in broadleaved woodland and other habitats with suitable trees, including gardens. Adults are attracted to light in small numbers. The recorded flight period is 30 April to 22 August. The peak period is late May and early June. The August records suggest there is a partial second generation, which is normal in Britain. The larval foodplants are poplar, Aspen and willow. Poplar and aspen are absent from many of its sites here and it appears likely that willow is its main foodplant; however, no larvae appear to have been found in Northern Ireland. It overwinters as a pupa.

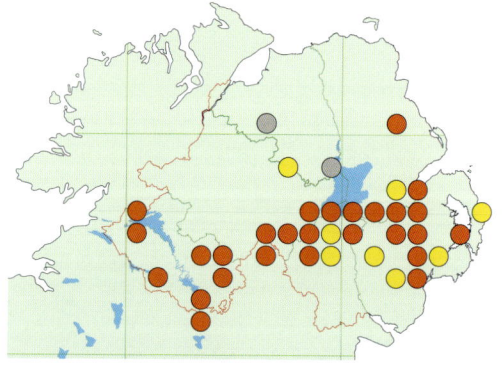

Lunar Marbled Brown

Drymonia ruficornis (Hufnagel) Category three.

Scarce. Recorded from a scattering of sites in Armagh and Fermanagh, and parts of Down and Tyrone adjacent to these counties. Apparently absent from east Down, Londonderry and most of Tyrone. It was last recorded in Antrim at Ballymacilrany by Wright in 1951, but it has not

Moth accounts

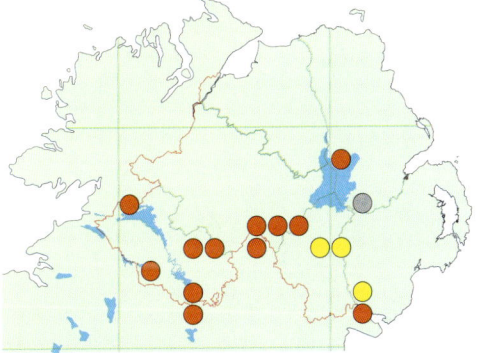

Lunar Marbled Brown.

Rostrevor Wood National Nature Reserve, County Down.

been seen in that county since. Most of the recent records have been from well-established woods in the west such as Crom, Garvary Wood and more recently, Aghalane in Fermanagh. Other sites include the Argory, Clare Glen, Slieve Gullion and Gosford Forest (Armagh); Derryleckagh Wood, Killowen Point and Rostrevor Wood (Down) and Crilly, Favour Royal and Rehaghy Mountain (Tyrone). These records have come from targeted and more intensive trapping in spring during its flight period. The distinctive adults look brownish-grey. The middle third of the forewing has a cross-band which is pale grey, edged with black wavy lines, and marked with a small, black crescent or comma. Lunar Marbled Brown is strictly a woodland species only found in mature semi-natural woodland containing oak, its sole foodplant. Adults appear at light in small numbers. In Northern Ireland, adults have been recorded from 20 March to 5 June, with the majority in April to mid-May. The larvae can be found during July and August on oak. It overwinters as a pupa.

Small Chocolate-tip
Clostera pigra (Hufnagel) Category three.

Scarce. This species is known from scattered localities in the north-west from west Fermanagh to north Antrim and especially in north Londonderry. It is absent from south Antrim, Armagh and Down except for a few sites on the fringes of the Mourne Mountains. Unusually, almost all the records of this moth are of larvae found during the daytime. Adults have only ever been recorded at light on a few occasions in Northern Ireland – at Carnmore Lough (Fermanagh) on 25 May 1997 and Altadavan (Tyrone) on 3 June 2001. Other sites include Garry Bog and Glenarm Forest (Antrim); Glennasheevar (Fermanagh); Banagher Glen and Benbradagh (Londonderry) and Gortin Glen (Tyrone). Adults have brownish-grey forewings marked with fine white cross-lines and a chocolate-coloured blotch at the apex. It is found on the edges of woodland and wooded bogs, wet heathland and damp moorland, especially on the margins of uplands. Adults are rarely attracted to light. The two confirmed records were of single individuals that had settled on the edge of the trap rather than inside it. In Britain, this species is bivoltine in the south (May/June and August/September), but univoltine (June/July) in the north and it is reported that there is just a single annual generation in Ireland. Larvae have been recorded in July and August in Northern Ireland on willows, particularly Eared Willow. The larvae feed at night and hide during the day between leaves spun together with silk. It overwinters as a pupa.

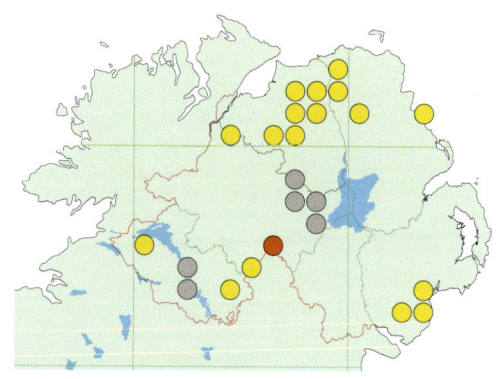

Buff-tip
Phalera bucephala (Linnaeus) Category one.

Widespread. This species has been recorded from all counties, but the distribution is mainly southern. It is frequent in Down, north Armagh and Fermanagh. Further north in Antrim, it occurs at a few sites in the south and around Lough Neagh, on the north coast around Ballycastle and on Rathlin Island. It is apparently rare in Londonderry and known only from Banagher Glen, Curran Bog, Kilrea and Traad Point. There have been no recent records from Tyrone away from the extreme south of the county. Adults are easily recognised by the silvery-grey forewings and the large buff patch that covers the front of the thorax and apex of the forewing. At rest the adult holds its wings tightly beside its body, giving a very convincing impression of a broken birch twig. The Buff-tip occurs in many habitats, including gardens, but is most frequent where there are stands of birch in open broadleaved woodland and on cutover bogs and other wetlands. Adults come frequently to light in moderate numbers. The larvae are gregarious until the latter instars and can sometimes be found sitting unconcealed on the branches of the foodplant. They are normally found between June and September. The recorded flight period is 16 May to 10 August. In Northern Ireland, larvae have been found on birch, Ash and willow. Other recorded foodplants are oak, Hazel, Alder and Beech. It overwinters as a pupa.

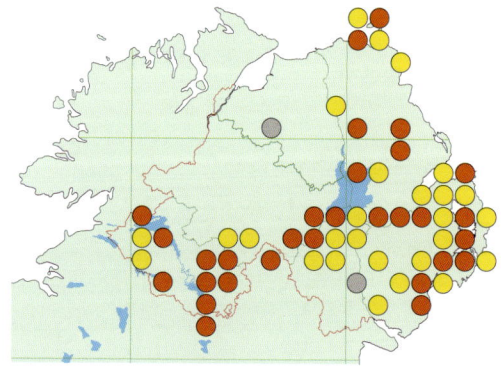

Figure of Eight
Diloba caeruleocephala (Linnaeus) Category three.

Scarce. Found in scattered localities mainly around Lough Neagh and Upper Lough Erne. Sites include Brackagh Moss, Oxford Island and Peatlands Park (Armagh); Aghalane, Crom and Knockninny (Fermanagh) and Rehaghy Mountain (Tyrone). In Londonderry, where it

Moth accounts

Small Chocolate-tip.

Altadavan, County Tyrone.

This small attractive species has rarely been taken as an adult in Northern Ireland.

Figure of Eight.

Peatlands Park, County Armagh.

was reported from Banagher Glen by Magurran in the 1970s and at Magilligan in the nineteenth century, the most recent record was in 1990 at Traad Point. Hillsborough Forest is the only known site in Down. It is equally rare in Antrim and only found in the Montiaghs Moss area in the extreme south-west. The monitoring data from the Rothamsted trap network indicates a decline in abundance of this species in Britain of 95 per cent between 1968 and 2002. Adults are dark grey-brown. The forewing has a white stigma which is shaped like the figure eight, from which the moth gets its name. There is also a chequered fringe and a dark tornal dash on the forewing. This species is associated with well-established lake shore scrub, broadleaved woodland and old mature hedgerows. Adults are attracted to light but are never seen commonly, even at many of its well-known sites. In Britain, the flight period lasts from late September to mid-November, but in Northern Ireland the recorded flight period is 6 to 23 October. The larvae have been found on Blackthorn in Northern Ireland; other known foodplants in Britain include Hawthorn and wild roses. It overwinters as an egg.

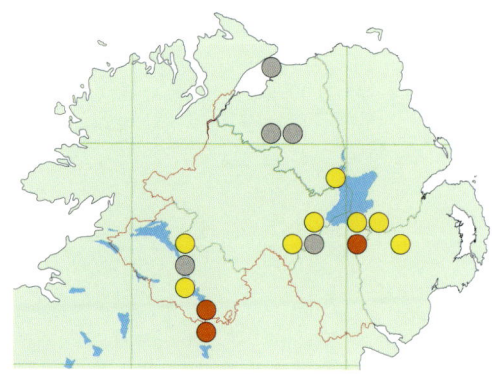

LYMANTRIIDAE

This is a family of medium-sized moths, some of which have long hairy forelegs that are stretched out in front when the moth is at rest. Four of the eleven species recorded in Britain are found in Northern Ireland. The most commonly encountered species has a day-flying male and a wingless female. The others are scarce and mainly recorded in southern counties. Two species appear to be recent colonists and spreading northwards. The adults (with the exception of the Vapourer) are attracted to light, but appear only sporadically in traps. They do not visit flowers or sugar.

The larvae of this family are quite attractive and feed openly on the leaves of a variety of trees and shrubs. The larvae tend to be rather hairy, with some species having distinctive dorsal tufts. Handling of caterpillars should be avoided as the hairs of some species may cause minor skin irritations in some individuals. The pupa is protected inside a tough silken cocoon that is often covered with hairs spun on the foodplant, or occasionally on the trunks of trees and fence posts.

Vapourer

Orgyia antiqua (Linnaeus) Category one.
Scarce. Although it has been recorded in all counties, the distribution of this species is largely southern. The only known sites north of Lough Neagh are Lisnagunoge and Rea's Wood (Antrim) and the Umbra (Londonderry). The majority of records have come from Armagh in particular wetland sites in the north of the county, including the Argory, Brackagh Moss, Oxford Island, Peatlands Park and Selshion Moss. It has also been found at lakes in the south of the county along the border with Monaghan. There have been only a few records from Down at Derryleckagh, Drum Lough, Lackan Bog and Murlough. In Fermanagh, there are recent records from Crom, Legatillida and Monmurry and Crilly and Glenkeen (Tyrone). Adult males have ochreous-red fore- and hindwings. There is a distinctive, white comma-shaped spot at the lower edge of the forewing and two dark cross-lines. The female is flightless with vestigial wings and remains at rest on her cocoon after emergence, waiting for passing males. The Vapourer is found chiefly on wooded cutover bogs, lakeside fens and wet woodlands, and less commonly on dunes. This species tends to be rather elusive even in localities where it is well-known and the adult males are never seen

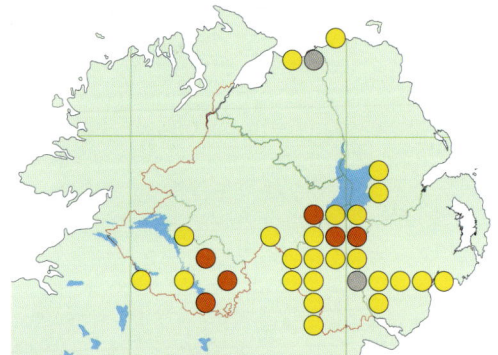

Vapourer.

Ballynakill Lake, County Galway.

in numbers. Males are active in sunshine and are often seen flying at great speed and often quite high up, where they could be mistaken for a butterfly. Most of the records are of larvae rather than adults. These can be found resting, exposed on the upper surface of leaves during the day. Its recorded flight period is 9 July to 12 October, although it is most commonly encountered during August and September. The larvae feed on a wide variety of trees and shrubs but have been found in Northern Ireland on birch, bramble, alder and willow. It overwinters as an egg.

Dark Tussock

Dicallomera fascelina (Linnaeus) SOCC Category four.

Rare. This species is currently known only from the Argory and Peatlands Park in north Armagh. It is found throughout the area encompassed by Peatlands Park including Annagarriff Wood, Derryadd Lough and Mullenakill Bog. It has not been recorded from any other county in recent years although old records exist for Tamnamore (Tyrone) and Tempo (Fermanagh). It may occur more widely, especially in upland areas in the west where recording has been less intensive. This is a large, grey moth that rests distinctively with the long hairy front legs stretched out in front. The wings are grey with fine, black speckling and black cross-lines suffused with orange. In Britain, the Dark Tussock occurs on bogs, moors and sand dunes, but in Northern Ireland the species has only been found on intact and cutover raised bogs. Adults appear sparingly at light even at well-known localities. The recorded flight period is 13 June to 21 July. The larvae can be found in August and September and, after overwintering, in the May and June of the following year. They are occasionally seen resting exposed on the tips of

The Butterflies and Moths of Northern Ireland

Pale Tussock.

Castlewellan Forest, County Down.

A fairly conspicuous species with long forelegs which are stretched out in front when at rest.

Heather stems in late summer prior to hibernation. The only reported foodplant in Northern Ireland is Heather, but in Britain, willow, Bramble, Hawthorn and birch are used in other habitats. It overwinters as a larva.

Pale Tussock

Calliteara pudibunda (Linnaeus) Category four.

Scarce. This species was virtually unknown in the region before 1997 and Baynes listed it only from the southern half of Ireland. The first record was in 1987 at Killowen Point in south Down. From 1997 on, it has been found at several sites in south Fermanagh, two in south Tyrone, two in Armagh and six in south Down. It has never been seen in Antrim or Londonderry. The majority of records come from regularly trapped sites including Carganamuck and Peatlands Park (Armagh), Castlewellan Forest, Murlough NNR and Rostrevor Wood (Down), Aghalane, Crom and Monmurry (Fermanagh) and Rehaghy Mountain (Tyrone). It is unlikely that this conspicuous species was overlooked at these sites in the past, and it appears that its range has

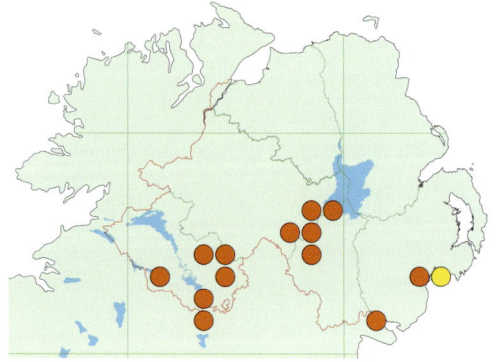

expanded northwards. This is a large, rather hairy-looking moth with grey forewings covered in fine black speckling. The male has a conspicuous brown band across the centre of the forewing. The female is much larger than the male and greyer. The Pale Tussock is found in damp woodland, mature hedgerows and scrubby open areas. Adults are attracted to light in small numbers, especially males, which are more frequently encountered than females. The recorded flight period is 2 May to 30 June. The larvae can be found from July until September on birch, Hazel and oak. It overwinters as a pupa.

Yellow-tail.

Castle Espie, County Down.

Yellow-tail

Euproctis similis (Fuessly) Category two.

Scarce. The Yellow-tail is a recent colonist that is now well established on the east and north coast of Down. It was first recorded in Ireland in 1934 from the Dublin City area and it is now locally common in eastern coastal counties. The first Northern Ireland specimen was seen at Newcastle (Down) in 1973 and since then it has spread further north, appearing in August 2004 at Breckenhill Dam (Antrim), its most northerly and inland locality. The known sites in Down include Ballydorn, Castle Espie, Castleward, Copeland Bird Observatory, Helen's Bay, Killard Point, Maghery Bog, Mill Bay, Murlough NNR, Rostrevor Wood, Seaforde and Tyrella. The adults have a silky white appearance. Both sexes have a white body with a deep yellow anal tuft,

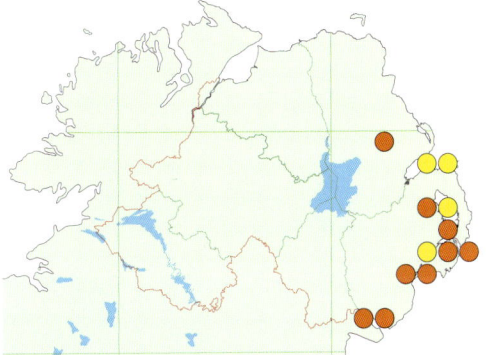

which is often elevated when the moth is at rest. The males have a small dark mark on the tornal aspect of the forewing. Adults are rarely found in numbers, even at well-known sites. Both sexes are attracted to light but appear only occasionally in traps. The recorded flight period is 6 July to 24 August. The colourful larvae, which have the characteristic hump on the first segment, are gregarious during early instars. They are best looked for during the latter part of summer, prior to hibernation, on Hawthorn, willow and oak. It overwinters as a small larva.

ARCTIIDAE

Sixteen of the 32 species in this family known from Britain and Ireland have been recorded in Northern Ireland. Fourteen of the species are resident; the other two are rare immigrants. The family is divided into two distinct subfamilies, the Lithosiinae (footman and allied moths) and the Arctiinae (tigers and ermines). The Lithosiinae are small, rather plain-looking moths with elongated forewings which in most species are folded tightly around the body, supposedly resembling the formal dress coats of Victorian servants. The adults of the majority of footman species have developed mouthparts and will visit flowers as well as sugar. Many species are also attracted to light, but few are encountered in daylight. The majority of the larvae of the footman species feed on lichens and algae growing on walls, fences and tree trunks. Pupation takes place among ground vegetation and leaf litter.

The species in the Arctiinae include the tigers and ermines. These are medium-sized and often brightly-coloured moths. None of the species have functional mouth parts and therefore the adults do not visit flowers or sugar. The species are mainly nocturnal and frequently attracted to light, although the adults of some species, especially Wood Tiger and Clouded Buff, are active during the day. The full-grown larvae of these species are large and most are conspicuously hairy; those of the Garden Tiger are known as 'woolly bears' or 'granny greybeards'. These come to the attention of the public, particularly curious children, in early summer, when they can be seen crawling on roads and pavements as they search for suitable pupation sites. Pupation takes place at ground level among leaf litter.

Round-winged Muslin
Thumatha senex (Hübner) Category one.

Widespread. Recorded from all counties except Londonderry, this species is generally distributed across southern counties, but apparently absent from most of Antrim. It was previously considered rare throughout Ireland but, with increased light trapping since 1990, it has been discovered at many sites in north Armagh, Down, Fermanagh and south Tyrone. Regular sites include Aughinlig, Brackagh Moss and Peatlands Park (Armagh), Hillsborough Forest (Down) and Crom (Fermanagh). There is no evidence that it has increased, rather it seems likely that it was overlooked as it is a small, inconspicuous moth which could be mistaken for a micro-lepidopteran. It may also be confused with Muslin Footman, but its wings are less rounded and it has a more transparent appearance. 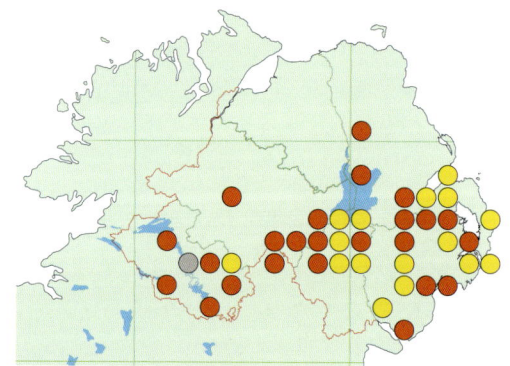 Round-winged Muslin is chiefly associated with bogs, fens, flushes and damp grassland, but adults can be encountered in other habitats, including gardens. The adults are attracted to light in small numbers. The recorded flight period is from 17 June to 2 September. The larvae are reported to feed on species of lichen, especially *Peltigera* species and ground-living mosses such as *Homalothecium sericeum*. It overwinters as a small larva, feeding occasionally throughout the winter months.

Muslin Footman
Nudaria mundana (Linnaeus) Category two.

Scarce. Recorded from Armagh, Down, Fermanagh and Tyrone. The most recent records are from eastern Down, Fermanagh and Tyrone including, Derryleckagh, Mill Bay, Hillsborough Forest, Killard Point, and Lackan Bog (Down); Brookeborough, Crom, Garvary Wood, Marlbank and Monawilkin (Fermanagh) and Fivemiletown (Tyrone). There have been just two confirmed records from Armagh since 1980, from Ballyrea and Lisadian. It is known from

Red-necked Footman.

Lough Alaban, County Fermanagh.

a scattering of sites in Tyrone, but there are no recent confirmed records from either Antrim or Londonderry. Donovan considered this species as being common and locally distributed. This opinion was not shared by Baynes, who claimed to have only a few records, including unlocalised records from both Antrim and Down. This is a small moth with transparent greyish forewings. There are two faint cross-lines on the forewings. Adults closely resemble Round-winged Muslin and they may be confused with this species. Muslin Footman is found in open woodland, bogs and rocky coasts. Adults are attracted to light in small numbers. The recorded flight period is 22 June to 18 August. The larvae feed on lichens growing on stone walls and small trees. It overwinters as a larva.

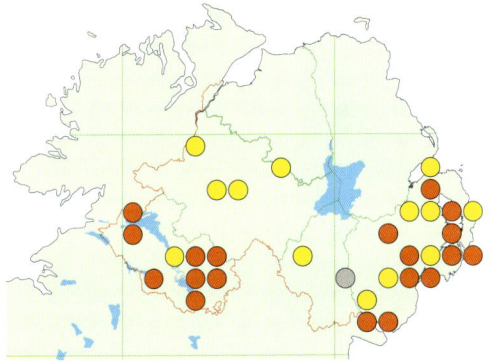

Red-necked Footman

Atolmis rubricollis (Linnaeus) Category four.

Scarce. Recorded from all counties except Londonderry. Until recently this species was considered rare in Northern Ireland, confined to a few localities in Armagh, Fermanagh and Tyrone. The increase in records may be due to the greater level of recording, but it genuinely may have increased and spread as conifer plantations have matured. The species strongholds are in the extensive woodlands, both conifer and broadleaved, in the west and south of Fermanagh. Further east it has been recorded from woodland habitat in scattered localities including the Argory and Peatlands Park (Armagh) and Belvoir Park (Down). There are also some records from coastal sites in Carlingford Lough (Down) and on the north Antrim coast at Portrush and White Park

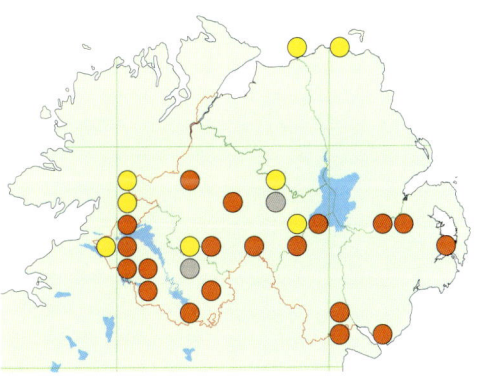

Bay, suggesting immigration or dispersal from established inland, or even southern Scottish, populations. Adults are very distinctive, having completely black wings and body apart from a conspicuous red collar. The forewings are long and held tightly over the body in typical footman fashion. This is a woodland moth associated especially with conifer plantations, but also broadleaved woods; however, a significant number of the Northern Ireland records are from other habitats. Groups of adults have been observed flying, in sunshine, around the tops of conifers or resting on low vegetation along woodland tracks in dull, cool conditions. Adults occasionally appear in light traps in very small numbers. Its recorded flight period is 8 June to 25 July. The larvae feed on lichens and algae growing on broadleaved and coniferous trees. It overwinters as a pupa, protected inside a cocoon.

Four-dotted Footman
Cybosia mesomella (Linnaeus) Category four.
Rare. Recorded once, when an adult was trapped in a mature Belfast garden in July 1991. This was the first and, to date, only record for Ireland. It is generally considered this record was of an immigrant as it has never been reported since; however, Four-dotted Footman is not generally known as a migrant species. In Britain, it is mainly a southern species although found as far north as north-east Scotland and also known from the Isle of Man. The adults have creamy white forewings with a yellow costa and outer edge. There is a small black dot at the costa and another on the inner margin of the forewing. In Britain, the Four-dotted Footman inhabits damp grassland, fens, heathland, moorland and open woodland. Adults become active at dusk and are attracted to light. It is reported that adults hide by day among low-growing vegetation and small bushes from which they can be easily flushed. The main flight period in Britain is from mid-June to the beginning of August. The larvae, which overwinter, feed on algae and lichens growing on the stems of heathers, willows and other shrubby plants.

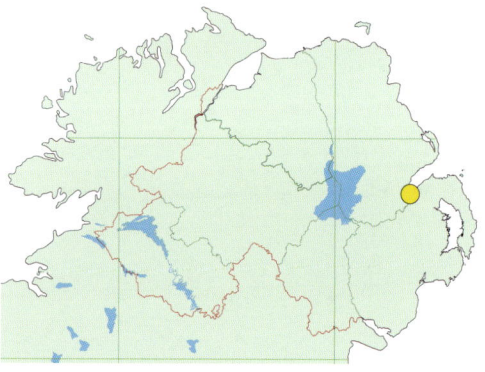

Scarce Footman
Eilema complana (Linnaeus) Category four.
Notable. Local and confined to coastal localities in Antrim and Down. In Down it has been recorded at Castleward, Copeland Bird Observatory, Kilclief and Murlough. Wright listed it from Antrim in 1964. The only known sites in this county are on Rathlin Island, where it appears to be common, especially around the western end of the island. There is one old record for Tyrone at Tamnamore by Greer, who described it as rare. Adults are silvery-grey with a yellow streak of uniform width running the whole length of the costa. The thorax is grey with an orange collar. Adults rest with their wings held tightly around the body. It bears a close resemblance to Common Footman but, in this species, the yellow streak on the costa of the forewing clearly tapers to a point near the apex of the forewing and at rest, the wings are held flat. In Northern Ireland, Scarce Footman is recorded from low, rocky coasts, coastal heathland, and mature sand dunes. In Britain, it is also found in fens and open woodland. Adults become active at dusk and are attracted to flowers, especially knapweeds and thistles. They also appear at light in small numbers. The recorded flight period is 15 June to 19 August, with a peak in July. The larvae feed on lichens growing on rocks, walls and plant stems. It spends the winter months as a larva.

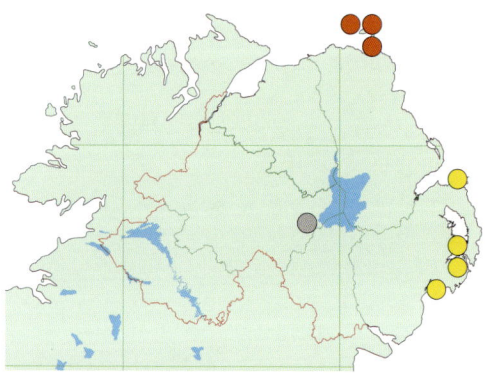

Moth accounts

Scarce Footman.

Rathlin Island, County Antrim.

A coastal species easily distinguished from the Common Footman (below) by the uniform length and shape of the yellow costal stripe and the rolled, resting position of the forewings.

Common Footman.

Crossgar, County Down.

This species, unlike the Scarce Footman, has been recorded from inland localities in County Down.

225

Buff Footman

Eilema depressa (Esper) Category four.

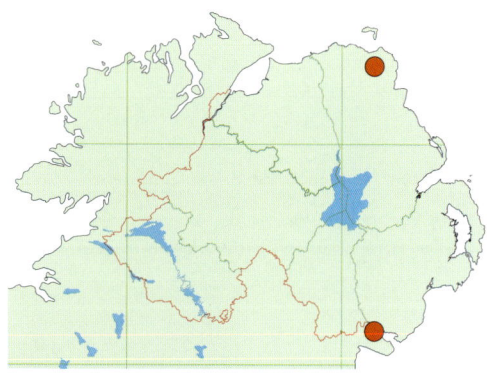

Rare. Recorded for the first time in Northern Ireland at Rostrevor Wood (Down) on 4 August 2004, where a total of three adults was taken at light. A single individual was also recorded at light from Breen Wood (Antrim) on 11 August 2004. Baynes describes this species as extremely rare in Ireland and reported it only from Killarney (Kerry). There are recent records from Kilkenny, Laois and Wicklow. The status of Buff Footman in Northern Ireland is unclear. The lack of earlier records, especially from the well-trapped woodland at Rostrevor, suggests the species is a recent colonist. The male has a greyish-white appearance to the forewings with a pale yellow stripe emanating from the basal area along part of the costa. There is also yellow fringing on the outer edge of the forewings. The larger female is normally slightly darker but with similar markings to the male. The hindwings are greyish-white. Baynes refers to a rare orange buff form (f. *unicolor* Banks) which has been recorded again recently at Killarney (Kerry). Buff Footman is a lichen-feeding species which seems confined in Ireland to ancient or long-established woodland. In Britain, it is single-brooded occurring in July and August. Adults are attracted to light in small numbers but none have been found by day. Several pupae were found on the trunk of an old Yew tree in Killarney National Park (Kerry) and successfully reared through to adults. The larval foodplants are lichens and algae found on the trunks and branches of a wide variety of trees including oak, birch, spruce and Yew. It overwinters as a small larva.

Common Footman

Eilema lurideola (Zincken) Category three.

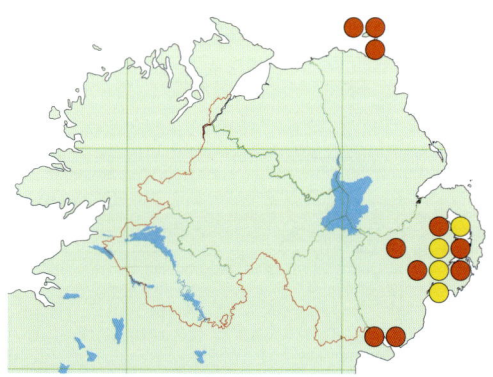

Notable. Reported from east and south Down and Rathlin Island (Antrim). It was not listed from the region by Baynes and the first record was in 1988 from Seaforde (Down). In Down, it is restricted to sites close to the shores of Strangford and Carlingford Loughs including Ballydorn, Castle Espie, Castleward, Mill Bay and Rostrevor Wood, although some spread inland is evident following records at Bohill Forest, Hillsborough Forest and near Killinchy. It appears widespread on Rathlin Island where it has been seen in reasonable numbers alongside Scarce Footman. The adults of Common and Scarce Footman appear similar, but can be told by the shape of the yellow streak on the costa of the forewings. In Common Footman this tapers to a point at the apex. Another distinction is that, at rest, the wings are held flat in Common Footman, but tightly rolled around the body in Scarce Footman. Adults have been taken at light in Northern Ireland, but usually only in ones and twos, apart from on Rathlin Island and at Rostrevor Wood, where they have been recorded in larger numbers. The recorded flight period is 15 June to 18 August. The larvae feed on lichens growing on rocks, trees and fence posts. It overwinters as a larva.

Four-spotted Footman

Lithosia quadra (Linnaeus) Category three.

Rare. This is a suspected immigrant recorded seven times in Northern Ireland. The most recent records were in 2004 from Crom (Fermanagh) on 17 July, Castleward (Down) on 24 July and Lisburn (Antrim) on 10 August. The other records are Tollymore Forest (Down) by Wright in 1973, an unknown locality in Down in 1950 and 1970 and the Belfast area in 1940. The species is a well-known immigrant and these records probably refer to migrant individuals as there is no evidence for a resident population in Northern Ireland; there are, however, resident populations in the south and west of Ireland. Four-spotted Footman is the largest of the footman moths recorded in Northern Ireland and is a distinctive species. The sexes differ in appearance and size, males being smaller than females. The male has bluish-grey forewings and an orange head and thorax. The female is entirely yellow apart from two dark spots on each forewing. Resident populations of this species in Britain and south-western Ireland inhabit broadleaved woodland. Adults are attracted to light in small numbers. The main flight period in Britain is July to September. The larvae, which overwinter, feed on lichens, especially dog lichens, from autumn until the following June.

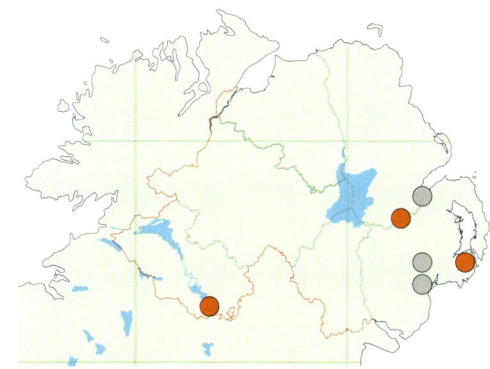

Wood Tiger
Parasemia plantaginis (Linnaeus) SOCC (P) Category two.
Scarce. This attractive day-flying moth has been recorded in every county but it is known from just a few sites in each. It seems to have declined in recent times probably due to loss of habitat, but it is also an elusive species that appears absent from many apparently suitable sites. On the north coast it occurs at Portstewart Dunes and the Umbra (Londonderry) and Cushleake Mountain, Fair Head, and the west end of Rathlin Island (Antrim). It was recorded by Stelfox in 1949 on the Garron Plateau (Antrim). The only known sites in Down are Leitrim Lodge and Murlough NNR. In Armagh, there are records from several areas within Peatlands Park and a few neighbouring bogs. The only Tyrone locality is Teal Lough. Finally, in Fermanagh, it is known from a number of sites including Braade, Belmore Mountain, Cavans Bog, Doon Forest, Knockmore and Monawilkin. Adults are distinctive and unmistakable. The forewings are black with elongate, creamy patches. The hindwings are yellow with black lines and spots. Although its name suggests an association with woodland it is, in fact, found on bogs, wet heathland, sand dunes and calcareous grassland. The males, which are often difficult to catch, are day-flying in warm sunshine, actively seeking out newly-emerged females. The females do not fly until late afternoon or after dark. Adults are frequently disturbed during the day from low vegetation, especially in dull overcast conditions. The recorded flight period is 5 June to 3 July. The larvae, which overwinter, feed from July until April of the following year on a wide variety of herbaceous plants.

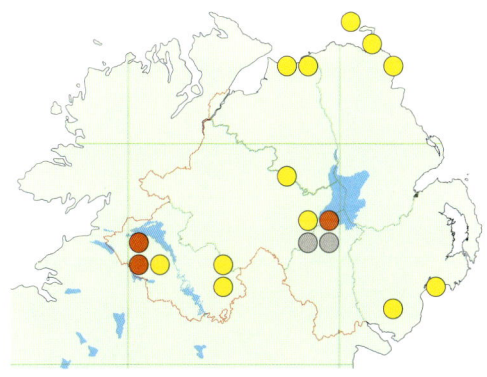

Garden Tiger
Arctia caja (Linnaeus) Category one.
Ubiquitous. This large, colourful and instantly recognisable species is widespread in all counties, but less abundant than it has been in the past. It has been recorded at several sites on Rathlin Island (Antrim) and at Copeland Bird Observatory (Down). Both sexes are similar in appearance. The forewings are dark chocolate brown with white streaks and patches, which are variable in size and shape. The hindwings are orange-red with a series of dark blue spots outlined in black. These are a distinctive feature of this species but are only displayed when the adults are disturbed. Some of the aberrations known in this species have been seen in Northern Ireland. For example, one in which most

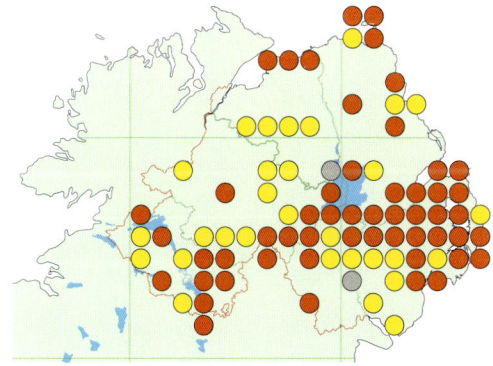

of the chocolate brown markings on the forewings have been replaced with white, has been seen at Killard Point (Down). Garden Tiger is found in a wide variety of habitats including woodland, bogs, coastal heaths and dunes as well as large gardens. The adults are seldom found by day but appear frequently at light. Many of the records are of larvae, which are commonly referred to as 'woolly bears' or 'granny greybeards'. These are often seen wandering across paths and roads in search of a suitable pupation site by members of the general public. The monitoring data from the Rothamsted trap network in Britain indicates a decline in abundance of this species of 30 percent between 1968 and 2002. The recorded flight period is 13 June to 2 September. The distinctive larvae feed from September until June of the following year on a wide variety of wild and garden plants. It overwinters as a larva.

Garden Tiger.

Derrylin, County Fermanagh.

A striking moth more frequently encountered as a larva than as an adult.

Clouded Buff
Diacrisia sannio (Linnaeus) Category two.
Notable. Restricted to a few localities in the west and north of Fermanagh at Blackhill Bog, Correl Glen, Garvary Wood and Glennasheevar. There is also an historical record from the Belfast area in the 1950s, and an adult was seen at Copeland Bird Observatory (Down) in 2004. This latter is considered an immigrant, possibly from Scotland, which is less than 40 kilometres away. In view of its short flight period and remote haunts, the species is probably under-recorded, especially in western counties. The Irish distribution is also western and southern. The adults are brightly coloured and unmistakable. The males, which are larger than the females, have yellow forewings with pink fringes and a reddish discal mark. Females have dark orange forewings with prominent dark veining and a smaller discal spot. The Clouded Buff is found on upland and lowland bogs and heaths. Adults are easily disturbed from the vegetation where they rest during the daytime. Males fly during the day but only in warm sunshine; the females are seen less often and tend not to fly before dark. The Northern Ireland records have all been in the two-week period from 12 to 26 June but in Britain the flight period lasts well into July. The larvae feed on Heather and other low-growing plants from August until May of the following year.

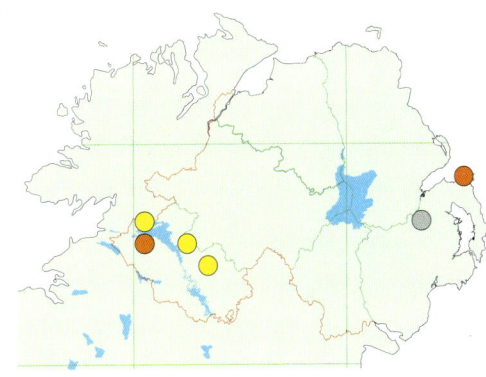

White Ermine
Spilosoma lubricipeda (Linnaeus) Category one.
Ubiquitous. This is one of the most commonly recorded arctiids in Northern Ireland. It has been seen in all counties and it is known from Copeland Bird Observatory (Down) and a number of places on Rathlin Island (Antrim). The majority of adults have white forewings but it is common to see individuals with cream or buff forewings. The wings have small, black dots, but there is much individual variation in the number and intensity of these markings. The head and thorax are both white and quite hairy and the abdomen has yellow and black markings. This species bears a close resemblance to Muslin Moth, but this has fewer black spots on the wings and a white and black abdomen. White Ermine occurs in many habitats with patches of coarse, tall vegetation including hedgerows, woods, rural and urban gardens. Adults are attracted to light and appear frequently in traps in moderate numbers. The recorded flight period is 1 May to 5 September. The larvae are polyphagous, feeding during August and September on a wide variety of wild and garden plants including docks and Common Nettle. It overwinters as a pupa.

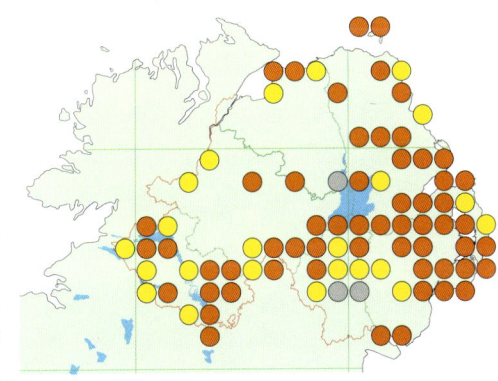

Buff Ermine
Spilosoma luteum (Hufnagel) Category one.
Widespread. Recorded from all counties and from Copeland Bird Observatory (Down) and Rathlin Island (Antrim). Buff Ermine is a distinctive and easily recognised moth. The forewings are buff with small black markings, which vary in number and intensity from moth to moth. In most individuals the black markings form a curved line across the forewings, but in others the markings may be absent. The head and thorax are buff and rather hairy. Females are generally paler than males but otherwise similar in appearance. This is a generalist species that can be seen in most habitats. Adults are attracted to light

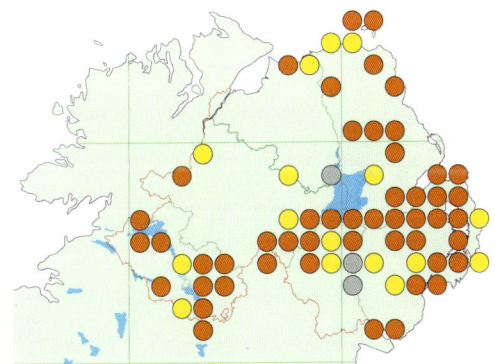

Clouded Buff.

Glennasheevar, County Fermanagh.

White Ermine.

Castlewellan Forest, County Down.

and appear frequently in moderate numbers in traps. The recorded flight period is 14 May to 4 August. The larvae feed during August and September on herbaceous plants including Common Nettle, and woody plants such as Honeysuckle and birch. It overwinters as a pupa.

Muslin Moth
Diaphora mendica (Clerck) Category one.

Widespread. A commonly recorded species across north Armagh and north Down but much more local elsewhere. It is apparently absent from south Down and much of the north and west of Antrim, Fermanagh, Londonderry and Tyrone. The northern limits to its distribution in Northern Ireland are currently Ballymena in Antrim and Banagher Glen in Londonderry. The forewings are creamy white in the male and silky white in the female. There is some variation in colouring, with some adults having a slight ochreous tint. The abdomen is white with a series of black marks along the dorsal surface. Adults resemble those of White Ermine, but are usually smaller and with less extensive black markings on the forewings. Muslin Moth is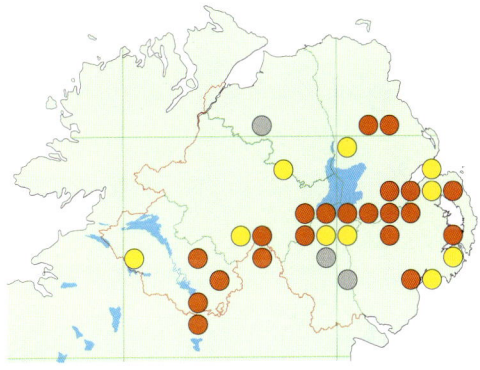
chiefly found in damp, open woodland, bogs, fens and marshy areas. Adults, especially males, are frequently seen at light but in small numbers. The recorded flight period is 11 April to 7 July. The larvae feed from July until September on a wide variety of herbaceous plants including docks and plantains. It overwinters as a pupa.

Ruby Tiger
Phragmatobia fuliginosa (Linnaeus) Category one.

Widespread. This is a local species mainly found in north Armagh and Fermanagh. It is rarely seen in large numbers even at well-known localities. It is unrecorded from Londonderry and apparently absent from much of Tyrone with the exception of Favour Royal and Knockaraven. In Antrim it is found on Rathlin Island, Clare Wood and White Park Bay on the north coast, and at several other sites in the county. There are scattered records from Down, mainly in the east and south of the county at Derryleckagh, Killard Point and Murlough NNR. Sites in the west include Garvary Wood, Glennasheevar and Legatillida (Fermanagh). Ruby Tiger is mainly associated with coastal and inland heaths. Adults are said to fly occasionally in warm sunshine,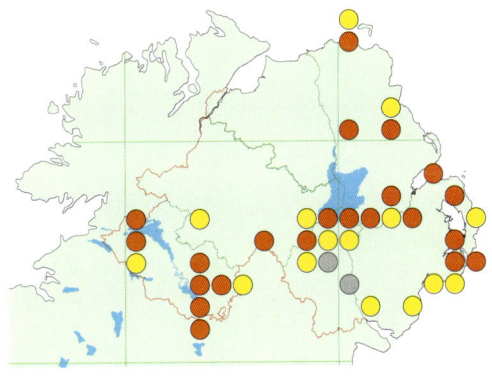
but they are more usually encountered at light in small numbers. In Britain, this moth is double-brooded in the south but single-brooded in the north of its range. The Northern Ireland phenology would suggest that it is double-brooded here with the first brood in May and June and the second from mid-July to the end of August. The recorded dates for the two broods are 17 May to 25 June and 17 July to 29 August. The larvae can be found from July until April of the following year on a variety of herbaceous plants including Heather, Common Ragwort, plantains and heathers. It overwinters as a larva.

Cinnabar
Tyria jacobaeae (Linnaeus) Category one.

Widespread. Generally distributed in Armagh, Down and Fermanagh and on the north coast including Rathlin Island (Antrim). Seemingly much scarcer in Tyrone and unrecorded from much of inland Antrim and Londonderry. Some of the biggest populations are in the west. All

The Butterflies and Moths of Northern Ireland

Cinnabar.

Murlough National Nature Reserve, County Down.

stages of the Cinnabar are colourful and very distinctive. The adults have greyish-black forewings with two large scarlet spots and a scarlet costal streak that runs almost to the apex. The hindwings are scarlet with the costa and cilia darkish-grey. Aberrations are occasionally seen and Donovan mentions specimens of ab. *flavescens* being taken in 1915 at Tempo (Fermanagh). There is also a specimen of this aberration, in which the scarlet is replaced by yellow, in a collection at the University of Ulster, Coleraine, collected near Forkhill (Armagh) in the mid-1980s. The yellow- and black-banded larvae are instantly recognised. Cinnabar occurs on sand dunes, grassland and heaths where its main foodplant, Common Ragwort, is found. Occasionally it will feed on other ragwort species, for instance, Marsh Ragwort in wetter habitats. Adults rest by day hidden in low vegetation but they are easily disturbed giving rise to the idea that they are day-flying. The moth appears frequently at light in small numbers. The recorded flight period is 6 May to 7 August. The larvae can be found easily by day from July to September communally on their foodplant. Their distinctive colouration deters birds and other predators and serves to remind them of their toxic nature. It overwinters as a pupa.

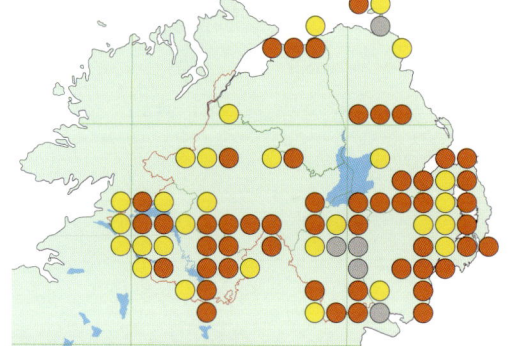

Moth accounts

Least Black Arches.

Castlewellan Forest, County Down.

NOLIDAE

The moths in this small family can easily be overlooked because of their resemblance to some micromoths. Four species are resident in Britain and Ireland and just one occurs in Northern Ireland. Adult colouration in this family is predominantly grey or white; all have rounded forewings with small raised scale tufts resembling small pyralids when viewed in close-up. They are seldom encountered in the daytime, but all are attracted to light.

The larvae are by nature quite hairy and feed on various trees and shrubs. The pupa is protected in a boat-shaped cocoon formed with fragments of bark and normally attached to the foodplant or occasionally to fences.

Least Black Arches
Nola confusalis (Herrich-Schäffer) Category two.
Widespread. Recorded from all counties, but mainly found south of a line from Belfast Lough to the north of Lower Lough Erne. It is not uncommon in north Armagh, Down and south Fermanagh but much more local in Antrim and Londonderry. Sites where it has been recently recorded include the Argory and Peatlands Park (Armagh); Belvoir Park, Castlewellan Forest, Helen's Bay, Murlough NNR and Seaforde (Down) and Crom, Garvary Wood and Legatillida (Fermanagh). It is known from a few scattered sites in Tyrone, including Cullinane, Davagh Forest and Rehaghy Mountain. In Antrim it has been recorded from Ahoghill, Derryola Bridge and Portmore Lough and, more recently, from gardens in Ballymena, Finaghy and Lisburn. There has

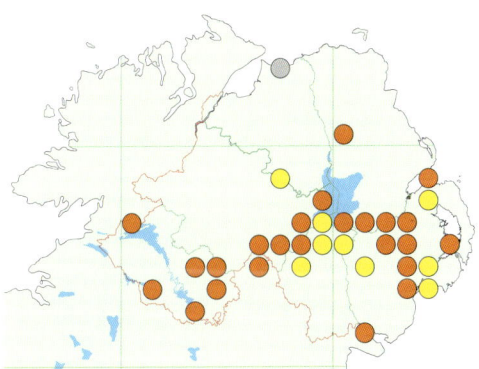

been just a single record from the north of the region at the Umbra (Londonderry) in 1970. This is a small micro-like moth with whitish forewings, tinged pale green and with a pale brown suffusion on the outer edge; the green fades to white with age. The median and postmedian cross-lines are black and scalloped; the latter curves strongly towards the costa. Least Black Arches is probably more widespread than the available records suggest. It is easily overlooked, especially by inexperienced recorders, due to its small size. Adults are attracted to light in small numbers and reportedly rest by day on the trunks of trees, branches and possibly the underside of foliage. It is mainly a woodland species, but it also occurs in other habitats including old orchards, parkland and occasionally, gardens. The recorded flight period is 30 April to 17 July. The larvae, which have never been observed in Northern Ireland, may be found from late July and throughout August on birch, oak, lime, Blackthorn and Beech. It overwinters as a pupa.

NOCTUIDAE

This is the largest family of macromoths in the northern Irish fauna with just over 200 recorded species. The Irish checklist includes 235 species and there are over 400 resident species in Britain. Adults are generally of medium size and rather thick-bodied; most have long forewings which are slightly overlapped at rest, thus giving the moth an elongated appearance. Most noctuid species are generally some shade of brown with various visible markings, the most noticeable being the elliptical and kidney-shaped marks in the central area of the forewing. The shape and structure of these markings are helpful in identification. The markings and the subdued colour of most species aids camouflage during the day. Many noctuids are capable of long-distance flight and some are regular migrants. They feed on nectar from flowers, honeydew and fermenting fruit and can be lured to artificial baits. Most species are single-brooded in Northern Ireland.

The vast majority of noctuids are attracted to light. Some species such as Large Yellow Underwing often occur in prodigious numbers in traps during the summer months, making examination of the contents a laborious task. The larvae of most species are generally smooth-skinned and rather mundane in appearance. They feed mostly on the roots of various grasses, herbaceous plants and also the leaves of trees and shrubs. The vast majority of larvae feed at night, but remain concealed during the day among leaves either on the foodplant or on the ground. Most species overwinter as pupae below ground or among leaf litter, but there are some that pass the winter months as larvae, feeding periodically when conditions are suitable. A small number of species hibernate as adults, reappearing in early spring.

Square-spot Dart
Euxoa obelisca (Denis & Schiffermüller) SOCC Category three.
Rare. This moth has only been reliably recorded in recent times from Rathlin Island (Antrim) and Copeland Bird Observatory (Down). It was discovered on Rathlin Island in 1988 and was found widely on the island in 2001. The first record from Copeland Bird Observatory came in 2003. There is an 1893 report of the species at Magilligan (Londonderry) by Campbell; however, Donovan refers to this as being a mistake. Adults have chocolate brown forewings with a slight reddish tinge, but this is prone to much variation. There is a black, dart-shaped mark on the forewing containing the pale reniform and orbicular stigmas. It closely resembles White-line Dart from which it is sometimes indistinguishable; however, Square-spot Dart lacks the

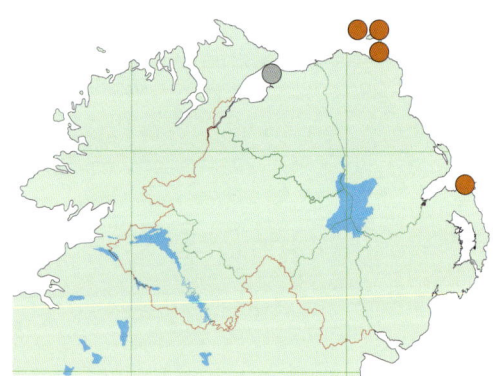

black arrowheads on the forewings and the pale shoulder patch is never as prominent as it is in White-line Dart. Square-spot Dart is a coastal species that favours rocky habitats. Adults come to light in moderate numbers and also visit the flowers of Common Ragwort and Heather. The Northern Ireland records have been between 20 July and 28 August, but it probably has a longer flight period than these few records indicate. In Britain, adults fly from early August to October. The larval history and the foodplant of this species are unknown in Northern Ireland.

White-line Dart
Euxoa tritici (Linnaeus) Category three.
Notable. A coastal species in Northern Ireland, recorded from scattered sites with suitable habitat on the coasts of Antrim, Down and Londonderry as well as Rathlin Island (Antrim). Sites include White Park Bay (Antrim); Ballykinler, Killard Point and Murlough NNR (Down) and Magilligan and the Umbra (Londonderry). There are inland records from Glenariff Forest (Antrim) and suburban gardens in the Greater Belfast area. There are historical records from Killymoon and Stewartstown (Tyrone) by Greer, but there are no recent reports from this county. The adults are extremely variable both in colour and markings, but they tend to match the habitat they live in; those from sandy sites are the palest. Most individuals have a pale, costal stripe and 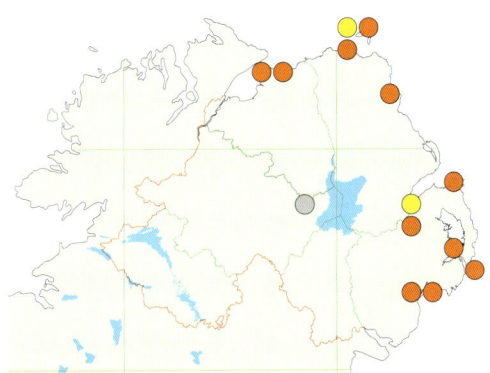 whitish cross-lines on the forewings. The stigmas are usually present but both vary in size. The most constant and reliable feature is the series of small dark, wedge-shaped markings near the outer margin of the forewing. White-line Dart is found on sand dunes, coastal heathland and moorland. Adults appear commonly at light and visit flowers especially Common Ragwort and Heather. The recorded flight period is 9 July to 25 September. The larvae feed mainly at night from late March to early July on small herbaceous plants including bedstraws and mouse-ears. It overwinters as an egg.

Garden Dart
Euxoa nigricans (Linnaeus) Category three.
Scarce. A very localised species largely confined to Down. It has been recorded in all counties, but there have been no recent records from Londonderry or Tyrone. Greer recorded it from Killymoon, Lough Fea, and Lough Neagh (Tyrone) in the 1920s. The most recent records have come from Belfast Harbour Estate, Copeland Island, Helen's Bay and Murlough NNR (Down) and suburban gardens in the Greater Belfast area. The monitoring data from the Rothamsted trap network indicates a decline in abundance of this species in Britain of 97 per cent between 1968 and 2002. This is quite a variable species, best distinguished from similar species by its relatively uniform appearance and the absence of both a pale costal streak and dark wedges on the forewings. Adults can 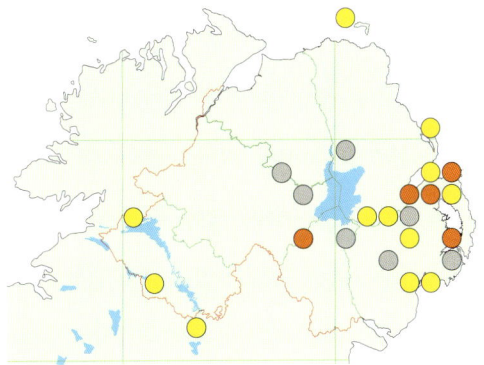 range in colour from pale to dark brown, but some individuals found in Northern Ireland appear almost black. Garden Dart occurs – as its name suggests – in gardens, but also in open, ruderal habitats, dunes and heaths. Adults are attracted to light in very small numbers and also feed at flowers, especially Buddleia and Common Ragwort. The recorded flight period is 30 June to 2 September. The larval foodplants are clovers, docks, plantains and other herbaceous plants. It overwinters as an egg.

Coast Dart

Euxoa cursoria (Hufnagel) Category three G.

Unknown. Baynes gives two old nineteenth-century records from Newcastle (Down) in 1895 and Magilligan (Londonderry) around 1875. The validity of these records is questionable, as no voucher specimens appear to exist for either record. There has been some recent effort in rediscovering the species, but this has not produced any confirmed records. This species is extremely variable in appearance and individuals may require confirmation by examining the genitalia. The forewing colour varies from pale straw to greyish-brown, but with a sandy tint. The reniform stigma is generally paler than the rest of the forewing with a distinct white outline. As its name indicates, Coast Dart is a coastal species that occurs in the slacks and on stable, well-vegetated yellow dunes of intact dune systems. In Ireland the range is western, from Donegal to Kerry. The most recent confirmed record is from Mayo. The adults are attracted to light and flowers such as Common Ragwort and Marram. The flight period in Britain extends from late July to late August and occasionally into early September. The larvae are nocturnal and feed in spring and early summer on herbaceous plants and grasses including Sea Sandwort, violets and Early Hair-grass. It overwinters as an egg.

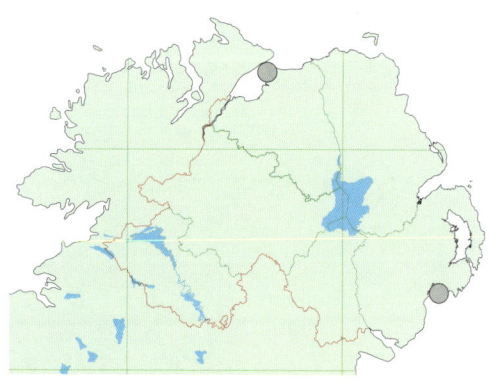

Light Feathered Rustic

Agrotis cinerea (Denis & Schiffermüller) Category four.

Unknown. Recorded once at Tullylagan (Tyrone) by Greer in June 1897. The voucher specimen for this record (which was the first for Ireland) is in the collections of the National Museum of Ireland in Dublin. It was also reported from Aghalee (Antrim) by Wright in 1958. This species appears to be rare throughout Ireland and apart from the Northern Ireland sites, is confined to Kerry in the south-west. Adults are variable in colour with individuals ranging from light brown to grey. The forewings have small reniform stigmas and dark dentate cross-lines with several dark spots along the costa. Light Feathered Rustic is associated with sparsely-vegetated grassy habitats such as abandoned industrial sites, coastal cliffs, shingle beaches and rocky hillsides. Males come to light more often than females which are seldom seen. Adults become active at dusk and visit flowers such as Red Valerian. The larvae reportedly feed on Wild Thyme from July to September. The flight period in Britain is May and June. It overwinters as a mature larva.

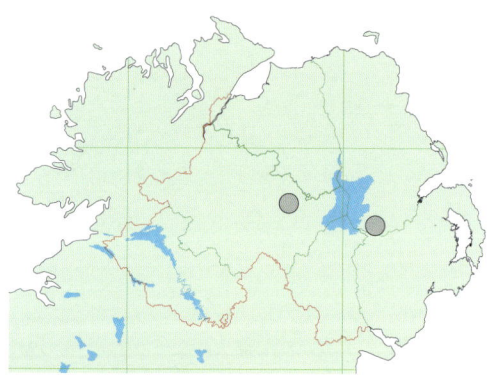

Archer's Dart

Agrotis vestigialis (Hufnagel) Category two.

Rare. Confined to a few coastal localities in Down and Londonderry. It has been recorded regularly from Ballykinler Dunes, Killard Point and Murlough NNR (Down) and Magilligan and the Umbra (Londonderry). An inland population (the only one in Ireland) existed on Lough Neagh, but the last record from here was in 1916 from Washing Bay (Tyrone) in the south-west corner of the lough. Although, like related *Agrotis* species, the adults are prone to much variation in colour and markings, this is the most distinctive of the Irish darts. The forewings vary from pale grey to dark brown with prominent dark markings. The stigmas are usually darker and there is

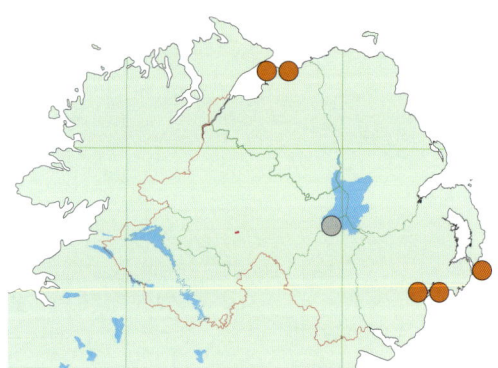

Archer's Dart.

Murlough NNR, County Down.

A colourful species which has been recorded from a few coastal localities in Down and Londonderry.

a series of dark, wedge-shaped markings near the outer margin of the forewing. Archer's Dart seems in recent times to be restricted to sandy habitats on the coast. The Lough Neagh population presumably was found on the sandy beaches of the lake, now largely disappeared due to the changes of the water levels. Adults are frequently seen at light in moderate numbers. Individuals can be found by day, usually in early morning, at rest on low-growing vegetation and on flowers, including Common Ragwort and Heather. The recorded flight period is 19 June to 25 September. The larvae, which overwinter, feed from September until the following May on herbaceous plants including bedstraws, stitchworts and Sea Sandwort.

Turnip Moth.

Crom, County Fermanagh.

Changes in agricultural practices and increased use of insecticides has probably contributed to a decline of this moth in some areas.

Turnip Moth
Agrotis segetum (Denis & Schiffermüller) Category two.
Scarce. The records are mainly concentrated in north Armagh at Aughinlig, Oxford Island and Peatlands Park and around Belfast, with just a few scattered records from Tyrone, at Favour Royal and Stillago. It appears to have shown a marked decline and its present strongholds are around the southern part of the Lough Neagh basin, but it is nowhere common. Adults show some variation in colour. The forewing is mainly light brown and the stigmas are usually darker. The most distinctive feature is the pearly white colour of the hindwings. Turnip Moth is found in cultivated areas especially arable farmland and gardens, but also natural habitats including woodland clearings and parkland. Adults are attracted to light but appear infrequently. There

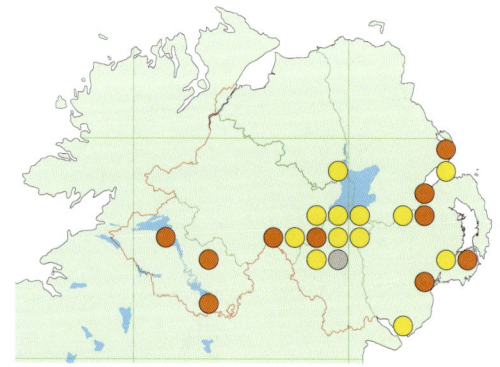

are two broods, one in spring to midsummer (6 May to 31 July) and a second in early autumn (4 September to 27 October). The larvae feed during the autumn until spring and again in summer on cultivated plants, particularly root vegetables (including turnip but also swede, cabbage and carrot). The effect of the use of insecticides and the reduction in the acreage of root crops has probably reduced the Northern Ireland population.

Heart and Club
Agrotis clavis (Hufnagel) Category four.
Unknown. There have been eight confirmed records from Londonderry (Castlerock and Derry City), Tyrone (Stewartstown), Antrim (Ballymacilrany) and Down (Belvoir Park and Murlough NNR). The lack of any recent records suggests this species is an occasional migrant or is now

most probably extinct. Elsewhere in Ireland, it is chiefly found in the south and east. The adults are variable in colour ranging from pale greyish-brown to dark brown with well-defined stigmas. It is similar in appearance to the much commoner Heart and Dart, but has a more rounded central dart and lacks the dark thoracic collar. Heart and Club is found in coastal dunes and similar dry habitats. Adults are attracted to light and will visit flowers such as Red Valerian. In Britain it flies from mid-June to early July, but the flight period in Northern Ireland is unknown. The larval foodplants are species of dock and possibly other herbaceous plants. It overwinters as a larva.

Heart and Dart
Agrotis exclamationis (Linnaeus) Category one.
Widespread. Recorded from all counties and on Rathlin Island (Antrim) but especially common in the east. It appears to be generally scarcer in western counties. Adults show some variation in the ground colour. The forewings are mainly light brown and the stigmas are usually slightly darker and well defined. There is also a dark thoracic collar visible on most individuals. Heart and Dart can be found in many sites but appears especially common in highly modified habitats such as gardens. Adults appear commonly at light, often in good numbers and are also attracted to the flowers of Common Ragwort, Red Valerian and Buddleia. Individuals reportedly rest concealed low down among ground vegetation and are seldom found during the daytime. The recorded flight period is extensive from 29 April to 2 September, peaking in late June. The larvae feed on a wide variety of plants. It overwinters as a fully grown larva.

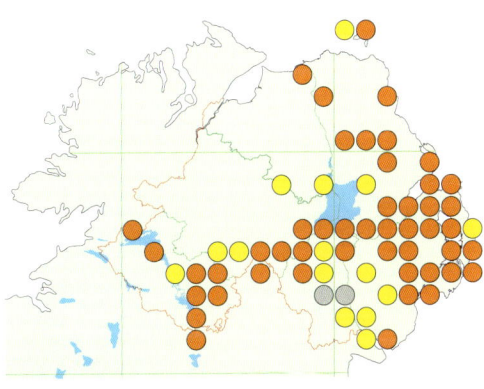

Crescent Dart
Agrotis trux (Hübner) Category four.
Rare. Following its discovery in 1973 at Copeland Bird Observatory (Down), this species went unrecorded in Northern Ireland until it was relocated on the same island in 2004. Adults were recorded successively between 14 and 17 July. A single adult was also taken at Castleward (Down) on 24 July in the same year. These records confirm its continued existence in Northern Ireland. It is rare in the rest of Ireland and confined to east and south coastal counties between Kerry and Louth. Adults show some variation in the ground colour. The forewings range from light to dark grey brown. The stigmas are usually paler than the ground colour and have well-defined black borders. Crescent Dart occurs in Britain on dunes, cliffs and low rocky shores, and so there would appear to be suitable habitat for it at Copeland Bird Observatory. Adults are attracted to light and also to sugar. The flight period in Northern Ireland is unknown. In Britain it flies during July and August. The larval habits and foodplants are not known in Northern Ireland. In Britain, the larvae have reportedly been recorded on Thrift and Rock Sea-spurrey. It overwinters as a fully-grown larva.

Dark Sword-grass
Agrotis ipsilon (Hufnagel) Category two.
Widespread. An immigrant species and short-term resident, recorded from all counties and Rathlin Island (Antrim); the majority of recent records emanate from Armagh and Down. It

has been regularly recorded from Helen's Bay, Hillsborough Forest, Killard Point and Rostrevor Wood (Down). Further west and north there are records from Crom and Garvary Wood in Fermanagh and the Umbra on the Londonderry coast. There are reports virtually annually. Like all *Agrotis* species, the adults vary in colour but this species is readily identifiable by size, as it is easily the largest member of the genus in the region. There is a conspicuous black dart on the forewings that extends distally from the reniform stigma. Adults are attracted to light and can turn up in any habitat at any time between April and November. The recorded flight period is 3 April to 27 November with the peak period in August and September. The larvae are considered polyphagous in the wild, but are seldom encountered.

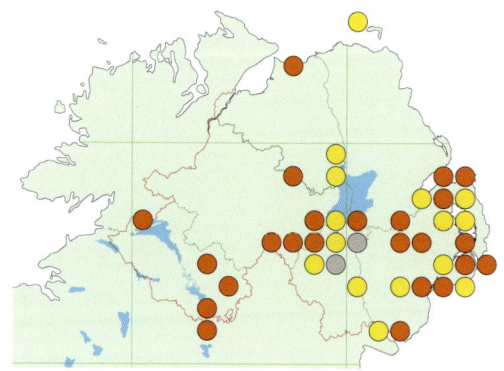

Shuttle-shaped Dart

Agrotis puta (Hübner) Category four.

Rare. This species has been recorded on three occasions in Northern Ireland; all the records are considered to have immigrant individuals. The first record was by Partridge in 1892 at Enniskillen (Fermanagh) and the second from Helen's Bay by A.F. O'Farrell in 1947. It was not reported again until May 2001, when a single adult was taken in a mature suburban garden in the Greater Belfast area. There has been just one other authenticated Irish record in Cork in August 1984. The other records mentioned in Donovan (from Louth and Westmeath) are considered incorrect. Adults are quite variable in colour, ranging from pale greyish-brown to a more uniform darker brown. The reniform stigma is darker than the ground colour and the cross-lines are very pale or indistinct. Adults are attracted to light and flowers, especially Common Ragwort and Buddleia. In Britain, there are at least two broods and adults can be encountered between April and October. The larval foodplants are Dandelion, docks and Knotgrass. It overwinters as a larva.

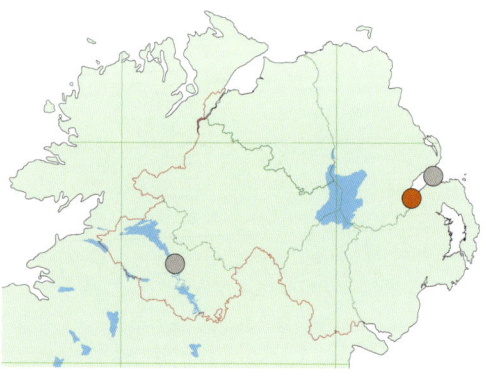

Flame

Axylia putris (Linnaeus) Category one.

Widespread. Recorded in all counties, but seems to be most common in the southern counties of Armagh, Down and Fermanagh. More local further north where there are recent records from Glenariff, Portglenone and Rathlin Island (Antrim). It has also been taken in several suburban gardens in Antrim and Londonderry. Frequently recorded sites in the south include Oxford Island and Peatlands Park (Armagh); Killard Point (Down); Aghalane, Crom, Garvary Wood and Legatillida (Fermanagh) and Rehaghy Mountain (Tyrone). Adults are distinctive and rest with their wings wrapped tightly around the body, resembling a piece of dead stem or a broken fragment of wood. The forewings are usually straw-coloured with a dark brown costal streak. Flame is found in a wide range of habitats including woodland margins, heathland, gardens and hedgerows. Adults are frequently encountered at light and occasionally visit flowers. The flight period is prolonged, from early May to early October, with a peak in June and July. The recorded flight period is 25 April to 11 November. The phenology does suggest a partial second brood in late September. The larvae feed at night on herbaceous plants including Common Nettle. It overwinters as a pupa.

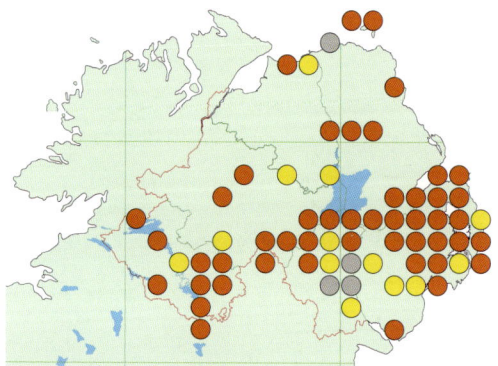

Moth accounts

Shuttle-shaped Dart.

Belfast, County Antrim.

A rare immigrant that is seldom encountered in Northern Ireland.

Flame.

Tattanellen Bog, County Tyrone.

A small dull-coloured moth which resembles a fragment of wood when at rest.

Portland Moth.

Belmullet, County Mayo.

Much effort has been expended to relocate this species in Northern Ireland, so far without success. It was last reported in 1978.

Portland Moth
Actebia praecox (Linnaeus) Category four.

Rare. Recorded from a few locations in Down and on the north Antrim coast, but it has not been seen anywhere in Northern Ireland for over 25 years and it is perhaps extinct. Some of its occurrences in Northern Ireland could relate to immigrants. The only accepted records are from Portballintrae and Portrush (Antrim) in 1903, Tollymore in 1973 and Murlough NNR in 1978, both in Down. No voucher specimen exists to validate the last record, but it is generally accepted as genuine. Foster described it as being abundant on the dunes at Murlough NNR during the early 1930s; Wright also reported it as common here in the 1950s. Its decline at this site may be due to changes in the vegetation and a decrease in the extent of bare sand, an essential requirement of the larvae. 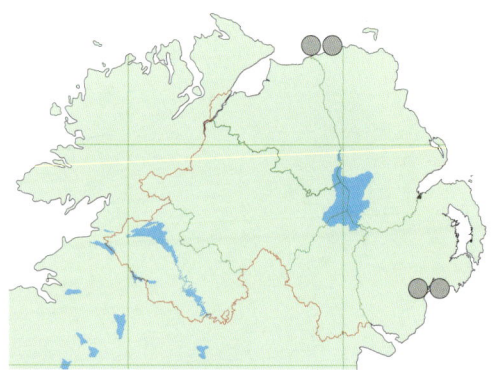 The species has been searched for on the dunes at Ballykinler just north of Murlough NNR, where some suitable habitat still exists, but without success. Elsewhere in Ireland, Portland Moth is found along the west coast in Donegal, Mayo, (where it was recently recorded in small numbers), Galway, Clare and Cork. This distinctive, long-winged moth is easily recognised by the shiny, silvery-green reflective colour of the forewings. All the Irish records are from large sand dune systems or machair. Adults are attracted to light and also flowers, especially Common Ragwort and Heather. There is insufficient data to give a reliable flight period in Northern Ireland. In Britain, it flies from mid-August and throughout September. The larvae have never been found in Northern Ireland. They are nocturnal and rest by day concealed in the sand near the foodplant. The larval foodplants are Creeping Willow and several other plants including chickweeds and Bird's-foot Trefoil. It overwinters as a larva.

Flame Shoulder
Ochropleura plecta (Linnaeus) Category one.

Widespread. Frequently recorded in all counties and on Rathlin Island (Antrim). This species is easily identifiable by the chestnut brown colouration and the pale costal streak that runs from the basal area of the forewing and fades out well short of the apex. This is further highlighted by the black streak that runs alongside it. Flame Shoulder can be found in many habitats including woodland, marshes and gardens. Adults appear frequently at light and also visit various flowers, particularly Common Ragwort. The recorded flight period is 1 May to 18 October, but it is most frequently recorded in June and July. The phenology indicates that the species is single-brooded here. The larvae feed throughout the summer until late autumn on a wide variety of 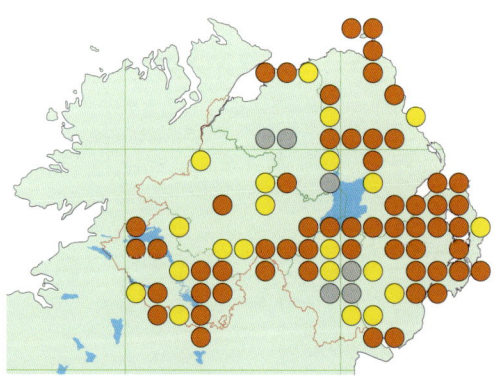 herbaceous plants including Ribwort Plantain, Groundsel and bedstraws. It overwinters as a pupa.

Northern Rustic
Standfussiana lucernea (Linnaeus) Category three.

Notable. This species is known from several sites on Rathlin Island (Antrim), where it appears common and widespread. There are also records from Helen's Bay and Killard Point (Down) in 1995 and 1999, respectively. It was recorded at Cushendall (Antrim) in 1972 and Wright reported it from Ballymacilrany, near Portmore Lough, in 1951 in south Antrim. This last record is the only confirmed inland occurrence. The forewing colour in the adults is variable, from pale ochreous-grey to dark brown. There is a series of indistinct bands on the forewings which is obvious on pale individuals, but hardly discernible on darker specimens. The Northern Rustic is primarily a coastal species found on rocky coasts and cliffs. Adults are attracted to light but never commonly. They visit the flowers of Heather and Red Valerian. The recorded flight period is 15 June to 16 August. The larval habits are unknown in Northern Ireland. In Britain, they feed at night from October until spring of the following year on grasses and herbaceous plants including Harebell, stonecrop and species of saxifrage. It overwinters as a small larva.

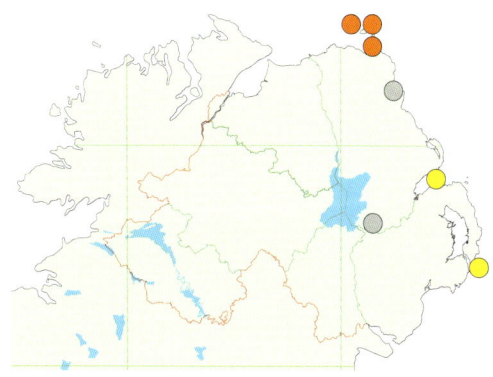

Large Yellow Underwing
Noctua pronuba (Linnaeus) Category one.
Ubiquitous. This is one of the most common and widespread moths in Northern Ireland, recorded from all counties and on Rathlin Island (Antrim). The forewing colour of the adults is quite variable, ranging from pale to dark reddish-brown. Dark individuals are usually males. The forewing is long and narrow with a small, but conspicuous spot, close to the apex. The hindwing is orange-yellow with a black border which is conspicuous in flight. Large Yellow Underwing can be seen in most habitats, but is particularly common in lowland grasslands and similar open habitats, including gardens of all types. Adults appear commonly at light, frequently in large numbers and midsummer catches are often dominated by this species. They also visit the flowers

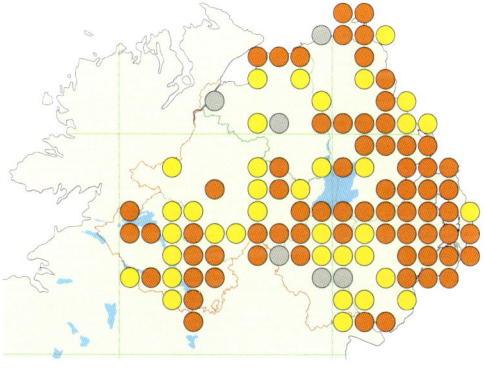

of Common Ragwort, Red Valerian and Buddleia. They are frequently seen during the day as they are often disturbed from their resting position in low vegetation. The recorded flight period is 23 May to 25 October, with the peak period for records being between mid-July and mid-August. The resident population is often increased by migrants. The larvae feed on grasses and herbaceous plants from September to April. It overwinters as a larva.

Lesser Yellow Underwing
Noctua comes (Hübner) Category one.
Ubiquitous. A common moth found in all counties and on Rathlin Island (Antrim). Adults show much individual variation in the colour and the intensity of the markings on the forewings. The ground colour of the forewings varies from pale brown to dark greyish-brown. The hindwings are orange-yellow with a broad black band and a central dark crescent, which separates it from the similar Large Yellow Underwing. Lesser Yellow Underwing occurs in most habitats, including gardens. Adults are attracted to light and often appear in large numbers. It rests by day concealed amongst low vegetation and is seldom seen. Adults also visit the flowers of Common Ragwort, Buddleia and Heather. It is single-brooded and the recorded flight

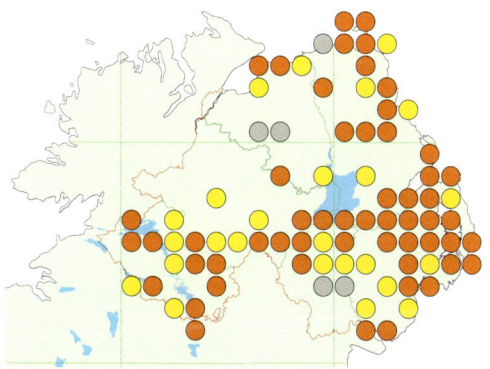

period is 10 June to 15 October with a peak in early August. The larvae are nocturnal and feed from September to April on Common Nettle, docks, Hawthorn, Bramble, Broom, willows and probably other trees and shrubs. It overwinters as a larva.

Broad-bordered Yellow Underwing
Noctua fimbriata (Schreber) Category two.

Scarce. There are records from all counties, but the majority are from the south-east. There has been an increase in inland records in the north and west suggesting it has been overlooked in many areas. It has also been recorded on Rathlin Island (Antrim). This is a richly-marked species. Forewing colour in the male varies from reddish-brown to olive brown, while the female is much lighter and pale brown or light greenish-brown. The stigmas are conspicuous and outlined with a fine whitish line. The hindwings are orange, with a broad black band that tapers towards the inner margin. It can be found in habitats with broadleaved trees, including parks and large gardens. Adults appear frequently at light, usually in small numbers. The recorded flight period is 23 June to 17 September. Adults aestivate after emergence and normally reappear in August, which is the peak period for records. The larvae feed from September to May of the following year on Common Nettle, Primrose, Bramble, Blackthorn, willows, and docks. It overwinters as a larva.

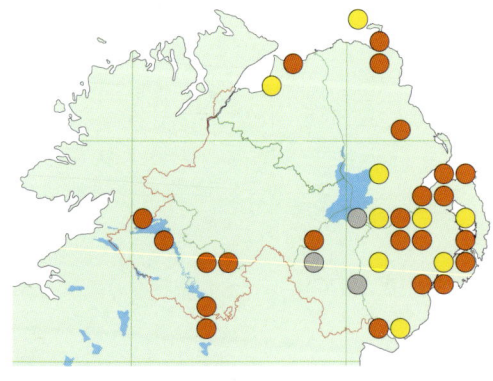

Lesser Broad-bordered Yellow Underwing
Noctua janthe (Borkhausen) Category one.

Ubiquitous. A common species found throughout Northern Ireland and on Rathlin Island (Antrim). Adults can be quite variable in colour. The forewings are usually purplish-brown suffused with grey. There is a small, fine, pale greenish-brown collar behind the head which is diagnostic. This colour extends part way down the costa, but fades with age. The stigmas are normally outlined with a pale fine line but are not always clearly defined. The hindwings are more yellow than the other species in this genus. Lesser Broad-bordered Yellow Underwing is found in most lowland habitats, including gardens. Adults are attracted to light and also visit flowers. This is one of the most frequently encountered moths in garden traps. The recorded flight period is 15 June to 15 October, with a peak in late July and early August. The larvae, which overwinter, feed from September to May on a wide range of herbaceous and woody plants including docks, Blackthorn, willows and Hawthorn.

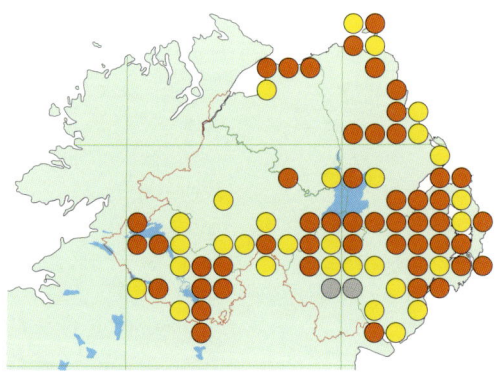

Least Yellow Underwing
Noctua interjecta Hübner Category three.

Scarce. This species has a curiously patchy distribution in Northern Ireland. It occurs at coastal localities in south Down, including Ballyquintin, Killard Point and Murlough NNR, and inland in Fermanagh, south Tyrone and north Armagh. Fermanagh sites include Correl Glen and Garvary Wood in the north and Crom in the south. Other sites include Peatlands Park (Armagh) and Tattenallen Bog (Tyrone). At many of its localities (which include some that have been trapped for many years), the species has only recently been discovered, indicating an expansion in range. Baynes makes reference to an early nineteenth-century record from Glenarm (Antrim), but there have been no recent records from here. This is a much smaller species than the other 'yellow underwings'. The forewings are normally reddish-brown or clay-coloured with poorly-defined stigmas and cross-lines and a dark subterminal line. The hindwings are

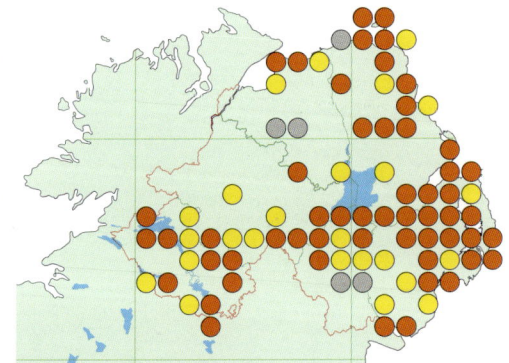

Lesser Broad-bordered Yellow Underwing.

Rostrevor Wood National Nature Reserve, County Down.

deep yellow with a broad black band and small discal lunule near the costa. The Least Yellow Underwing has been recorded from bogs, coastal heaths, dunes and especially broadleaved woodland. Adults are attracted to light in small numbers and also to the flowers of Common Ragwort and Lavender. The recorded flight period is 5 July to 25 September. The larvae, which overwinter, feed on herbaceous plants including Meadowsweet and grasses before hibernation, switching to willows and Hawthorn in the following spring.

Double Dart

Graphiphora augur (Fabricius) Category one.

Widespread. Recorded from all counties, but most commonly seen in Armagh at sites such as Brackagh Moss and Oxford Island and in Down at Belvoir Park and Helen's Bay. It is also frequent at Crom, Garvary Wood and Monmurry in Fermanagh, where most of the recent records originate. It was recorded at Copeland Bird Observatory (Down) in 2003. Other recent sites include the Ecos Park and Portmore Lough (Antrim); Magilligan (Londonderry) and Coolcush and Tattanellen Bog (Tyrone). The monitoring data from the Rothamsted trap network indicates a decline in abundance of this species in Britain of 97 per cent between 1968 and 2002. The forewings have a rather plain appearance and vary in colour from pale to dark brown with small black markings. The cross-lines 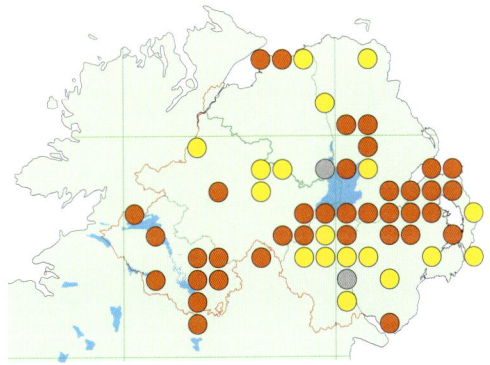 are scalloped and poorly-defined, occasionally reduced to a series of small dots. This species is found in many habitats, but especially broadleaved woodland, hedgerows, fens and bogs. Adults are attracted to light, usually in small numbers. The recorded flight period is 1 June to 15 August. The larval foodplants are willows, birch, Blackthorn and Hawthorn. It overwinters as a larva.

The Butterflies and Moths of Northern Ireland

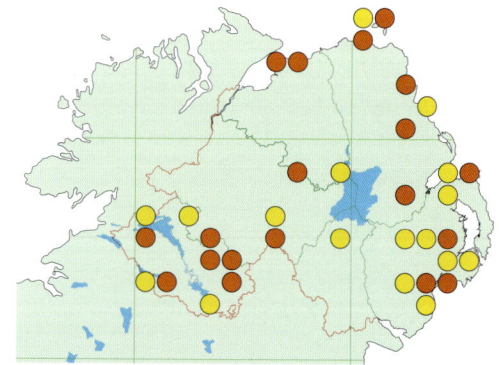

True Lover's Knot.

Lackan Bog, County Down.

Autumnal Rustic

Paradiarsia glareosa (Esper) Category two.

Widespread. This species has a patchy distribution, but there are records from all counties and it is also present on Rathlin Island (Antrim). Sites for it include Clare Wood, Glenariff Forest and Slievenacloy (Antrim); Turmennan Fen (Down) and Correl Glen and Garvary Wood (Fermanagh). Loughgall is the only locality in Armagh where it was last taken, in 1988. The forewings range in colour from pale grey to brownish-grey. The best distinguishing feature is the conspicuous black marking between the stigmas which is obvious even on the darkest individuals. The Northern Ireland records of this species come from lowland and upland heaths and moors, open woodland and fens. Adults appear sparingly at light, but also will visit flowers after dark, especially Heather and Common Ragwort. As its name suggests, this species flies late in the year. The recorded flight period is 3 August to 1 October. The larvae feed after dark on Heather, birch, willows and grasses. It overwinters as a small larva.

True Lover's Knot

Lycophotia porphyrea (Denis & Schiffermüller) Category one.

Widespread. A common species found throughout Northern Ireland and on Rathlin Island (Antrim) where it can be abundant. Commonly recorded sites include Glenariff Forest

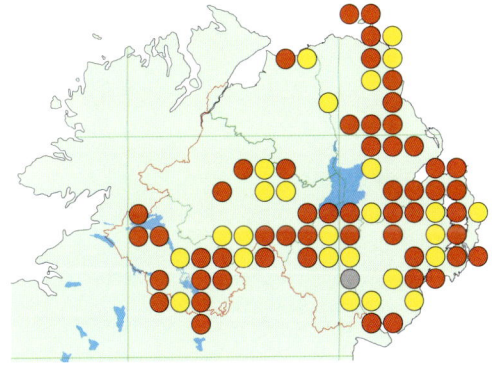

Autumnal Rustic.

Glenmore, Ballygawley, County Tyrone.

(Antrim); the Argory, Oxford Island and Peatlands Park (Armagh); Killard Point, Lackan Bog and Murlough NNR (Down); Crom and Garvary Wood (Fermanagh) and Rehaghy Mountain (Tyrone). Adult colouration and patterning can show some variation within individual populations. The forewings are normally deep reddish-brown with white stigmas and cross-lines; some individuals can have a mottled grey appearance. True Lover's Knot is found on upland and lowland heaths and bogs and, occasionally, in gardens. Adults appear frequently at light in moderate numbers and also visit flowers, particularly of Heather. Although mainly nocturnal, it can occasionally be found in late afternoon, especially in sunshine. The recorded flight period is 19 May to 8 September. The larval foodplants are Heather and Bell Heather. It overwinters as a larva.

Pearly Underwing
Peridroma saucia (Hübner) Category three.

Scarce. Each year since 1996 between one and seven individuals of this immigrant from southern Europe has been seen in Northern Ireland. Before this date, the species was much less commonly recorded. This may reflect a real increase, but it is more likely to be due to the increased level of moth recording. Records are concentrated in southern counties in Armagh, Down and Fermanagh, but there has also been one record from Rathlin Island (Antrim). Adult colouration is quite variable ranging from light buff to dark brown with a pale thoracic crest. The large reniform stigma is well defined and varies in colour relative to that of each individual moth. The hindwings are translucent with pale brown veining. Most records are in early autumn in September and October. The recorded flight period is 10 June to 1 November. Adults come often to light and will also visit flowers. This is an immigrant species only and it is highly unlikely that the larvae of this species could successfully overwinter in Northern Ireland.

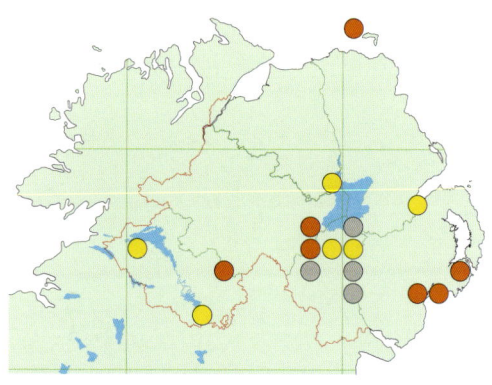

Ingrailed Clay
Diarsia mendica (Fabricius) Category one.

Ubiquitous. Recorded throughout Northern Ireland and on Rathlin Island (Antrim). Adults are extremely variable in colour and in the extent of the markings, even within a population from a single locality. The forewings range from pale straw to a light brown or chestnut colour. The stigmas are not always clearly defined. Two dark marks are usually visible between the anterior and posterior cross-lines, but may be indistinct on some individuals. The hindwings are pale yellowish-grey, with a small discal spot. The Ingrailed Clay occurs in many habitats, but particularly woodland, heathland and moorland. It has been recorded from gardens, but uncommonly. Adults appear regularly at light and also visit flowers and honeydew. The recorded flight period is 16 May to 10 September with a peak in midsummer. The recorded larval foodplants are Bramble, Bilberry, Blackthorn, Hawthorn, Heather, willows and possibly herbaceous plants such as Primrose and violets. It overwinters as a larva.

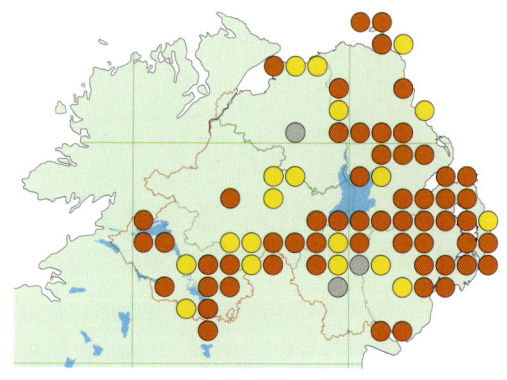

Barred Chestnut
Diarsia dahlii (Hübner) Category three.

Scarce. Barred Chestnut is found on the coast of Down and in a few sites in Armagh and Fermanagh. In Down, there have been recent records from Castlewellan Forest, Helen's Bay, Murlough NNR and Rostrevor Wood. Away from the Down coastline it is recorded from just two sites in north Armagh (the Argory and Ardress House) and two in north-west Fermanagh (Correl Glen and the Magho Cliffs). The adults are sexually dimorphic. The males are usually lighter than females, but darker examples can also be found. The forewings are broad with a noticeably convex costa and vary in colour from reddish-brown to dark brown. The stigmas are not well defined and the cross-lines are often faint, apart from the pale, wavy subterminal line. Barred Chestnut is mainly associated with acid woodlands especially adjacent to heaths and bogs. Adults are attracted to light in small

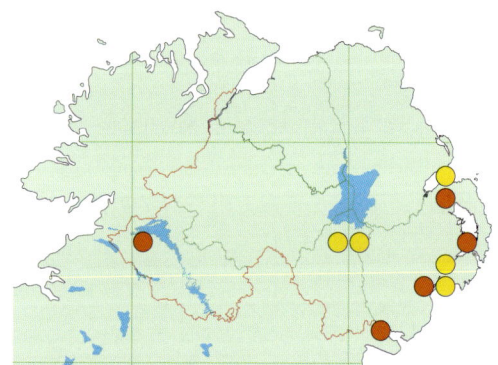

Barred Chestnut.

Rostrevor Wood National Nature Reserve, County Down.

A scarce species that appears occasionally at light. Most of the recent records emanate from coastal sites in Down.

numbers and also to flowers, such as Heather and Common Ragwort. This is a late summer species with a recorded flight period of 30 July to 12 September. The larval foodplants are birch, Bilberry and possibly willows and Bramble. It overwinters as a larva.

Purple Clay
Diarsia brunnea (Denis & Schiffermüller) Category one.
Widespread. Widely distributed throughout all counties, but more frequently encountered in north Armagh, Down and Fermanagh than elsewhere. Regularly recorded sites include the Argory and Brackagh Moss (Armagh); Helen's Bay and Rostrevor Wood (Down); Crom and Garvary Wood (Fermanagh); the Umbra (Londonderry) and Rehaghy Mountain (Tyrone). The forewings are broad with a purplish-brown sheen in fresh individuals. There is a square black mark above the conspicuous, straw-coloured reniform stigma. Purple Clay is associated most closely with broadleaved woodland, although it has been recorded from some suburban gardens and several bogs. Adults come frequently to light in small numbers and will also visit flowers and sugar. The recorded

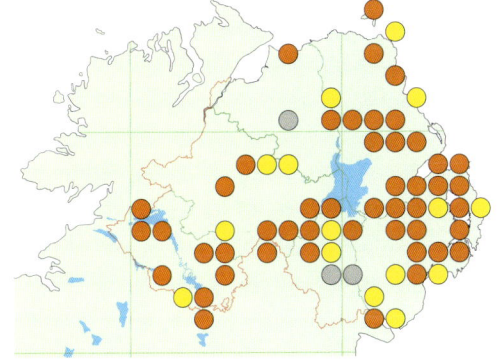

flight period is 7 June to 19 August with a peak in early July. The larvae feed from September to late spring on Bilberry, Foxglove, Bramble, willows and birch. It overwinters as a larva.

Small Square-spot
Diarsia rubi (Vieweg) Category one.
Ubiquitous. Common and widespread in southern counties where it is regularly recorded from sites such as Brackagh Moss and Peatlands Park (Armagh); Killard Point (Down) and Crom and

Setaceous Hebrew Character.

Ballykinler Dunes, County Down.

This adult was found on Ragwort in early morning close to an actinic trap.

Garvary Wood (Fermanagh). It is apparently absent from much of Tyrone (although it is frequently recorded from Rehaghy Mountain), Londonderry and upland areas of Antrim. Adults are variable in colour, ranging from light to dark reddish-brown. The stigmas are similar to the base colour with a fine paler outline. There is a conspicuous black square-shaped mark between both stigmas. The cross-lines are dark and evenly curved towards the costa. Small Square-spot is found in many open and lightly-wooded habitats, particularly wet sites. Adults come frequently to light and also visit flowers, especially Bramble, Heather and Common Ragwort. The phenology indicates that this species is double-brooded at many of the Northern Ireland sites, as it is in southern Britain. Two peaks are apparent in the pattern of records, in May/June and in August/early September, but with considerable overlap in the two broods in July. It is also possible that there is just a single generation at some sites, as is the case in northern Britain. The recorded flight season is 1 May to 23 October. The larvae overwinter and feed from autumn to spring on small shrubs and herbaceous plants including Heather, Dandelion and docks.

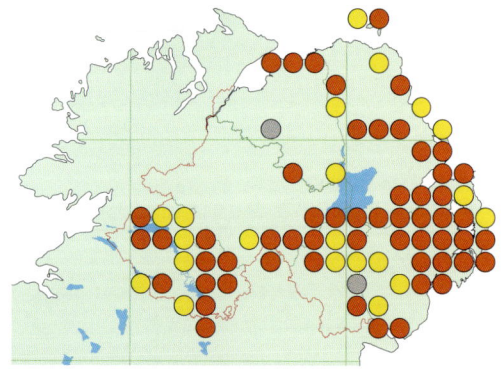

Setaceous Hebrew Character
Xestia c-nigrum (Linnaeus) Category three.
Scarce. This species was frequently encountered in the mid-1970s in north Armagh, but it appears to have declined in this area. The majority of recent records have been from sites in north and east Down including Ballykinler Dunes, Belfast Harbour Estate, Castleward, Helen's Bay and Murlough NNR. It is known from only a few sites in each of the other counties. Adults are normally greyish-brown, but this is prone to variation. The stigmas are not well defined. There is a creamy triangular patch extending along the costa of the forewing, which is bordered behind

by a conspicuous black mark. There is also a small black spot near the apex of the forewing. The hindwings are white with darker veining. Setaceous Hebrew Character is found primarily in lowland habitats including woodland, marshy areas and heathland and, occasionally, gardens. Adults are attracted to light and they also visit flowers including Buddleia and Common Ragwort. Individuals can be found at rest by day. The recorded flight period is 21 June to 17 October in two distinct broods; one from late June to early August and the second from late August to mid-October. The small number of annual records makes it difficult to establish whether the resident population here is reinforced with immigrants. The larvae of the second brood overwinter. The larvae feed on herbaceous plants including Common Nettle and species of willowherb.

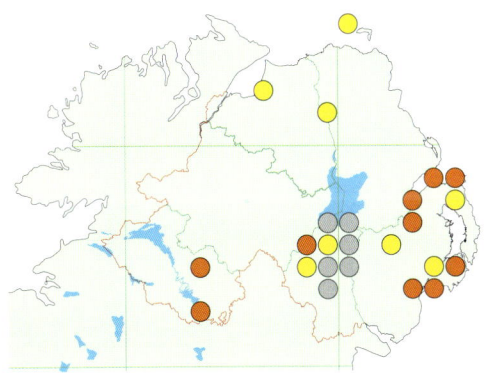

Double Square-spot
Xestia triangulum (Hufnagel) Category one.
Widespread. Common in the southern counties, but apparently more local in the north and west. It has been recorded from Rathlin Island (Antrim). Adults are ochreous to greyish-brown with two small, black, square-shaped markings finely outlined in a paler colour. There is also a small black mark near the apex of the forewing. Double Square-spot is typically a woodland species, although it is also found on heaths and bogs and, in the Greater Belfast area, in suburban gardens. Adults are attracted to light and flowers. This is a single-brooded species. The recorded flight period is 8 June to 9 September. The larvae, which overwinter, feed from September to the following spring on Blackthorn, Hawthorn, and possibly some herbaceous plants such as Primrose and Cow Parsley. It overwinters as a larva.

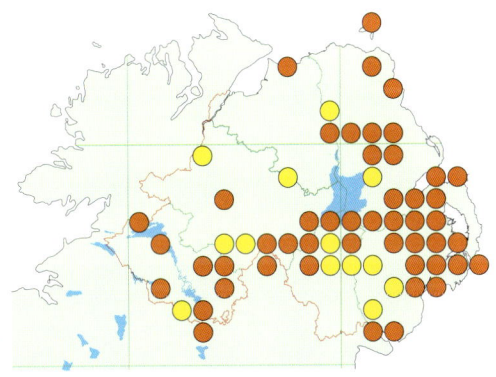

Dotted Clay
Xestia baja (Denis & Schiffermüller) Category one.
Ubiquitous. This common moth has been recorded in all counties but is especially common in Down, north Armagh and south Fermanagh. It has also been recorded on Rathlin Island (Antrim). Adults are generally reddish-brown with poorly-defined stigmas. There are two small black marks near the apex of the forewing which are diagnostic. The postmedian line is broken into a series of small black dots. Dotted Clay is recorded from woodland, heaths and bogs. It has also been recorded from several suburban gardens in Antrim and Down. Adults are attracted to light and also visit flowers, especially Common Ragwort. This is a single-brooded species and adults have been recorded between 15 June and 31 August. The larvae feed from September to late spring on shrubs and tall herbaceous plants including Common Nettle, birch, Bramble, willows, Bog Myrtle and Blackthorn. It overwinters as a larva.

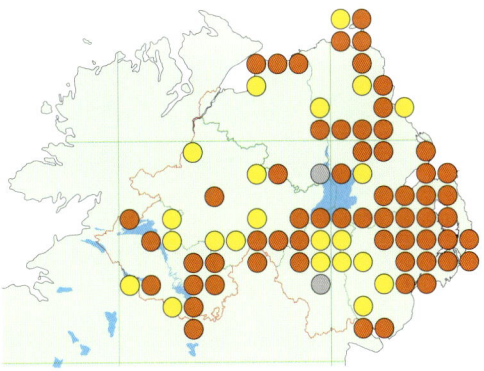

Neglected Rustic
Xestia castanea (Esper) Category three.
Notable. Donovan regarded this species as rare in Ireland, but in the last twenty years it has been discovered at scattered sites in several counties. It is known from Murlough Bay (Antrim); the Argory and Peatlands Park (Armagh); Correl Glen, Magho Cliffs and Monastir Gorge (Fermanagh) and Ballynasollus and Davagh Forest (Tyrone). There have been no recent

confirmed records from either Down or Londonderry. Adults are normally reddish-brown but some individuals are often much lighter. The reniform stigma on the forewing is usually poorly defined, but it often has an obvious darker mark on the basal aspect. The cross-lines are faint and indistinct on most individuals. Square-spot Rustic is similar in appearance but has a paler hindwing and lacks the dark mark on the stigma. Neglected Rustic is primarily a heathland species, although it has also been recorded from woodland and bogs. Adults are attracted to light and to the flowers of Heather. The Northern Ireland records have been between 18 August and 10 September. The larval foodplants are Heather and Bell Heather. It overwinters as a larva.

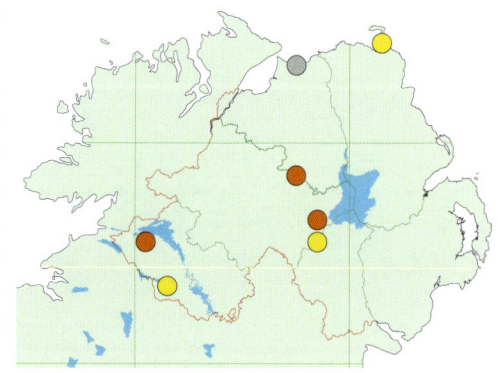

Six-striped Rustic

Xestia sexstrigata (Haworth) Category one.

Widespread. There are many records from the southern counties, especially Armagh and Down, but very few from the north and west although this is probably a function of recording effort. To the north of Lough Neagh, it is currently known from Ballymena, Glenarm and Rathlin Island (Antrim) and Duncrun, Magilligan and the Umbra (Londonderry). Cullinane, Favour Royal and Rehaghy Mountain are the only sites in Tyrone with recent records. Adults have light to darkish-brown forewings with poorly-defined stigmas and six darker cross-lines which are usually distinct. The hindwings are pale ochreous-brown with darker veining. Six-striped Rustic is associated with damp woodland, bogs and fens. Adults come frequently to light and also visit flowers especially Common Ragwort. The recorded flight period is 3 July to 12 September with a very sharp peak in late July and early August. Nothing is known about the larval habits or foodplants in Northern Ireland. In Britain, the recorded foodplants are Bramble, plantains and Hedge Bedstraw. It overwinters as a larva.

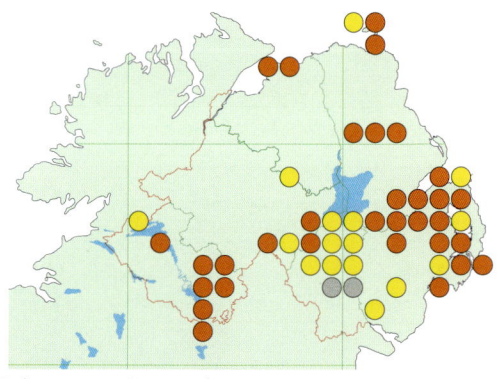

Square-spot Rustic

Xestia xanthographa (Denis & Schiffermüller) Category one.

Ubiquitous. Recorded in all counties and on Rathlin Island (Antrim). The colouration of the adults is quite variable, ranging from light to dark reddish-brown. The pale square reniform stigma is quite obvious in most colour forms. The postmedian line is broken into a series of small dots. The hindwings are light brown with dark veining. Square-spot Rustic can be found in lowland habitats including woodland, most types of grassland, bogs and in many suburban gardens. Adults come frequently to light and also visit the flowers of Common Ragwort, Buddleia and Heather. Adults have been reported between 5 May and 20 October, but infrequently outside the peak period in August and early September. The larvae, which overwinter, feed mainly on grasses and other low-growing herbaceous plants.

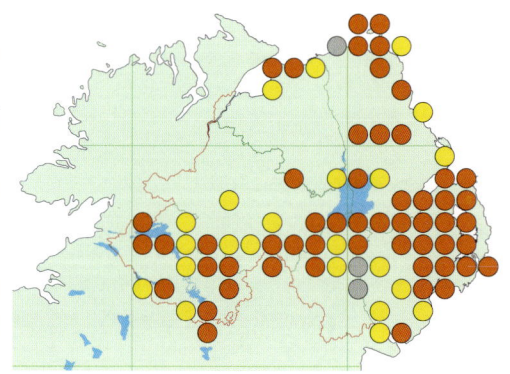

Heath Rustic

Xestia agathina (Duponchel) Category two.

Scarce. This species occurs in north Antrim, east Down and at scattered sites in north Armagh, Fermanagh and Tyrone. Sites include Peatlands Park and the Argory (Armagh); Ballykinler Dunes, Lackan Bog, Murlough NNR and Seaforde (Down); Correl Glen and Monmurry Bog

Heath Rustic.

Lackan Bog, County Down.

(Fermanagh) and Annaloughan Bog, Davagh Forest and Rehaghy Mountain (Tyrone). Adults are mainly reddish or greyish-brown with small but conspicuous, pale stigmas. There is a broad, black, wedge-shaped mark emanating from the basal area of the forewing and extending to the postmedian line and a series of small, dark, wedge-shaped marks along the inner edge of the subterminal line. In Northern Ireland, Heath Rustic is found on moorland, heathland and cutover bogs, usually close to trees. It is said to prefer stands of mature Heather. Adults come to light in small numbers but can also be found, especially in the morning, resting in Heather plants. The recorded flight period is 2 August to 1 October, with a peak in mid-August and early September. The larvae feed from October to May on young Heather. It overwinters as a small larva.

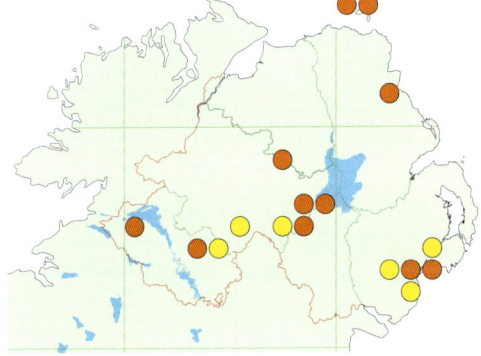

Gothic

Naenia typica (Linnaeus) Category one.

Scarce. Donovan described this species as being common in suitable localities throughout Ireland, a statement that Baynes disputed, based on a lack of information and records available at the time. In Northern Ireland it has been recorded from all counties, but it seems to be absent from many areas. It has most often been recorded in the Greater Belfast area and north Down, as well as parts of Armagh. It is known otherwise from just scattered sites in Antrim, south Down and south Fermanagh. The only recent record from Londonderry is from the dunes at Portstewart. Adults are a dark greyish-brown with distinctive brownish-

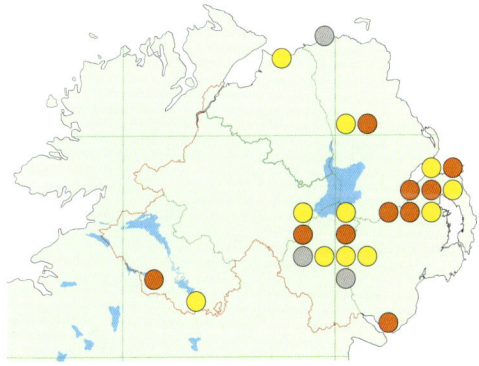

Green Arches.

Rehaghy Mountain, County Tyrone.

A colourful species recorded frequently from broadleaved woodlands in southern counties.

white veining and prominent jagged cross-lines; the stigmas are outlined in white. The Gothic favours weedy and damp habitats. In Northern Ireland, records have come from suburban gardens, cutover bogs, dunes and fens. Adults come to light very sparingly, but also visit flowers. They rest during the day concealed in vegetation. The recorded flight period is 23 June to 25 August, with a peak in July. The larval foodplants include Blackthorn and species of willowherb and willows. It overwinters as a larva.

Great Brocade
Eurois occulta (Linnaeus) Category four.
Rare. This has been recorded twice in Northern Ireland. The first record was in 1875 from Derry City (Londonderry) and the second in August 1997 at Dundrum (Down). There would appear

to have been only two other Irish records, one from County Sligo published by Kane and mentioned in Donovan, and the other from Glenageary (Dublin) in 1960. This is a fairly large moth. The forewings are predominately dark grey with variable amounts of black. The stigmas are outlined in white and the postmedian and subterminal lines are pale grey. In Britain, Great Brocade is resident in the central highlands of Scotland, but an immigrant elsewhere. The resident populations are found on sheltered moorland and woodland. In Britain, adults are on the wing in July and August. The Dundrum record was on 7 August but the exact date of the Londonderry record is unknown. The larval foodplants are Bog Myrtle, birch and willows. It overwinters as a larva.

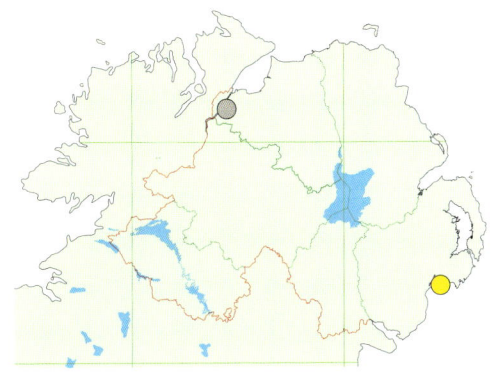

Green Arches
Anaplectoides prasina (Denis & Schiffermüller) Category one.
Widespread. This species is not uncommon in suitable habitat in southern counties, but has not been recorded in recent times north of Breen Wood (Antrim). It occurs most frequently in sites with extensive broadleaved woodland including Portglenone Wood (Antrim); Castlewellan Forest and Rostrevor Wood (Down); Crom and Florencecourt (Fermanagh) and Rehaghy Mountain (Tyrone). It has not been recorded in recent times from Londonderry but there are records for Knocktarna, near Coleraine, in the 1960s. The freshly-emerged adult is a richly-marked moth that is readily distinguished from other noctuids. The forewings are mottled bright green mixed with brown, white and black. The stigmas are large and paler than the ground colour and the cross-lines are conspicuous. Some individuals are more richly marked than others. Green Arches is associated with well-wooded habitats especially semi-natural broadleaved woodland. It can also be found, but in smaller numbers, in mature gardens and bogs. Adults come frequently to light. By day they remain concealed in low-growing vegetation or at rest on the trunks of trees, where they blend cryptically with their surroundings. The flight period extends from 28 May to 10 August, peaking in late June and early July. The larvae feed from August to April on Honeysuckle, Primrose, Bilberry and Bramble. It overwinters as a larva.

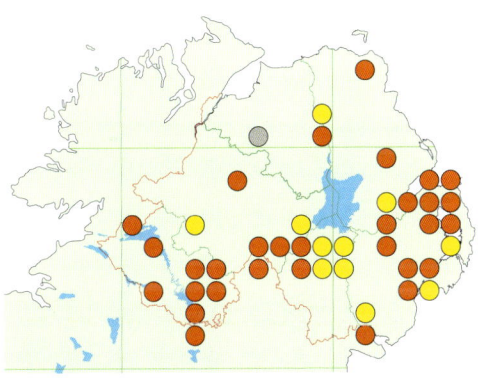

Red Chestnut
Cerastis rubricosa (Denis & Schiffermüller) Category three.
Scarce. Mainly found in north Armagh, south Down and Fermanagh. It is apparently absent from most of Antrim with the exception of Montiaghs Moss and Rathlin Island. There are no records for Londonderry. Brackagh Moss (Armagh); Rostrevor Wood and Seaforde (Down); Garvary Wood and Legatillida (Fermanagh) and Rehaghy Mountain (Tyrone) are some of its most commonly recorded sites. It is probably overlooked in some areas due to its early flight period. Adults have reddish-brown forewings that are suffused with grey along the costa. The cross-lines are usually weak and indistinct. In Northern Ireland it has been found most often in damp, broadleaved woodland, wooded cutover bogs and sheltered moorland. Adults come frequently to light and will also visit willow catkins. This is a spring-flying species and the recorded flight period is 17 March to 30 May. Little is known about the habits and ecology of Red Chestnut

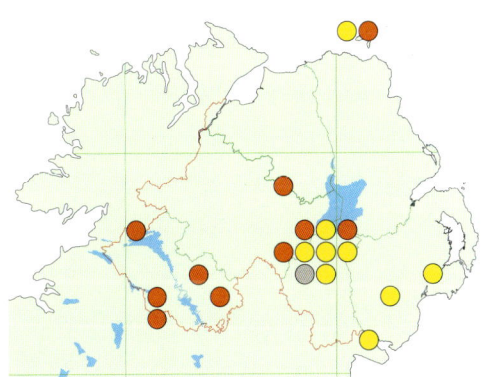

in Northern Ireland. In Britain, the recorded foodplants are herbaceous plants, shrubs and small trees, including Bilberry and species of bedstraw and willows. It overwinters as a pupa.

Beautiful Yellow Underwing
Anarta myrtilli (Linnaeus) Category one.
Scarce. This attractive day-flying moth has been recorded from suitable habitat (lowland bogs) in all counties. Adults have been seen regularly at Peatlands Park (Armagh) which is the most reliable site for this species. Other sites include Sharvogues Bog (Antrim) and Drumnahavil Bog (Armagh). This is a small and distinctively marked moth, usually seen flying at great speed just above the Heather; however, it can easily be overlooked, even by experienced observers. During dull overcast conditions, it rests concealed among Heather stems, where its cryptic colouration makes it extremely difficult to detect. The forewings are a rich brownish-red with a small central white mark and white antemedian and postmedian lines. The outer margin is chequered brown and white.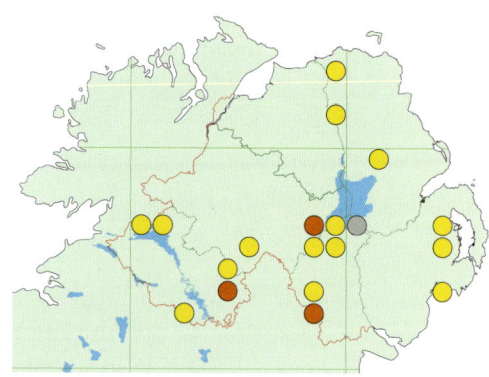
The hindwings are dark yellow with a broad black border and thin white outer fringe. There are two generations in southern Britain and one in the north. The data indicates that there is just a single generation in Northern Ireland as the recorded flight period is 3 May to 9 August. Most sightings have been in May and June. The distinctive green and white larvae have been swept from stands of Heather (which is one of the principal foodplants) at Peatlands Park. The larvae feed on the developing shoots of Heather and Bell Heather. The overwintering stage is unknown.

Nutmeg
Discestra trifolii (Hufnagel) Category four.
Rare. A suspected immigrant that was recorded at Mill Bay on Carlingford Lough in south Down on 9 August 2004. There is currently no evidence that the species is resident anywhere in Ireland. It was first reported in Ireland by W.E. Hart early in the twentieth century at Kilderry (Donegal). Baynes described it as rare in Ireland, mentioning additional records from Timoleague (Cork) and Lough Caragh (Kerry). It has since been recorded from Ballyteigue (Wexford) in 2004. There are no Irish specimens in the National Museum collections in Dublin. The ground colour of the adults is greyish-brown. The reniform stigma is conspicuous and darker than the ground colour in the lower half of the forewing. There is also a double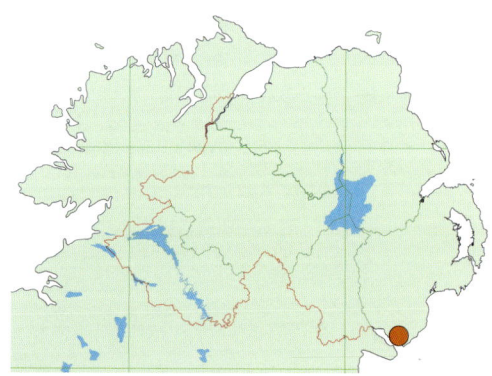
projection on the pale subterminal line which is clearly evident. In Britain, the Nutmeg is found in many habitats including open woodland, fens, coastal sites, gardens, waste ground and river valleys. Adults are attracted to light and also to flowers. There are two generations in southern Britain, but only one in June and July from central England northwards. The larvae are nocturnal, feeding on goosefoots, orache and probably other herbaceous plants.

Shears
Hada plebeja (Linnaeus) Category three.
Widespread. This species is found widely on the coast of Northern Ireland from Dundrum (Down) round to Magilligan Point (Londonderry) and on Rathlin Island (Antrim). Inland it is found to the south of Lough Neagh at the Argory and Aughinlig (Armagh) and more generally in Fermanagh from well-recorded sites such as Crom, Garvary Wood and Legatillida.

Beautiful Yellow Underwing.

Peatlands Park, County Armagh.

The adults of this small moth are active on bright sunny days. It is often overlooked even by experienced recorders.

Adults are greyish with conspicuous white cross-lines and a series of small white marks along the outer edge. There are pale scissor-like markings beside the stigmas which give the moth its common name. The Shears is associated with a wide range of habitats such as woodland, bogs and coastal heaths. Adults are attracted to light in small numbers and also to flowers such as Viper's Bugloss and Red Valerian. It reportedly rests on tree trunks and fence posts by day although, to the best of our knowledge, no adults have been found in Northern Ireland during the day. The recorded flight period is 19 May to 23 July. The larval foodplants are herbaceous plants including Smooth Hawk's-beard and hawkweeds. It overwinters as a pupa.

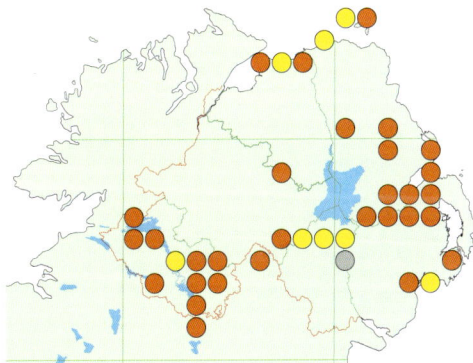

The Butterflies and Moths of Northern Ireland

Grey Arches.

Lackan Bog, County Down.

Grey Arches

Polia nebulosa (Hufnagel) Category one.

Widespread. Found in all counties but is more commonly met with in Armagh, Down and Fermanagh. It is seemingly absent from most of Antrim and only known from three sites in the county – Breen Wood, Magheramorne and Portglenone Wood – although the scarcity of records could be attributed to the lack of recording effort in the northern part of the county. Adults are large with a marbled appearance but are variable both in colour and markings. The pale colour form f. *pallida* predominates but darker individuals have been taken at Rostrevor Wood. The colour of the forewings ranges from ochreous-white through to dark brown with a mixture of grey, brown and white markings. The stigmas are clearly defined and outlined in black. This species is found mainly in woodland habitat. Adults come frequently to light and also visit flowers. They apparently rest by day concealed amidst ground vegetation or on tree trunks and fence posts. Grey Arches is single-brooded and recorded here between 16 June and 13 August. The recorded larval foodplants are birch, willows, Honeysuckle, Hazel and Bramble. It overwinters as a larva.

White Colon

Sideridis albicolon (Hübner) Category four.

Unknown. Restricted to the dunes at Murlough NNR (Down) and Magilligan (Londonderry), but it has not been seen in Northern Ireland since 1978 and its current status here is unknown. It was first recorded at Magilligan by Campbell in 1875 and seen again here in 1971. Wright recorded it twice between 1951 and 1953 at Murlough NNR and it was reported again in 1978. The lack of recent records from either of these well-studied sites suggests it may have become extinct in Northern Ireland. The forewings are variable in colour, ranging from pale brown through to dark grey. The stigmas and cross-lines are poorly defined. There are two small white dots (colon-shaped) on the distal half of the wing, giving the species its common name. It is similar in appearance to the Cabbage Moth but this is usually darker and has the reniform stigma clearly outlined in white. It also has a curved spur on the tibia and darker hindwings. In Ireland, White Colon appears to be found only on sand dunes; in Britain it also occurs inland. Adults are attracted 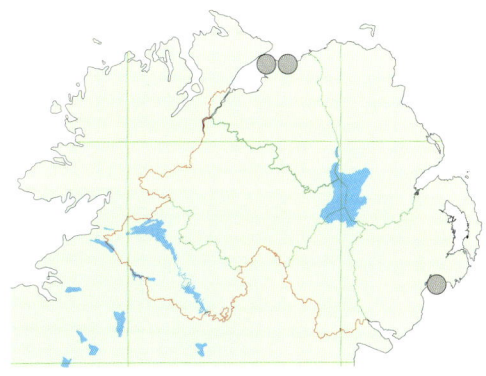 to light and are said to conceal themselves by day under sandy overhangs and among Marram grass. The flight period is not known in Northern Ireland. In Britain, it is mainly single-brooded, in late May and June. The larval foodplants are not known in Northern Ireland, but recorded foodplants in Britain include Common Restharrow, species of goosefoot and orache, Sea Bindweed and Sea Sandwort. It overwinters as a pupa.

Cabbage Moth
Mamestra brassicae (Linnaeus) Category one.
Widespread. Generally distributed in Armagh and Down, but more local elsewhere. This lowland species appears to have become scarce in areas where it once was common. This may be related to changes in farming practice, the use of pesticides and the reduction in the cultivation of cabbages in gardens and allotments. The forewings of the adults are normally reddish-brown with a marbled appearance. The reniform stigma is outlined in white. A diagnostic feature is the curved spur on the tibia which separates it from other similar species. The Cabbage Moth is most common in farmland and gardens where brassicas are grown, but it also appears in natural habitats especially woodland. Adults are attracted to light and flowers, especially Red 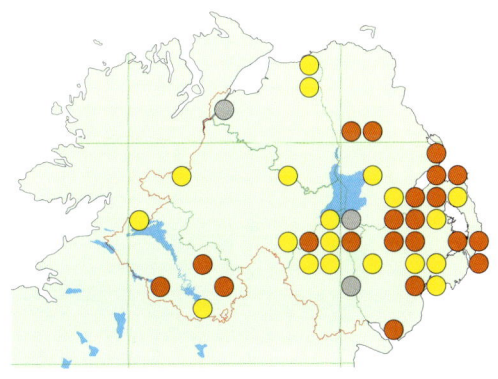 Valerian. The larvae are nocturnal and polyphagous, but they have a preference for cultivated brassicas. The recorded flight period is prolonged, from 1 April to 23 August with a peak in late July and early August. There is an indication from the pattern of records that there may be two generations in Northern Ireland. The larvae can be found throughout the summer and autumn. Most individuals overwinter in the pupal stage.

Dot Moth
Melanchra persicariae (Linnaeus) Category three.
Rare. The current status of this species in Northern Ireland is unknown. The only recent record is from Rathlin Island (Antrim) in the 1980s, but there is no corroborating specimen. Donovan, who regarded this species as being local and not common in Ireland, gave records for Antrim and Fermanagh. In Down, Wright apparently recorded it from the Newcastle area in 1973 although it does not appear on the site list from Murlough NNR. A record from Rostrevor Wood is given in Baynes who said it appeared to be widely distributed. The forewings are almost black with a conspicuous white reniform stigma. The cross-lines are obscure with the exception of the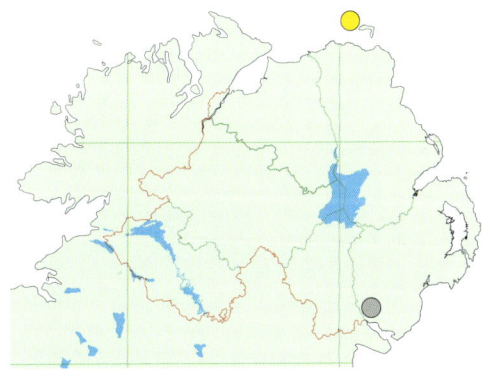

Dot Moth.

Killarney National Park, County Kerry.

This species is readily identified by the blackish appearance of the forewings and the white reniform stigma. There have been no confirmed records since the 1980s.

subterminal line, which is made up of a series of pale ochreous dots. In Britain, Dot Moth is recorded from many habitats including gardens, but in Ireland it seems to be found mainly on bogs and heaths. There is a single generation in Britain from late June to August. The larval habits and foodplants are not known in Northern Ireland. In Britain, the larvae reportedly feed from the late summer to autumn on a wide range of woody and herbaceous plants including Common Nettle, White Clover, Ivy, Hazel and willows. It overwinters as a pupa.

Broom Moth

Melanchra pisi (Linnaeus) Category one.

Widespread. Recorded from many localities in all counties and on Rathlin Island (Antrim) and Copeland Bird Observatory (Down). Adults show some variation in colour. The forewings are generally reddish-brown. The most obvious diagnostic feature is the pale white subterminal line and the similarly coloured blotches at the inner edge of the forewing. The stigmas are not clearly defined. Broom Moth occurs in many habitats including open woodland, bogs, heaths and suburban gardens. Adults come frequently to light and also visit flowers and sources of honeydew after dark. The recorded flight period is prolonged, from 2 May to 8 September, with the majority of records between mid-May and the end of June. The larval foodplants are Broom, Bracken, birch, Heather and Bramble. It overwinters as a pupa.

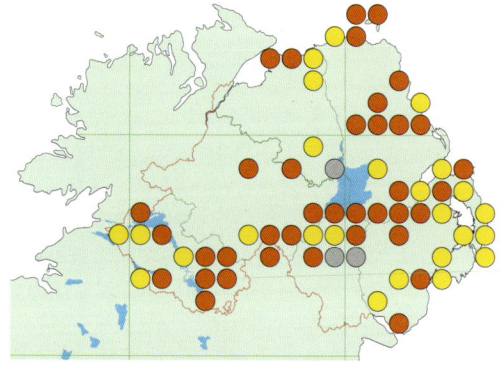

Beautiful Brocade

Lacanobia contigua (Denis & Schiffermuller) Category four.

Unknown. There have been no confirmed records for this moorland species for many years. Watts (1894) refers to a specimen being taken in the Mourne Mountains (Down) in July 1893

Moth accounts

and Johnson (1902) described it as scarce in the same locality. Foster apparently recorded a single adult in Donard Demesne (Down). In his county list, Wright noted it for Antrim and also listed it from Newcastle (Down) in 1973 on a record card submitted to the Biological Records Centre at Monks Wood. Beautiful Brocade is associated with heathland and moorland. It is possible that a small population may well still exist in the Mourne Mountains as no trapping has been carried out there for many years. In Britain the flight period is June-July. The larval foodplants include birch, oak, Bog Myrtle and Heather. It overwinters in the pupal stage.

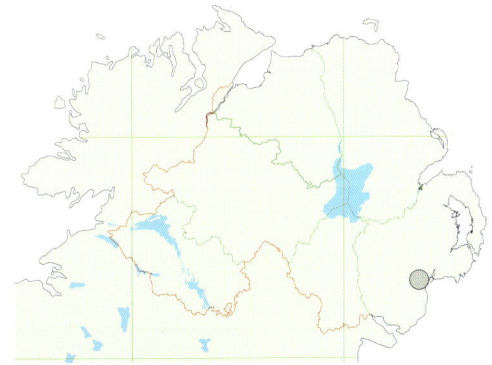

Pale-shouldered Brocade.

Banbridge, County Down.

The Butterflies and Moths of Northern Ireland

Pale-shouldered Brocade
Lacanobia thalassina (Hufnagel) Category one.
Widespread. This species occurs throughout Northern Ireland and on Rathlin Island (Antrim). Aughinlig and Peatlands Park (Armagh); Helen's Bay (Down); Garvary Wood and Legatillida (Fermanagh) and Rehaghy Mountain (Tyrone) are among the most reliable sites for this species. This is a medium-sized moth with reddish-brown forewings suffused with lighter brown and creamy white scales. The base of the wings has a conspicuous dark streak and there is a black bar connecting both the antemedian and postmedian cross-lines. The postmedian cross-line has two projections that extend to the outer margin of the wing. The stigmas are similar to the ground colour but edged with a fine pale line. This species is usually found in woodland, but it also occurs in fens, bogs and heathland, providing there are trees present. It is seen frequently at light and also visits flowers. The adults apparently rest by day on the trunks of trees and, occasionally, fence posts. The recorded flight period is 29 April to 8 September with a peak in early June. The larvae feed from July to September on trees and woody plants including oaks, Hawthorn, Broom and Honeysuckle. It overwinters as a pupa.

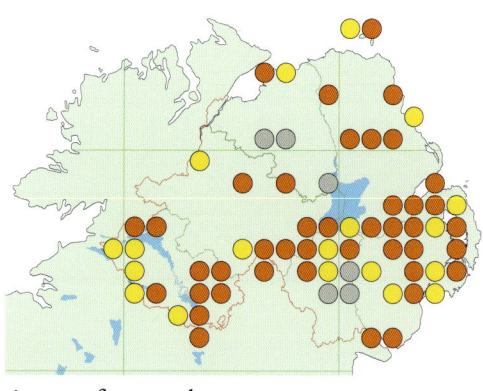

Dog's Tooth
Lacanobia suasa (Denis & Schiffermüller) Category three.
Rare. Recorded only from Down and Antrim. There are just four recent records: from Portmore Lough (Antrim) on 21 July 1998; Castle Espie (Down) on 26 June 2001 and Mill Bay (Down) on 6 and 15 July 2003. The only other records are from Aghalee (Antrim) in the 1950s by Wright and Kircubbin (Down) by Foster. Adults are normally greyish-brown, but individuals can show some variation in colour. The diagnostic feature is the double tooth-shaped projection extending to the outer margin of the forewings. In Northern Ireland, Dog's Tooth has been recorded from saltmarshes and damp, marshy pasture. Its foodplants suggest it could be found in ruderal habitats. Adults are attracted to light, but are seldom encountered, or perhaps overlooked in a few cases. In Britain this has two generations in the south (May to early July and late July to September) but only one in the north in June to July. The Northern Ireland records span the period from 26 June to 21 July, suggesting it is single-brooded here. The larvae have not, to our knowledge, been found in Northern Ireland. In Britain, they are reported to feed throughout the summer to early autumn on various herbaceous plants, including Greater Plantain and goosefoots. It overwinters as a pupa.

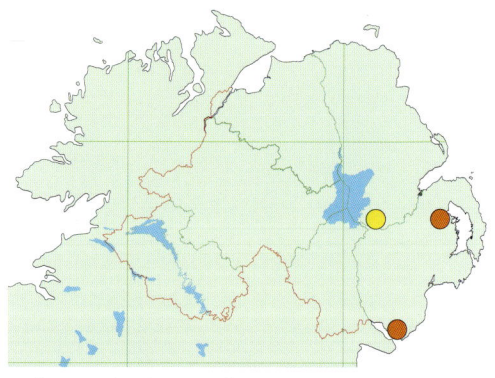

Bright-line Brown-eye
Lacanobia oleracea (Linnaeus) Category one.
Widespread. A common species in the south-east and throughout most of Fermanagh and Antrim including Rathlin Island. It is apparently much scarcer in Londonderry where it is largely restricted to coastal localities such as Magilligan, Portstewart Dunes and the Umbra. In Tyrone it has been recorded recently from several sites including Crilly, Fivemiletown, Lough Fea, Rehaghy Mountain and Stillago. The forewings are normally dark reddish-brown. The orbicular stigma is small and finely outlined in white and the reniform stigma is dark orange. There is often a dusting of white scales near the basal part of

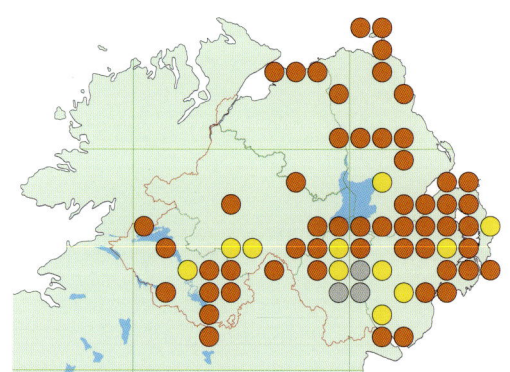

Glaucous Shears.

Ballynagilly, County Tyrone.

This moorland species appears to have its stronghold in the west.

the wing. The subterminal line is white, with two tooth-shaped projections that extend to the outer margin of the wing. This species occurs in many lowland habitats, including suburban gardens. Adults come to light and also visit flowers, especially Red Valerian. It has a prolonged flight period, with a single annual generation, from 15 April to 4 September. The peak period for records is July and early August. The recorded larval foodplants are Common Nettle, willowherbs and Hazel. It overwinters as a pupa.

Glaucous Shears

Papestra biren (Goeze) Category two.

Scarce. An attractive species known from all counties except Londonderry. Most of the records are from Fermanagh where it has been taken frequently at Garvary Wood, Knocknalosset, Legatillida and the Marlbank area. Outside Fermanagh it has occurred on Rathlin Island (Antrim); at Loughgall (Armagh); Aughintober (Down) and Ballynagilly (where it has been recorded in good numbers) and Davagh Forest (Tyrone). Adults are heavily marked and mostly dark grey in colour, although becoming lighter towards the outer margin; this can vary among individuals. The stigmas are conspicuous and normally lighter than the ground colour and outlined in white. The antemedian and postmedian lines are also paler than the forewing colour. Glaucous Shears is essentially a moorland species found on upland and lowland bogs and heaths but clearly also strays into neighbouring habitats. Adults are attracted to light and also visit flowers. Individuals rest by day on the shaded sides of large stones, old fence posts and tree trunks. Individuals have been found in Northern Ireland on stones and ground vegetation by day. The recorded flight period in Northern Ireland is 7 May to 3 July but there have been few records in June and only one in July. The larval habits and foodplants are not known in Northern Ireland. In Britain, the recorded foodplants are Meadowsweet, Bog Myrtle, willows, Bilberry and Heather. It overwinters as a pupa.

Broad-barred White
Hecatera bicolorata (Hufnagel) Category three.
Scarce. Confined mainly to Down and north Armagh, apart from apparently isolated populations at Portstewart and the Umbra on the north coast of Londonderry and at Beihy in west Fermanagh. It is recorded from the White Mountain on the northern (Antrim) side of Belfast Lough and at Magheramorne (Antrim). There are also records from gardens in Belfast, Lisburn (Antrim) and Helen's Bay (Down). In north Armagh it has been found around Loughgall and Portadown and at Peatlands Park. Localities in Down include Killard Point, Murlough NNR and Seaforde. The adult moth is pale grey with a dark central band which gets broader towards the costa. The stigmas are paler against the darker band. Broad-barred White is recorded from dunes, gardens, old quarries and abandoned urban sites. Adults appear at light in small numbers and also visit flowers, especially Red Valerian and Viper's Bugloss. It has been found by day resting on stone walls in coastal localities. The recorded flight period is 25 May to 1 August. The larvae feed on the flowers of composites especially hawkweeds and hawk's-beards. It overwinters as a pupa.

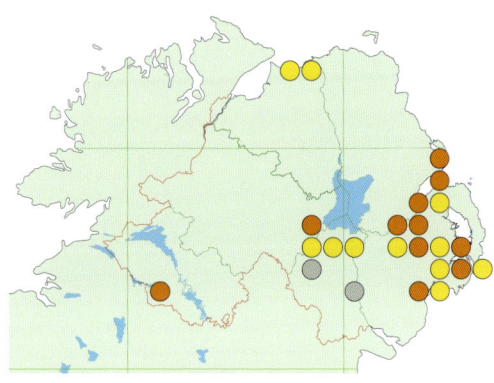

Campion
Hadena rivularis (Fabricius) Category one.
Widespread. The Campion has been recorded frequently at many of the regularly trapped sites in Armagh, Down and Fermanagh. It is also known from sites in mid-Antrim around Ballymena and, further north, on Rathlin Island. There are many records from Copeland Bird Observatory and from Derryleckagh and Helen's Bay (Down). It is commonly recorded from Brackagh Moss and Oxford Island (Armagh) and Crom, Garvary Wood and Legatillida (Fermanagh). This is a richly-marked species. The forewings are purplish-brown, often tinged with pink. The orbicular and reniform stigmas are finely outlined in white and both are connected by a white line. The antemedian and postmedian lines have a purplish hue and are quite distinctive. The Campion is associated with damp woodland, bogs, fens, coastal grassland and, occasionally, gardens. Adults appear at light in small numbers and also visit flowers, especially campions. It is a single-brooded species recorded between 20 May and 22 August. The larvae feed on the developing seeds of White, Sea and Red Campion and Ragged Robin. It overwinters as a pupa.

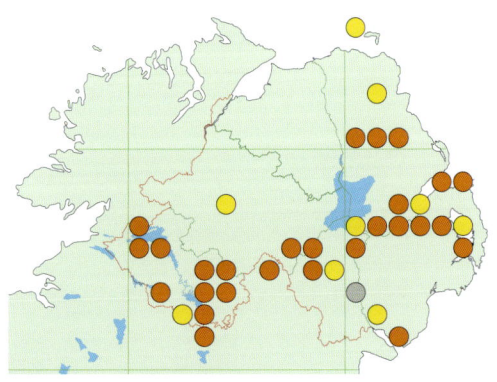

Tawny Shears and Pod Lover
Hadena perplexa (Denis & Schiffermüller) Category two.
Notable. Confined to a few sites on the coasts of Down and north Antrim including Rathlin Island. There are recent records from Murlough Bay, Portballintrae and White Park Bay (Antrim) and Ballyquintin, Copeland Island, Killard Point and Murlough NNR (Down). In Ireland this species is represented by the subspecies *capsophila* which is called the Pod Lover. Adults of this subspecies are darker in appearance than those of the nominate form. The

Tawny Shears and Pod Lover.

Killard Point National Nature Reserve, County Down.

forewings are dark brown with the reniform stigma outlined with a fine whitish line. The cross-lines are indistinct and the subterminal line is wavy and pale in colour. The Pod Lover occurs in Northern Ireland on dunes and rocky coasts. Adults appear in small numbers at light and visit flowers, especially campions and Red Valerian. The recorded flight period is 22 May to 22 August. The larvae have not, to our knowledge, been found in Northern Ireland. In Britain they reportedly feed during July and August on the seed heads of campions, especially Sea Campion, and also Rock Sea-spurrey. It overwinters as a pupa.

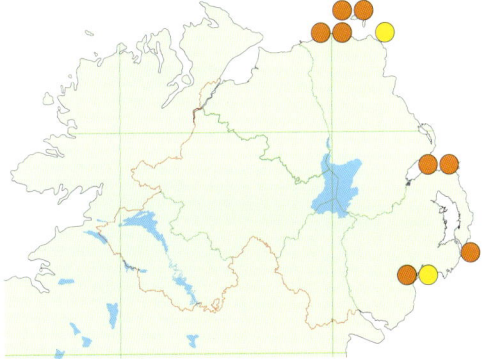

Marbled Coronet
Hadena confusa (Hufnagel) Category three.
Scarce. Recorded from scattered localities in all counties, but with concentrations of records in Fermanagh and north Antrim. The majority of recent records are from regularly trapped sites in Fermanagh including Derrylin, Eshywulligan, Garvary Wood, Glennasheever and Legatillida. It has obviously been overlooked in this county in the past. It has also been taken on the north Antrim coast at Murlough Bay and White Park Bay, at several sites on Rathlin Island and, further south, in Glenariff Forest. This is a distinctive moth not easily confused with other species. The forewings are a mixture of black and white with a chequered fringe and a small white apical blotch. The orbicular and reniform stigmas are fused together to form a large white

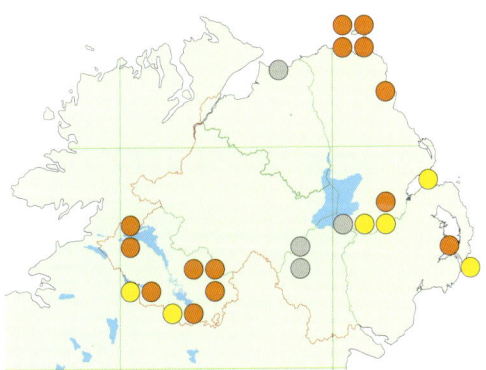

265

The Butterflies and Moths of Northern Ireland

Marbled Coronet.

Brookeborough, County Fermanagh.

A brightly-marked noctuid which has a more widespread distribution than previously thought.

blotch. Marbled Coronet is found on calcareous grassland, both inland and on the coast. There are also some records from woodlands. Adults are attracted to light in small numbers. It is also attracted to flowers especially Red Valerian. The recorded flight period is 29 May to 20 August. The larval foodplants are the ripening seed pods of Sea and Bladder Campion. Ragged Robin would seem to be the foodplant in the west as most campions are mainly coastal in Northern Ireland with the exception of Red Campion. It overwinters as a pupa.

Lychnis
Hadena bicruris (Hufnagel) Category two.
Scarce. Recorded in the Greater Belfast area and in north Down, north Armagh and at a few sites in south Fermanagh. The only places it is known from north of Lough Neagh are

Antler Moth.

Knockbarragh, Rostrevor, County Down.

Portrush and Rathlin Island (Antrim). It was taken commonly at Copeland Bird Observatory (Down) in 2003. Adults are similar in appearance to the Campion but lack the pinkish tinge evident on this species. The forewings are greyish-brown to dark brown with a pale irregular subterminal line. The orbicular and reniform stigmas are outlined in white and well defined. Lychnis is recorded from low rocky coasts, woodland and, especially, suburban gardens. Adults come occasionally to light and also to flowers after dusk. The recorded flight period is 27 May to 27 July. Like other closely-related species, the larvae feed on the seeds of various species of campions including garden varieties. It overwinters as a pupa.

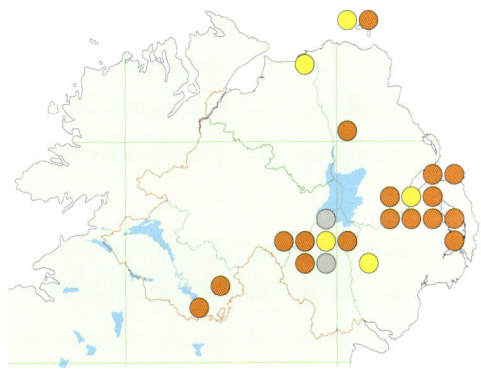

Antler Moth

Cerapteryx graminis (Linnaeus) Category one.

Widespread. Recorded from all counties including Rathlin Island (Antrim). The species is most frequently recorded at coastal sites, particularly the Magilligan and Umbra dunes (Londonderry) and Killard Point (Down). Inland localities include Slievenacloy (Antrim); Pigeon Rock and Rostrevor Wood (Down) and Cuilcagh Mountain (Fermanagh). It may have disappeared from some of its inland, lowland localities, for example, in Armagh, where it has not been recorded since 1996. This is a distinctive, easily recognised species. The forewings are reddish-brown with a conspicuous white marking (shaped like the antler of a deer) and black longitudinal streaks. There is considerable variation in the colour of the wings and the extent of the markings.

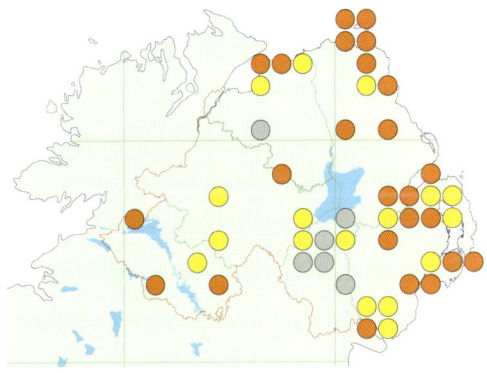

Antler Moth occurs in lowland and upland rough grassland, dunes, heaths and bogs and open woodland. It may be under-recorded in upland areas in Northern Ireland. Adults come frequently to light and will also visit flowers, especially Common Ragwort, where they have been found sitting on the flower heads early in the morning. The recorded flight period is 3 July to 18 September. The larval foodplants are coarse grasses including Sheep's Fescue, Mat-grass and Purple Moor-grass. It overwinters as an egg.

Hedge Rustic
Tholera cespitis (Denis & Schiffermüller) Category four.
Notable. Recorded from a few widely-scattered sites in Armagh, Down, Fermanagh and Tyrone. It has been recorded several times in successive years at Murlough NNR (Down), Legatillida (Fermanagh) and Davagh Forest (Tyrone). At other sites it has only been recorded on single occasions. The monitoring data from the Rothamsted trap network indicates a decline in abundance of this species in Britain of 97 per cent between 1968 and 2002. This is a rather dark, stocky moth. The forewings are dark brown with paler orbicular and reniform stigmas which are outlined with yellow. The subterminal line is pale and irregular. Hedge Rustic has a preference for grassy habitats both inland and in coastal localities. Adults appear sparingly at light, but also visit flowers, especially Common Ragwort. The recorded flight period is 10 August to 25 September. The larvae have not been recorded in Northern Ireland. In Britain, they feed after dark from March to July on Mat-grass and probably other species of grasses. It overwinters as an egg.

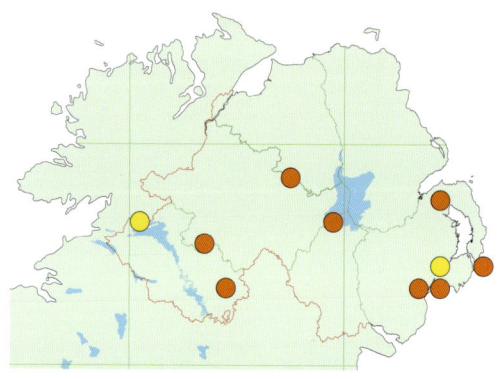

Feathered Gothic
Tholera decimalis (Poda) Category three.
Notable. Currently known from just a few sites in Down and Tyrone. It was recorded regularly in the Loughgall area (Armagh) in the 1970s, but it has not been in the county since 1980 and it is feared extinct. The majority of recent records have been from Killard Point (Down) where there appears to be a stable population. The only other sites it has been recorded at since 1990 are Clogher in Tyrone, and two sites in Down, Carryduff and Derryleckagh. This is a thickset species. The males have feathery antennae. The forewings are dark brown and marked with a series of pale lines radiating from the reniform stigma which is conspicuous and finely outlined in pale yellow. The subterminal line is irregular, creamy white in colour with a series of small internal black wedges running across the wing. This moth is found in rough grassland and marshy areas. Adults are attracted to light, especially the males. The recorded flight period is 14 July to 7 September. The larval habits are unknown in Northern Ireland. In Britain it is nocturnal, feeding from March to July on Mat-grass and Sheep's Fescue. It overwinters as an egg.

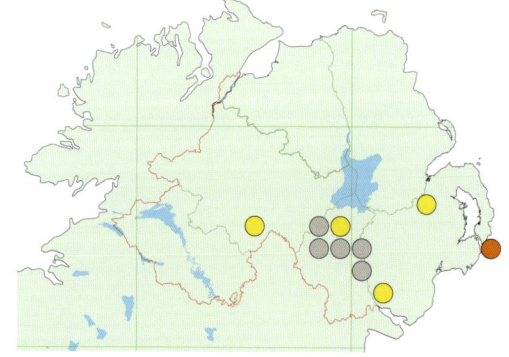

Pine Beauty
Panolis flammea (Denis & Schiffermüller) Category three.
Scarce. This species is almost certainly more widespread than the map suggests, especially in the north. It is probably overlooked in many localities owing to its early flight period. There have been recent records from scattered localities in southern counties including the Argory (Armagh), Murlough NNR and Seaforde (Down), Garvary Wood and Legatillida (Fermanagh)

Opposite page
Pine Beauty.

The Argory, County Armagh.

The intricate patterning on the forewings helps the moth to blend naturally with the developing buds of pine trees.

and several localities in the Greater Belfast area. As this moth can be a pest in commercial pine plantations, Forest Service conducted a survey for the species throughout Northern Ireland in 1990 using pheromones. It was found to be widespread in the coniferous forests, although not in numbers that would constitute a serious problem; however, these records are not mapped. Adults are distinctive and show little variation in markings. The forewings are rich red-brown with fairly narrow darker brown cross-lines. The reniform stigma is enlarged and darker than the ground colour and outlined in white. The outer margin of the forewing has a series of fine white streaks. This is an early spring species. Pine Beauty is found in commercial pine plantations and mixed woodland with pines. Adults come sparingly to light and rest by day among the branches, shoots and on the trunks of the foodplant. The larvae feed during June and July on the foliage of pines. The recorded flight period is 8 March to 31 May. It overwinters as a pupa.

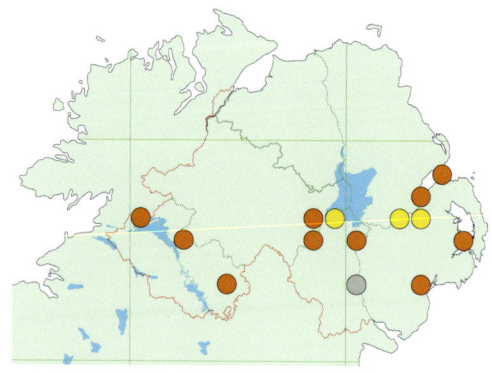

Small Quaker
Orthosia cruda (Denis & Schiffermüller) Category one.
Widespread. Recorded in all counties but very local in the north where it is apparently absent from large areas, although this may largely be due to the lack of early trapping. It is frequently encountered in sites to the south of Lough Neagh such as the Argory and Peatlands Park (Armagh), Crom and Garvary Wood (Fermanagh). The species is on the wing early in the year when little recording is done and this may exaggerate its rarity in the northern counties. The adults are small in comparison to related species and rather plain in appearance. The forewings vary from greyish-brown to reddish-brown, usually with darker reniform stigmas and small dark dots towards the costa. The apex of the forewing is well rounded. The subterminal line is normally a series of small pale dots. This is a woodland species that can also be seen occasionally in heaths, fens and some suburban gardens. Adults come frequently to light and will also feed at willow catkins. The recorded flight period is 11 March to 22 May. The larval foodplants are oak, birch, Hazel and willows. It overwinters as a pupa.

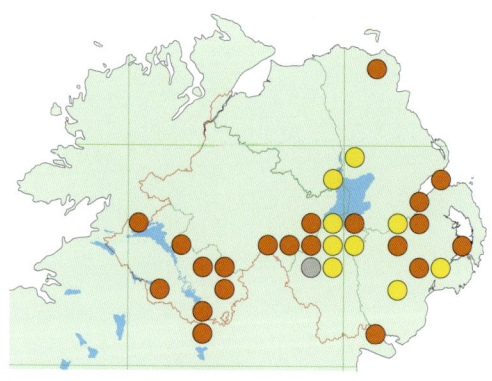

Northern Drab
Orthosia opima (Hübner) Category three.
Rare. This species had not been recorded in Northern Ireland for many years. Baynes describes it as being 'uncommon, but widely distributed'. Donovan listed records from Belfast (Antrim); Lough Neagh (Armagh); Lough Neagh and Lagan Marshes (Down) and Lough Neagh, Stewartstown and Tamnamore (Tyrone). This species has been targeted in recent fieldwork surveys at some of its former haunts and at other potentially suitable areas around the Lough Neagh basin without any success. Like many *Orthosia* species this is a rather hairy moth. The forewing colour ranges from pale to greyish-brown with a central band which broadens towards the costa. The orbicular and reniform stigmas are conspicuous and finely outlined in yellowish-white. The subterminal line is straight and dark and edged with white. In Britain it is found in open, damp habitats but the habitat used in Northern Ireland is unknown. Adults are attracted to light and are said to

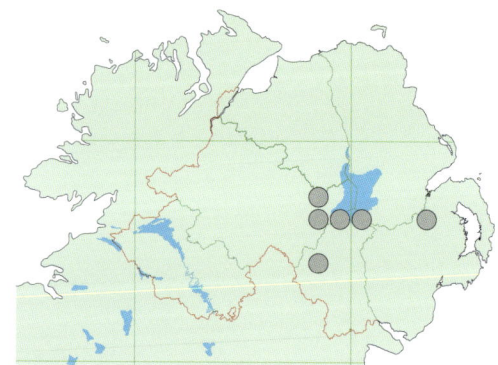

Lead-coloured Drab.

Crom, County Fermanagh.

visit willow catkins and Blackthorn blossom. This is a spring species and the few Northern Ireland records have been in April. The main flight period in Britain is April and May. The larval foodplants include willow, birch and Common Ragwort. It overwinters as a pupa.

Lead-coloured Drab
Orthosia populeti (Fabricius) Category three.
Rare. There are reliable records of this species from only two sites in Northern Ireland. It was recorded by Wright at Tollymore Forest (Down) in the early 1970s, but has not been reported from here since. Its other confirmed locality is Crom (Fermanagh), where it was discovered in 1997; records from Crom have been virtually annual since. It is considered likely that this species is under-recorded, especially in the west. Partridge (1895) and Allen (1907) both

claimed to have taken larvae at Lough Erne near Enniskillen (Fermanagh); those found by Partridge were reportedly reared through to adults. Donovan considered the records to be questionable and Baynes was also of a similar opinion. This is a dark greyish-brown moth with fine wavy cross-lines which are poorly defined. The orbicular and reniform stigmas are finely outlined in white and are always visible. Confusion is possible between this species and the much commoner Clouded Drab. Lead-coloured Drab has been trapped at Crom in or close to broadleaved woodland. Adults appear at light in small numbers and they reportedly visit willow catkins. Lead-coloured Drab is an early spring species recorded between 22 March and 21 April in Northern Ireland. The larvae feed during May and June on species of poplar including Aspen, which is a common tree in the Crom woodlands. It overwinters as a pupa.

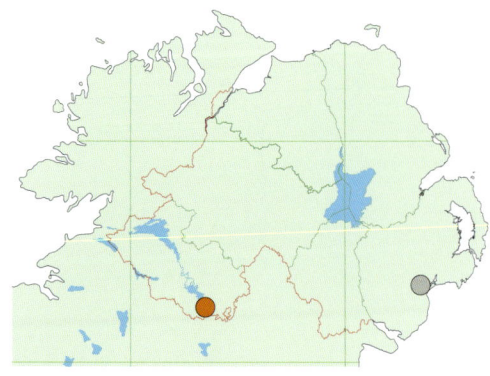

Powdered Quaker
Orthosia gracilis (Denis & Schiffermüller) Category two.
Scarce. Widely distributed across southern counties but absent from the northern half of Antrim, and the north and west of both Londonderry and Tyrone. Regular sites include Brackagh Moss and Oxford Island (Armagh); Helen's Bay (Down); Crom, Garvary Wood and Legatillida (Fermanagh) and Rehaghy Mountain (Tyrone). In Antrim there are recent records from Montiaghs Moss and Killybegs Bog. The only known site in Londonderry is Traad Point. Adults show a wide variation in colour, ranging from a pale sandy or greyish-brown to a darker orange-brown. The orbicular and reniform stigmas are obvious and generally darker than the ground colour. There is a series of small dark dots arranged in a curve below the reniform stigma. The subterminal line is usually pale and fairly straight. Darker forms tend to have a lighter hindwing. Powdered Quaker has been recorded from cutover bogs, fens, wet woodland and also gardens. Adults are frequently attracted to light and will also visit willow catkins and Blackthorn blossom. The recorded flight period is 17 March to 27 May. The larval foodplants include willow, Aspen, Blackthorn, Bog Myrtle, Meadowsweet and Purple Loosestrife. It overwinters as a pupa.

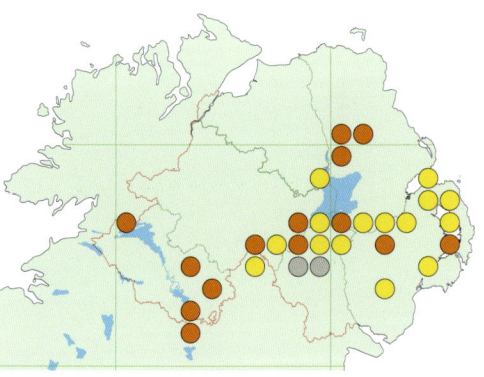

Common Quaker
Orthosia cerasi (Fabricius) Category one.
Widespread. Common and widely distributed across southern counties, but appears to be less frequent further north. Recently recorded sites include Breen Wood (Antrim); Oxford Island, Peatlands Park and Slieve Gullion (Armagh); Bohill Forest, Castlewellan Forest and Rostrevor Wood (Down); Aghalane and Garvary Wood (Fermanagh); the Umbra (Londonderry) and Ballynasollus (Tyrone). Adults vary considerably in colouring within and between populations. The forewing colour ranges from greyish-brown to reddish-brown with many intermediate forms. The orbicular and reniform stigmas are large and conspicuous and finely edged in yellowish-white. The subterminal line is fairly straight and also yellowish-white. This species occurs in lowland broadleaved woodland and other habitats with trees, including gardens and parks. Adults are frequently seen at light in moderate numbers and will also visit willow catkins and Blackthorn blossom. The recorded

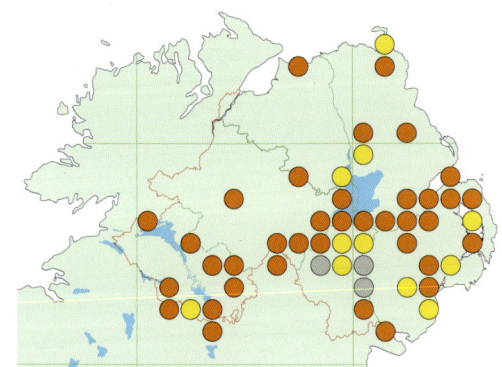

Powdered Quaker.

Crom, County Fermanagh.

flight period is 30 January to 8 June with a peak in mid-March to mid-April. The larvae feed during May and June on oak, willows, birch, Hawthorn and Hazel. It overwinters as a pupa.

Clouded Drab
Orthosia incerta (Hufnagel) Category one.

Widespread. Found in all counties, but relatively scarce in the north and west. Adults are extremely variable in colour and markings with many named aberrations and forms. The forewing colour ranges from blackish- or reddish-brown through to pale grey. The orbicular and reniform stigmas are finely outlined with white. Both these, along with the subterminal line, are more conspicuous in darker forms. The much rarer Lead-coloured Drab is similar in appearance but slightly smaller and has a more rounded forewing. Clouded Drab is a spring species found commonly in a wide range of woodland habitats and suburban gardens. Adults are frequently seen at light and will also visit willow catkins and Blackthorn blossom. The adults emerge in late winter and

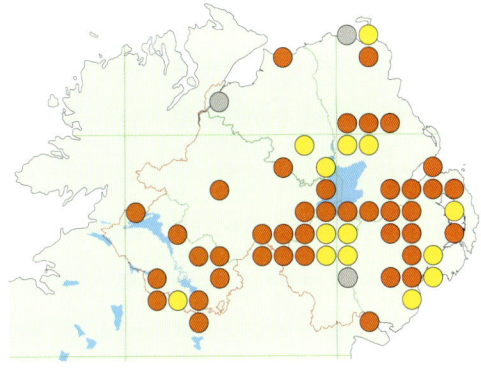

have been recorded up to mid-June, but are commonest in March, April and early May. The recorded flight period is 29 January to 18 June. The larvae feed during May and June on a variety of trees including oak, willow, birch, Hawthorn, Hazel and possibly other broadleaved trees. It overwinters as a pupa.

Twin-spotted Quaker
Orthosia munda (Denis & Schiffermüller) Category one.

Widespread. Largely restricted to the southern half of Northern Ireland. Regular sites include the Argory (Armagh); Belvoir Park (Down) and Crom and Garvary Wood (Fermanagh). It is

Hebrew Character.

Rehaghy Mountain, County Tyrone.

The Hebrew-like markings on the forewings are diagnostic.

seemingly absent from most of Tyrone and Londonderry with the exception of Crilly, Favour Royal and Rehaghy Mountain (Tyrone) and Traad Point (Londonderry). Similarly, in Antrim, the few records have been from the extreme south of the county with the exception of a recent record from Breen Wood and a 1940 record from White Park Bay. Adults are variable both in colour and markings and there are many named forms. The forewing colouration ranges from light ochreous-brown to dark greyish-brown. The orbicular and reniform stigmas are not always clearly defined and the subterminal line is also faint. A diagnostic feature is the two conspicuous black spots near the tip of each forewing. Twin-spotted Quaker is mainly a woodland moth but it can also be encountered in gardens and other well-wooded habitats. Adults come frequently to light and also to willow catkins. Its early flight period may account for the scarcity of records as there seems to be much apparently suitable habitat, especially in Fermanagh. The recorded flight period is 3 March to 18 May. The larvae feed during May and June on willow, oak, Aspen, Ash and possibly other broadleaved species. It overwinters as a pupa.

Hebrew Character
Orthosia gothica (Linnaeus) Category one.
Widespread. Present at a large number of sites in all counties and on Rathlin Island (Antrim). The vernacular name is derived from the appearance of the dark mark in the middle of the

Moth accounts

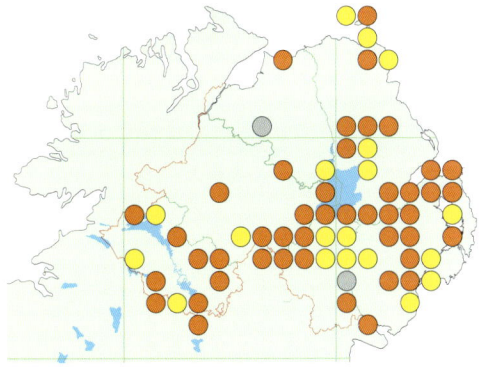

Brown-line Bright-eye.

Lackan Bog, County Down.

forewing which resembles a character from the Hebrew alphabet. The appearance of this mark is variable but it is usually discernible and provides for easy identification of this species. The forewings also vary in colour from pale sandy to blackish; most individuals are reddish-brown. Hebrew Character is a generalist species that can be found in virtually every type of habitat, both rural and suburban. Adults come frequently to light in moderate numbers. It is one of the most frequently encountered species in garden traps in early spring, even in the coolest of nights when little else is flying. There is a single generation in spring and early summer. Adults have been recorded between 15 January and 8 September, with the peak period mid-March to mid-May. The larvae feed on a wide variety of trees including oak, birch, Hawthorn and willows. It overwinters as a pupa.

Brown-line Bright-eye
Mythimna conigera (Denis & Schiffermüller) Category two.
Scarce. Baynes considered this species as common and widespread in Ireland and Wright listed it from all the counties of Northern Ireland. Since the mid-1990s it has been recorded widely in Down, at a few sites in Fermanagh, at Portmore Lough and Rathlin Island (Antrim) and at the Umbra on the north coast of Londonderry. Sites include Killard Point, Lackan Bog, Murlough NNR and Seaforde (Down) and Crom, Monmurry and Randalshough (Fermanagh). The forewings are rust-brown with darker brown anterior and postmedial lines.

The subterminal line is indistinct. There is a white tear-shaped mark at the base of the reniform stigma which is noticeable and diagnostic. The Brown-line Bright-eye is found primarily in coarse grassland on dunes, along the edges of woods and in clearings, bogs and suburban gardens. Adults are attracted to light and reportedly visit the flowers of campion, Common Ragwort and Red Valerian. There is a single generation in mid- to late summer. The recorded flight period is 13 June to 10 August. The larval foodplants are species of grass including Cock's-foot and Common Couch. It overwinters as a larva.

Clay
Mythimna ferrago (Fabricius) Category one.

Scarce. Although Wright listed this species from all counties, it appears to have a pronounced eastern distribution in Northern Ireland. Apart from single records from Derrylin (Fermanagh) and from Banagher Glen (Londonderry) in the late 1970s, it has not been found in recent times outside Antrim, Armagh or Down. It has been recorded regularly from Helen's Bay, Killard Point and Murlough NNR (Down) and at Aughrim on the White Mountain (Antrim). It also appears to be widespread on Rathlin Island (Antrim). Forewing colour in the adult ranges from light sandy brown to a darker rust-brown. The postmedian line is a strongly-curved line of small black dots. The stigmas are inconspicuous except for a small white mark on the reniform stigma.

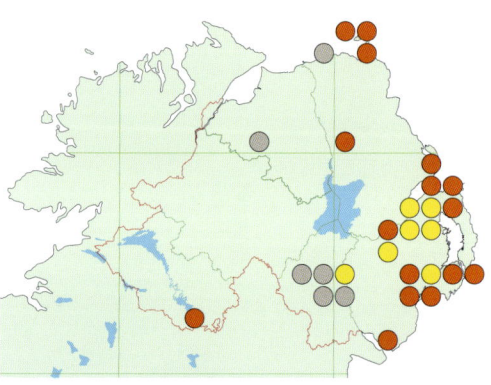

The Clay is found in dune grassland, heaths, woodland margins and clearings and, occasionally, gardens. Adults appear at light sparingly and visit flowers. The recorded flight period is 22 June to 22 August. The larvae, which overwinter, feed from September to May on Cock's-foot and meadow grasses.

Southern Wainscot
Mythimna straminea (Treitschke) Category three.

Notable. Considered by Baynes to be 'rare and local', this was recorded for the first time in Northern Ireland in 1997 at Helen's Bay (Down). It has been recorded on several occasions since at other Down sites – Bohill Forest, Castle Espie, Corbet's Wood – and at Crom (a total of four records between 2000 and 2002) in Fermanagh and on Rathlin Island (three records from two localities, East Light and Ushet, in 2001) in Antrim. This species probably occurs more widely than current records suggest. The other Irish records are from the southern half of the island apart from one in Donegal. The forewings and thorax are a pale straw colour. The greyish line between the head and thorax, along with the obviously straight edge to the forewing, helps to distinguish it from other similar species. Markings on the forewings include a prominent brown streak that runs from the basal area towards the centre of the wing and a curved line of black dots. Adults are attracted to light and also to flowers of grasses. Southern Wainscot inhabits wetlands, particularly those which have an abundance of Common Reed. It appears also to wander into other habitats. The Northern Ireland records have been between 30 June and 13 August. The larval habits are not known in Northern Ireland. In Britain, the larvae are nocturnal feeding from August to May on Common Reed and Reed Canary-grass. It overwinters as a small larva.

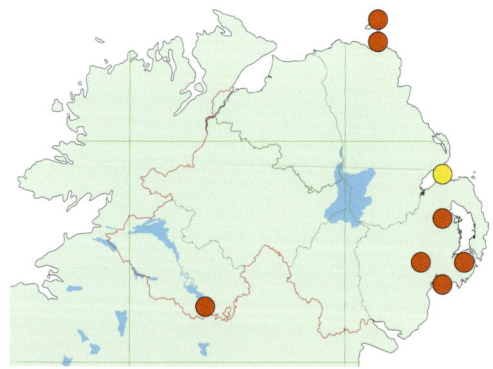

Southern Wainscot.

Castle Espie, County Down.

There were no records of this moth before 1997. It has been found at several widely spread localities since.

Smoky Wainscot

Mythimna impura (Hübner) Category one.

Ubiquitous. This is the most frequently recorded of the wainscot moths in Northern Ireland. It is generally distributed in all counties and it occurs at Copeland Bird Observatory (Down) and on Rathlin Island (Antrim). Adults are straw-coloured with pronounced whitish veins, a postmedian line, two small black dots and a dark basal streak extending to the median area of the forewing. The dark grey hindwings are the diagnostic feature as most other wainscots have paler hindwings. The hindwing has dark, prominent veining, a small discal spot and a white outer fringe. Smoky Wainscot is a grassland species with a preference for rank and damp types.

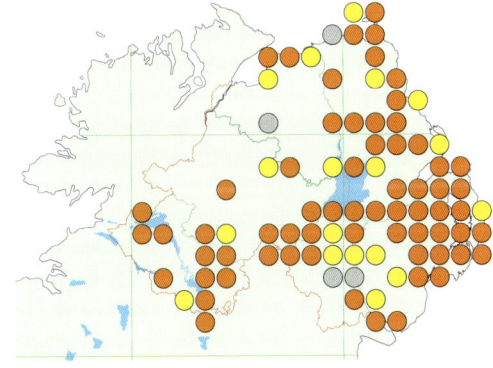

It can be found on bogs and fens (often in good numbers), also dunes, woodland clearings and rides and gardens. There is a single generation each year with a recorded flight period of 8 June to 7 October. The larvae are nocturnal, feeding from September until May of the following year, on Cock's-foot, Common Reed and other grasses. It overwinters as a small larva.

Common Wainscot

Mythimna pallens (Linnaeus) Category one.

Widespread. This species occurs throughout Armagh and Down, but it is much more local elsewhere and scarcer than the preceding species although it is often found with it, but usually in small numbers. It has been recorded from scattered sites in south Fermanagh, mid-Antrim and on Rathlin Island (Antrim). The adults superficially resemble those of the other wainscots, but the pale, straw-coloured wings are narrower. The reniform stigma is reduced to a small black dot and the postmedian line is replaced by two small black dots. The fringe is ochreous in colour. The hindwings are white with darker veining and a similarly coloured fringe. The females are slightly darker and most closely resemble Smoky Wainscot, but this has a darker basal streak and dark hindwing. Common Wainscot inhabits grassland, especially drier types, but can be found in many habitats, including gardens. Adults appear frequently at light and they will also visit flowers. The recorded flight period is 21 June to 30 September. The larvae are nocturnal, hiding among grassy tussocks during the day and feeding from September until May chiefly on Cock's-foot, Common Couch and Annual Meadow-grass. It overwinters as a larva.

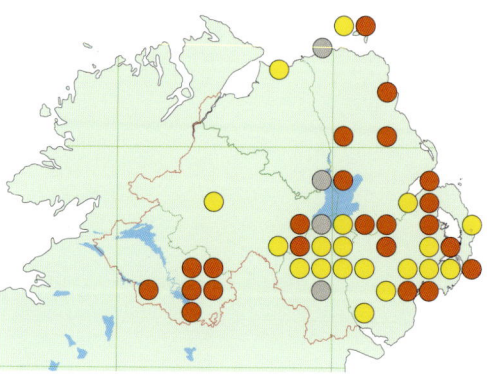

Shore Wainscot

Mythimna litoralis (Curtis) Category four.

Rare. Recorded only from Antrim, Down and Londonderry. Donovan described this species as being generally distributed along the Irish coast. There are historical records for Castlerock, Magilligan (where it was described by Greer and Campbell as abundant) and Portstewart (Londonderry). Baynes also mentions a record by Wright from Dundrum (Down). There were no other records until 1998 when a single adult was taken at the Umbra (Londonderry) on 9 August. It was also rediscovered nearby at Magilligan in 2001 when several adults were trapped at light. This is a distinctive moth, readily identified by the conspicuous white streak which runs from the basal area of the forewing to almost the outer margin and white hindwings. Shore Wainscot is principally associated with coastal dune systems in Ireland. Adults are attracted to light in moderate numbers. There is insufficient data to determine its phenology in Northern Ireland. In Britain it is normally single-brooded, appearing in late June and flying until late August. The larva is apparently nocturnal, concealing itself by day in the sand and feeding from August to May on Marram. It overwinters as a small larva.

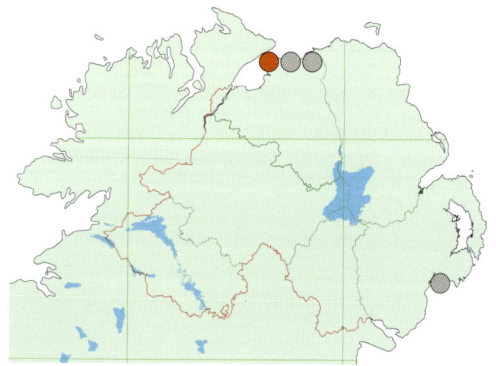

White-speck

Mythimna unipuncta (Haworth) Category four.

Notable. The White-speck is an immigrant that has been recorded on twelve occasions. Originally a North American species, this moth has colonised southern Europe and it is from here that the individuals seen in Northern Ireland almost certainly originate. All of the

Shore Wainscot.

The Umbra, County Londonderry.

Recorded mainly from the north coast. The conspicuous white streak on the forewing distinguishes it from similar species.

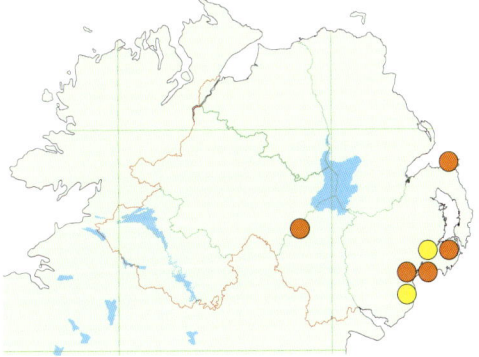

specimens, with the exception of one, have been found on the east coast of Down. It was first recorded in 1988 at Annalong. The other records were 1996 and in each year from 2000 to 2004. The most seen in one year is eight in 2000 – two were trapped at Castleward, one at Moy (Tyrone) and a total of five on two different occasions at Murlough NNR. The most recent records are from Murlough NNR and Copeland Bird Observatory in 2003. Adults are reddish or sandy-brown in colour with fine black speckling and a distinctly pointed forewing. Some individuals are paler in colour with little or no speckling. There is a fine basal white line that extends to a central white dot. There is also a dark apical streak, but this is greatly reduced in some individuals. The hindwings are darkish-grey with a paler fringe. Adults are attracted to light and are also said to visit Ivy blossom. The Northern Ireland records have been between 17 September and 11 October. The species is continually-brooded in southern Europe where adults can be seen throughout the year, but it is unlikely that they would be able to survive a winter in Northern Ireland. The larvae are said to feed on grasses.

Shoulder-striped Wainscot
Mythimna comma (Linnaeus) Category two.
Notable. Since 1990, this species has been recorded from Down and Londonderry. It was reported twice in the early twentieth century in Armagh by Johnson from Acton Glebe and Armagh City, but it has not been seen in the county since. The recent records have come from Dundonald, Helen's Bay, Killard Point, Murlough NNR and Seaforde (Down) and the Umbra (Londonderry). Adults are mainly straw-coloured with paler veining. The forewings are marked with a black basal streak (which extends to the middle of the wing) and a white median dot. There is a series of fine black streaks in the outer region of the forewing and a pale band running along the costa. The hindwings are brown with a paler fringe, but occasionally lighter in some individuals. In Northern Ireland, Shoulder-striped Wainscot has been recorded in dune slacks and grassland and gardens. In Britain, it is also found in fens and damp woodland. Adults hide during the day among grass tussocks and other dense vegetation. They are attracted to light and flowers. The recorded flight period is 10 June to 9 August. The larvae are nocturnal, feeding during August and September on Cock's-foot and probably other grasses. It overwinters as a larva.

Chamomile Shark
Cucullia chamomillae (Denis & Schiffermüller) Category four.
Rare. The status of this species is unclear as there have been few records and none have been reported since 1994. An adult was taken at light at Belvoir Park (Down) on 26 May 1991; in June of the following year a larva was recorded from Ballymacormick Point (Down). The only other accepted record is from Ballymacilrany (Antrim) by Wright in 1951. Adults have long narrow forewings which are greyish-brown and pointed at the apex. This, and related species, rest with their wings held tightly to the body. All have a raised projection on the front of the thorax that recalls a dorsal fin of a shark. The females are darker than the males. Chamomile Shark has a mainly coastal distribution in Ireland, but is also found on waste ground and marginal edges of arable land inland.

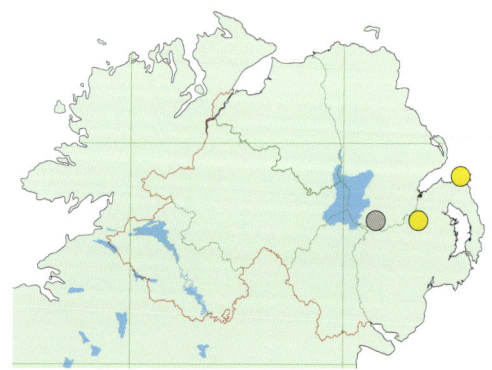

Adults are attracted to light and flowers, especially campions. It reportedly can be seen by day resting on fence posts and tree trunks. In Britain this species has a single generation in April and May, but there is insufficient data to determine its phenology in Northern Ireland. The larval foodplants are Scentless Mayweed, Chamomile and related composites. It overwinters as a pupa.

Shark
Cucullia umbratica (Linnaeus) Category one.
Scarce. Recorded on the east and north coasts between Dundrum (Down) and Portstewart (Londonderry) and on Rathlin Island (Antrim). Inland, it is known from north Armagh, Fermanagh and the Greater Belfast area. Sites with recent records include Glenariff Forest, Portrush and White Mountain (Antrim); the Argory, Aughinlig, Ballyrea and Portadown

Moth accounts

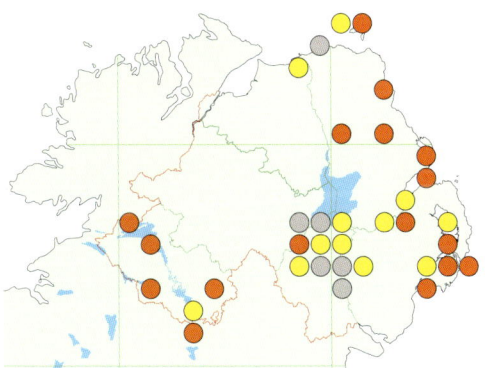

Shark.

Killard Point National Nature Reserve, County Down.

The most frequently recorded of the two Cucullia *species found in Northern Ireland.*

(Armagh) and Helen's Bay, Killard Point and Seaforde (Down). In Fermanagh, it has been found since 2000 at Cuilcagh, Derrygonnelly, Legatillida and Garvary Wood. This is the only widespread species of *Cucullia* in the region, the only other being the very rare Chamomile Shark. The forewings of the Shark appear greyer and more streaked than in that species. The apex of the wing is also arched and pointed. The antemedian and postmedian lines are usually indistinct and there are small black streaks near the apex of the forewing which do not continue into the fringe. There is also a small black broken line at the outer margin of the forewing. The Shark has been taken in bogs, heaths, gardens and urban sites with patches of tall, ruderal vegetation.

281

The Butterflies and Moths of Northern Ireland

Sprawler.

Rehaghy Mountain, County Tyrone.

This large, easily identified moth is on the wing in autumn. It is restricted to a few sites in Fermanagh and Tyrone.

Adults come frequently to light and are reported to visit the flowers of Red Valerian and Honeysuckle. It is said to rest by day on the trunks of trees and, occasionally, fence posts. The recorded flight period is 18 May to 25 September. The main larval foodplants are sow-thistles, hawk's-beards and hawkweeds. It overwinters as a pupa.

Minor Shoulder-knot
Brachylomia viminalis (Fabricius) Category three.
Notable. This species has been recorded from Ballymacilrany and Shane's Castle (Antrim), Murlough NNR (Down) and Banagher Glen (Londonderry), but has not been seen at any of these sites since 1990. There have been recent records (since 1995) from the limestone uplands of west Fermanagh including Cuilcagh, Gortmaconnell Rock, Legacurragh and Legalough. In

2004 it was found at Glenarm (Antrim). Adults are generally light grey, but the colour varies between individuals. There is a short black basal streak on the forewing. The orbicular and reniform stigmas are paler than the ground colour and finely outlined in white. The subterminal line is pale white and irregular. There is also a small brown patch on the costa near the apex of the forewing and a fine black broken line at the outer margin. Minor Shoulder-knot is associated with damp woodland, heaths and marshy areas. Adults are attracted to light and flowers, in particular, Viper's Bugloss and campions. It hides by day among the undergrowth. The recorded flight period is 26 June to 18 August. Nothing is known about the larval habits of this species in Northern Ireland. In Britain, its main larval foodplant is Grey Willow but other willows and Aspen may also be used. It overwinters as an egg.

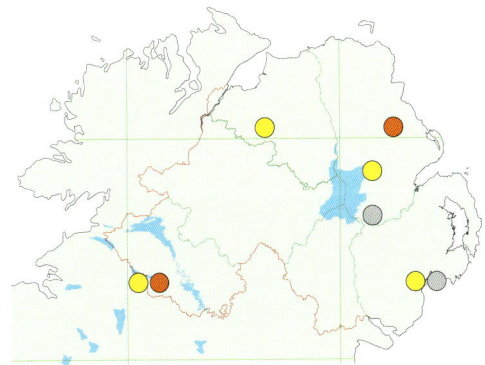

Sprawler
Asteroscopus sphinx (Hufnagel) SOCC Category four.
Rare. This large attractive moth was unknown in Northern Ireland before 1992 when it was discovered at Crom in Fermanagh. It has subsequently been found nearby at Aghalane (Fermanagh) in 2000 and at three sites in south-east Tyrone, Crilly, Lemnagore Wood on the Caledon Estate and Rehaghy Mountain. It has been seen regularly at both Crom and Rehaghy Mountain. This is a fairly large moth with a rather hairy thorax. The forewings are brownish-grey with a few darker longitudinal and thoracic streaks. The basal streaks are darker brown. Its size and late flight period make confusion with any other species unlikely. It has been found in Northern Ireland in mature broadleaved woodland, particularly old estate woodland. Adults come to light, usually after midnight in very small numbers, but do not feed. They rest by day on branches, tree trunks and old fence posts. There is one annual generation in autumn. The recorded flight period in Northern Ireland is 23 October to 12 November. The larvae have not been reported in Northern Ireland. In Britain, it feeds during May and June on a range of broadleaved trees including oak, Hazel, willow and Hawthorn. It overwinters as an egg.

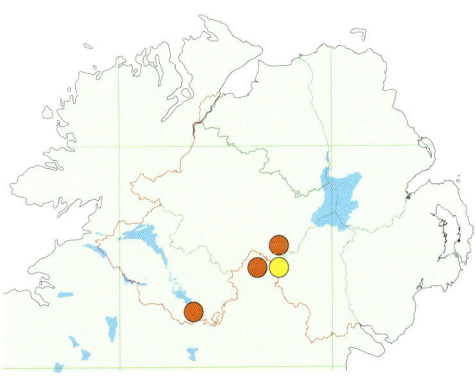

Brindled Ochre
Dasypolia templi (Thunberg) Category three.
Notable. This is a coastal species, recorded in recent times mainly from Down. There have been reports since 1990 from Down at Killinchy (1990), Castleward (2000), Copeland Bird Observatory (2003) and Murlough NNR (2000 and 2003). The only records since 1980 away from Down are from several sites on Rathlin Island (Antrim) where it was reported first in 1989 and again in 2000. There are also historical records from Glenarm (Antrim) (1902), Limavady (Londonderry) (1969) and Tollymore Forest (Down) (1969). The male has yellowish-brown forewings; the females are generally much darker. There is a rough appearance to the forewings and both the thorax and abdomen are quite hairy. The orbicular and reniform stigmas are paler than the ground colour. Brindled Ochre has been recorded on dunes, coastal heaths and unmanaged rough grassland. There is one generation a year, emerging in autumn. Once mated, the female enters

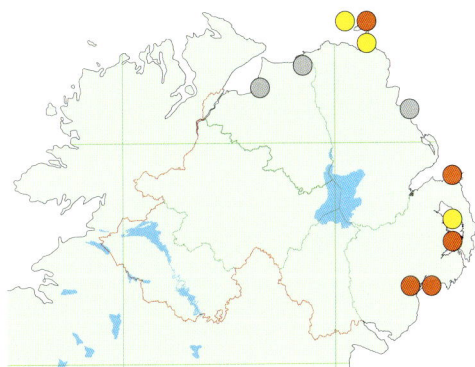

hibernation, reappearing in early spring. Adults are attracted to light, but they do not feed. The recorded flight period is 23 September to 1 November. There have been no records of overwintered females in Northern Ireland. The larval habits are unknown in Northern Ireland. In Britain, it feeds from late spring until early August in the roots of Hogweed and Wild Angelica.

Northern Deep-brown Dart

Aporophyla lueneburgensis (Freyer) Category four.
Notable. Apart from nineteenth-century records from Castlerock, Kilderry and Magilligan in Londonderry, this is currently known only from Rathlin Island (Antrim) (where it appears widespread), Ballynasollus (Tyrone) – where four adults were recorded in 1995 – and several sites on the coast of south Down. Here, it was trapped at Rostrevor Wood in 2000 and Ballykinler Dunes in 2002. Additionally, it was recorded from Murlough NNR and Tollymore in 1978 and 1988, respectively. This is a dark grey moth with a blackish central band which widens towards the costa and is finely edged in white. The orbicular and reniform stigmas are lighter than the central band and outlined in white. The hindwings are white with a series of black dots which often form a curved line. Northern Deep-brown Dart was formerly regarded as a subspecies of Deep-brown Dart, but is now widely accepted as a distinct species. Adults have been found on rough grassland and heaths in rocky areas on the coast and just inland. Both sexes are attracted to light and to flowers, including Heather and Common Ragwort. The recorded flight period is 24 August to 30 September. The larvae, which overwinter, feed from September until May of the following year on Heather, Bilberry and Bird's-foot Trefoil.

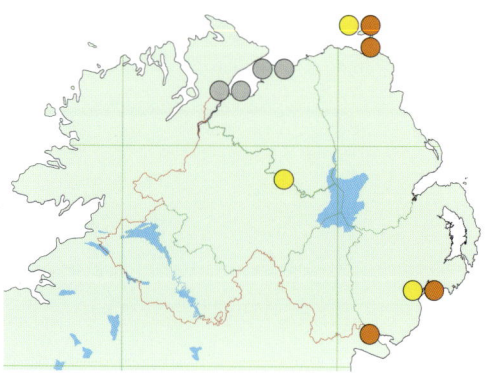

Black Rustic

Aporophyla nigra (Haworth) Category two.
Scarce. This moth seems largely confined to Armagh, Down and Fermanagh, although its current distribution may mostly reflect recording effort. It occurs mainly on the coast in Down including at Ballykinler Dunes, Murlough NNR, Rostrevor Wood and Seaforde. The Umbra is the only known site in Londonderry and it is almost as rare in Antrim, occurring only on Rathlin Island and at Magheramorne. Crilly, Favour Royal and Rehaghy Mountain are the only sites in Tyrone. In Fermanagh, Crom and Garvary Wood are the most commonly recorded sites. This is a very dark moth and the adults appear to be completely black. The forewings vary from very dark brown to black. The antemedian and postmedian lines are black. The subterminal line is irregular and curves towards the costa. The distal aspect of the reniform stigma is white appearing like a small mark in the central area of the forewing. The hindwings in the male are white with fine darker veining. Black Rustic is an autumnal species found in a variety of habitats including coastal heaths, moorland, woodland clearings and, occasionally, rural gardens. Adults come frequently to light and apparently visit flowers and blackberries. The recorded flight period is 25 August to 9 November. The larvae feed on a variety of plants including Heather and clovers. It overwinters as a small larva.

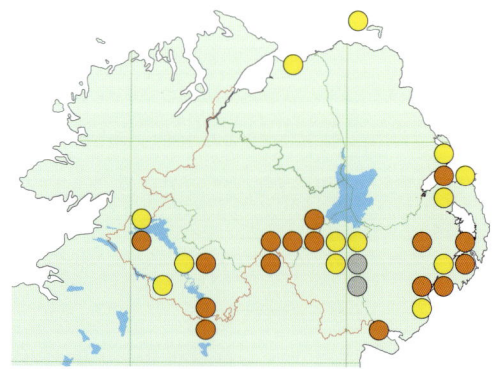

Pale Pinion

Lithophane hepatica (Clerck) Category one.
Scarce. Common and widespread in a band running across southern counties. Unrecorded in the north, the northernmost locality being Traad Point (Londonderry) on the north-west shore

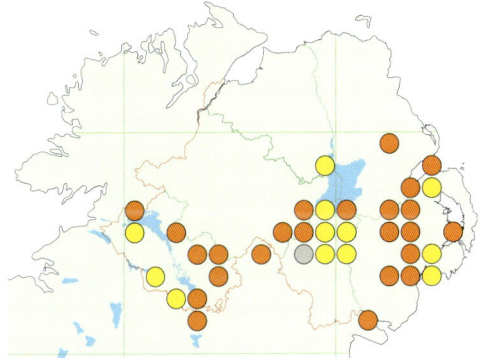

Pale Pinion.

Rehaghy Mountain, County Tyrone.

of Lough Neagh, where it was recorded in 1990. Reliable sites include the Argory, Aughinlig and Peatlands Park (Armagh); Helen's Bay and Rostrevor Wood (Down); Crom and Garvary Wood (Fermanagh) and Rehaghy Mountain (Tyrone). Adults are straw-coloured, with darker shaded areas running along the length of the forewing. The thorax has rounded shoulders and a prominent crest, both diagnostic features. The antemedian and postmedian lines are unclear, reduced to a few black dots. This species keeps its wings tightly wrapped around its body and resembles a piece of weathered wood when at rest. Pale Pinion is primarily a woodland species in Northern Ireland but it also occurs in small numbers on wooded bogs and fens. Adults appear at light in small numbers and also visit Ivy flowers and fermenting fruit in the autumn and sallow catkins in early spring. The recorded flight period is 23 September to 6 November and again after hibernation from 24 February to 4 July. The adults are most commonly encountered in the spring after hibernation. The larvae feed from May until July on willow, oak, birch, Bramble and possibly other trees and shrubs.

Grey Shoulder-knot
Lithophane ornitopus (Hufnagel) Category two.
Notable. This has a very restricted distribution and has been found mainly in east Tyrone and north Armagh. The known sites included the Argory, Brackagh Moss and Gosford Forest (Armagh) and Rehaghy Mountain (Tyrone). It has also been recorded from Glenarm (Antrim), Murlough NNR and Rostrevor Wood (Down) and Crom and Killadeas (Fermanagh). Adults are normally whitish-grey, with a series of small irregular darker patches across the wing. The diagnostic feature is a small black basal streak at the base of the forewing which is forked at the end and the central thoracic crest. Grey Shoulder-knot is an autumnal species favouring broadleaved woodland habitats with mature oaks. Adults are attracted to light, but are seldom found in any 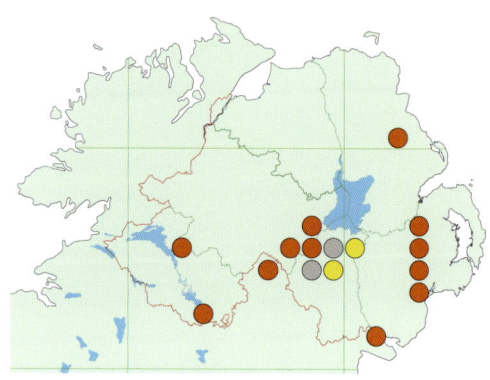 numbers. The adults are more commonly encountered in early spring after hibernation. In the autumn they have been noted feeding at Ivy blossom, berries and rotting fruit. They apparently rest by day on the trunks of trees and old fence posts. The recorded flight period is 23 September to 7 November and again, after hibernation, from 23 February to 23 April. The larval foodplant is oak. It hibernates as an adult.

Blair's Shoulder-knot
Lithophane leautieri (Boisduval) Category four.
Rare. This species was recorded for the first time in Northern Ireland from a garden at Tyrella (Down) on 31 October 2004. There was a second record from a mature garden at Aughinlig (Armagh) on 15 November 2004. These are the only two confirmed records at present, although considering the abundance of its foodplants, it is likely to appear in other localities in the near future. The only other Irish records are from Wicklow. Adults have long, narrow grey forewings with two conspicuous black basal streaks. There are other smaller streaks located centrally on the forewing. The size and intensity of these markings can vary between individuals. The thorax is normally a lighter grey than the forewings and slightly crested. This species is quite distinctive and unlikely to be 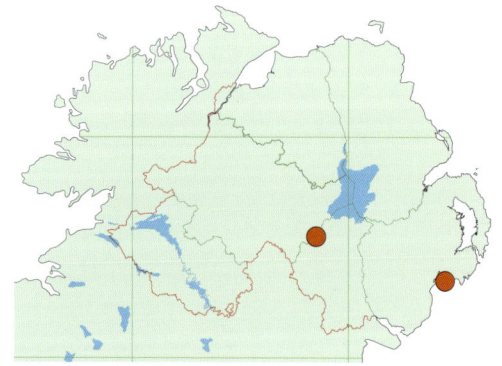 confused with any other autumnal species. Adults are likely to be encountered from late summer onwards. In Britain, it has become an established resident and widely distributed occurring on the Isle of Wight (where it was first discovered in 1951) northwards through southern England, the Midlands, Wales to Scotland. Adults are attracted to light but are seldom seen by day. Its preferred habitat is conifer plantations, but it is also found in gardens and other areas where its foodplants occur. The larvae feed on ornamental conifers including Monterey, Lawson's and Leyland Cypress.

Red Sword-grass
Xylena vetusta (Hübner) Category one.
Scarce. Recorded in all counties, but most frequently recorded across the southern half of Northern Ireland. It has been reported in north Antrim from Rathlin Island and Breen Wood

Moth accounts

Grey Shoulder-knot.

Rehaghy Mountain, County Tyrone.

Blair's Shoulder-knot.

Tyrella, County Down.

This photograph shows the first adult ever recorded in Northern Ireland.

287

Red Sword-grass. Lackan Bog, County Down.

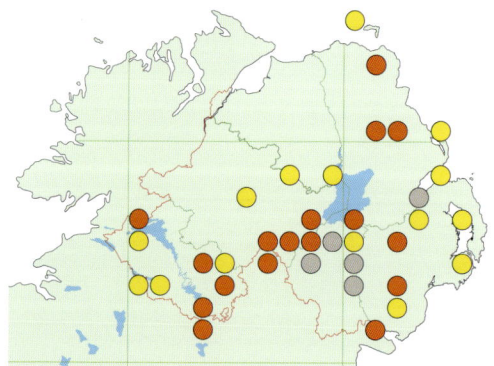

(where it was found in 2003). The species is probably still under-recorded in the north and west. Adults are distinctive and easily recognised when at rest as they wrap their wings tightly around their body, thus resembling a broken or dead piece of wood. The forewings are straw-coloured with a reddish flush along the dorsal half of the wing. There is a small white mark in the central area of the forewing and an irregular subterminal line. The hindwings are fuscous with a small discal lunule. Red Sword-grass has been found in damp woodland, bogs, fens and moorland in Northern Ireland. Adults come regularly to light and will feed on fermenting berries, fruit, and Ivy blossom in autumn. The recorded flight period is 15 September to 5 November and again after hibernation from 1 January to 13 May. The larvae feed on herbaceous plants, including Bog Myrtle, dock, Yellow Iris and grasses.

Sword-grass
Xylena exsoleta (Linnaeus) SOCC (P) Category three.
Unknown. Baynes stated this species was distributed throughout Ireland, but was less common than Red Sword-grass. Donovan was of the same opinion. This appears to overstate the case, at least in the north, where it appears always to have been rare. There are only three acceptable records, all from Armagh, from Acton Glebe in 1899, Drumbanagher in 1940 and Loughgall in 1953. There are historical records for Enniskillen (Fermanagh), Londonderry and Favour Royal (Tyrone). The current status of the Sword-grass here is unknown. Whether it was ever resident is also impossible to determine as the old records may refer to immigrants. In Britain, the species

has declined in the south and remains common only in Scotland. Adults are similar in appearance to the more common Red Sword-grass when at rest. The forewings are straw-coloured and lack the reddish colouration found on Red Sword-grass. There is no information on the habitat of this species in Northern Ireland. In Britain it is found on moorland and open woodland. In the autumn prior to hibernation, adults are attracted to light, fermenting berries and Ivy blossom and in early spring to Blackthorn blossom and willow catkins. The flight period is not known in Northern Ireland. In Britain, adults are seen in September and October and again after hibernation from March to May. The larval habits and foodplants are not known in Northern Ireland. In Britain the larvae feed on Common Restharrow, Creeping Cinquefoil, Groundsel and docks.

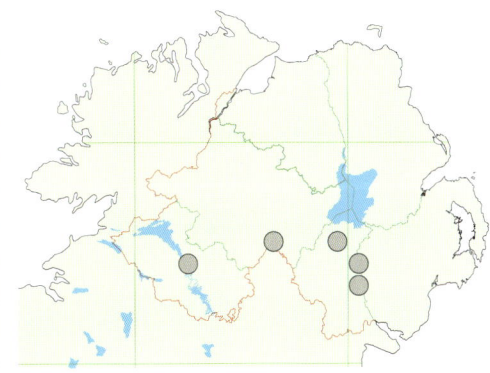

Early Grey
Xylocampa areola (Esper) Category one.
Widespread. Widely distributed and frequently trapped at sites in the south, especially Fermanagh and the Greater Belfast area, but seems to be more local further north and west and confined to a few coastal localities in north Antrim and Londonderry. Adults vary in colour from light to dark grey. Some individuals have a slight pinkish tinge to the forewings. There is a small black basal streak and scattered darker patches across the forewing. The subterminal line is generally grey and irregular. The hindwings are pale grey with a small discal spot. The largest populations of Early Grey are found in broadleaved woodland, but it can be taken in many other habitats with trees, including suburban gardens. Adults come frequently to light and will also visit willow catkins. The adults are occasionally found at rest in early morning on tree trunks close to the light trap. They can also be found occasionally on fence posts and rocks, by day. There is a single generation annually, emerging in late winter and early spring. The recorded flight period is 4 February to 28 May. The larvae can be found from April to June on Honeysuckle. It overwinters as a pupa.

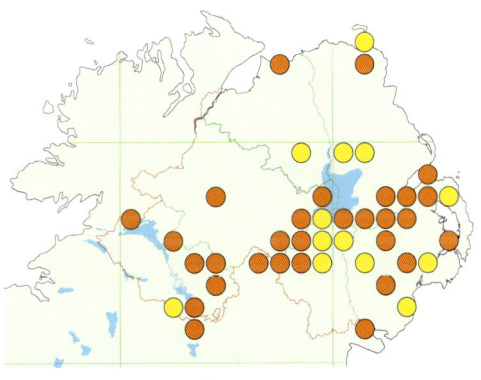

Green-brindled Crescent
Allophyes oxyacanthae (Linnaeus) Category one.
Widespread. Widely recorded in Armagh and Down but apparently more localised in the west and north. It has been trapped regularly in the appropriate season at several sites in both Fermanagh and Tyrone including Aghalane, Crom, Garvary Wood and Monmurry (Fermanagh) and Benburb, Crilly and Rehaghy Mountain (Tyrone). It is probably still under-recorded in most of the north as a result of its late flight period. The northernmost record is from Ballymena in mid-Antrim. Adults are predominately brown with a green metallic sheen to the forewings. They are marked with a black basal streak and antemedian line and brown postmedian line. Green-brindled Crescent has been recorded in clearings and margins of broadleaved woods, along mature hedges and in gardens. Adults appear commonly at light and are said to feed at Ivy blossom and fermenting berries. The recorded flight period is the 20 August to 7 November. The peak period for records is late September to mid-October. The larval foodplants are Blackthorn, Hawthorn and willows. It overwinters as an egg.

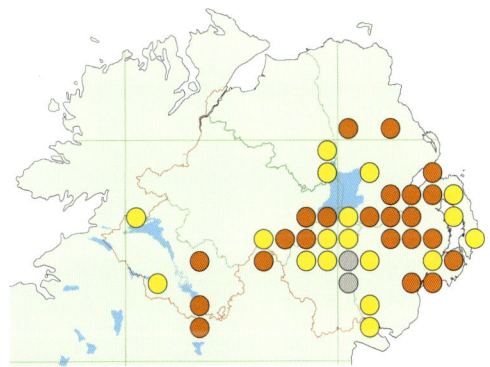

Early Grey.

Lackan Bog, County Down.

Green-brindled Crescent.

Castlewellan Forest, County Down.

Merveille du Jour.

Rehaghy Mountain, County Tyrone.

A striking moth that appears in early autumn. The mottled green markings camouflage the moth against lichen-covered branches and tree trunks.

Merveille du Jour
Dichonia aprilina (Linnaeus) Category one.
Scarce. This distinctive species has been recorded in all counties, but from relatively few, well-scattered sites; at many of these the records have been sporadic. The most reliable sites appear to be in north Armagh at the Argory, Aughinlig, Loughgall and Peatlands Park. In Fermanagh there are records for Aghalane, Crom, Derrylin and Garvary Wood. It has recently been discovered at Breen Wood in north Antrim. There are also recent records for Down from Belvoir Park, Castlewellan Forest Park, Hillsborough Forest and Helen's Bay. It is probably still under-recorded especially in broadleaved woodland in the west and north. This attractive and richly-marked moth is unlikely to be confused with any other species flying in early autumn. There is some

variation between individuals in the amount of colour and patterning. Freshly-emerged adults have pale green forewings, but the green colour fades fairly quickly. However, unlike many other green species some of the pigment is always retained. There is a series of black spots across the central area of the forewing and a few black marks on the top of the thorax. Both stigmas are paler than the ground colour. Merveille du Jour has been recorded in semi-natural and old estate broadleaved woodland and, occasionally, gardens. Adults come frequently to light in small numbers and apparently visit Ivy blossom and various overripe berries. The recorded flight period is 22 September to 10 November. The larvae feed in spring and early summer on the opening leaves and flowers of oaks, resting by day in crevices on the branches and the trunk. Pupae have been found at Belvoir Park (Down) and Enniskillen (Fermanagh). It overwinters as an egg.

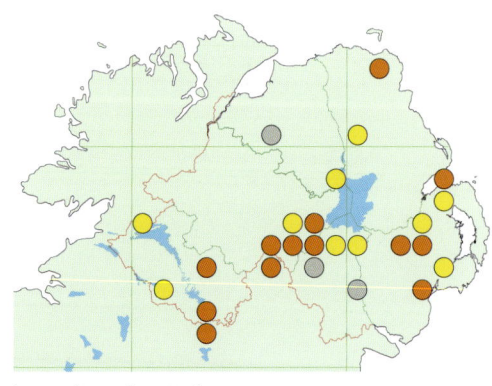

Brindled Green
Dryobotodes eremita (Fabricius) Category three.

Scarce. Baynes considered Brindled Green a rarity in Ireland and 'of northern and north-eastern distribution'. It appears on Wright's list from Aghalee (Antrim) in the 1950s and Greer recorded it at sugar in 1916 near Stewartstown (Tyrone); however, recent records indicate it is confined to southern counties and not found north of the southern shore of Lough Neagh. It appears sporadic in occurrence even at well-trapped sites such as Crom (Fermanagh). It has recently been rediscovered at Rostrevor Wood (Down), where it was described as abundant in the early twentieth century. Other recent sites include the Argory and Loughgall (Armagh); Castleward, Castlewellan Forest and Seaforde (Down) and Aghalane, Brookeborough, Derrylin, Garvary Wood and Marble Arch (Fermanagh). Adults are generally a dull greenish-grey, sometimes flecked with black or brown. There is a small whitish blotch adjoining the orbicular stigma which is bordered with black on the distal edge. The hindwings are light brown with a small discal spot. The Northern Ireland localities are all in, or close to, established broadleaved woodland. Adults come sparingly to light and will visit fermenting blackberries. The recorded flight period is 31 July to 7 November. The larvae have not been recorded in Northern Ireland. In Britain, they feed inside the developing buds of oak and possibly Hazel, although the latter is not found at most of its Northern Ireland sites. It overwinters as an egg.

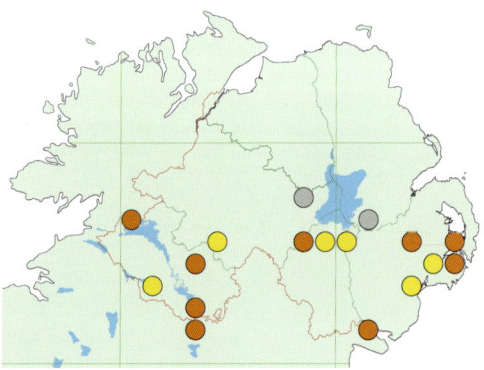

Dark Brocade
Blepharita adusta (Esper) Category two.

Scarce. Present in the south and west especially west Fermanagh and north Armagh, but only known from one site in Down (Saintfield). There appear to be no recent records from Londonderry. Antrim sites include Ballinderry, Magheramorne, Portmore Lough and Rathlin Island. In Tyrone, it has been recorded at Aughentaine and Rehaghy Mountain in the extreme south of the county and at Lough Fea in the north. Adults are quite variable, ranging from reddish-brown to dark brown. The orbicular and reniform stigmas are paler and outlined in white. A black bar usually connects the antemedian and postmedian cross-lines. The subterminal line is irregular and yellowish-white with two tooth-shaped

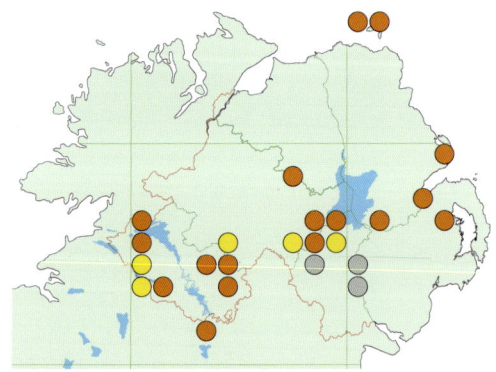

Brindled Green.

The Argory, County Armagh.

projections. This moth has been recorded in damp woodland, heathland, fens and bogs. Adults appear in small numbers at light and they also visit flowers. The recorded flight period is 23 May to 20 August. The larvae feed from July to September on grasses and other plants associated with wetland areas such as Bog Myrtle and Alder. It overwinters as a mature larva.

Grey Chi

Antitype chi (Linnaeus) Category two.

Scarce. In Ireland this species has not been recorded south of Dublin and Clare. It is recorded from all the counties in Northern Ireland, but the records are concentrated in the Greater Belfast area and around Lough Neagh. It is known from two sites on the north coast in Londonderry (Limavady and the Umbra), on Rathlin Island (Antrim), and three sites in Fermanagh (Aghalane, Garvary Wood and Legalough). Other recent localities include Ballymena and Slievenacloy (Antrim) and Belvoir Park (Down). Adults are pale grey with a dark mark or bar in the middle of the forewing. The orbicular and reniform stigmas are similarly coloured and outlined with a fine dark line. The hindwings are white in the male, light grey in the female with darker veining. Grey Chi has been recorded in grassland, bogs and heaths, both upland and lowland. It is not confined to natural habitats as it has been taken frequently in gardens and other habitats in urban areas. Adults appear at light in small numbers and may be found by day at rest on stone walls and rocks. The recorded flight period is 3 August to 30 September. The larval habits are unknown in Northern Ireland. In Britain, the principal foodplants are considered to be docks and sorrels. It overwinters as an egg.

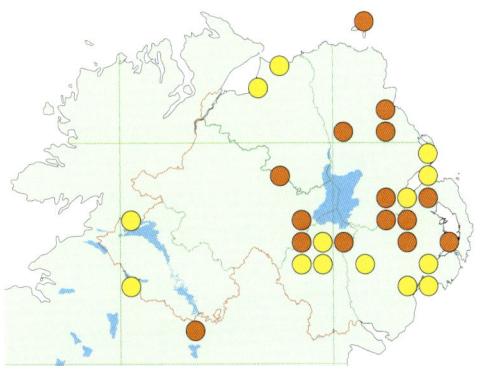

Feathered Ranunculus

Polymixis lichenea (Hübner) Category four.

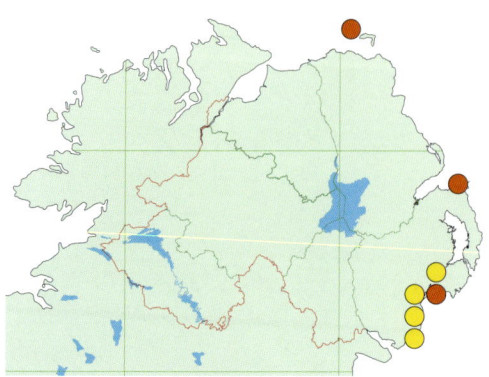

Notable. An autumnal species that is restricted to a few localities on, and close to, the coast of south Down and Rathlin Island (Antrim). It was reported for the first time in Northern Ireland by Wright in the Newcastle area in 1973. In Down, it has been recorded from the coast at Annalong, Ballykinler, Murlough NNR and Tyrella and inland at Castlewellan Forest, Corbet's Wood and Seaforde. It was found at the West Light on Rathlin Island in 2001. There are no recent records from any other counties. Elsewhere in Ireland, it is confined to the south and east coasts from Kerry to Dublin. The forewings are greyish-green and appear mottled when the adult is at rest. The antemedian and postmedian lines are darker than the ground colour. The orbicular and reniform stigmas are paler than the ground colour and outlined in white. The males have feathered antennae. The hindwings are white in the male but grey in the female with a small discal spot. On the coast, Feathered Ranunculus has been found on dunes and heaths. The habitat at the few inland localities is not known. Adults appear at light in small numbers and also visit flowers, especially Ivy blossom and overripe blackberries. The recorded flight period is 24 August to 11 October. The larval foodplants are not known in Northern Ireland. In Britain, the recorded larval foodplants include Hound's-tongue (this is rare in Northern Ireland and restricted to the east cost of Down), Thrift and stonecrop. It overwinters as a larva.

Satellite

Eupsilia transversa (Hufnagel) Category two.

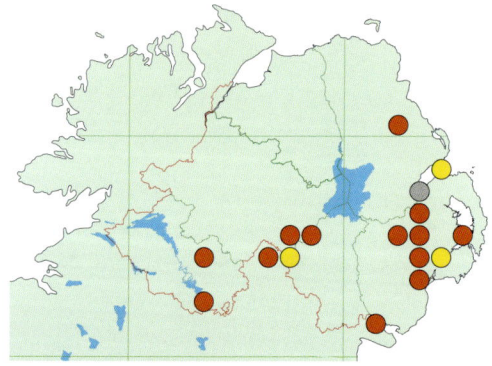

Scarce. Recorded from all counties except Londonderry. It is most frequent in Down, where it is mainly reported from the coast. It is apparently much scarcer in Armagh, Fermanagh and south Tyrone although it has been recorded recently from Caledon Estate and Creeve Lough in Tyrone. There is just one recent record from Antrim but the lack of early- and late-season trapping, when the adults are on the wing, may account for its absence in seemingly suitable localities. Regular sites include the Argory (Armagh); Helen's Bay and Rostrevor Wood and Crom (Fermanagh). It was recorded a number of times in 1937 and 1938 by Neal and Denis Rankin in Belfast Castle Estate (Antrim). The forewings have a satiny sheen on freshly-emerged individuals. The colour varies from light to dark reddish- or orange-brown. The antemedian and postmedian lines are darker than the ground colour. The reniform stigma appears as a small white or orange dot which is diagnostic. The Satellite has been recorded in broadleaved woodland (especially old estate woodland), heathland and suburban gardens. Adults come to light in very small numbers and also visit Ivy blossom, willow catkins and, occasionally, overripe berries. There is one annual generation appearing in early autumn. Both sexes overwinter and re-emerge in late winter during mild spells. The recorded flight period is 30 September to 16 December and again, after hibernation, from 10 February to 23 April. The larvae feed during May and June on oak, birch, willow and probably other trees and shrubs.

Chestnut

Conistra vaccinii (Linnaeus) Category one.

Widespread. Recorded from all counties, but most commonly encountered in the south and east. Baynes described it as abundant, but this is not the case at present as it appears to be absent from

Dark Chestnut.

The Argory, County Armagh.

some well-trapped localities in each county. Regularly recorded sites include the Argory (Armagh), Helen's Bay (Down), Crom and Monmurry Bog (Fermanagh) and Rehaghy Mountain (Tyrone). Adults can be quite variable in colour and pattern with some individuals being more richly marked than others. The forewing colour ranges from light to dark reddish-brown. The orbicular and reniform stigmas are clearly defined in some forms, but obscure or reduced to a dark dot in others. Many of the adults in Northern Ireland populations have a pale band near the termen. It closely resembles the Dark Chestnut but that species has more pointed forewings. The Chestnut can be found in a wide range of habitats including woodland, bogs, heaths and suburban gardens. It is attracted to light in moderate numbers although exceptionally large numbers have been recorded at several sites, such as the Argory (Armagh). Adults also visit Ivy blossom, ripe berries and willow catkins. The recorded flight period is 22 September to 14 May. Adults can be seen throughout the winter, emerging during mild spells. The larvae feed from May until June on oak, birch, Hawthorn and Blackthorn. It overwinters as an adult.

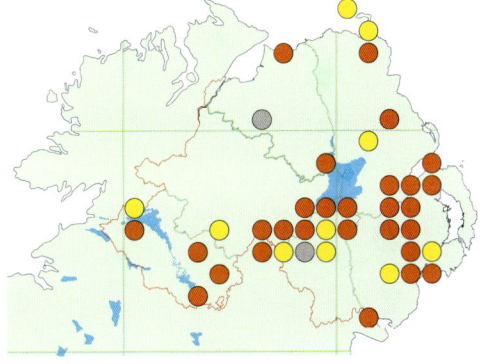

Dark Chestnut

Conistra ligula (Esper) Category three.
Notable. The Dark Chestnut is seemingly not as common in Northern Ireland as the closely-related Chestnut; however, this species does not have such a prolonged flight period so there may be some under-recording. It has been recorded from fewer than fifteen sites. Most records come from Aughinlig (Armagh) (1975-2001) and Traad Point (Londonderry) (1989 and 1990). The only other post-2000 records are from the Argory (Armagh), Monmurry Bog (Fermanagh) and Rehaghy Mountain (Tyrone). This moth is very similar to the Chestnut but can be distinguished

with care by the more pointed forewings and the generally darker appearance. The forewings are dark reddish-brown with inconspicuous cross-lines and stigmas. The hindwings are dark. The Northern Ireland records of Dark Chestnut have come from scrubby woodland, gardens, bog and fens. Adults are attracted to light and to flowers of Ivy, fermenting berries and willow catkins. The recorded flight period is 21 September to 20 January. The females are mated and lay eggs in late winter. The larvae hatch in spring as the leaves of the foodplant open. They feed initially on shrubs including Blackthorn and Hawthorn, willows and oak, then on herbaceous plants such as Dandelion.

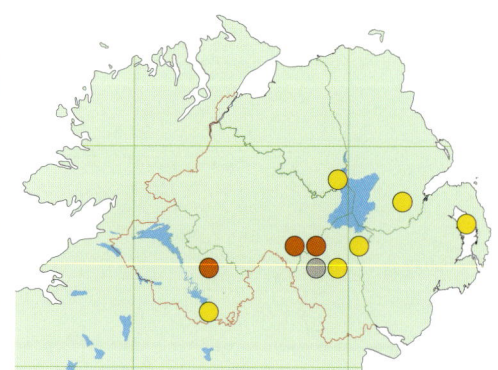

Brick
Agrochola circellaris (Hufnagel) Category two.
Scarce. Widespread throughout north Armagh and adjacent parts of south Tyrone. It also occurs at scattered sites in Down and Fermanagh. There have been no records north of Traad Point (Londonderry) on the north-west corner of Lough Neagh. Recent localities include Finaghy and Montiaghs Moss (Antrim); Aughinlig (Armagh); Castlewellan Forest, Copeland Island and Corbet Lough (Down); Crom (Fermanagh) and Favour Royal and Rehaghy Mountain (Tyrone). Adult colouring in this species is prone to individual variation from light to dark reddish-brown. The antemedian and postmedian cross-lines vary in intensity; on some individuals they are hardly noticeable. The subterminal line is irregular and either red or yellowish-brown. The lower part of the reniform stigma appears black. The Brick is an autumnal species most commonly recorded in broadleaved woodland, but also in wetlands, hedgerows and, rarely, gardens. Adults appear at light in very small numbers. It has also been documented that they often visit Ivy blossom and fermenting berries. Adults usually remain well concealed in vegetation; however, individuals have been disturbed from undergrowth by day in Northern Ireland. There is one generation a year in autumn. Adults have been seen between 31 August and 14 November. The larval habits and foodplants in Northern Ireland are not known. In Britain it feeds from April to June on Wych Elm, poplars, willows and Ash. It overwinters as an egg.

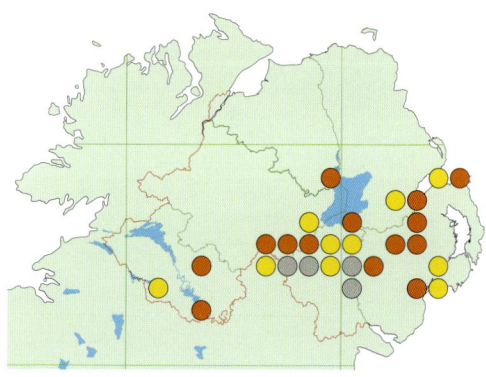

Red-line Quaker
Agrochola lota (Clerck) Category two.
Scarce. This autumnal species is found mainly in north Armagh, north Down and south Antrim as far north as Capanagh. Apparently absent from the northern part of Antrim and most of Londonderry and Tyrone away from Lough Neagh. There have been a few records from Fermanagh and south of Newcastle in Down. Reliable sites include the Argory, Aughinlig, Brackagh Moss, Loughgall and Peatlands Park (Armagh); Crom and Monmurry (Fermanagh) and Rehaghy Mountain (Tyrone). Adult colour ranges from pale to dark lead grey. The antemedian and postmedian lines are reduced to small dark dots. The orbicular and reniform stigmas are indistinct, but a small dark spot is clearly visible in the median area of the forewing. The subterminal line is pale and bordered with a fine red line. Red-line Quaker is found commonly in open woodland, heaths, fens and, occasionally, suburban gardens. Adults come to light in small numbers. Like

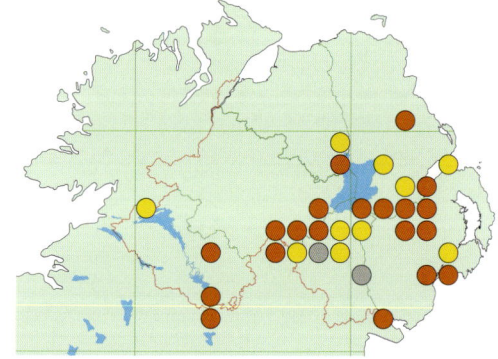

Flounced Chestnut.

Rehaghy Mountain, County Tyrone.

All the recent records have come from Rehaghy Mountain.

many other autumnal species it feeds at Ivy blossom and fermenting berries. The recorded flight period is 21 September to 13 December. The larvae feed on willows. It overwinters as an egg.

Yellow-line Quaker
Agrochola macilenta (Hübner) Category two.
Scarce. This species has a similar distribution and flight period to the Red-line Quaker, but it occurs further north to Breen Wood in north Antrim. It also is found in south Down at Derryleckagh and Rostrevor Wood. This species is probably still under-recorded in quite a few areas due to its late flight period. Adults vary in colour, ranging from light to dark yellowish-brown. The antemedian and postmedian lines are reduced to small dark dots. The subterminal line is fairly straight but angled towards the apex. It is paler than the ground colour and outlined with a fine reddish line. The stigmas are not well defined, but a small dark spot is normally visible in the distal area of the reniform stigma. It is similar in appearance to the Brick, but can be distinguished from that species by the absence of the oblique mark on the costa near the apex of the forewing. The Yellow-line Quaker occurs mainly in open woodland, fens and heaths. Adults are attracted to light in moderate numbers and visit flowers and overripe berries. The recorded flight period is 22 September to 26 December. The larval foodplants are poplar, Beech, Hawthorn, willows and oak. It overwinters as an egg.

Flounced Chestnut
Agrochola helvola (Linnaeus) Category three.
Rare. Since 1986 this species has been recorded from only two sites in Northern Ireland, Loughgall (Armagh) and Rehaghy Mountain (Tyrone). It has been seen frequently at Rehaghy

Mountain since it was first discovered at the site in 1996; the most recent record from here was in 2002. The Loughgall records date from 1986 and 1988. The historical accounts suggest this species used to be more widespread. Donovan gave records from Armagh, Down, Londonderry and Tyrone. Baynes described it as uncommon and local, but stated that Wright considered it present in all the Northern Ireland counties and described it as common in Aghalee (Antrim). Foster states that it was 'abundant on heather from the end of August and September at Dundrum [Down]'. Whether this represents a decline, or reflects a change in trapping effort in the appropriate habitat is unclear. This is an attractive and richly-marked species. The forewings are dark reddish-brown with pale beige antemedian and postmedian lines. The central band is broad and normally a reddish- or purplish-brown colour. The subterminal line and orbicular and reniform stigmas are pale beige and clearly marked. The recent records have come from open woodland and a semi-rural garden. Adults appear sparingly at light and will also visit the blossom of Ivy and fermenting berries. The recorded flight period is 22 September to 27 October. The larval habits and foodplants are unknown in Northern Ireland. In Britain, the larvae feed on oak, birch, willows, Bilberry and Heather. It overwinters as an egg.

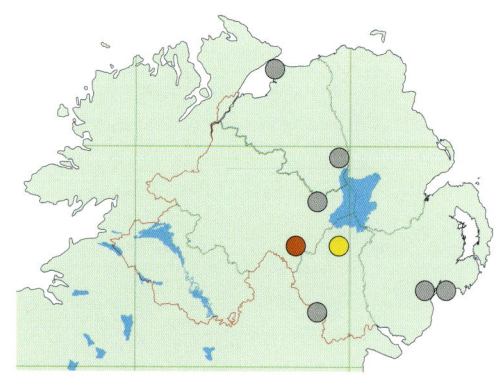

Brown-spot Pinion
Agrochola litura (Linnaeus) Category four.
Rare. This species was added to the Northern Ireland list based on a single adult taken at Copeland Bird Observatory (Down) on 25 September 2004. An earlier specimen has since been found which was taken at Dundrum (Down) on 25 October 1996. These are the only confirmed records for Northern Ireland. Adults are reddish-brown with conspicuous black marks along the costa and a dash near the apex of the forewing. The orbicular and reniform stigmas are usually obvious and darker than the ground colour and finely outlined in light brown. The hindwings are dark greyish-brown. Brown-spot Pinion is associated with deciduous woodland, heathland, fens, scrub and suburban habitats. Adults come to light and are also attracted to Ivy blossom and fermenting blackberries. There is insufficient data to determine its phenology in Northern Ireland. In Britain, there is a single generation from August to October. The larvae feed at night on Meadowsweet, Common Sorrel and Bladder Campion and also deciduous trees such as oak, willows and Hawthorn. It overwinters as an egg.

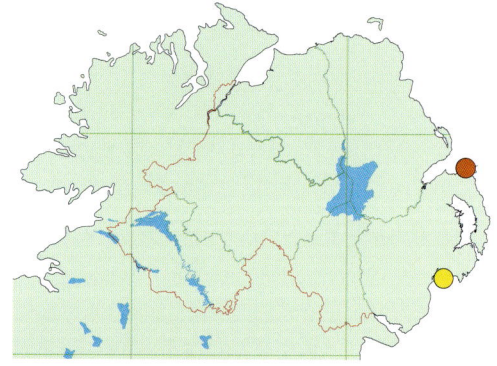

Beaded Chestnut
Agrochola lychnidis (Denis & Schiffermüller) Category two.
Notable. This autumnal species has been recorded from scattered localities in north Armagh, east and south Down and north-west Fermanagh. In Armagh, there are many records from Loughgall and Aughinlig, but not since 1984 and 1996, respectively. The known localities in Down are Ballykinler Dunes, Castleward, Dundrum, Helen's Bay, Murlough NNR, Rostrevor Wood and Seaforde. The most recent records from any of these sites were in 2004. The only locality in Fermanagh is Correl Glen where it has been recorded once, in 2000. Adult colouration ranges from light to dark reddish-brown, usually with an orange tinge on most individuals. The forewings are rather pointed and the orbicular and reniform stigmas are

darker, slightly elongated and finely outlined in pale orange-brown. There is also a small dark mark near the costal margin at the apex of the forewing. The strength of the markings on the forewings is prone to much variation. Beaded Chestnut has been taken most often in woodland in Northern Ireland and less often on dunes and in gardens. Adults are attracted to light in small numbers and will also visit Ivy blossom and overripe berries. The recorded flight period is 2 September to 28 October. The larvae feed on low-growing herbaceous plants especially buttercups, clovers, chickweeds and grasses. It overwinters as an egg.

Centre-barred Sallow
Atethmia centrago (Haworth) Category one.
Scarce. Widely distributed across southern counties and in scattered localities in north Antrim (Glenariff Forest and Murlough Bay) and Londonderry (the Umbra). These northern records are all since 1998, and it is probably still under-recorded in the north and west and inland parts of Antrim. In the west there are recent records for Aghalane, Crom and Monmurry (Fermanagh) and Ballynasollus and Favour Royal (Tyrone). Adults are a light orange colour which darkens slightly towards the outer margin. The forewings have a slightly scalloped hind margin and a dark central band that is normally pinkish-brown and outlined with a fine pale line. There is a darker subcostal dot but this is not always clearly defined. The orbicular and reniform stigmas are 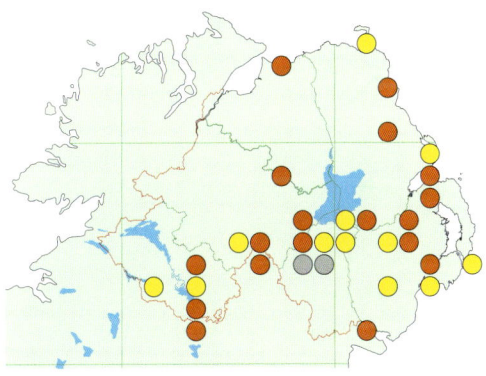 usually faint. This is essentially a woodland species that favours areas where Ash is well established, but it also occurs in other wooded habitats, including gardens. Little is known about the habits of this species in Northern Ireland. Adults appear sparingly at light and show little interest in flowers. The recorded flight period is 5 August to 27 September. The larvae are reportedly nocturnal and feed from April to the beginning of June on Ash. They stay concealed by day at the base of the foodplant or in a small crevice in the bark. It overwinters as an egg.

Lunar Underwing
Omphaloscelis lunosa (Haworth) Category three.
Scarce. Largely confined to Down, where it is commonly encountered at many well-trapped localities, including Ballykinler Dunes, Copeland Bird Observatory, Helen's Bay and Murlough NNR. Elsewhere it is much scarcer, existing in just a few widely scattered sites. In the west it has been recorded only from Crom (Fermanagh) in 1997 and Rehaghy Mountain (Tyrone) in 1996 and 2002. Further east it occurs at the Argory, Gosford Forest and Oxford Island (Armagh), and near Aghalee and on Rathlin Island (Antrim). The base colouration varies considerably between individuals. The forewings are rather long and narrow, ranging from light grey to brownish-grey to dark brown. The antemedian and postmedian cross-lines are paler than the ground colour. 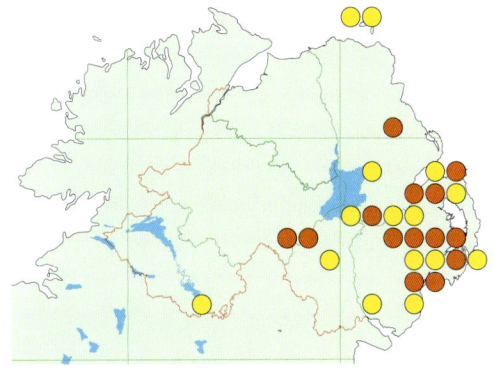 The veining on the forewings is conspicuous and usually lighter than the forewing colour. The subterminal line is a series of black dots. The orbicular and reniform stigmas are darker and edged with a fine pale line. The hindwing has a distinctive crescent. Lunar Underwing is found in grassland both inland and on the coast. Most of the Northern Ireland records are from well-

Pink-barred Sallow.

Annaloughan Bog, County Tyrone.

wooded localities, suggesting a need for either shelter or shady conditions. Adults come frequently to light in moderate numbers and also to Ivy blossom. The moth reportedly conceals itself by day, but becomes active around dusk, where it will often rest on stone walls, fence posts or high up on grass stems. The recorded flight period is 7 September to 14 October. The larvae, which overwinter, feed from October to May of the following year on grasses.

Pink-barred Sallow

Xanthia togata (Esper) Category two.

Widespread. This is a common and widespread species recorded from all counties including Rathlin Island (Antrim). Frequently recorded sites include Montiaghs Moss (Antrim); the Argory, Brackagh Moss and Peatlands Park (Armagh); Ballykinler Dunes and Belvoir Park (Down); Crom and Monmurry (Fermanagh); the Umbra (Londonderry) and Rehaghy Mountain (Tyrone). Adults have deep orange-yellow forewings with darker reddish-brown patches along the costa. The central band or bar is pinkish-brown. The antemedian and postmedian lines are reduced to a series of purple dots. The head is pinkish-brown, contrasting with the deep yellow thorax. It closely resembles and often occurs with the Sallow, but that species shows no purple pigmentation on the head. This species can be found in damp woodland, fens, heaths and other wetland habitats. Adults come to light in small numbers and visit flowers and overripe berries. The recorded flight period is 3 August to 6 November. The larvae have not been found in Northern Ireland. In Britain, they feed initially on willow catkins and, later, on low-growing herbaceous plants. It overwinters as an egg.

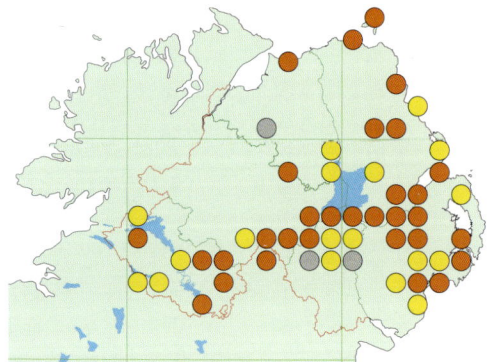

Poplar Grey.

Rehaghy Mountain, County Tyrone.

Sallow
Xanthia icteritia (Hufnagel) Category one.

Widespread. This is a common and widespread species across southern counties, appearing frequently at many well-trapped localities. It would seem to be more scarce further north and confined mainly to coastal localities including the Umbra (Londonderry), where it has been frequently recorded. It has also been found on Rathlin Island (Antrim). Adults are extremely variable both in colour and markings. The colour ranges from pale yellow to a deeper orange-yellow. Some individuals (f. *flavescens*) have entirely yellow forewings with a small dark discal spot. The purplish central band, which is often fragmented, is widest at the costa and extends almost to the apex of the forewing. It closely resembles Pink-barred Sallow, but can be distinguished by the colour of the head, which is pinkish-brown in Pink-barred Sallow and yellow in Sallow. Individuals show some variation in the amount of purple shading on the forewings. The subterminal line is irregular comprising of a series of well-defined black dots. Sallow is found in wet woodlands, the margins of wetlands and other habitats with willows. Adults are attracted to light and are also said to feed on Ivy flowers and fermenting berries. The recorded flight period is 26 July to 19 October. The larvae have not been reported in Northern Ireland. In Britain they feed from March to June, initially on the willow catkins and, later, on low-growing herbaceous plants. It overwinters as an egg.

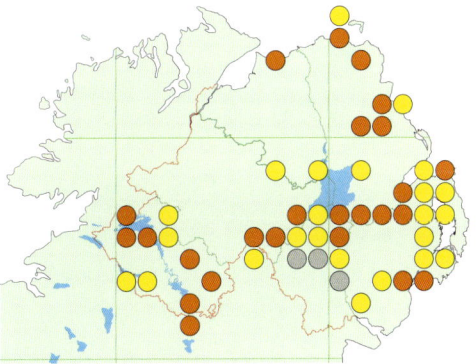

Poplar Grey
Acronicta megacephala (Denis & Schiffermüller) Category one.

Scarce. Generally distributed and not uncommon in a narrow band running from Strangford Lough (Down), west through north Armagh to west Fermanagh. There have been no recent records from the northern counties. Since 2000 it has been recorded in suburban gardens in Finaghy and north Belfast (Antrim); the Argory (Armagh); Belvoir Park and Derryboye (Down); Crom and Gortmaconnell Rock (Fermanagh) and Caledon (Tyrone). Adults have a fairly broad, well-rounded forewing. The colour ranges from pale to dark grey with darker grey antemedian and postmedian lines; however, in some individuals these are not always clearly defined. The orbicular and reniform stigmas are pale grey outlined with black. The subterminal line is white, but obscured for most of its length by a darker suffusion in some individuals. Poplar Grey has a similar appearance to Knot Grass, but the latter has darker hindwings. As its name implies, the Poplar Grey is found in habitats where poplars are present. Adults are attracted to light and flowers. It rests during the daytime on the trunks and branches of trees. The recorded flight period is 1 May to 5 August. The larvae feed from July until September on poplar and willows. In view of the general scarcity in Northern Ireland of poplars, it seems more likely that willows are the principal foodplant here. It overwinters as a pupa.

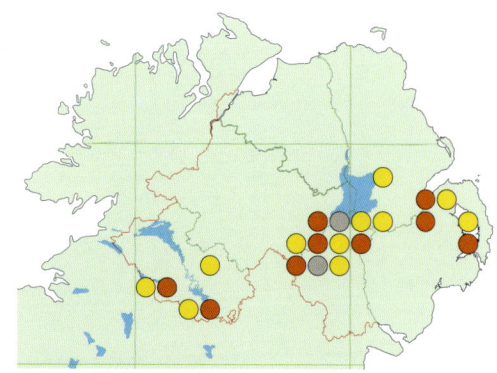

Miller

Acronicta leporina (Linnaeus) Category three.

Scarce. The majority of recent records of this local species are from Armagh and parts of Fermanagh. Reliable sites in Armagh include Aughinlig, Brackagh Moss, Derryadd Lough, Drumnahavil, Loughgall and Peatlands Park. It has also been taken at Finaghy (Antrim); Belvoir Park and Tollymore Forest (Down) and Rehaghy Mountain (Tyrone). There have been no records from anywhere north of Lough Neagh since it was recorded at Banagher Glen (Londonderry) in 1978. Adults are mainly pale grey with fine, dark speckling but darker individuals are occasionally seen. Markings include a small, fine, dark basal streak and a dark arrowhead mark close to the thorax. The orbicular and reniform stigmas are faint except for the black crescents which are conspicuous and form part of the reniform stigmas. Miller has been found on wooded bogs, heaths and fens, in damp woodland and once, in a garden. Adults appear sparingly at light and will also visit flowers from dusk onwards. The recorded flight period is 17 June to 26 July. The larvae, which are quite hairy and conspicuous, feed on birch, Alder, willow, oak and poplar. It overwinters as a pupa.

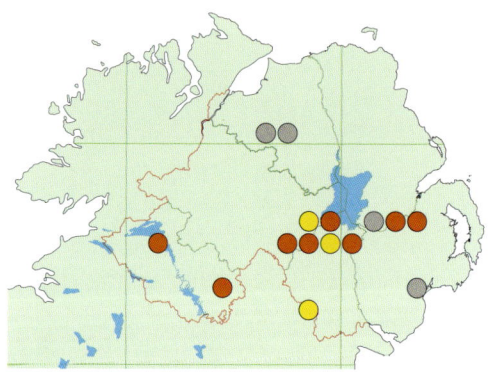

Alder Moth

Acronicta alni (Linnaeus) Category three.

Scarce. This species was first recorded in Northern Ireland by Langham at Tempo (Fermanagh) in 1942; the specimen is now in the Ulster Museum. There were no other reports until 1982 when it was taken at Drumherriff (Armagh) and then a few days later at Cottage Farm near Omagh (Tyrone). Recording since 1982 has revealed that the Alder Moth has a mainly western distribution in Fermanagh, south Tyrone and north Armagh. Within this area it is recorded from many of the regularly trapped sites – for example, Crom and Garvary Wood (Fermanagh) and Rehaghy Mountain (Tyrone) – and also Gosford Forest (Armagh), Altaclanabryan and Gortmaconnell Rock (Fermanagh) and Fardross Forest (Tyrone). Most of the records have involved a few

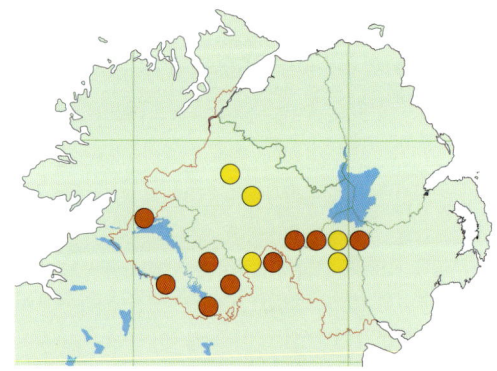

Moth accounts

Alder Moth.

Rehaghy Mountain, County Tyrone.

This moth is probably under-recorded, especially in western counties, where there is plenty of suitable habitat.

individuals, but an exceptional total of 17 was reported on 6 June 2004 at Brackagh Moss (Armagh). Alder Moth is thought to be under-recorded in Northern Ireland, especially in the west where suitable habitat exists. The forewings are mainly grey with a long darker dorsal streak, obscure grey antemedian and postmedian lines and a long, black tornal streak. The orbicular and reniform stigmas are grey and finely edged in black. The distinctive dark patches on the forewings in contrast with the light greyish thorax make it difficult to confuse with any other species. Alder Moth is found in the damper parts of well-established woodlands. Adults are attracted to light. The recorded flight period is 2 May to 7 July. The strikingly patterned, yellow- and black-banded larvae can be found during July and August on Alder but also birch, willow and oak. Despite their conspicuous colouring, the larvae have only been found in Northern Ireland on a few occasions. It has been stated that they prefer to rest and feed in the higher branches of trees, which might account for the lack of records. When fully mature, they will rest and feed quite openly on the surface of the leaves. It overwinters as a pupa.

Grey Dagger

Acronicta psi (Linnaeus) Category one.

Widespread. Recorded in all counties. In the south it is generally distributed, but it is known from only a scattering of localities in the north and west. It is probably more widespread than current records suggest, especially in parts of mid-Antrim where there appears to be suitable habitat. This is a rather distinctive moth with pale to dark grey colouring. There is a black basal streak which is clearly forked at the apex and two dagger-like tornal marks on the forewings. These markings readily distinguish it from other closely-related species. The orbicular and reniform stigmas are paler and outlined in black. Grey Dagger can be found in woodland and other habitats with plenty of trees, including suburban gardens. Adults appear at light frequently,

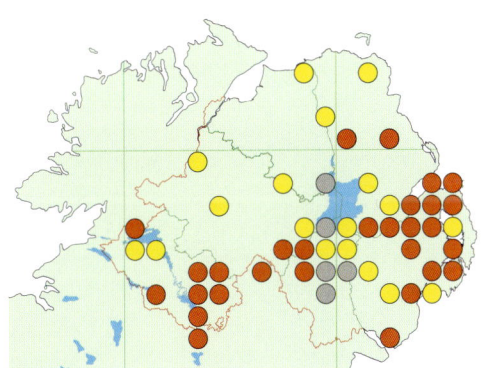

303

The Butterflies and Moths of Northern Ireland

but in small numbers. The recorded flight period is 12 May to 17 August. Individuals are occasionally reported in September and October in Britain, which represents a partial second brood. There is a single late record from Northern Ireland on 4 October which may suggest a similar occurrence here. The colourful and conspicuous larvae are not uncommon and easily found throughout their development. They feed from July to early October on many species of trees and shrubs including birch, oak, Hawthorn and Blackthorn. It overwinters as a pupa.

Light Knot Grass
Acronicta menyanthidis (Esper) Category four.
Rare. Baynes described this species as scarce and local, being confined to the northern half of Ireland. There are historical records from Aghalee (Antrim); Churchill, Drumman More Lough and Maghery (Armagh) and Lough Fea (Tyrone). The latter record was of a larva; Greer described the larvae as common on Heather on bogs near and around Lough Neagh (Tyrone) in the 1920s. Despite regular trapping in suitable localities around the Lough Neagh area, there were no confirmed records of this rare heathland species until 2003, when a single adult was taken in a light trap at Mullenakill NNR in Peatlands Park (Armagh) on 23 May. There are records elsewhere in Ireland for counties Donegal, Galway, Kildare, Mayo, Offaly, Roscommon and Sligo. Adults have a stocky appearance. The forewings are grey with conspicuous black markings and cross-lines. The postmedian cross-line is finely edged with white and both stigmas are conspicuous against the paler background colour. Light Knot Grass is associated with the damper parts of heathland and moorland. Adults are attracted to light and rest by day on rocks and fence posts. There is insufficient data to determine its flight period in Northern Ireland. In Britain, it occurs as a single generation from late May until mid-July. The larval foodplants include Bog Myrtle, Heather, Bilberry and willows. It overwinters as a pupa.

Knot Grass
Acronicta rumicis (Linnaeus) Category one.
Widespread. Recorded from all six counties and on Rathlin Island (Antrim). Most records are from the south, especially north Armagh, north Down and Fermanagh and it has rarely been recorded in the north and west. The most recent record from Londonderry was at Ballymaclary in 1993, but there is reason to believe that it is still extant in the county. This is a medium-sized moth with mottled forewings in a mixture of light and dark grey. Adults show some variation in colour throughout its range here. The diagnostic features are the pale white spots on the inner edge of the forewing and the broken white subterminal cross-line. Knot Grass can be found in a variety of habitats including lush, dry and damp grassland, bogs, woodland and suburban gardens. Adults appear frequently at light and will also visit flowers. During the day they have been found resting on tree trunks and fence posts. The colourful larvae are often encountered by day, low down in the vegetation. There appears to be a single annual generation here with a recorded flight period of 10 May to 27 July. A September record suggests a partial second brood may occasionally be produced. The main foodplants are Hawthorn, Bramble, Common Sorrel, plantains and docks. It overwinters as a pupa.

Opposite page
Grey Dagger.

Murlough NNR, County Down.

Coronet
Craniophora ligustri (Denis & Schiffermüller) Category four.
Rare. This was recorded in Northern Ireland for the first time at Crom (Fermanagh) on 2 July 2000. Despite regular trapping at this site, it has not been reported again. There is an old record from Glenarm (Antrim) in 1874, but the finder is unknown. Elsewhere in Ireland, it seems confined to the Burren in Clare and Galway. This is a rather distinctive moth with a marbled appearance. The forewings are dark brown (almost black) with a suffusion of white on the thorax. The orbicular stigmas are finely outlined in white. The subterminal line is black, but it is masked by conspicuous white blotches near the apex of the forewing. In Britain, Coronet is found in calcareous woodland and scrub, fens and lightly-wooded valleys and hillsides, but the precise habitat the Fermanagh specimen was found in is unknown. Crom has extensive areas of mature broadleaved woodland, scrub and fen. Adults appear sparingly at light and reportedly rest by day on tree trunks and fence posts. The larval habits and foodplants are not known in Northern Ireland. Its main flight period in Britain is June and July. The principal larval foodplants are Ash and Wild Privet, but it also uses Alder and Hazel. It overwinters as a pupa.

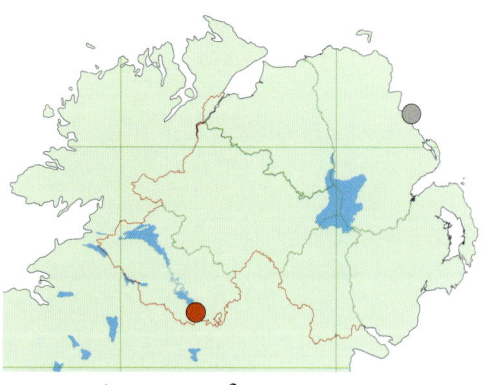

Marbled Beauty
Cryphia domestica (Hufnagel) Category one.
Widespread. Widely recorded in the east of Northern Ireland, and unknown west of the Lower Bann and Blackwater rivers. It has been recorded frequently and regularly from sites in the Greater Belfast area and north Down including Finaghy (Antrim) and Belvoir Park and Helen's Bay (Down). It is also known from Rathlin Island (Antrim). There have been no confirmed records from either Fermanagh or Tyrone. This is a small, rather attractive moth, mottled grey and white in appearance with black antemedian, postmedian and subterminal lines. The basal area and the thorax are pale grey. The orbicular and reniform stigmas are finely outlined in white and the forewings have a chequered fringe. Marbled Beauty occurs in woodland, suburban gardens and coastal sites. Individuals have been found at rest during the day on rocks and old gable walls near the coast. Adults come to light in small numbers and also to flowers after dusk. The recorded flight period is 8 June to 17 September. The larval foodplants are lichens growing on old walls, fences and buildings. It overwinters as a larva.

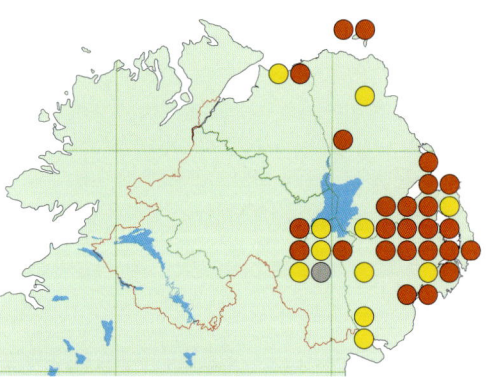

Copper Underwing
Amphipyra pyramidea (Linnaeus) Category three.
Widespread. This autumnal species seems to be almost entirely confined to southern counties. It is known from many localities in Armagh and Down including most regularly trapped sites. In Fermanagh it has been recorded from Aughalun, Castle Coole, Correl Glen, Crom, Legatillida, Monastir Glen, Monmurry and Reilly Wood. Altadavan, Favour Royal and Rehaghy Mountain are all recently recorded sites in Tyrone. The only known sites north of Lough Neagh are Glenarm (Antrim) and the Umbra (Londonderry). The forewings are dark brown with pale cross-lines. The antemedian and postmedian lines are pale straw, the latter being more conspicuous. The orbicular

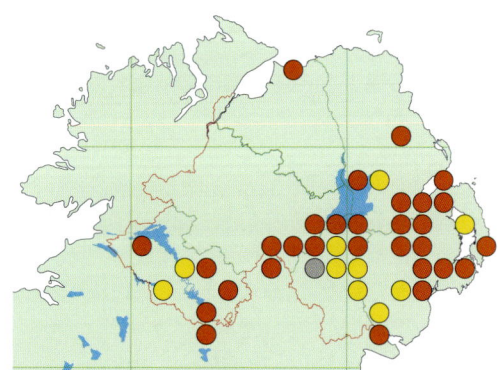

Moth accounts

Coronet.

Crom, County Fermanagh.

An extremely rare moth which has been reported only once in Northern Ireland despite regular trapping at this locality.

Marbled Beauty.

Peatlands Park, County Armagh.

307

and reniform stigmas are dark brown and surrounded by a fine, pale line. The outer edge of the forewing is very slightly scalloped. The hindwings are copper-coloured with darker veining. This is unlikely to be confused with any other species found in Northern Ireland; however, there is a very similar species, Svensson's Copper Underwing *A. berbera*, which has been reported from Dromore Forest (Clare), and which may be present in Northern Ireland. The two species differ in the patterning on the hindwing. Copper Underwing is mainly associated with woodland and old estates. Adults are only occasional visitors to light, but are readily attracted to flowers, sugar and overripe blackberries. They conceal themselves by day in old hollow trees and dark corners of derelict sheds and outhouses. The recorded flight period is 26 July to 23 October. The larvae have been recorded on Ash in Northern Ireland. Other recorded foodplants in Britain include oak, Bramble, Honeysuckle, birch, Hawthorn, Blackthorn and Wild Privet. It overwinters as an egg.

Mouse Moth

Amphipyra tragopoginis (Clerck) Category one.
Scarce. The Mouse Moth is mainly found in Down and north Armagh. It is much more local in other areas, although it has been recorded regularly from the Belfast area. It is only known from three sites in south Fermanagh at Aghalane, Crom and Derrylin and one in Tyrone, Aughentaine. The only recent record in the north-west is from Magilligan on the north coast of Londonderry. Adults have a rather drab, plain appearance. The forewings are greyish or mousy-brown and highly reflective when freshly emerged. The orbicular and reniform stigmas are reduced to a series of dark dots. Mouse Moth has been recorded in dune grassland, fens, heaths, broadleaved woodland and suburban gardens. Adults appear sparingly at light, but are also attracted 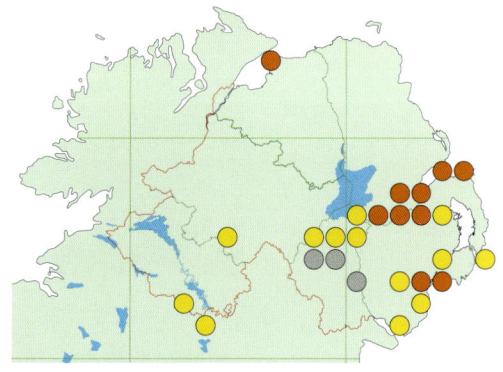 to flowers. They rest by day in the darker areas of old disused buildings and under sheds. It prefers to run rather than fly when disturbed, heading quickly for cover; this scuttling behaviour, in the manner of a mouse, probably accounts for the moth's vernacular name. The recorded flight period is 18 July to 17 October. The larval foodplants are herbaceous plants and woody shrubs including Mugwort, Teasel, Hawthorn and willows. It overwinters as an egg.

Old Lady

Mormo maura (Linnaeus) Category two.
Scarce. Recorded in all counties, but only known from a few sites in each. Since 2000 it has been recorded from the Argory and Peatlands Park (Armagh) and Bleanalung Bay, Crom and Legatillida (Fermanagh). In northern counties the only records are from Breen Wood, Glenshesk and White Park Bay (Antrim). This is a large distinctive species with a broad forewing which is a rich chocolate brown marked with paler scalloped lines, appearing like a shawl. There is a darker central band, two black spots near the basal area and a paler brown patch at the apex of the heavily-scalloped forewing. Old Lady occurs in damp and wetland habitats including wooded valleys, fens and raised bogs and it also, though rarely, appears in suburban gardens. An association with water is 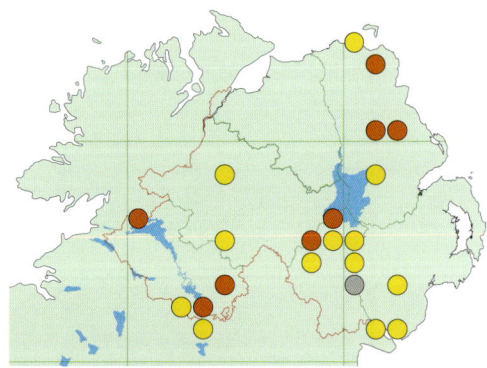 apparent in most of the records. Adults show little interest in light and as a result are seldom found in traps. When it does appear at light, it seems to show a preference for actinic light rather than mercury vapour traps, since many of the adults captured have been taken in Heath traps.

Old Lady.

Maghery Bog, County Down.

A large unmistakable moth that appears infrequently at light. Most of the records emanate from sites close to water.

Adults have also been recorded at sugar in Northern Ireland at the Argory (Armagh). It seems also to have a liking for sap runs and honeydew. According to Donovan, it could be found commonly in Sand Martin burrows in riverbanks during the day. Adults can also be encountered in old buildings (especially boathouses) and under stone bridges. The recorded flight period is 4 July to 2 September. The larvae feed from September to May on the foliage of trees and shrubs such as Blackthorn, Hawthorn, Ivy and birch. It overwinters as a small larva.

Brown Rustic

Rusina ferruginea (Esper) Category two.

Scarce. Generally distributed in Down and north Armagh, but more localised in the west and north. In the south and west there are recent records from Dundrum, Hillsborough Forest, Mill Bay and Murlough NNR (Down) and Hanging Rock and Legatillida (Fermanagh). Further north in Londonderry, it has been recorded from Banagher Glen (in the late 1970s), Binevenagh Forest, Magilligan and the Umbra. There are also records for Glenarm, Montiaghs Moss and White Mountain in Antrim. This is a rather plain, dark brown moth often with a conspicuous series of small white dots along the costa and dark brown cross-lines. In some individuals the cross-lines are faintly marked but there is usually some trace visible. The males are easily

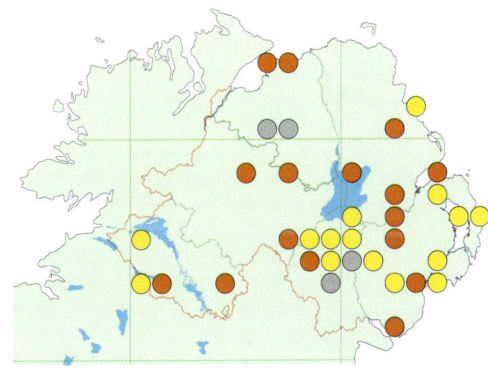

identifiable by their feathered antennae. Brown Rustic is recorded mainly from woodland habitats, but it also occurs in some suburban gardens. Adults are attracted to light and also flowers. The recorded flight period is 26 May to 13 July. The larvae feed on vetches, plantains, docks and possibly other herbaceous plants. It overwinters as a larva.

Straw Underwing
Thalpophila matura (Hufnagel) Category three.
Notable. This is a mainly coastal species in Northern Ireland but there are historical inland records for Armagh City, Drumman More Lough, Lurgan and Poyntzpass (Armagh); Tempo (Fermanagh) and Killymoon and Lissan (Tyrone). It has been recorded frequently at Killard Point and Murlough NNR (Down) and in the same county it was found inland from Rostrevor in 1999. Since 2000 it has been taken on the north coast on Rathlin Island (Antrim) and at Duncrun, Magilligan and the Umbra (Londonderry). Adults are deep reddish-brown with pale antemedian and postmedian lines which are edged with black. The orbicular and reniform stigmas are similar to ground colour and edged with a fine black-and-white line. The diagnostic feature is the light straw hindwings 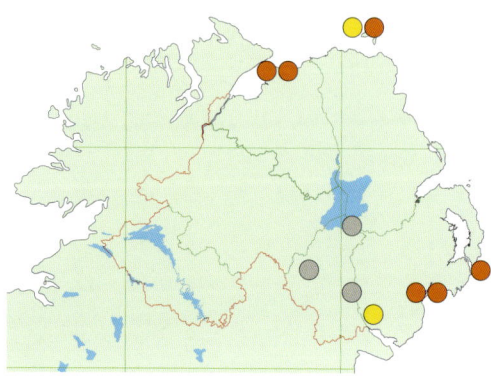 with a dark brown border. Straw Underwing has been taken on dunes, coastal grassland and heathland. It is probably restricted here to the coast for climatic reasons, as it is not a coastal species in the south of its range. Adults come frequently to light in moderate numbers and also visit flowers, especially those of Common Ragwort. The recorded flight period is 3 July to 25 August. The larvae feed on grasses from autumn to May. It overwinters as a larva.

Small Angle Shades
Euplexia lucipara (Linnaeus) Category one.
Widespread. Generally distributed and frequently encountered across southern counties, but more localised further north. It is recorded from both Copeland Bird Observatory (Down) and Rathlin Island (Antrim). Northern localities include Ballymena, Ballymoney and Capanagh Wood (Antrim) and Banagher Glen (Londonderry). This moth is easily recognised at rest as the posterior end of the forewing is creased and raised. The only other species to rest like this is the much larger Angle Shades. The forewings are russet brown with a suffused pinkish-brown band. The most conspicuous feature is the large golden blotch near the apex of each forewing. Small Angle Shades is found in a variety of habitats including woodland, parkland, heaths and bogs and often in gardens.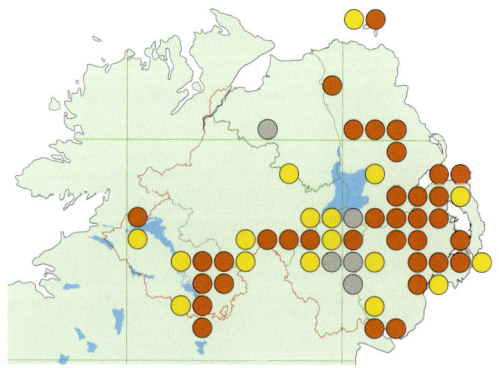
Adults appear frequently at light and at flowers. Its cryptic appearance allows it to rest with relative safety during the day concealed among leaves and other ground vegetation where it is difficult to detect. The recorded flight period is 17 May to 10 August. There are also two late September records indicating a partial second generation. The larvae feed on a wide range of herbaceous and woody plants but especially wild and cultivated ferns including Bracken. Other recorded foodplants include willowherbs, Foxglove, birch, willow and oak. It overwinters as a pupa.

Angle Shades
Phlogophora meticulosa (Linnaeus) Category one.
Widespread. Recorded in all counties and on offshore islands. Gaps in the distribution probably reflect lack of coverage rather than genuine absence. The shape of this moth when it

Angle Shades.

Banbridge, County Down.

is at rest is unusual and distinctive. The leading edge of the forewing is creased and raised and the outer margin is strongly scalloped. Adults have a similar resting posture to Small Angle Shades. The forewings range in colour from olive green through to rosy red. It can be found in most lowland habitats and it is a frequent garden visitor. Adults are attracted to light and flowers and are often found by day at rest on fences, walls and low vegetation. Angle Shades has been recorded in every month of the year in Northern Ireland. There are two peaks in the records, one in June, the second in September, suggesting two generations. The number of records is greatest in early autumn and this probably includes immigrants. The larvae feed throughout the year on a variety of herbaceous plants and shrubs, including Bramble, broom, oak and birch. It overwinters as a larva.

Olive

Ipimorpha subtusa (Denis & Schiffermüller) SOCC Category three.

Rare. Olive has been recorded since 1996 at four sites in Northern Ireland, Gosford Forest (Armagh), Belvoir Park (Down), Crom (Fermanagh) and Favour Royal (Tyrone). It has always been considered rare in Ireland as Donovan stated it was confined to Fermanagh, where it was first recorded from the Enniskillen area around 1893-94 and again in 1907. There are, however, other historical records from Antrim, Tyrone and Kildare. It may have been overlooked in other parts of Fermanagh where there is plenty of suitable habitat. Adults are normally olive brown with pale, almost straight, antemedian and postmedian lines. The latter curves inwards at the end towards the costa. The orbicular and reniform stigmas are large and conspicuous and finely outlined in cream.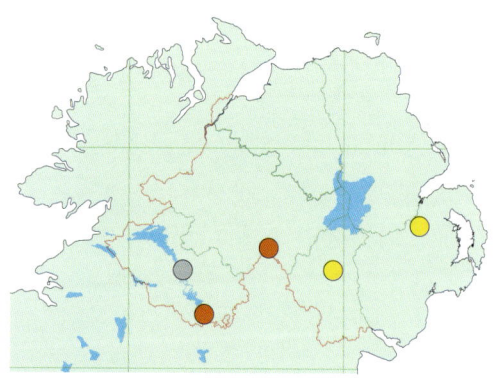
Olive is a species of damp, semi-natural and planted broadleaved woodland with Aspen and poplars. The recent Northern Ireland records have all been from old estates with a long history of woodland cover. Adults usually appear at light sparingly; however, at the Gosford Forest site the moth was found in surprisingly large numbers. Individuals can be rather restless and can damage themselves quite quickly in traps. The recorded flight period is 23 July to 5 September. The larvae have been recorded on Aspen in Northern Ireland. Aspen is absent from some of the known sites, so it seems likely that other poplars are being used. It overwinters as an egg.

Angle-striped Sallow

Enargia paleacea (Esper) Category four.

Rare. This species was recorded for the first time in Northern Ireland at Legatillida, in eastern Fermanagh on 7 August 2002. There have been no other records although the site is regularly trapped. This would suggest that this individual was an immigrant. In Britain, this is a northern species resident in the Scottish Highlands and in a wide area across the north and midlands of England and parts of Wales. It has a long history of random vagrancy in southern England. Adults are variable in colour ranging from light to deep yellow or orange-brown. The antemedian and postmedian cross-lines are darker than the ground colour. The former has a conspicuous, almost right-angled bend, which is a useful identification feature that helps to separate it from Dun-bar and Sallow.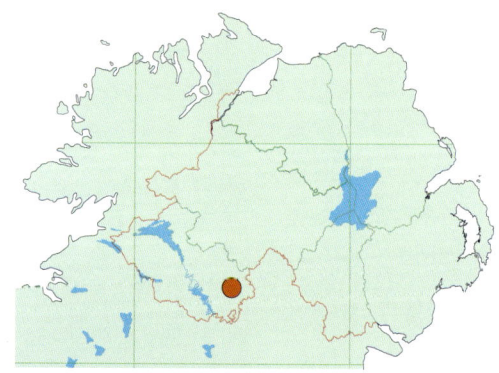
Angle-striped Sallow is associated with mature birch woodland in both upland and lowland sites. Adults are attracted to light, particularly the males. In Britain, it flies during August to mid-September. The larval foodplants are birch and possibly Aspen. It overwinters as an egg.

Suspected

Parastichtis suspecta (Hübner) Category four.

Unknown. The current status of this species in Northern Ireland is uncertain as there have been no recent confirmed records. There are historical records from Churchill (Peatlands Park) (Armagh) and Killymoon and Tamnamore (Tyrone). Kane mentioned a record from Londonderry, but considered it dubious. Elsewhere in Ireland it has been recorded sporadically in counties Cork, Dublin, Kerry and Wicklow. This is a rather variable species. The forewing ground colour ranges from greyish-brown to reddish-brown. The antemedian, postmedian and subterminal lines are in most cases reduced to a series of small dots. The orbicular and reniform stigmas are similar to the ground colour of the forewing; both are outlined with a fine, pale line. Suspected is mainly associated with woodland but it also inhabits fens and heathland where birch is present. Greer described seeing this species abundantly after dark in August on the bogs at Lough Neagh, when Heather was in bloom. He referred to it as being of a shy and retiring nature, settling deep into the Heather and not easily seen. Other than this, little is known about its habitat requirements in Ireland. Adults show some attraction to light and also flowers, especially Common Ragwort and Heather. There is no information to provide a flight period in Northern Ireland. In Britain, it flies during July and August. The larval habits and foodplants are not known in Northern Ireland. In Britain, it feeds throughout April and May on birch and possibly willow. It overwinters as an egg.

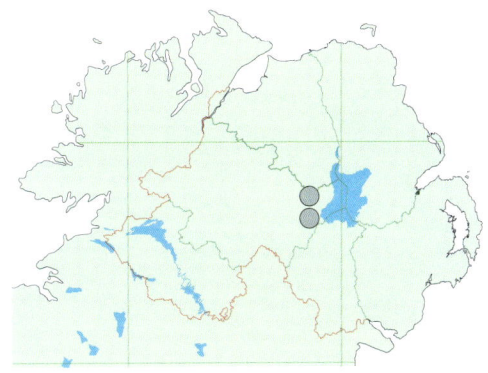

Dingy Shears
Parastichtis ypsillon (Denis & Schiffermüller) Category four.
Rare. Only recorded from a few sites in the east in Antrim, Armagh and Down. It was recorded at Glenarm (Antrim) in 1874 according to Baynes. There is a specimen in the Magilligan Field Centre (Londonderry) labelled 'August 1969 Magilligan Point'. It was not seen again until 1991 when it was discovered at Dunmurry (Antrim). There have been seven records since from Finaghy (Antrim) in 2001 and 2003; Oxford Island (Armagh) in 2001, 2003 and 2004; Derryboye (Down) in 2003 and most recently, Aughinlig (Armagh) in 2004. The coincidence of timing of the records suggests a recent expansion in range or immigration, although there is no history of the latter in Britain. Adults are variable in colour, ranging from light to dark greyish-brown with paler antemedian and postmedian cross-lines. The orbicular and reniform stigmas are similar to the ground colour and finely outlined with a paler line. There are two tooth-like marks in the central area of the forewing. The subterminal line is irregular with two small, dark wedge-shaped marks near the central area. In Britain, Dingy Shears is associated with damp woodland, fens and marshy areas. The records from Derryboye and Oxford Island are from this type of habitat, but the Aughinlig, Dunmurry and Finaghy specimens were trapped in gardens. Adults become active from dusk onwards, but come sparingly to light. Sugar has also proved to be a alternative attractant to light; adults were taken at Oxford Island in 2004 using this method. The recorded flight period is 6 July to 8 August. In Britain, the main flight period is late June to mid-August. The larval foodplants are willows and poplars. It overwinters as an egg.

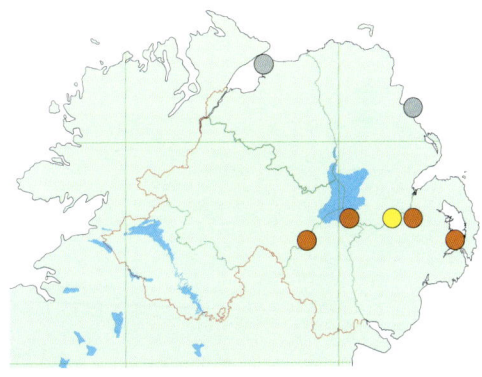

Dun-bar
Cosmia trapezina (Linnaeus) Category one.
Widespread. Recorded in all counties and generally distributed in the south-east, but more scattered across the northern half and in Fermanagh. Antrim localities include Ballymena, Breen Wood and Portglenone Wood. It was taken in 1988 on Rathlin Island (Antrim), but not thereafter. Colouring in this species is very variable, but most individuals are easily recognised by the prominent central band. The forewings range from reddish-brown to light ochreous-brown. The median area tends to be darker than the ground colour, creating a broad band. There is also a small dark discal spot present on all colour

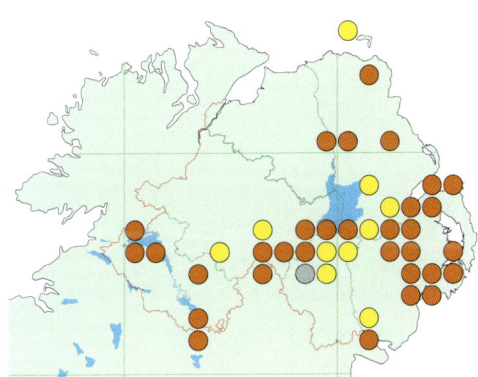

Dun-bar.

The Argory, County Armagh.

forms. The antemedian line is fine and runs obliquely towards the costa. The postmedian line curves in towards the costa. The colour of the lines varies with the ground colour of the forewings, but all are clearly defined. This species can be found in damp woodland and other habitats with trees, including gardens. Adults rest by day in ground foliage; they come frequently to light and occasionally visit the flowers of Common Ragwort. The recorded flight period is 5 July to 8 September. The larvae feed on a wide range of trees and shrubs including oak, birch, Hazel, Hawthorn, Blackthorn and willow. The larvae are also partly carnivorous, feeding on other moth larvae. It overwinters as an egg.

Dark Arches

Apamea monoglypha (Hufnagel) Category one.
Ubiquitous. This common and frequently encountered moth is recorded from all counties and on offshore islands. Individuals show considerable variation in colour. The forewings range from greyish-brown to dark brown. The forewings have a short dark basal streak and lighter, finely outlined stigmas. The antemedian and postmedian lines are light brown and irregular. The subterminal line has double-toothed projections and a series of small black dashes. Dark Arches is a generalist species that may be encountered in many habitats. Adults are frequently encountered at light (often in good numbers) and it is one of the most regular visitors to garden-based traps. They are also attracted to flowers. By day, they usually conceal themselves but, occasionally, individuals can be seen resting on fence posts, walls or tree trunks. The recorded flight period is 26 May to 28 September. The larval foodplants are Cock's-foot, Common Couch and other grasses. It overwinters as a larva.

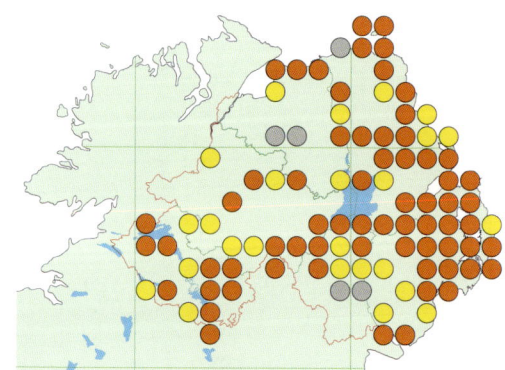

Light Arches.

Clare Glen, County Armagh.

Light Arches

Apamea lithoxylea (Denis & Schiffermüller) Category one.

Widespread. Generally distributed through all counties and on Copeland Bird Observatory (Down) and Rathlin Island (Antrim). It is especially frequent in Armagh and Down. Adults are pale straw-coloured and white but slightly darker towards the terminal border, with a series of fine dark streaks or lines. There is a brownish mark running obliquely from the costa towards the central area of the forewing. The postmedian line is reduced to a series of small black dots. It is similar in appearance to Reddish Light Arches but individuals of that species are smaller and darker overall. This is a grassland species that occurs in open areas within many lowland habitats including woodland clearings, coastal grasslands and gardens. Adults are attracted to light and flowers, but not in the same numbers as Dark Arches. The adults mainly rest by day, concealed amongst ground vegetation or, occasionally, on walls, fence posts and tree trunks. The recorded flight period is 18 June to 25 August. The larvae feed during the autumn until spring on various grasses. It overwinters as a larva.

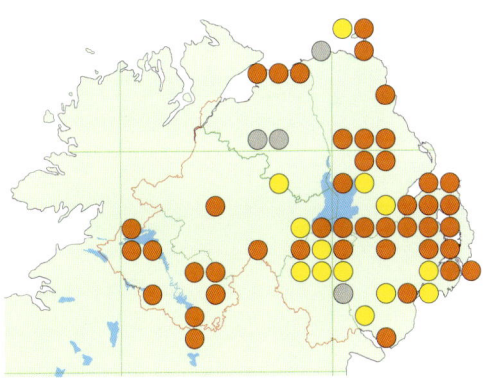

Reddish Light Arches

Apamea sublustris (Esper) Category four.

Rare. The modern records indicate this species is confined to Fermanagh. There are three historical records from Glenarm (Antrim) in 1875, Favour Royal (Tyrone) in 1902 and Down. Baynes gave no details regarding the Down record other than the year (1943) and recorder (O'Farrell). It has been recorded twice in Fermanagh at Monawilkin (23 June 1988) and

Gortmaconnell Rock (18 July 2000). Efforts have been made to rediscover the species at both these sites, so far without success. These records are supported by voucher specimens. There have been records from counties Clare (particularly the Burren), Dublin, Galway, Kerry, Limerick, Meath, Offaly, Sligo, Westmeath and Wicklow. Adults are similar in appearance to the Light Arches but are usually more reddish-brown with a darker thorax and wedge-like marks on the outer margin of the forewing. The postmedian line is reduced to a series of small black dots. The subterminal line is faint and poorly defined. The hindwings are light brown with dark veining and a small discal spot which is not always clearly defined. The Fermanagh sites conform to typical habitat of this species, unimproved calcareous grassland. Adults have been recorded in June and mid-July in Northern Ireland although there is insufficient data to give an accurate flight period. The larval foodplants in the wild are unknown. It is reported to overwinter as a larva.

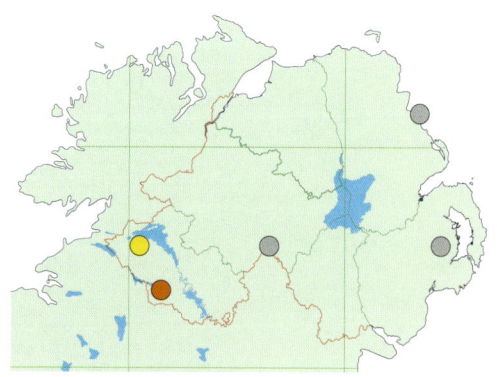

Crescent Striped
Apamea oblonga (Haworth) Category four.
Unknown. Crescent Striped is confined to the east coast of Ireland from Antrim and Down to Waterford. There have been no reliable records from Northern Ireland since 1977 when it was reported from Murlough NNR (Down). No specimen exists to validate this record. There are early twentieth-century records from an unspecified locality in Antrim (most probably Belfast) in 1903 and from Rostrevor in 1932. The lack of recent records (despite targeted trapping at former and potential sites) suggests it may now be extinct. Loss of its habitat may be a contributing factor to its apparent decline. Adults have rather broad forewings which are greyish-brown with a smooth appearance. The orbicular and reniform stigmas are similar to the ground colour, sometimes finely outlined but not always clearly defined; in some cases the reniform stigma takes the form of a small colon mark. There is a series of small black dots along the termen. The subterminal line is irregular and poorly defined. Crescent Striped is found in Britain in saltmarshes, mudflats and generally brackish areas. Only small fragments of these habitats remain in Northern Ireland and any surviving populations of the species are likely to be small and isolated. Adults are reportedly attracted to light and flowers, especially Sea Campion and thistles. The flight period for Northern Ireland is unknown. In Britain, the main flight period is late June to early August. The larval habits are unknown in Northern Ireland. In Britain, the larvae overwinter and feed on the roots of various saltmarsh grasses from late summer until June.

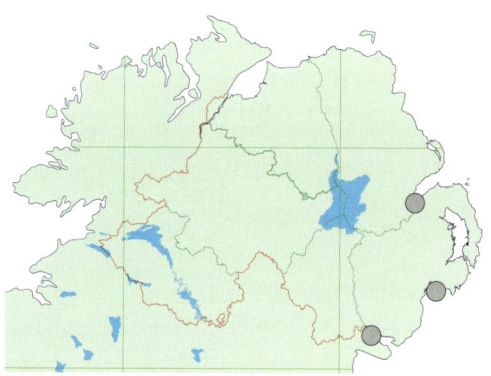

Clouded-bordered Brindle
Apamea crenata (Hufnagel) Category one.
Ubiquitous. Recorded from all counties and commonly met with in most. It is also recorded from Rathlin Island (Antrim). Apparently absent from much of Tyrone (with the exception of Rehaghy Mountain) and inland in Londonderry, although it has been recorded from Duncrun and the Umbra on the coast. It is probably overlooked rather than genuinely absent from the areas lacking records. Clouded-bordered Brindle exists in two different colour forms. In the first, adults are buff with darker reddish patches along the costa and dark basal and tornal streaks. The thorax is dark and double-crested. The stigmas are similar to the ground colour but are finely outlined with a pale line. The darker form f. *combusta* has a similar appearance,

Clouded-bordered Brindle.

Portadown, County Armagh.

but the ground colour is a dark reddish-brown. The stigmas are evident but the other markings are less obvious. This species is similar in appearance to the Clouded Brindle which has less pointed forewings and has a wedge-shaped tornal mark. The Clouded-bordered Brindle is found in woodland rides and clearings and a wide range of other habitats including grassland and mature gardens. Adults come frequently to light and also visit flowers. It remains concealed during the day among ground vegetation and is seldom found unless by chance. The recorded flight period is 1 May to 10 September. The larvae feed from August to April on Cock's-foot and other grasses. It overwinters as a larva.

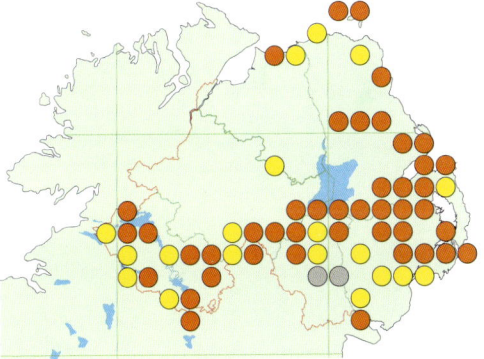

Clouded Brindle

Apamea epomidion (Haworth) Category three.
Notable. Known from all counties. Wright listed it for Londonderry but it has not been seen there for many years. Donovan gave records for Lough Erne (Fermanagh) and Altadavan and Favour Royal (Tyrone). The majority of recent records are from Fermanagh where it has been regularly trapped at Crom. It has also been found at Aughalun, Glen Wood and Marble Arch. It is present at just one or two localities in other counties, including Finaghy and Rathlin Island (Antrim); Loughgall and Peatlands Park (Armagh); Belvoir Park and Seaforde (Down) and Crilly (Tyrone). The forewings of this species have a marbled appearance and range from light to dark reddish-

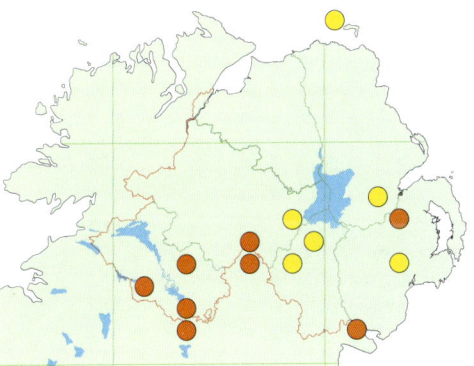

The Butterflies and Moths of Northern Ireland

Clouded Brindle.

Rostrevor Wood, County Down.

brown. The thorax is darker than the wings and slightly crested and on each side there is a small black basal streak. The orbicular and reniform stigmas are outlined in black. The subterminal line is pale and contains two dark wedge-like markings. Clouded Brindle has been recorded in, or close to, mature broadleaved woodland and, occasionally, suburban gardens. Adults come sparingly to light and visit flowers. The recorded flight period is 2 June to 30 July. The larval foodplants are Cock's-foot and probably other grasses. It overwinters as a larva.

Confused

Apamea furva (Denis & Schiffermüller) SOCC Category three.

Rare. The only recent records of Confused are from Rathlin Island (Antrim). It was first recorded on the island in 1988 and in 2001 was found to be present at seven localities between the east and west lights. Baynes describes it as being rare and local but widely distributed in Ireland. Donovan gave records for Derry City and Magilligan (Londonderry) and Lissan (Tyrone). It is primarily a northern species in Britain and not uncommon in Scotland. Its presence on Rathlin Island is therefore not unexpected and other populations may exist along the north coast and in the uplands of Antrim and Londonderry. Elsewhere in Ireland it has been reported from Clare, Cork, Donegal, Dublin, Kerry, Sligo and Waterford. The ground colour of the forewings is a dull dark brown. The cross-lines are poorly defined, with the exception of the subterminal line which is white. The reniform stigma is roughly square and often partially outlined in white. There are two dark bands on the underside of the forewing, near the outer margin. It is similar in appearance to Dusky Brocade, but Confused has a wider forewing and more square reniform stigma; the subterminal line has a slight double projection and there is

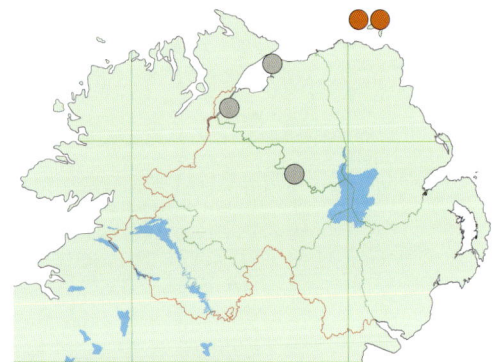

also a fine black basal streak. On Rathlin Island the species has been recorded in heathland. It also is found on moorland and dunes in Britain. Adults are attracted to light and also flowers. There is insufficient data to give an accurate flight period since adults have only been recorded at the end of July in Northern Ireland. In Britain, it flies from July to September. The larval habits are not known in Northern Ireland. In Britain, it feeds on the roots and lower parts of grasses. It overwinters as a larva.

Dusky Brocade
Apamea remissa (Hübner) Category one.
Widespread. Generally distributed in all counties and also recorded from Rathlin Island (Antrim) and Copeland Bird Observatory (Down). Adults have rather broad forewings which range in colour from greyish-brown to brownish-grey. The basal area of the forewing has a black streak. The orbicular and reniform stigmas are lighter than the ground colour of the forewings and contrast conspicuously with the darker coloured patch between antemedian and postmedian lines. There is a series of small white marks along the edge of the costa. The subterminal line has a slight double projection. It closely resembles Confused which has a narrower forewing, square reniform stigmas and two dark bands on the underside of the forewing. Dusky Brocade is found in a wide 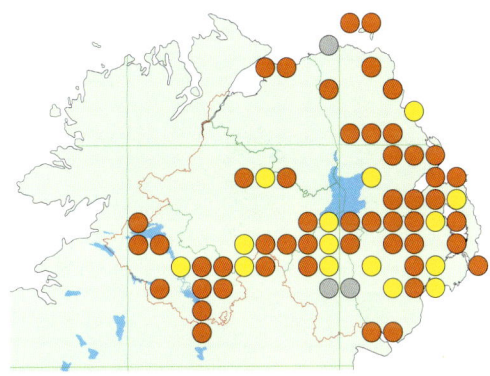 variety of habitats especially woodland, fens, bogs and suburban gardens. Adults become active after dark and appear commonly at light and flowers. It rests during the day, hidden in low-growing vegetation and is seldom seen. The recorded flight period is 21 May to 24 August. The larvae feed from autumn until spring on various grasses. It overwinters as a larva.

Small Clouded Brindle
Apamea unanimis (Hübner) Category three.
Scarce. Largely confined to the south-east, especially the Lough Neagh basin and across central Down. Sites in this area include Montiaghs Moss (Antrim); the Argory, Brackagh Moss and Loughgall (Armagh) and Lackan Bog (Down). Outside Armagh and Down there are few known sites, though it occurs at Aghalane, Crom and Monmurry in Fermanagh, Rehaghy Mountain in Tyrone and Glenshesk in north Antrim; the last is the only recent occurrence from north of Lough Neagh. It is seemingly uncommon in the rest of Ireland with records only from counties Dublin, Louth and Westmeath. This is a rather small species and individuals vary in colour from light to dark brown or reddish-brown. There is a short black basal streak along the thorax and 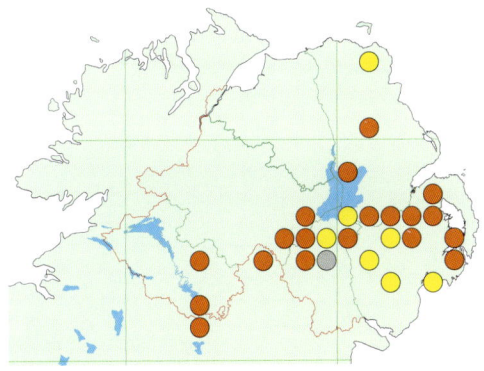 another at the costal edge and four small white marks on the costa near the apex of the forewing. The reniform stigma, which is outlined in white, is the most conspicuous feature of this moth. Small Clouded Brindle is found in damp open woods, fens and cutover bogs. Adults have also appeared in gardens. They come to light in small numbers and also visit the flowers of some grasses and rushes. The larval foodplants are large grasses including Reed Canary-grass. The recorded flight period is 25 May to 4 August. It overwinters as a fully-grown larva.

Rustic Shoulder-knot
Apamea sordens (Hufnagel) Category two.
Scarce. Baynes described this species as being very common and widely distributed, a view which does not reflect its present distribution in Northern Ireland. It has a discontinuous

distribution, occurring in Fermanagh and in a band running from Belfast south-west to north Armagh. The most recent records are from Magheramorne Quarry (Antrim) and the Greater Belfast area, where it has been recorded from a few suburban gardens. There is a 1938 record from Portballintrae on the north Antrim coast. In Armagh the majority of records are pre-1980, with the exception of two records, one from Brackagh Moss in 1992 and the other from Portadown in 1993. Other known localities include Belvoir Park and Helen's Bay (Down) and Boho, Garvary Wood, Marlbank and Monawilkin (Fermanagh). It appears to be absent from Tyrone and Londonderry, which is surprising considering the abundance of its foodplants. Most individuals are sandy

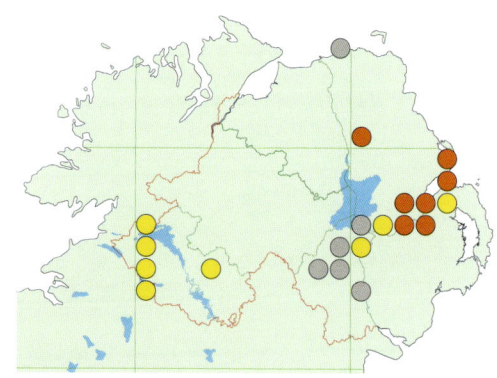

brown with a slightly reddish tint, but this is subject to some variation. There is a distinctive black basal streak, slightly forked at the tip. The orbicular and reniform stigmas are similar to the base colour and finely outlined in white. The Rustic Shoulder-knot is a grassland species. Most of the Northern Ireland records are from gardens, but it has been found in woodland, limestone grassland and rough coastal grassland. Adults come to light in small numbers and occasionally visit flowers. During the day it remains concealed low down among vegetation and is seldom encountered. The recorded flight period is 15 May to 18 July. The larvae feed from August until March on Cock's-foot and probably other grasses. It overwinters as a larva.

Slender Brindle
Apamea scolopacina (Esper) Category four.
Rare. This has been recorded in Rostrevor Wood (Down), but is unknown anywhere else in Ireland. It was first found during a moth workshop on 30 July 1999; there have been other reports in 2000, 2001 and 2004. Slender Brindle is widespread in Britain, occurring in much of England and Wales as far north as Cumbria. Adults are straw-coloured with darker patches on the crest of the thorax, costal edge and outer edge of the forewing. The reniform stigma is white and quite conspicuous. The antemedian and postmedian cross-lines are slightly scalloped and black in colour. This is a woodland moth primarily found in open spaces in mature broadleaved woodland. The habitat at Rostrevor Wood is therefore typical of its habitat in Britain. Adults are

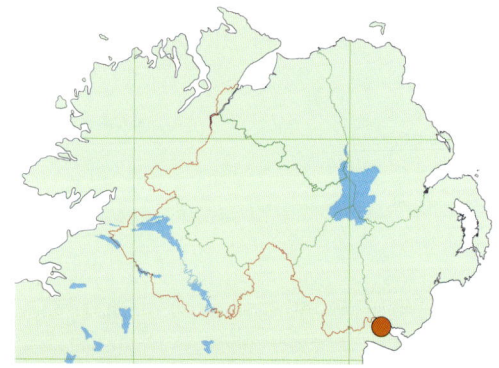

attracted to light and flowers and apparently rest by day concealed among oak leaves on the woodland floor. The Rostrevor records have been in the period from 23 July to 25 August. The main flight period in Britain is late June to mid-August. The larvae feed on the stems and leaves of various woodland grasses and wood-rushes including Great Wood-rush which is abundant at Rostrevor. It overwinters as a larva.

Double Lobed
Apamea ophiogramma (Esper) Category one.
Scarce. Common and widespread across southern counties particularly Down, north Armagh and south Fermanagh. It is rare north of Lough Neagh and the only records are from the dunes at Portstewart (Londonderry) and Rathlin Island (Antrim). To the south of Lough Neagh, it has been recorded regularly from the Argory and Oxford Island (Armagh); Belvoir Park and Lackan Bog (Down) and Crom (Fermanagh). This is a rather conspicuously marked species. The forewings are quite narrow and generally sandy in colour with a large dark, reddish-brown patch which extends from the costal area to the base of the forewing and incorporates the

Moth accounts

Slender Brindle.

Rostrevor Wood National Nature Reserve, County Down.

This is the only known site in Ireland for this woodland species.

Doubled Lobed.

Lackan Bog, County Armagh.

A wetland species that is frequently recorded in small numbers from a few well-trapped sites in southern counties.

orbicular and reniform stigmas. The reniform stigma is light brown and clearly defined within this darker area. The postmedian line is made up of a series of small black dots. There is a distinctive wedge-shaped streak in the tornal area of the forewing. The thorax has a fine black line and is crested. This species is found in wetlands, especially fens and cutover bogs and also wet woodland. It has also been taken in gardens. Adults come to light in small numbers and visit flowers. The recorded flight period is 28 June to 31 August. The larvae feed from August until June on various grasses. It overwinters as a larva.

Marbled Minor
Oligia strigilis (Linnaeus) Category four G.
Rare. There are only three confirmed records of this species: the Argory (Armagh) on 1 July 1999; Gorteen (Fermanagh) on 3 July 2002 and Cuilcagh Mountain (Fermanagh) on 19 July 2002. Marbled Minor cannot be reliably distinguished in the field from Tawny Marbled Minor and correct determination of either species requires examination of the genitalia. The forewing colour is variable ranging from brown to almost blackish, with a black bar which connects both central cross-lines. There is a conspicuous, broad pale band towards the outer area of the forewing; however, this is not present in all individuals. Some adults are considerably darker, almost melanic and many of the markings are difficult to distinguish. Marbled Minor inhabits grassland and grassy areas along rides and in open clearings in woodland. Adults have only been recorded in July in Northern Ireland. In Britain the normal flight period is late May to July. Adults are attracted to light and apparently to sugar and honeydew. The larval habits are not known in Northern Ireland. In Britain, it feeds from August to May on large grasses including Cock's-foot, Common Couch and Reed Canary-grass. It overwinters as a larva.

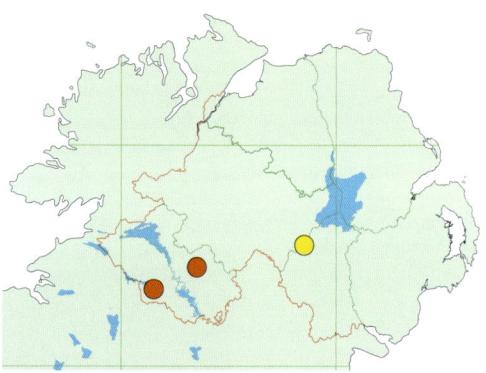

Rufous Minor
Oligia versicolor (Borkhausen) Category four G.
Rare. Up until 2001 the status of this species was unknown, the only confirmed record being from Lissan (Tyrone) in 1903 by Thomas Greer. It was rediscovered in 2001 when a single adult was trapped at Rostrevor Wood (Armagh) on 23 July. It was also discovered in Fermanagh at Legatillida in 2003 and Monmurry in 2003 and 2004. The specimens were confirmed as this species by examination of the genitalia. Due to the difficulty in identification, the Irish distribution of Rufous Minor remains unclear, but there have been confirmed records from Clare, Galway and Kerry. Adults are dark reddish-brown with large, conspicuous, pale orbicular and reniform stigmas which are outlined in white. The antemedian and postmedian cross-lines are obvious and there are two black bars in the median area of the forewing. The outer area of the forewings is normally paler than the base colour. All the *Oligia* species are similar and certain identification requires examination of the genitalia. Records cannot be substantiated without a voucher specimen. Rufous Minor is a grassland species mainly associated with woodland clearings, coastal grasslands and cliffs. Little is known of its habits or ecology in Ireland. Adults are attracted to light. The recorded flight period is mid-July to late August. The larval foodplants are unknown in the wild.

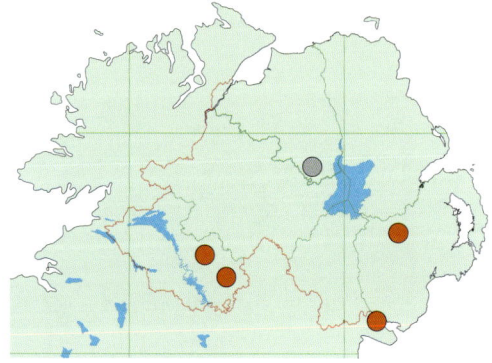

Tawny Marbled Minor
Oligia latruncula (Denis & Schiffermüller) Category four G.

Widespread. Only reliably recorded from Armagh, Down, Fermanagh and Tyrone, although there are numerous claimed records which have not been verified. Like other *Oligia* species, specimens of this moth require confirmation by examination of the genitalia. Confirmed records have come from the Argory, Aughinlig, Brackagh Moss, Derryadd Lough, Drumbanagher and Drumman More Lough (Armagh); Castleward, Castlewellan Forest, Rostrevor Wood and Seaforde (Down); Derryvore and Gortmaconnell Rock (Fermanagh) and Altadavan, Crilly and Favour Royal (Tyrone). This species bears a close resemblance to Marbled Minor which has led to some confusion. Both species probably exist more widely than current records suggest – a view which was shared by Baynes. However it is not known how many of the records have been identified by critical examination. The forewings are generally dark brown or reddish-brown with a copper-coloured central and outer band. The orbicular and reniform stigmas are not always clearly defined. Tawny Marbled Minor is a grassland species which occurs in a variety of grassy habitats including gardens. Adults come frequently to light in small numbers. The recorded flight period is late May to mid-August. The larvae feed internally during April and May on Cock's-foot and probably other grasses. It overwinters as a larva.

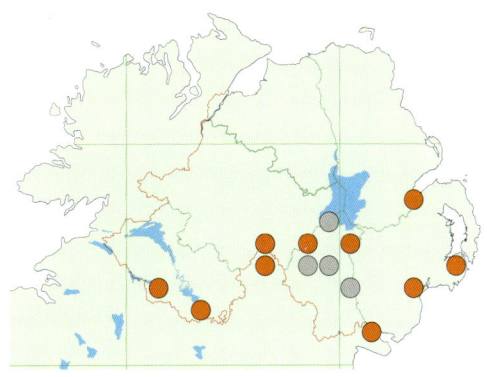

Middle-barred Minor
Oligia fasciuncula (Haworth) Category one.

Widespread. This is the only one of the locally occurring *Oligia* species that can be certainly identified in the field. It is known from many sites in all counties and it has been recorded on both Rathlin Island (Antrim) and Copeland Bird Observatory (Down). Adults are quite variable in colour. The forewing colour ranges from light brown or straw to brick red. The most distinctive feature is the darker central band which is finely outlined in white, especially on the distal side. The males have a distinct anal tuft. Middle-barred Minor is a species of lush vegetation and damp grassland on the edge of woodland, fens and other wetlands. It is also commonly found in gardens. Adults are attracted to light in small numbers and to flowers, especially those of Common Ragwort and umbelliferous plants. The recorded flight period is 27 May to 2 September. The larvae, which overwinter, feed from August to May on grasses.

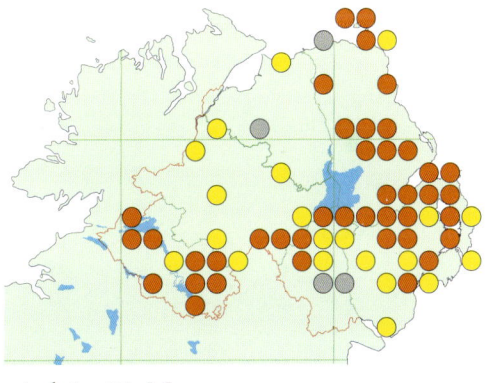

Cloaked Minor
Mesoligia furuncula (Denis & Schiffermüller) Category one.

Scarce. Recorded in all counties but never commonly encountered and known from just a few sites in each. Since 2000, records have come from throughout its known range including Ecos Park and Rathlin Island (Antrim); Killard Point, Lagan Meadows and Rostrevor Wood (Down); Crom (Fermanagh) and Magilligan (Londonderry). Adults appear more slender than closely-related species. The general colour is variable; most are pale straw brown, but individuals from inland sites tend to be darker. The forewings have fine whitish cross-lines, but these are not always clearly defined. The orbicular and reniform stigmas are often

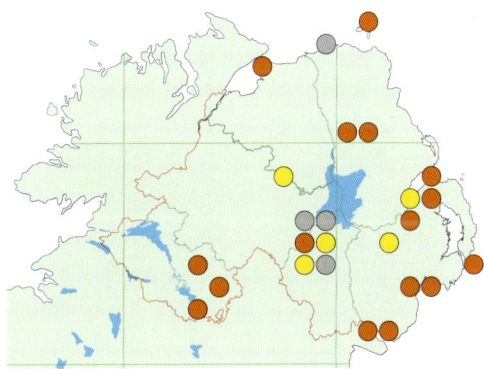

inconspicuous but, where visible, are faintly outlined in white. Cloaked Minor can be found in sites with light, well-drained soils. There are records from sand dunes and coastal heaths. There is no obvious pattern to the inland sites as there are records from a wide variety of localities. Adults are attracted to light and flowers, especially Common Ragwort. The recorded flight period is 10 July to 28 August and it overwinters as a larva. The larval foodplants are grasses.

Rosy Minor
Mesoligia literosa (Haworth) Category one.
Widespread. Recorded in all counties and on Rathlin Island (Antrim). It is generally distributed in Armagh and Down, especially along the Lagan Valley and inland around Lough Neagh, but appears to be more local in other counties. It is known from all the dune systems on the north coast. This is a distinctive, small moth and most individuals can be told by the combination of size and colour. Adults vary little in colour and are mainly a mixture of red, brown and grey. The median cross-lines are straight and conspicuous. The termen area is reddish-brown. The subterminal line is pale, edged with red. It can be found in open grassy habitats including fens, coastal grassland, dunes and gardens. The adults are attracted to light and to flowers, in particular, Common Ragwort and

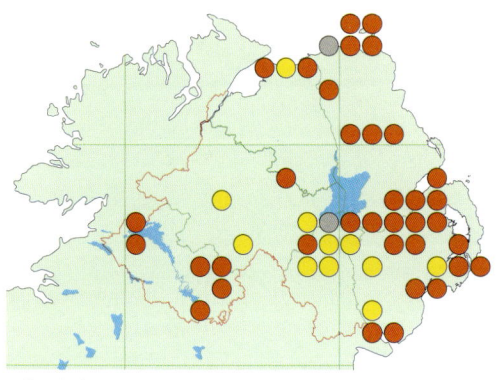

Marram. The recorded flight period is 19 June to 25 September. The larvae feed from August to May on Cock's-foot, Marram and probably other grasses. It overwinters as a larva.

Common Rustic
Mesapamea secalis (Linnaeus) Category three G.
Two species of *Mesapamea* – Common Rustic *M. secalis* and Lesser Common Rustic *M. didyma* – are found in Northern Ireland. Neither can be identified in the field and the genitalia provide the only diagnostic features. Few individuals of either species have been certainly identified in Northern Ireland so the map shows the distribution of both species combined. The two species are ubiquitous and together they account for some of the most frequently recorded moths in Northern Ireland. Adults are extremely variable in colour, but the one reliable feature is the reniform stigma which is always evident and, in most cases, finely outlined in white. The forewing colour ranges from light to greyish-brown through to dark, almost blackish-brown. The reniform stigma is large, usually lobed,

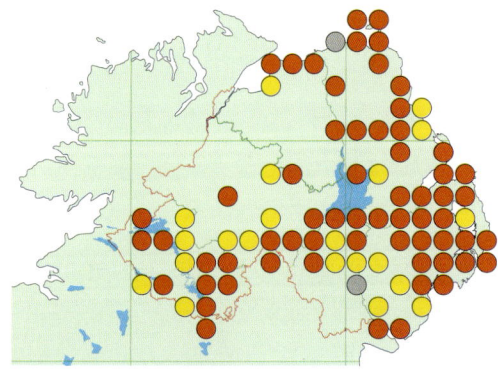

with a smaller white-lined stigma inside. The orbicular stigma is generally inconspicuous. The antemedian and postmedian lines vary in intensity of colour, but are irregular and strongly indented. The termen area is normally darker and clearly seen on paler individuals. Other similar species, such as Small Clouded Brindle, have a dark basal streak on the forewing and a paler hindwing. The 'common rustics' are found in a wide variety of habitats including grassland, coastal heaths, woodland and gardens. Adults rest concealed among ground vegetation by day and become active after dark. They come frequently to light and to flowers, especially Common Ragwort, Buddleia and grasses. The recorded flight period of both species is 31 May to 1 October. The larvae feed from autumn until May on Cock's-foot and other grasses. It overwinters as a larva.

Lesser Common Rustic
Mesapamea didyma (Esper) Category three G.
As stated under the previous species, the two species of *Mesapamea* in Ireland can only be identified by examination of the genitalia. As this is rarely done, it is difficult to establish the

Lesser Common Rustic.

Murlough NNR, County Down.

precise distribution of either species. Lesser Common Rustic was recognised as a distinct species in 1983 and Rathlin Island (Antrim) produced the first confirmed Northern Ireland record. Although the *M. secalis/didyma* aggregate is ubiquitous in Northern Ireland, the relative abundance and the exact distribution of each species is still not clear. The forewing colour range is similar to *M. secalis*; however, black individuals with a white reniform stigma are probably, but not certainly, Lesser Common Rustic. There is considerable overlap in the habitat usage, foodplants and phenology of the two species. The recorded flight period of both species is 31 May to 1 October. The larvae feed on Cock's-foot and other grasses. It overwinters as a larva.

Small Dotted Buff

Photedes minima (Haworth) Category two.

Widespread. Records are widely distributed across Armagh, Down and Fermanagh. In the north it is confined to coastal localities and Rathlin Island (Antrim). Sites in the north include Kebble on Rathlin Island, Glenariff Forest and Larne (Antrim) and Duncrun and Magilligan (Londonderry). This is a rather small and delicate-looking moth. The forewings are pale straw-coloured with a darker patch on the costa near the termen. The antemedian and postmedian lines are reduced to a series of small dark dots. This species is found in wetland habitats including cutover bogs, fens, wet woodland and upland damp grassland and moorland. Adults become active from dusk onwards and appear frequently at light in small numbers. On account of their small size the

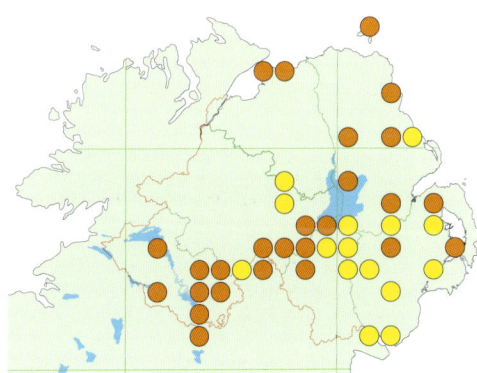

adults can be easily overlooked or misidentified by the inexperienced observer. The recorded flight period is 16 June to 20 August. The larval foodplant is Tufted Hair-grass. It overwinters as a larva.

Small Wainscot
Photedes pygmina (Haworth) Category one.

Ubiquitous. Small Wainscot has been recorded from all areas of Northern Ireland and any gaps in the north and west probably reflect poor coverage rather than genuine absence. It has also been recorded from Rathlin Island (Antrim). Adults are small and rather stocky in appearance. The forewing is quite variable in colour, ranging from straw to orange or reddish-brown. There is a small darker central streak that extends from the basal area of the forewing to the termen and a line of dark dots which curve towards the costa. This species is characteristic of wetland habitats especially fens, cutover bogs, wet woodland and lake shores; however, it can be encountered in other habitats and adults frequently appear in garden traps. Adults become active at dusk and appear at light in small numbers, but they do not feed in the adult stage. The recorded flight period is 22 June to 15 October. The larval foodplants are sedges and possibly rushes. It overwinters as a larva.

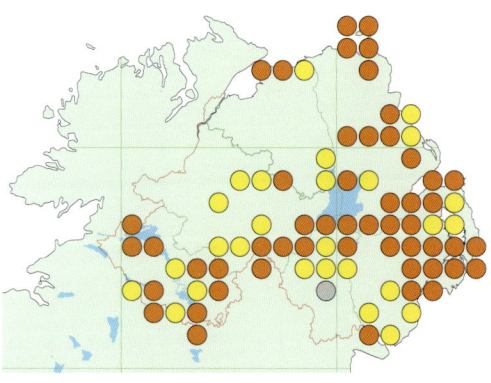

Flounced Rustic
Luperina testacea (Denis & Schiffermüller) Category one.

Widespread. Baynes considered this species mainly coastal in its distribution. Recent records, however, show it to be widespread inland, especially in Armagh and Down, and extending continuously from the east coast as far west as Clogher in Tyrone. It is also found along the north coast between Ballycastle (Antrim) and Magilligan (Londonderry), on Rathlin Island (Antrim) and at two sites in west Fermanagh. Adults are variable in colour, ranging from light brown to very dark brown. The orbicular and reniform stigmas are usually slightly paler than the ground colour and finely outlined with black. The antemedian and postmedian lines are pale brown and edged with black. There is a small black bar connecting both lines. Flounced Rustic is associated with grassland habitats, especially those in coastal areas and on well-drained soils. Adults are attracted to light, but can also be found resting on grass stems after dark. The recorded flight period is 14 July to 6 October. The larvae feed from September until June on Common Couch and probably other grasses. It overwinters as a larva.

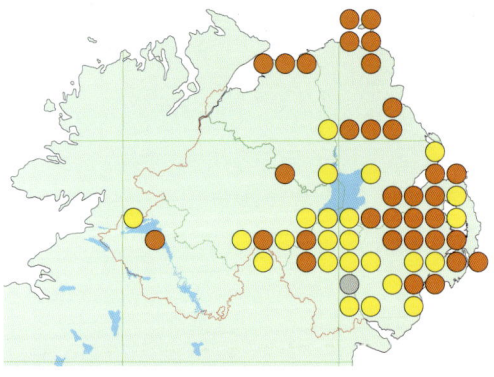

Notes on the three *Amphipoea* species in Northern Ireland

The three species of Ear moths are difficult to identify by visual characteristics alone. For this reason, it is recommended that a specimen should be retained for examination of the genitalia to confirm identification. This is especially important when recording in new sites. Many of the current records of all three species have not been fully authenticated and as a result we have mapped them all together. Validated records for each species have been mentioned in the appropriate text.

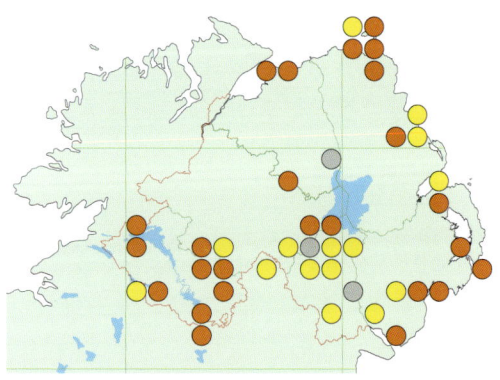

Ear Moth.

Ballykinler Dunes, County Down.

Large Ear
Amphipoea lucens (Freyer) Category three G.
Scarce. The distribution of this and the other 'ears' are not well known due to the difficulty in determination. Large Ear has been confirmed from all counties; most verified reports are from the west, especially Fermanagh. Sites include Carmacullagh, Correl Glen, Garvary Wood and Monmurry (Fermanagh) and Davagh Forest (Tyrone). There are also recent definite records from the Argory, Aughinlig and Canary (Armagh) and Killard Point and Mill Bay (Down). Records from Drumman More, Loughgall and Richhill (Armagh) collected in the 1970s are supported by correctly determined voucher specimens. The three ear moths overlap in size and markings and firm identification requires the examination of the genitalia. Hybridisation may occur between species and sampling of a number of adults at each site may be needed for certain identification. The forewings are long and narrow; the colour can be variable but is usually a rich reddish-brown. The orbicular stigma is usually discernible. The reniform stigma is large and conspicuous and usually orange but occasionally white or yellow. The antemedian and postmedian lines are darker and clearly visible. Little is known about the habits of Large Ear in Northern Ireland. Although Baynes described this species as mainly coastal, many of the confirmed records are from inland sites. Adults are attracted to light and flowers, especially rushes. The recorded flight period is 7 August to 25 October. The larvae feed from May to July on Purple Moor-grass and Common Cottongrass. It overwinters as an egg.

Crinan Ear
Amphipoea crinanensis (Burrows) Category three G.
Scarce. The true distribution of this species is difficult to ascertain due to the problem in identifying the ear moths. There are verified records of this species from Aughinlig, Canary

(1974), Drumbanagher (pre-1940), Drumman More Lough (1966 and 1967), Loughgall and Lurgan (pre-1933) (Armagh); Ballykinler Dunes and Murlough NNR (Down); Monmurry (Fermanagh); Magilligan Dunes (Londonderry) and Rehaghy Mountain (Tyrone). This species is almost impossible to differentiate from Ear Moth. The orbicular and reniform stigmas are usually orange. The cross-lines are generally darker than the base colour. Crinan Ear is not necessarily restricted to coastal localities as some of the confirmed records indicate. Adults are attracted to light and flowers particularly those of Heather and Common Ragwort. The recorded flight period is 10 August to 25 September. The larval foodplants in the wild are unknown. It overwinters as an egg.

Ear Moth

Amphipoea oculea (Linnaeus) Category three G.

Scarce. Like the two other species of ear moth present in Northern Ireland, the precise distribution of this species is not known. There have been confirmed records of Ear Moth from Ballykinler Dunes, Killard Point, Murlough NNR (Down); Correl Glen, Crom, Eshywulligan and Legatillida (Fermanagh); Magilligan Dunes (Londonderry) and Davagh Forest and Rehaghy Mountain (Tyrone). Adults most closely resemble the Crinan Ear and many specimens are indistinguishable in the field. The forewings of this species are slightly broader and reddish-brown in colour. The reniform stigma is usually creamy white, but it can also be orange on some individuals. The cross-lines are darker and usually conspicuous. Ear Moth has been found on a wide range of habitats including coastal dunes, heathland and woodland. Adults are attracted to light and flowers, especially Common Ragwort and thistles. The recorded flight period is 3 July to 29 September. The larval foodplants are various grasses. It overwinters as an egg.

Rosy Rustic

Hydraecia micacea (Esper) Category one.

Widespread. Known from a wide range of sites in all counties and both Copeland Bird Observatory (Down) and Rathlin Island (Antrim). Adults have rather long forewings which are slightly pointed at the apex. The general colour varies from light to dark rosy brown. The orbicular and reniform stigmas are similar to the ground colour and finely outlined with a darker line. The antemedian line is strongly convex; the postmedian line is well defined and curved towards the costa. Rosy Rustic occurs in a wide range of habitats including woodland rides, marshy areas, fens, bogs, waste ground and gardens. Adults are attracted to light in small numbers and also visit flowers. The recorded flight period is 3 July to 5 November. The larval foodplants are dock, plantain and possibly horsetails and Yellow Iris. It overwinters as an egg.

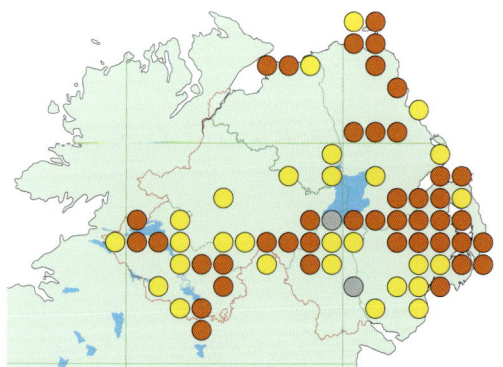

Frosted Orange

Gortyna flavago (Denis & Schiffermüller) Category two.

Widespread. Recorded continuously from north and central Down across north Armagh to east Tyrone. It is more sporadic elsewhere although it is known from several sites in records from Fermanagh, south Londonderry and mid-Antrim. This species is probably more widespread in the north, but limited trapping in late summer and the autumn would explain the lack of records from this region. There has been a single record on the north coast from Portrush (Antrim). This is an attractive and distinctive species. The ground colour of the forewing is greyish-brown with a conspicuous golden yellow central band. The orbicular and reniform stigmas are finely outlined with a darker line. There is also a suffusion of orange at the apex and a small blotch at the basal area of the forewing. Frosted Orange has been found in damp woodland, marshy areas, fens and, less frequently, gardens. Adults come to light in small numbers, usually well after dark. The recorded flight period is 5 August to 15 October. The larval foodplants are Foxglove, thistles, Common Ragwort and probably other herbaceous plants. It overwinters as an egg.

Haworth's Minor
Celaena haworthii (Curtis) Category two.
Widespread. Locally distributed through all counties. It is unrecorded, but probably overlooked, in much of the north and west, especially Tyrone and Londonderry. It is recorded from many bogs and fens in north Armagh while in Fermanagh, it is found in several widely-scattered sites including Correl Glen, Crom and Legatillida. In Antrim it occurs mainly in the far north at Breen Wood, Breesha Plantation and Clare Wood and on Rathlin Island. Haworth's Minor is a small moth with dark reddish-brown forewings which appear square at the ends. The stigmas are whitish and clearly defined, the orbicular being quite small. There are two fine white lines projecting from the distal aspect of the reniform stigma to the outer edge of the forewing. This moth is found on bogs, fens and moorland. Adult males will often fly in late afternoon sunshine, but most are seen at light after dusk. Individuals have been found at rest on the stems of Heather after dark. They are also attracted to flowers, especially Heather and Common Ragwort. The recorded flight period is 11 July to 23 September. The recorded larval foodplants are cotton grasses and rushes. It overwinters as an egg.

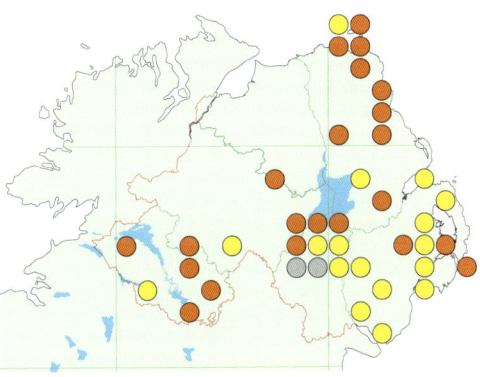

Crescent
Celaena leucostigma (Hübner) Category two.
Widespread. Recorded from all counties, but rather patchily distributed in most. It has been recorded from Rathlin Island (Antrim). This is a rather dark, medium-sized moth. The ground colour ranges from light to deep mahogany brown. There are small and conspicuous pale yellow reniform stigmas, but the antemedian and postmedian lines are not clearly defined. There is a series of small white dashes along the costal edge and in the darker colour form there is a wide light band running obliquely across the rear of the forewing. Crescent is a wetland species which has been found most frequently in fens and cutover bogs; however, individuals can also appear in traps well away from its normal habitat, including gardens. Adults are attracted to light in small numbers and also to the flowers of rushes. It can reportedly be found after dark at rest on the stems of Common Reed. The recorded flight period is 3 June to 4 October. The larvae feed from March to July on Yellow Iris and probably other wetland plants. It overwinters as an egg.

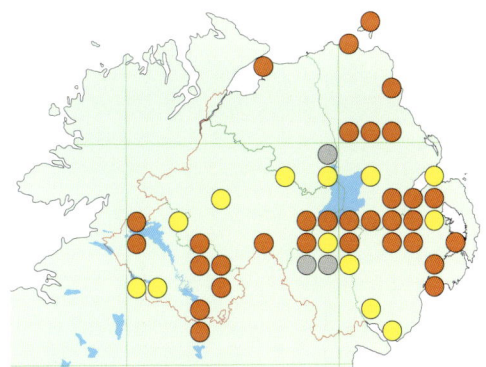

Bulrush Wainscot
Nonagria typhae (Thunberg) Category two.
Scarce. Widely distributed and commonly encountered in the southern counties, especially the bogs and fens around the Lough Neagh basin. It has been recorded from just one site, the

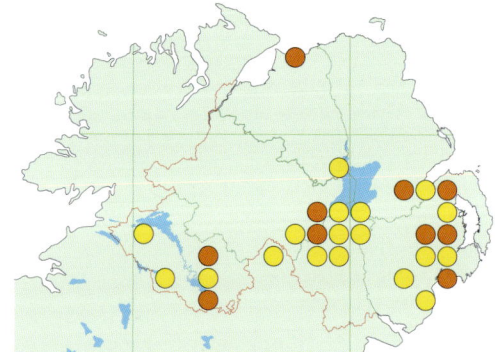

Bulrush Wainscot.

Ballykinler Dunes, County Down.

Umbra (Londonderry), to the north of Lough Neagh. It was found here in 2003, and it seems likely that it awaits discovery at other sites in Antrim, Londonderry and Tyrone. This is a large, rather plain-looking moth with a well-rounded forewing. The sexes differ in appearance. The males are reddish-brown whilst the larger females are pale straw-coloured. In both sexes the veining is conspicuous and there is a series of small black dashes or marks along the terminal area of the forewing. Bulrush Wainscot is found in lowland wetlands (fens, cutover bogs and lake shores) with stands of its foodplant. Adults come to light in small numbers and often rest by day concealed among the stems of bulrushes. It is often found well away from its normal breeding area and is therefore very quick to colonize new localities. The recorded flight period is 22 July to 13 October. The larvae feed inside the stems of Bulrush and Lesser Bulrush. It overwinters as an egg.

Rush Wainscot
Archanara algae (Esper) Category four.

Rare. Only known from Crom in Fermanagh where it was discovered on 22 August 2000. Since then it has been recorded at the same site on four other occasions on 24 August, 2 and 15 September 2000 and 25 August 2002. There have been no other records despite regular trapping in the same locality over the last two years. Whether it exists elsewhere remains unknown, but it is considered likely that there are undiscovered populations especially around Upper Lough Erne. It is uncommon generally in Ireland, known only from the Burren area of Clare, western Galway, Kildare and Roscommon. Adult males are orange-brown while the slightly larger females are straw-coloured. The forewings are rather square and pointed at the apex. The stigmas and cross-lines are represented by a series of dark dots. Rush Wainscot is found in lowland wetlands with large stands of emergent vegetation. Adults are attracted to light in moderate numbers and can be found after dark at rest on vegetation. The dates of the Northern Ireland records fall within the flight period quoted for Britain. The larval foodplants are Bulrush, Lesser Bulrush, Common Club-rush and Yellow Iris. It overwinters as an egg.

Large Wainscot
Rhizedra lutosa (Hübner) Category two.

Scarce. Commonly encountered to the south of Lough Neagh in Armagh and east Tyrone. Sites in this area include the Argory, Brackagh Moss, Oxford Island and Peatlands Park (Armagh) and Favour Royal and Rehaghy Mountain (Tyrone). It is more locally distributed in south Fermanagh where it has been recorded recently from Aghalane, Crom and Monmurry Bog. Murlough NNR and Turmennan Fen are the most recent records for Down. There have been single records from Antrim – Montiaghs Moss – and Londonderry – Traad Point on the north-west corner of Lough Neagh. Adults are pale straw-coloured with long narrow forewings which are rather pointed at the apex. The antemedian and postmedian lines are replaced by a series of small black dots, the latter having a distinct curve towards the costa. It closely resembles the Bulrush Wainscot but this has broader and more rounded forewings. The Large Wainscot occurs in wetland habitats including fens and cutover bogs. Adults have also appeared in gardens and non-wetland habitat. The adults come to light in small numbers. It rests by day low down among the stems of Common Reeds and wetland grasses. The recorded flight period is 21 August to 16 November. The larvae feed from April until July in the roots and stems of Common Reed. It overwinters as an egg.

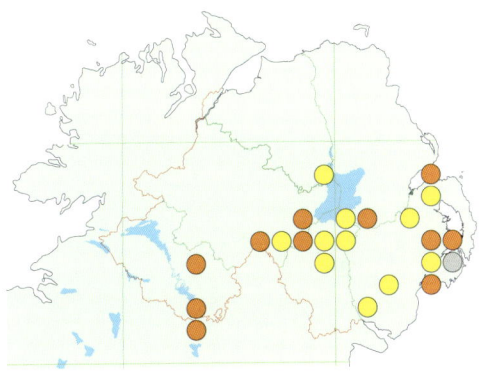

Treble Lines
Charanyca trigrammica (Hufnagel) Category two.

Notable. This species has been recorded in Antrim, Armagh, Down, Fermanagh and Tyrone, but it is very local in each. Since 1990, it has been recorded from a small number of sites and most recent records originate from Crom in Fermanagh. In Tyrone, it has been recorded from Crilly in good numbers on two separate occasions. In north Armagh it was found at a number of sites in the 1970s, including Drumman More Lough, but has not been seen anywhere in the county since 1984. Newcastle is the only known site in Down. The Antrim localities are Dunmurry and Portrush. Adults are a pale grey with a broad

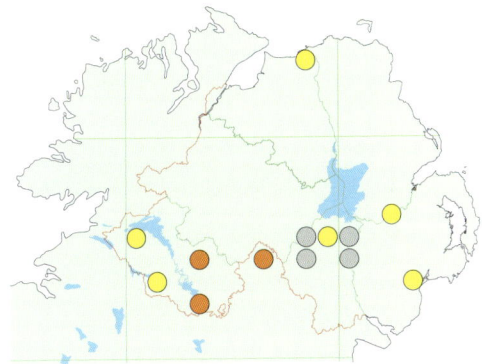

forewing which is pointed at the apex. It derives its common name from the three fine brown lines which run across each forewing. Treble Lines is found in clearings and the edges of woods, rough pasture and hedgerows. Adults become active from dusk onwards and are attracted to light in small numbers. The recorded flight period is 18 May to 28 June. The larvae feed at night on herbaceous plants including plantain and Common Knapweed. It overwinters as a larva.

Uncertain
Hoplodrina alsines (Brahm) Category two.
Widespread. This species has a strongly south-eastern distribution. It is restricted almost entirely to Armagh and Down apart from seemingly isolated records in Fermanagh and Tyrone and from Slievenacloy and Rathlin Island (Antrim). There have been no recent records from Londonderry. It appears regularly in traps in gardens in the Greater Belfast area, Helen's Bay and Killard Point (Down). The colour of the adults ranges from light to dark brown. The wings are rough-scaled and appear granular. The orbicular and reniform stigmas are darker than the ground colour and outlined with a fine pale line. The antemedian line is dark, extending obliquely from the costa to the dorsum; the postmedian line is curved towards the dorsum. It closely resembles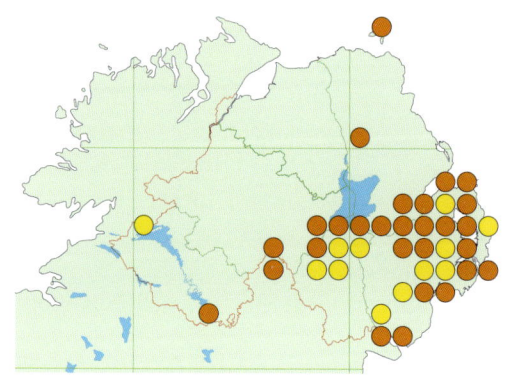
Rustic and with some specimens, the genitalia should be examined to determine identification. This is essentially a lowland species found in a diverse range of habitats including woodland, heaths, unimproved grassland and gardens. Adults come to light in small numbers. The recorded flight period is 23 June to 11 September. The overwintering larvae feed nocturnally from September until April on plantains, docks, dead nettles and Primrose.

Rustic
Hoplodrina blanda (Denis & Schiffermüller) Category two.
Widespread. This species seems to have several disjunct populations in Northern Ireland. It is found in two main areas — one along the north coast of Antrim and Londonderry, including Rathlin Island (Antrim) and a second, running discontinuously from Belfast Lough around the coast of east and south Down and inland along the Lagan Valley to the south shore of Lough Neagh. There is also an isolated record from Crom in Fermanagh. The coastal populations seem to be the strongest. It was taken many times at different localities on Rathlin Island in both 2001 and 2003; these adults were much darker than the usual paler form and resembled specimens from the Scottish islands. It was also frequently encountered during the 1996 moth survey of Killard Point (Down). The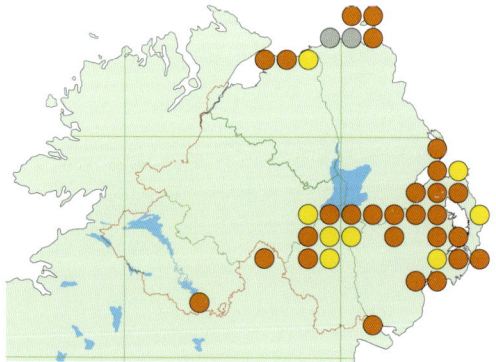
existence of old records from Fermanagh and Tyrone suggests the distribution may be wider than the records indicate. Adults appear smooth-scaled with a glossy appearance and vary from light grey to dark greyish-brown. The orbicular and reniform stigmas are similar to the ground colour and outlined with a fine pale line. The antemedian line is not clearly defined but represented by a series of sparse black scales or dots. The postmedian line comprises a series of black dots. It closely resembles Uncertain but that species is less variable in colour, has less sheen to the forewings and a browner hindwing. Individuals should be retained for verification, especially from unconfirmed sites. Rustic has been found on grassland and heaths on the coast and inland, also woodland and gardens. Adults come frequently to light and also to the flowers of Common Ragwort and Buddleia. The recorded flight period is 7 June to 25 August. The larvae feed from September to May on low-growing herbaceous plants including plantains and docks. It overwinters as a larva.

Vine's Rustic

Hoplodrina ambigua (Denis & Schiffermüller) Category four.

Rare. Recorded in Northern Ireland for the first time at the West Lighthouse on Rathlin Island (Antrim) on 28 August 2001. The adult was determined by Paul Waring. It seems highly likely that this insect was an immigrant rather than a resident since no adults have been reported since; however, this species could be easily overlooked due to its close resemblance to both Rustic and Uncertain. Baynes makes no reference to its occurrence in Ireland in his 1964 catalogue, but there are published records for Youghal (28 August 1976) and Fountainstown (1977 and 1978) (Cork). The forewings are brownish-grey, lightly dusted with pale grey. The orbicular and reniform stigmas are conspicuous, slightly darker than the ground 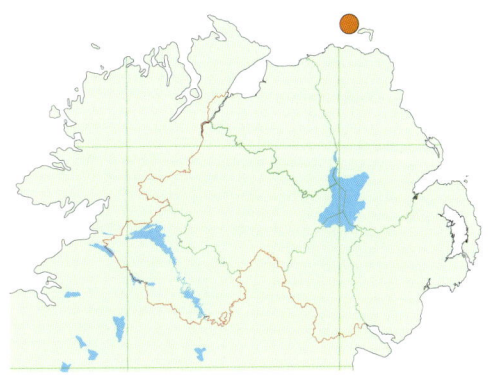 colour and outlined with a fine pale line. The hindwings are white with a faint sheen. In Britain, it is widespread along the south coast and across southern England. Adults are attracted to light and also to flowers. There are two generations in Britain, May to July and August to September. The larvae feed on docks, Dandelion, Primrose and other herbaceous plants.

Small Mottled Willow

Spodoptera exigua (Hübner) Category four.

Rare. This is an occasional immigrant to Ireland. It has been recorded ten times in Northern Ireland. It was first recorded in July 1903 at Raughlan (Armagh) on the south shore of Lough Neagh. It was not reported again until 1998 when three adults were taken in two separate traps on 3 July, one at Murlough NNR and one at Dundrum village (Down). The other records have been from Aughinlig (Armagh), 7 August 2000 and 5 August 2003; Drumbrughas North (Fermanagh), 2 August 2001; Killard Point (Down), 4 September 2003; Killynick (Fermanagh), 25 July 2000 and Mill Bay (Down) on 9 August 2004. This is a rather plain species. The forewings are narrow and range from pale brown to greyish-brown with darker speckling. The orbicular 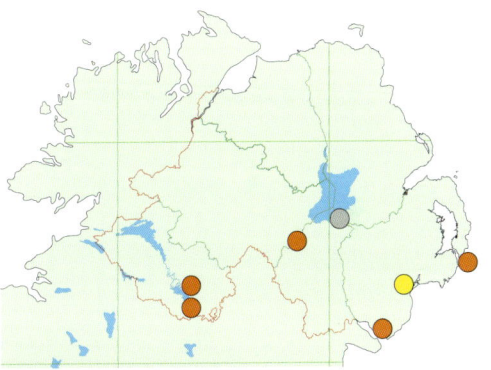 and reniform stigmas are conspicuous, usually pinkish-orange in colour and outlined with a fine pale line. The long forewings are wrapped tightly round the body when the moth is at rest. Small Mottled Willow is a cosmopolitan species which is regarded as a pest in many parts of its range. Adults are attracted to light and may appear in any month but are usually seen in the late summer and early autumn. In Northern Ireland, the records have been between 3 July and 4 September. It seems unlikely that the larvae would survive to maturity in Northern Ireland. In Britain, there have been several unproven claims of successful larval development.

Mottled Rustic

Caradrina morpheus (Hufnagel) Category two.

Scarce. An eastern species commonly recorded in Armagh and Down, but scarce in Antrim and Tyrone and unrecorded from Fermanagh and Londonderry. The westernmost locality is at Stillago in south-eastern Tyrone. It has been recorded at only three localities in mid- and north Antrim – Ballymoney, Ecos Park and on Rathlin Island. Adults range from light to dark greyish-brown. Fresh adults often have a slightly glossy appearance. The orbicular and reniform stigmas are not sharply defined but have a soft or blurred appearance. The cross-lines are darker than the base colour, but clearly visible. The hindwing is greyish-white. Mottled Rustic occurs in many habitats including woodland, bogs, rough grassland and scrub. It is a frequent visitor to gardens, particularly in Greater Belfast and along the Lagan Valley. Adults come frequently to light and less commonly to flowers. The recorded flight period is 14 May to 8 September. The larvae are nocturnal and feed from August to November on a variety of low-growing plants, including Common Nettle, docks, goosefoots, willows and Teasel. It overwinters as a larva.

Pale Mottled Willow
Caradrina clavipalpis (Scopoli) Category three.
Scarce. Baynes described this species as common and widely distributed but this is not apparent from recent records. These indicate it is present only across the southern half of Northern Ireland, south of a line from Belfast to Enniskillen. It was recorded regularly in north Armagh in the 1970s and 1980s. Since 2000 it has been recorded from a number of sites including Crom and Monmurry (Fermanagh), Aughinlig (Armagh), where it has been recorded continuously since 1999 until 2003, Gilnahirk and Tyrella in Down. A record from Portballintrae (Antrim) in 1937 remains the only one from the north of the region. Adults have a mottled appearance and long narrow forewings. The colouration ranges from greyish-brown to darker

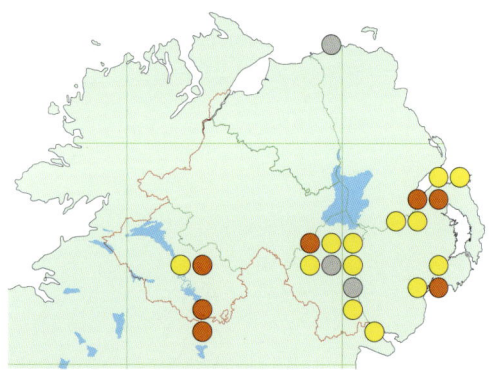

brown. There are four small marks along the costal edge of the forewing and both stigmas are inconspicuous and reduced to small dots. The hindwings are pearly white at the base, becoming darker towards the outer margin. Pale Mottled Willow has been found on coastal dunes, heaths, grassland and gardens. It also occurs in urban habitats. Adults come frequently to light and visit flowers. Adults have been recorded between 31 March and 25 October. It seems to be double-brooded as there are two peaks, one from early July to mid-August and a second from early September to the end of October. The larvae feed on the seeds of grasses and cereals including those in stores. It overwinters as a larva.

Anomalous
Stilbia anomala (Haworth) Category three.
Scarce. Recorded from all counties except Armagh, but mainly found on or close to the coast. It is known from regularly trapped coastal localities including Rathlin Island (Antrim), Ballykinler Dunes and Murlough NNR (Down) and the Umbra (Londonderry). There have been inland occurrences at Slievenacloy (Antrim); Belcoo, Gortmaconnell Rock, Legalough and Monawilkin (Fermanagh) and Ballynagilly (Tyrone). In Down it has been taken at Seaforde and from a suburban garden in the Greater Belfast area. There are historical records for Lough Fea (Tyrone) by Greer and the Mourne Mountains (Down) by Watts. The monitoring data from the Rothamsted trap network indicates a decline in abundance of this species in Britain of 93 per cent between 1968 and 2002. This is

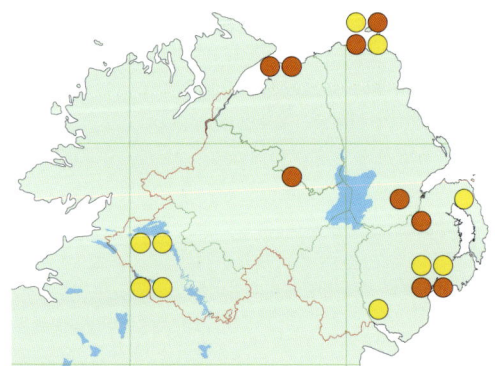

a fairly distinctive moth with a silky appearance when freshly emerged. The forewings are narrow at the basal area becoming much wider near the apex. The colour is generally grey but there can

Moth accounts

Anomalous.

Ballykinler Dunes, County Down.

A mainly coastal species in Northern Ireland that appears at light in small numbers.

be a little variation between individuals. There is a black patch between the orbicular and reniform stigmas and three small black marks near the apex of the forewing. Anomalous is recorded from coastal heaths, moorland and limestone grassland. Adults become active at dusk and are attracted to light in small numbers. Individuals have been found at rest on Heather and other low vegetation after dark. The recorded flight period is 11 August to 11 October. The larval foodplants are Wavy Hair-grass and Tufted Hair-grass. It overwinters as a larva.

Bordered Sallow
Pyrrhia umbra (Hufnagel) Category three.
Rare. Recorded in recent times from just three sites, Killard Point on the Down coast and Aughinlig and Loughgall in north Armagh. It has been recorded at the last two sites on only one occasion each on, respectively, 10 August 1978 and 1 August 1992. These records probably refer

to immigrants. It was first recorded at Killard Point on 4 July 1996 during an intensive moth survey which produced six more records during the same month. It was taken again at Killard Point on 16 July 2002 and 21 July 2004. It has still to be determined whether there is a resident population at this site (one of its foodplants, Common Restharrow, flourishes at this locality), or whether these are migrants. Langham recorded it at Tempo in 1902 and Donovan gives Armagh (date unknown) and Strangford, 1929. The forewings are a dark orange-yellow with fine dark brown cross-lines. There is also an obliquely angled cross-line which demarcates the junction between the rest of the forewing colour and darker suffusion at the outer edge. In Northern Ireland this moth has been found on sandy coastal grassland and in gardens. Adults are attracted to light and to flowers. In Britain the main flight period is June and July. Recorded larval foodplants in Britain include Common Restharrow and Sea Sandwort. It overwinters as a pupa.

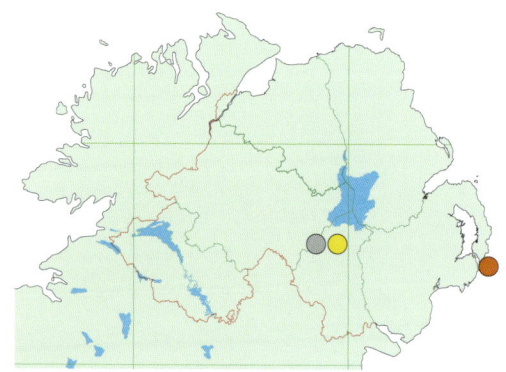

Scarce Bordered Straw

Helicoverpa armigera (Hübner) Category four.

Rare. This species occurs in Britain and Ireland as either an adult moth or as larvae imported on cultivated fruit and vegetables. It has been recorded on four occasions in Northern Ireland. All the records have been in Down. These were at Dundrum in 1954, Killard Point on 19 August 1996, Island Hill on 23 September 2000 and Banbridge in 2003. The last record involved two larvae found in flowers bought from a supermarket on 22 January. The other records were of adults in late summer on east coast sites which strongly suggest they were immigrants. Adults are variable in colour ranging from sandy brown to purplish-brown. The orbicular stigma is replaced by a small dark central dot. The reniform stigma is variable and either lighter or darker than the ground colour. The colour intensity of the markings on the forewings varies between individuals. The hindwings are cream-coloured with a broad dark terminal band. Scarce Bordered Straw is found worldwide in warm areas where it can be a serious pest of cultivated crops. It is resident in southern Europe and North Africa where adults can be seen throughout the year. In Britain and Ireland adults are normally seen in the late summer or autumn. They are attracted to light and flowers.

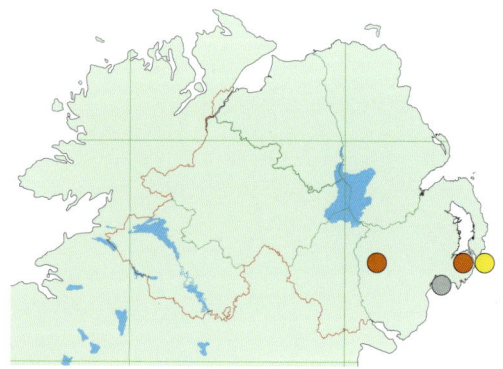

Bordered Straw

Heliothis peltigera (Denis & Schiffermüller) Category four.

Rare. This migrant moth, a resident of southern Europe, North Africa and eastern Asia, appears regularly in southern Britain each year during the summer months. It is less frequently seen further north and in Ireland. It has been recorded on only three occasions in Northern Ireland: at Tempo (Fermanagh) in 1902, Seaforde (Down) on 22 October 1988 and Helen's Bay (Down) on 1 June 1994. The ground colour of the forewings (which is apparently affected by temperature experienced during development) is variable, ranging from straw to sandy brown. The reniform stigma forms part of a triangular blotch which is conspicuous and extends out to the costa. The orbicular stigma is pale and often reduced to a small black dot. There is also a small dark patch on the costa near the apex of the wing. The hindwings are creamy white with heavy, darker veining, and a broad dark terminal band and

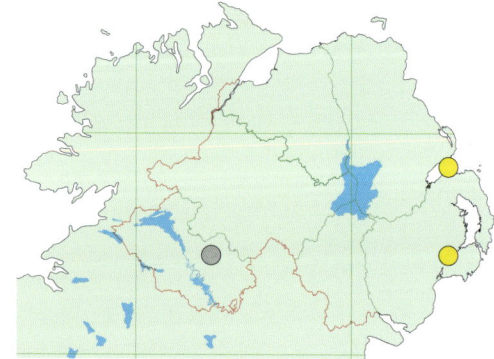

a central mark. Adults have been recorded throughout the year in Britain. The recorded larval foodplants are garden marigolds, Groundsel, Common Restharrow and species of thistle.

Marbled White Spot
Protodeltote pygarga (Hufnagel) Category four.
Notable. Known from five sites – the Argory and Peatlands Park (Armagh); Murlough NNR (Down); Crom (Fermanagh) and Rehaghy Mountain (Tyrone). It was first discovered in Northern Ireland at Murlough NNR on 19 June 1999 and then at Rehaghy Mountain on 29 June 2000. It is considered that the Marbled White Spot has been overlooked due, in part at least, to its resemblance to a micromoth and it probably will be found in other localities in Northern Ireland. Apart from the few sites in Northern Ireland, this species is confined to south-west Ireland in Cork, Kerry (at Killarney where it is common) and north Tipperary. Adults are normally brown, appearing mottled. The orbicular and reniform stigmas are similar to the ground colour 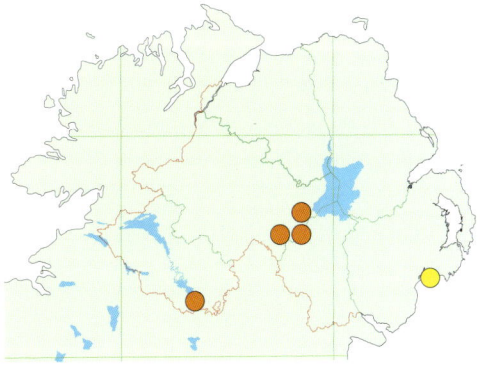 but conspicuous as they are finely outlined in white. The diagnostic feature is the white suffusion in the tornal area of the forewing. The fringe is chequered brown and white. In Britain, it is found mainly in woods and heaths on acid soils; the habitat at the Northern Ireland sites conforms to this description. Adults are attracted to light in small numbers and may also fly during the day when disturbed from vegetation. All the Northern Ireland records have been in the period 8 to 29 June. The main flight period in Britain is late May to July. The larval habits and foodplants are not known in Northern Ireland. In Britain it feeds on grasses, mainly Purple Moor-grass and False Brome. It overwinters as a pupa.

Bordered Straw.

Lisburn, County Antrim.

This rare migrant has only been recorded on a few occasions in Northern Ireland.

Silver Hook
Deltote uncula (Clerck) Category two.

Scarce. Recorded in all counties, but only in scattered sites outside Armagh and Down which appear to be its strongholds. It is present on many of the larger cutover bogs and fens in these two counties including Brackagh Moss and Peatlands Park (Armagh) and Drumawhy, Inishargy and Lackan Bog (Down). It is apparently more locally distributed in Fermanagh with recent records for Aghalane, Legatillida and Monmurry. Across the north it seems to be local. Northern sites include Garry Bog (Antrim) and Barmouth and Lough Bran (Londonderry). Wright also recorded it from Aghalee (Antrim) in 1954 and it has recently been found in the same part of the county at Montiaghs Moss (2003) and Portmore Lough (2000). 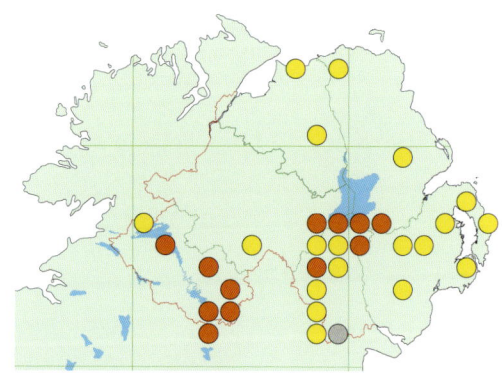 This small moth is not easily confused with any other species. The forewings are olive brown with a cream streak, suffused with grey, running along virtually the entire length of the costa. There is also a spur-like projection edged in white, running obliquely from the streak and pointing downwards towards the apex. This species is found on cutover raised bogs, fens, heathland and moorland. Adults are often disturbed from grassy tussocks and low vegetation during the day. They are attracted to light in small numbers and to flowers. The recorded flight period is 17 May to 27 July. The larval foodplants are wetland sedges and coarse grasses. It overwinters as a pupa.

Green Silver-lines
Pseudoips prasinana (Linnaeus) Category three.

Scarce. Baynes considered this species to be rare in Northern Ireland and not found outside Down, but recent fieldwork has shown it to be more widespread than was previously thought. It is now known to be present in the Greater Belfast area and along the Lagan Valley, in north Armagh and in well-trapped sites in Fermanagh. Recent records have come from the Argory (Armagh); Castleward and Hollymount (Down) and Aghalane, Altaclanabryan, Ballyreagh Bog and Legatillida (Fermanagh). Surprisingly, it has not been reported in recent times from northern counties as there would appear to be much suitable habitat. North of Lough Neagh it occurs at Ballymena (Antrim) and Banagher (Londonderry). There are old records of larvae from Derry City (Londonderry), Donard Demesne and Rostrevor (Down). 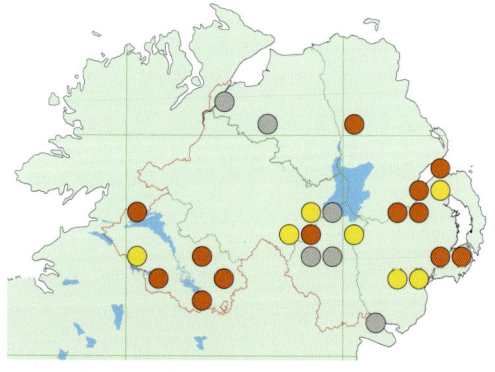 This is a very distinctive species not easily confused with any other moth. The forewings are bright green with three white lines which run obliquely across the wing. The thorax is a mixture of green and white; the legs and antennae are pink in the male. The hindwings are silky white. This is a woodland moth and it is present in many of the large broadleaved woodlands within its range; however, the records from the Belfast area in particular, show that it is not restricted to large expanses of woodland, and colonies can persist in suburban areas where groups of mature trees exist. The recorded flight period is 17 May to 4 July. The larval foodplants are broadleaved trees including oak, Beech, Aspen, birch and Hazel. It overwinters as a pupa.

Oak Nycteoline
Nycteola revayana (Scopoli) Category three.

Rare. There have been few confirmed records of this small moth. Historically, it has been recorded from Tyrone, at Favour Royal in 1902 by Kane and Stewartstown in 1921 by Greer. It has not been found in these areas since. Currently, it is known from just three sites, the

Oak Nycteoline.

Rostrevor Wood National Nature Reserve, County Down.

This woodland species is one of the smallest of the noctuid moths.

Argory (Armagh), Rostrevor Wood (Down) and Crom (Fermanagh). It was found at Rostrevor Wood in 1999 and recorded again in 2000. The populations at Crom and the Argory were discovered in 2000 and 2003, respectively. Adults show great variation in colour and markings. The forewings are strongly arched near the base of the wing. The colour ranges from greyish-brown to dark brown. The females tend to be darker than the males. The patterning on the forewings also shows some degree of variation. The greyish-brown form with small darker patches seems the most common one at Rostrevor Wood. The antemedian and postmedian lines are clearly defined. The hindwings are grey and vary in shade according to the colour form. Oak Nycteoline is a woodland species found in native or mature estate, broadleaved woodland with oak. Its early flight period and resemblance to a micromoth may account for its apparent scarcity at some suitable sites. It seems highly probable that other undetected populations await discovery in other mature woodland sites, especially in the west. Adults are attracted to light and to autumn flowers and berries. It has been stated that adults have a preference for hibernating in Ivy, old Yew and Holly trees. It can be beaten from the lower branches of oak or Hazel in the autumn. The adults appear in late summer and again after hibernation in early spring. The Northern Ireland records have been between 23 July and 23 September and in spring between 5 and 8 April. There is an historical record of larvae having been found in Northern Ireland on willow. Oak is probably its main foodplant in Northern Ireland as it is present at all currently known sites.

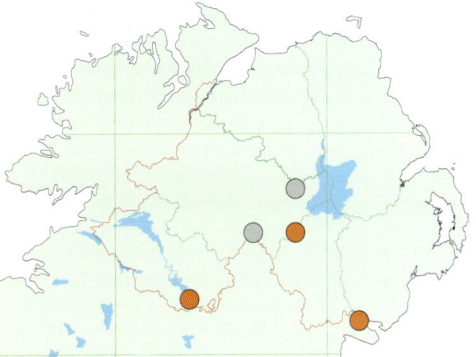

Nut-tree Tussock.

Glenarm, County Antrim.

Nut-tree Tussock

Colocasia coryli (Linnaeus) Category one.

Scarce. Recorded in all counties, but found mainly in Fermanagh and south Tyrone. It also occurs at a number of localities in Down, but is unrecorded in most of Antrim, Armagh and the west and north of Londonderry and Tyrone. Helen's Bay (Down); Crom, Garvary Wood and Legatillida (Fermanagh) and Rehaghy Mountain (Tyrone) have produced the majority of records. Sites in Down include Belvoir Park, Castlewellan Forest, Lackan Bog and Saintfield. This is a rather hairy moth with broad, well-rounded forewings which are a mixture of grey and brown. The basal area of the forewing to the median area is normally dark brown (except at the base of the wing which is grey) in all colour forms. The remainder of the forewing is generally paler grey.

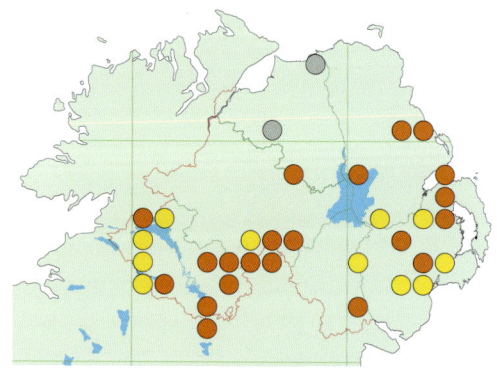

The stigmas are reduced to a small spot outlined in black and occasionally white in the centre. The fringe is normally chequered with a mixture of brown and grey. As the name suggests, Nut-tree Tussock is a woodland species associated often, but not exclusively, with Hazel. It is recorded from mature broadleaved woodland and gardens. Adults (generally males) come frequently to light in small numbers, the females are seldom seen. The recorded flight period is 22 April to 15 July. In southern England it is double-brooded, but there appears to be just a single annual generation here. The larvae have been found in Northern Ireland on oak; other known foodplants include Beech, Hazel and birch. It overwinters as a pupa.

Slender Burnished Brass
Thysanoplusia orichalcea (Fabricius) Category four.
Rare. This immigrant from the Mediterranean has been recorded once at Seaforde (Down) on 30 October 1989. There have been only three other Irish records, from Timoleague (Cork) in 1946, Connemara (Galway) in 1964 and Fountainstown (Cork) in 1978. This is a very distinctive moth, which can be confused only with the resident Burnished Brass. The forewings are mainly brown with a bright metallic brass-coloured patch running from the basal area to the apex of the wing. It can be told from Burnished Brass by the narrower forewing and the brassy patch which tapers to a point (rather than widening out) near the basal area of the forewing. In Britain, this is an irregular migrant to southern England that rarely reaches the north and west. Mid-August to October is the peak period for the British records. Adults are attracted to light and flowers. The larvae feed on various low-growing plants, including members of the daisy and cabbage families. They are occasionally found on imported plants, but there is no evidence that larvae occur in the wild here.

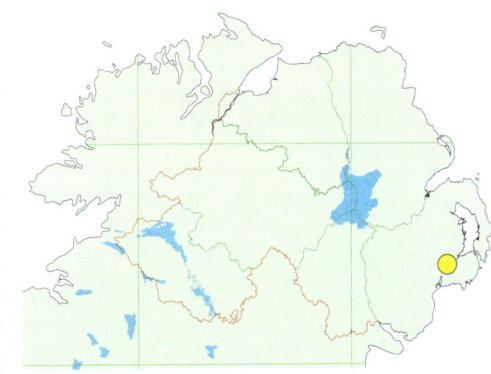

Burnished Brass
Diachrysia chrysitis (Linnaeus) Category one.
Widespread. Recorded from all counties, on Rathlin Island (Antrim) and Copeland Bird Observatory (Down). The gaps apparent in the north and west are probably due to under-recording rather than genuine absence. This is an easily recognised species that can be confused only with the rare migrant Slender Burnished Brass. The ground colour of the forewings is light brown with a bright brassy metallic suffusion extending from the basal area to the apex of the wing. The brassy patch widens towards the apex of the forewing which distinguishes it from Slender Burnished Brass. The head and front of the thorax is an orange-brown colour. When viewed in profile the thorax has a raised tuft or projection. This moth can be found in most habitats but especially woodland, fens, bogs, and suburban gardens. Adults come frequently to light and apparently also to the flowers of Red Valerian. There is a single annual generation which peaks between mid-June and early August. Adults have been recorded between 6 June and 11 October. The larvae feed on Common Nettle and Marjoram. It overwinters as a small larva.

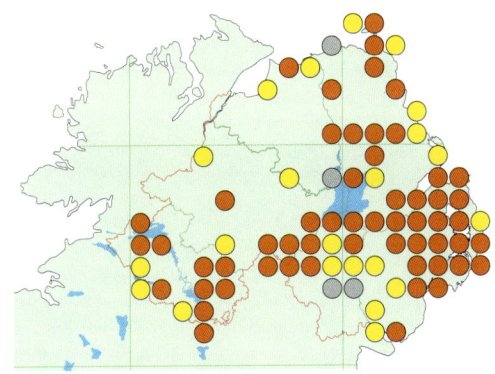

Golden Plusia
Polychrysia moneta (Fabricius) Category four.
Unknown. (Not mapped). Added to the Northern Ireland list based on a single record of an adult found in a MV trap at Loughgall Manor House (Armagh) on 13 September 1971. Although, the specimen was not retained there is no reason to doubt the authenticity of the

The Butterflies and Moths of Northern Ireland

Burnished Brass.

Peatlands Park, County Armagh.

The bright brassy patch on the forewings varies in intensity depending on the resting position of the moth. The sheen is less obvious when it is resting upside down among vegetation.

record, as its finder is a highly regarded and experienced lepidopterist. The only other Irish records come from Seapoint (Dublin), where it was first recorded in 1939, and Glenageary (Dublin) where it was recorded regularly by Baynes. In Britain, it is mainly recorded from gardens and parks. The flight period is late June until early August. It is possible that it may have been introduced with garden Delphinium species, on which the larva feeds.

Gold Spot
Plusia festucae (Linnaeus) Category one.
Widespread. Frequently encountered in all counties and on Rathlin Island (Antrim). This colourful moth is difficult to confuse with any other species. The head and front of the thorax is orange. The forewings are a rich golden brown with a definite metallic sheen. There are three

Gold Spot.

Argory Moss, County Armagh.

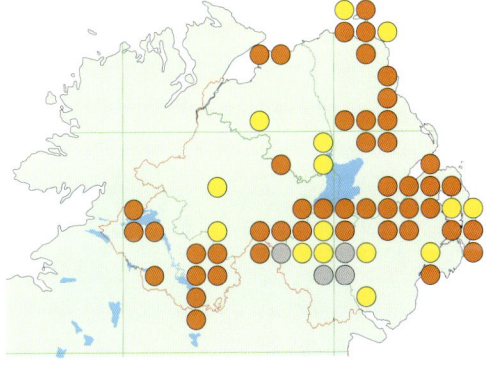

distinct silver blotches, two in the median area which are occasionally fused and a third, more elongated mark, near the apex of the forewing. There is a series of smaller yellow marks adjacent to this silver flash. In profile the thorax has a raised tuft or projection. Gold Spot is recorded from damp woodland, fens and bogs, where it is commonly encountered at light. The adults have occasionally been discovered by day, resting in low-growing vegetation. The recorded flight period is prolonged, extending from 30 May to 15 October. Two broods are normal in southern Britain, but the plot of records in Northern Ireland indicates that there is only a single generation here. The larvae feed during July and August on wetland sedges, Yellow Iris and bur-reeds. It overwinters as a larva.

Lempke's Gold Spot
Plusia putnami (Grote) Category four.
Unknown. This species is included in the Northern Ireland checklist based on a specimen taken at Drumman More Lough (Armagh) in 1963 by the late Phyllis Nelson, which was confirmed by the late John Heath. The specimen is in the Phyllis Nelson collection in the Armagh County Museum. To our knowledge, the record has not been published. There are only a few Irish records from Cork and Kerry. In Britain it occurs in fens, marshes and other similar places. The flight period is July-August in a single generation. Adults are attracted to light and flowers. The larval foodplants are Wood Small-reed, Purple Small-reed and possibly other grass species. It overwinters as a small larva.

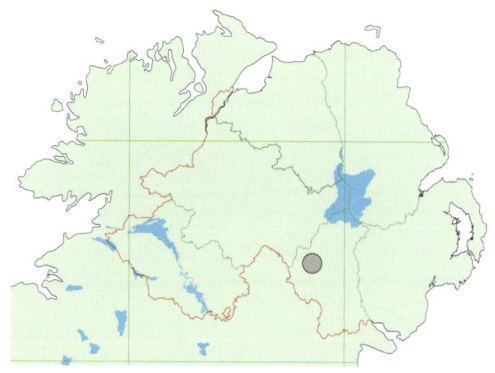

Silver Y
Autographa gamma (Linnaeus) Category one.
Ubiquitous. One of the most widespread and commonly recorded moths in Northern Ireland. It is recorded from all counties and on the larger offshore islands. The presence of this species each year is entirely dependent on immigration of adults, as it cannot overwinter successfully here. Early migrants will produce a locally-bred population. Adults show much variation in colour and markings. The forewings have a metallic sheen and are a mixture of greys and browns becoming paler towards the apex. Immigrants generally tend to be paler in colour than locally-bred adults. The median area of the wing contains a diagnostic silver Y-shaped mark. This mark is usually intact but, rarely, is broken. It is similar in appearance to Plain Golden Y, Beautiful Golden Y and Scarce Silver Y. The former two tend to have a redder colouration and the latter is a more blackish-grey colour. Silver Y is a highly mobile immigrant moth that can be found in virtually any habitat, but is most frequent in flowery grassland. Individuals are frequently seen during the day, especially in coastal localities, feeding at flowers. Adults also come to light in moderate numbers and will also visit garden flowers. Adults have been recorded from 15 April to 12 November. There are two peaks in the phenology, one in June, the second in late August. The larvae feed on bedstraws, clovers, Common Nettle and probably other low-growing herbaceous plants.

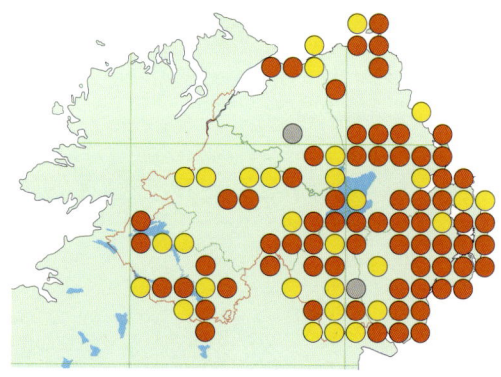

Beautiful Golden Y
Autographa pulchrina (Haworth) Category one.
Widespread. A commonly recorded moth which is generally distributed in most counties. It is recorded from both Copeland Bird Observatory (Down) and Rathlin Island (Antrim). Fresh adults have a distinct deep reddish-brown colour. The antemedian line is golden yellow, edged with purple. The postmedian line is faint and runs parallel to the costa. The reniform stigma is purplish-brown, finely outlined in gold and shaped like a figure 8. There is a small pale Y-mark in the median area of the forewing which is generally split into two. Plain Golden Y is less colourful and lacks any markings around the costa. This species can be found in most habitats, both upland and lowland, especially near stands of tall herbs in woods, grassland, scrub, hedgerows and gardens. Adults come frequently to light and also visit flowers. Individuals have

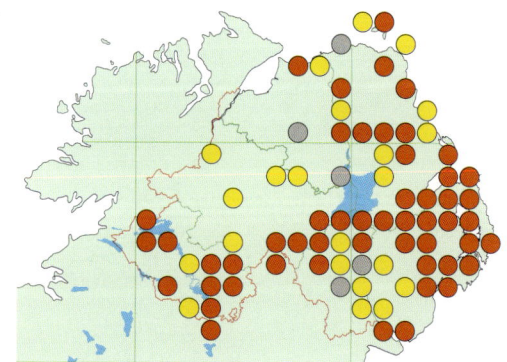

Beautiful Golden Y.

Banbridge, County Down.

occasionally been found by day often resting upside down low down among ground vegetation. The flight period is prolonged from 20 May to 11 September. The larvae feed on Common Nettle, Common Ragwort, Honeysuckle and probably other low-growing herbaceous plants from August to May. It overwinters as a larva.

Plain Golden Y
Autographa jota (Linnaeus) Category one.

Widespread. Generally distributed across the southern counties, but much more local in Antrim, Londonderry and Tyrone. It is recorded from both Rathlin Island (Antrim) and Copeland Bird Observatory (Down). Adults are a light reddish-brown with a slightly pink tinge. Superficially, they resemble the adults of the Beautiful Golden Y, but they are duller. There is a broad darker area running along the dorsum making the wing appear two-tone. The Y-shaped mark is greatly reduced and normally broken and is not as obvious as it is in the Beautiful Golden Y. The Plain Golden Y is found in most lowland habitats. Adults come frequently to light and also visit flowers. The recorded flight period is 9 June to 2 September, peaking in early July. The larvae feed on Common Nettle, Honeysuckle and probably other low-growing plants. It overwinters as a larva.

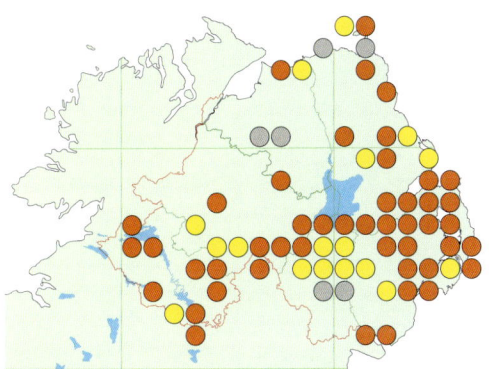

Gold Spangle
Autographa bractea (Denis & Schiffermüller) Category one.

Widespread. Common in north Down, Greater Belfast and west along the Lagan Valley to north Armagh and in Fermanagh. Sites are more scattered elsewhere. It has been recorded from

The Butterflies and Moths of Northern Ireland

Scarce Silver Y.

Peatlands Park, County Armagh.

This species is mainly associated with lowland bogs and heaths in the western counties.

Rathlin Island (Antrim). Adults are readily identifiable by the generally reddish-brown appearance and the large prominent golden metallic blotch in the median area of the wing; this spot can vary in size and colour — in a few individuals the gold blotch has been silver. The hindwings are pale grey with fine dark veining. It shows a preference for open habitats including woodland clearings, damp meadows by streams, upland pasture and moorland. Adults are active from dusk onwards and are attracted to light in small numbers and also to flowers. The recorded flight period is 13 June to 17 August. The larvae feed on low-growing plants including Common Nettle, Ground Ivy, Bilberry and Honeysuckle. It overwinters as a larva.

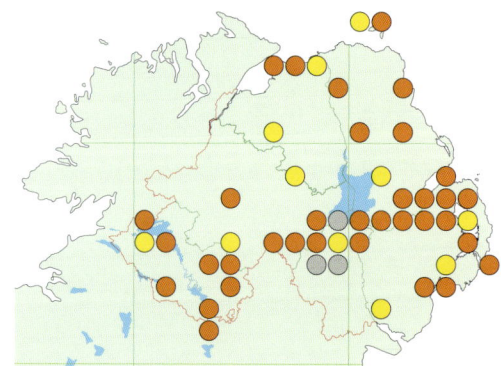

Scarce Silver Y

Syngrapha interrogationis (Linnaeus) Category three.
Scarce. This shows a distinctly western distribution in Fermanagh and central Tyrone. Sites include Correl Glen, Cuilcagh and Knocknalosset (Fermanagh) and Rehaghy Mountain and Tattanellen Bog (Tyrone). It is also recorded in north Armagh at Peatlands Park and the Argory and a few nearby sites. Creightons Wood and Kilrea Bog are the only known sites in Londonderry. Bohill Forest and Lackan Bog are its only localities in Down although it probably does occur in the Mourne Mountains. Breen Wood is the most recent record for Antrim. Adults have a mottled appearance which is composed of various shades of grey suffused with black, especially in the median area which appears darker than the rest of the forewing. As in all Y moths, the silver mark can vary

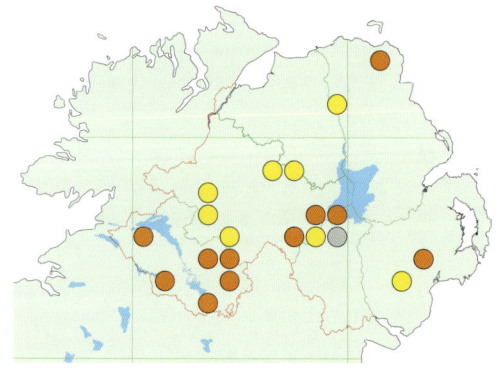

Dark Spectacle.

Knockbarragh, Rostrevor, County Down.

in size and shape between individuals. Scarce Silver Y is recorded from lowland raised bogs and upland heaths. It has also been taken on calcareous grassland sites in Fermanagh, although these are presumably stray specimens. Adults are attracted to light but are seldom seen in more than ones and twos. It is reported to fly occasionally in sunshine and said to rest on rocks or low down among vegetation during the day. The recorded flight period is 27 May to 26 August. The larvae, which overwinter, feed on Heather and Bilberry.

Dark Spectacle
Abrostola triplasia (Linnaeus) Category two.
Widespread. Generally distributed across southern counties but becoming more local further north. It is recorded from sites along the north coast and Rathlin Island (Antrim), but is seemingly absent from much of inland Antrim, Londonderry and Tyrone. Reliable sites for this moth

include Crom, Garvary Wood, Legatillida and Monmurry (Fermanagh). The forewings are narrow and normally very dark-brown, but slightly paler in the basal region. The upper part of the thorax is straw-coloured and appears as a raised projection. The orbicular and reniform stigmas are similar to the ground colour, but finely outlined in black. The postmedian cross-line is reddish-brown and there are three black parallel streaks near the apex of the forewing. It is similar to Spectacle, but the adults of that species are generally paler and the cross-lines are not reddish-brown. Dark Spectacle can be found in many habitats including bogs, heaths, moorland, damp marshy areas and woodland. It appears regularly in garden traps especially in the Greater Belfast area. Adults come to light in small numbers and also visit flowers after dusk. The recorded flight period is 7 May to 5 September. The larvae feed on Common Nettle. It overwinters as a pupa.

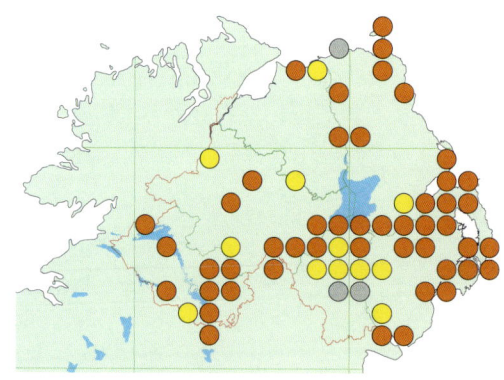

Spectacle

Abrostola tripartita (Hufnagel) Category one.

Widespread. Spectacle has been recorded in all counties and on both Rathlin Island (Antrim) and Copeland Bird Observatory (Down). It is generally distributed across southern counties but more sporadic further north. This species takes its name from the spectacle-shaped markings on the thorax which are visible when viewed from the front. The forewings are pale grey with a darker band across the median area. Both stigmas are conspicuous but the orbicular stigma is larger. It is similar in appearance to Dark Spectacle but the cross-lines are not reddish-brown. Spectacle is found in a wide range of habitats including woodland clearings, bogs, fens and mature gardens. Adults come frequently to light and flowers, especially Red Valerian. The flight season is prolonged in a single generation from 1 May to 8 September. The larvae feed on Common Nettle. It overwinters as a pupa.

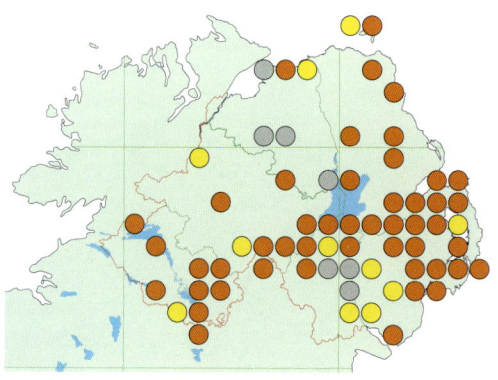

Clifden Nonpareil

Catocala fraxini (Linnaeus) Category four.

Rare. This immigrant moth is on the Northern Ireland list based on a specimen that reportedly flew through a window into a house in Derry City (Londonderry) in September 1896 and documented by Kane (1901). This is the only occurrence of this species in Northern Ireland. There has been only one other Irish record, in 1845 at Kingstown (modern Dun Laoghaire) by Rev. Joseph Greene. Clifden Nonpareil is a migrant in Britain and small numbers are seen most years. Populations were temporarily established in Norfolk and Kent in the mid-twentieth century. This is a large conspicuous grey moth with a distinct broad light blue band across the black hindwings. It reportedly rests by day on tree trunks and has a liking for sugar. In Europe it inhabits broadleaved woodland. The larval foodplants are Aspen and other species of poplar.

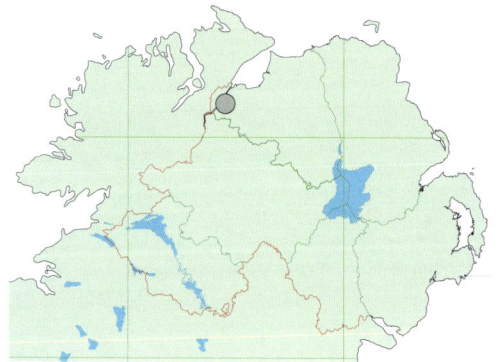

Mother Shipton

Callistege mi (Clerck) Category one.

Scarce. Recorded in all counties, but local in most. It is most frequent in east Down, north Armagh and Fermanagh. Surprisingly, in view of its habitat preferences, it is absent from most

Moth accounts

Mother Shipton.

Monawilkin, County Fermanagh.

An inconspicuous day-flying moth found in dry grasslands. It is often overlooked by inexperienced recorders.

of the north coast, except for Fair Head and Rathlin Island (Antrim). Elsewhere in Antrim it is recorded from Glenarm and Slievenacloy. The only sites in Londonderry are in the east of the county at Bracaghreilly, Carndaisy, Kilrea, and Springhill, although there are old records from Magilligan Dunes, where it was described by Campbell (1893) as abundant. Tyrone sites include Moneygal Bog, Mullaghmore Bog and Rehaghy Mountain. Strangely, it was not listed by either Wright or Nairn (1977) for Murlough NNR but a single larva was found at this site in September 1990 and an adult was recorded in 2000. This is a small day-flying moth which is most active in sunshine. The forewings are brown with a cream line that forms a shape said to resemble the

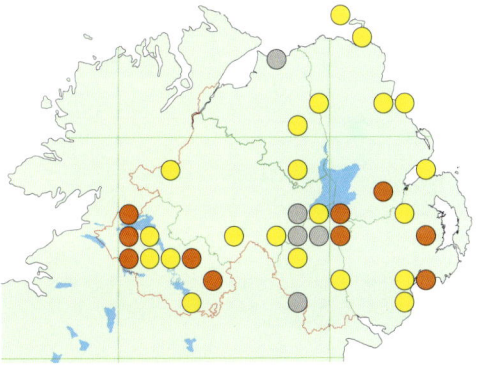

profile of a legendary Yorkshire witch. The antemedian and postmedian lines are white; the former is straight and quite oblique, the latter is dentate, forming a loop that connects with the antemedian line. The subterminal line is pale and quite straight. Mother Shipton can be found in open, dry grassland including flowery meadows, dry banks on cutover bogs, woodland rides and clearings and parkland. Its small size and dull colours may cause it to be overlooked, especially by the inexperienced observer. The adults rest on flowers and leaves and are often difficult to approach as they fly at the least sign of disturbance. The recorded flight period is 10 May to 10 July. The larvae feed on clovers, Bird's-foot Trefoil and various coarse grasses.

Burnet Companion
Euclidia glyphica (Linnaeus) Category one.
Scarce. Only found in limestone areas of Armagh and Fermanagh. In Armagh, it is found in the centre of the county at a restricted number of sites close to Armagh City, including Milford railway cutting, Navan Fort, Richhill and Thompson's Quarry. In Fermanagh, most of the sites

are in the north-west of the county, around Derrygonnelly and it is also known from Crom, Ederney and Maguiresbridge. This is a nervous and often difficult moth to approach, especially in warm sunny conditions. The forewings are greyish-brown with broad dark brown cross-lines outlined with a pale line. There is also a small dark patch near the costa. The hindwings are greyish-brown in the basal region. The outer area is orange-brown with a dark brown band running almost to the top of the wing. Burnet Companion is a small day-flying moth that is most active in bright sunshine. Adults are found in dry, unimproved, limestone grassland, on roadside verges (often with the Dingy Skipper) or, occasionally, wet meadows. The recorded flight period is 6 May to 3 July. The larvae feed on clovers and trefoils. It overwinters as a pupa.

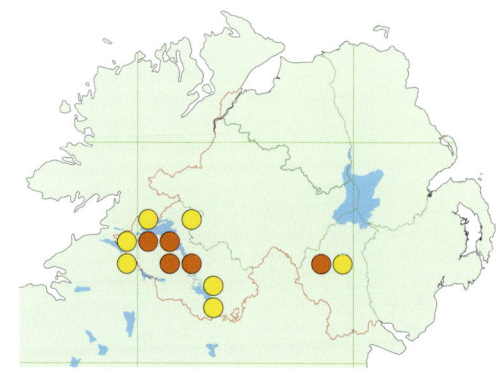

Herald

Scoliopteryx libatrix (Linnaeus) Category two.

Widespread. Generally distributed across the southern part of Northern Ireland, but much more local north of Lough Neagh. The only site on the north coast is Limavady (Londonderry). Regularly recorded sites include Brackagh Moss, Loughgall and Peatlands Park (Armagh); Lackan Bog (Down); Crom, Legatillida and Garvary Wood (Fermanagh) and Rehaghy Mountain (Tyrone). This is a distinctive, easily recognised species with deeply-scalloped brown and orange forewings, that resembles a withered leaf when at rest. The ground colour is light brown with pale orange median blotches which extend up to the basal area of the forewing and across the top of the thorax. There is also a tiny white spot in the central area of the forewing and a conspicuous, white postmedian cross-line. The Herald is found mainly in woodland, hedgerows, fens, marshy areas and, less frequently, in gardens. Adults have been seen in nearly every month of the year. The new generation of adults appears in August and September and enters hibernation, reappearing in spring with some surviving to early summer. Prior to hibernation, the adults feed on flowers, particularly Ivy and fermenting berries. Hibernation sites include old buildings. The peak period for sightings is May and June when adults are attracted to light. The larvae have been found on willow in Northern Ireland; other known foodplants include Aspen and other poplar species.

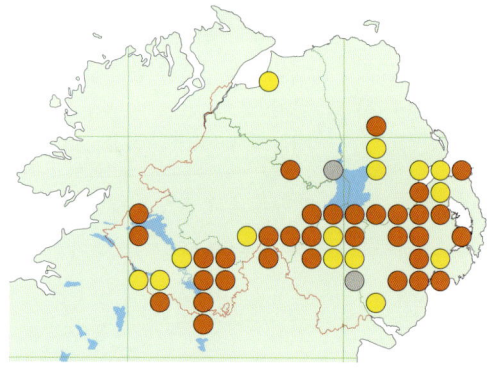

Small Purple-barred

Phytometra viridaria (Clerck) Category one.

Notable. Recorded from every county apart from Tyrone, but at only a few sites in each. There are two old records from Londonderry, the last in 1970. The only Antrim locality is Fair Head. The only known sites in Down are in the Mourne Mountains at the Annalong valley, Kinnahalla and Leitrim Lodge. In Armagh, it was recorded at Brackagh Moss in 1993 and at Peatlands Park in 1991 and 2002. There are four documented sites in Fermanagh at Cavans Bog, Dresternan Lough, Monawilkin and Lough Namanfin. The largest number of records is from Monawilkin, and it was last seen here in 2000. It is probably more widespread than the records would suggest as its resemblance to a micromoth makes it easy to overlook by all but experienced recorders. This is a small, inconspicuous day-flying moth that is most active in sunshine. The forewings are a mixture of olive and purple-brown.

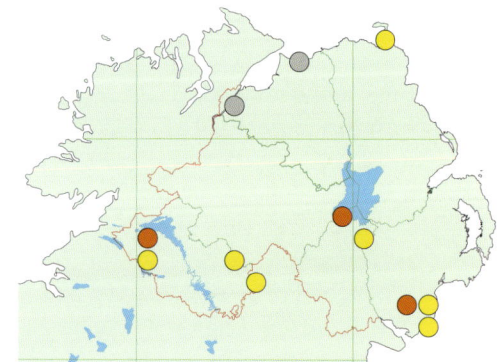

Herald.

Lackan Bog, County Down.

There are two distinct purple cross-bands near the outer edge of the forewing. The hindwings are similar in colour to the forewings. It can be found in heathland, both upland and lowland, cutover bogs and limestone grasslands. The adults normally fly rapidly in sunshine, but will settle usually on bare ground or short vegetation. They are generally wary and difficult to approach. During dull weather they are more difficult to find as they tend to retire deep into the vegetation. The recorded flight period is 6 May to 3 July. The larvae feed on Common Milkwort. It overwinters as a pupa.

Straw Dot

Rivula sericealis (Scopoli) Category one.

Widespread. Recorded in all counties and well distributed in all. There are some gaps apparent in the north and west where it is still under-recorded. Recent sites include Divis Mountain, Glenariff Forest, Montiaghs Moss, Portglenone Wood and Rathlin Island (Antrim), Magilligan Dunes and the Umbra (Londonderry). The forewings are pale straw or yellow-brown, becoming darker brown towards the outer edge. There is a small dark brown dot in the central area near the costa. The hindwings are light brown with darker veining. Straw Dot inhabits fens, heaths and the damper parts of woodland clearings and grasslands. It has also been trapped in several suburban gardens. Adults are active from dusk onwards and are attracted to light in small

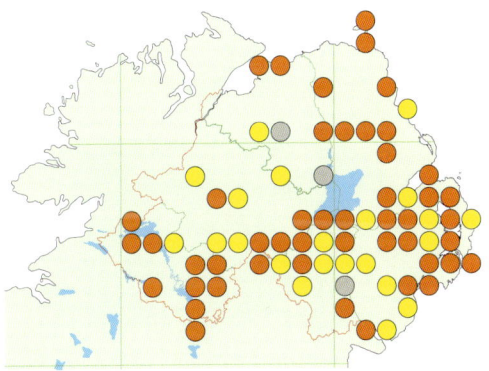

The Butterflies and Moths of Northern Ireland

Beautiful Snout.

Rehaghy Mountain, County Tyrone.

numbers. The recorded flight period is 18 June to 24 August. The larvae feed on Purple Moor-grass and other grasses. It overwinters as a larva.

Beautiful Snout

Hypena crassalis (Fabricius) Category three.
Notable. Unknown in Northern Ireland before 1988 when it was found at Crom in south Fermanagh. Since then it has been discovered at three other Fermanagh sites (Aghalane, Correl Glen and Monmurry Bog) and several sites in south Tyrone (Altadavan, Crockacleaven Lough and Rehaghy Mountain) and north Armagh (the Argory, Aughinlig and Gosford Forest) and from a suburban garden in Belfast. It is also known from Bohill Forest in Down. This is a distinctive species. Males and females differ in appearance. The forewings in the male are rich purplish-brown to the postmedian area with a fine white line and a dark apical streak. The female is more of a chocolate brown, becoming white towards the termen. There is a short white streak running from the basal area along the dorsal aspect of the wing and a short dark brown apical streak. The hindwings are dark brown in the male but a little paler in the female; both have darker veining and a small discal spot. Beautiful Snout is found in woodland, heathland and moorland. Adults come to light in very small numbers and reportedly rest by day among ground vegetation or on tree trunks. When disturbed, they fly erratically for some distance before settling again. The recorded flight period is 3 June to 26 August. The larvae feed on Bilberry and possibly Heather. It overwinters as a pupa.

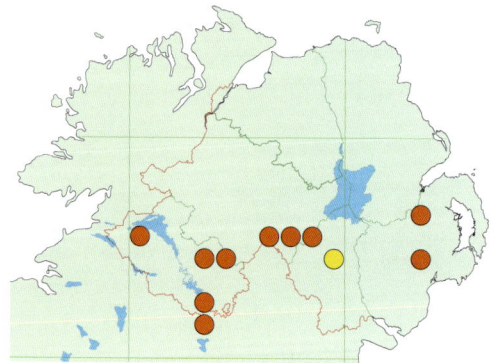

Snout
Hypena proboscidalis (Linnaeus) Category one.

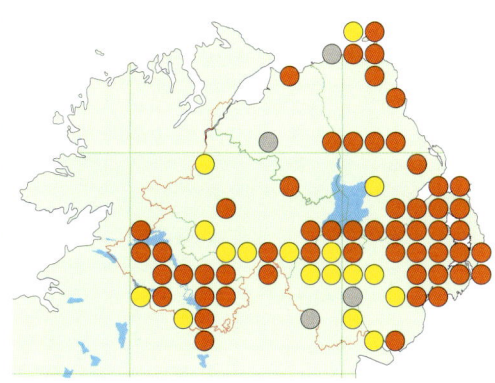

Widespread. Recorded from all counties and on Rathlin Island (Antrim). Adults are distinctive because of the long, projecting, inverted palps from which it gets its name. The ground colour ranges from light greyish-brown to darker purplish-brown. The forewings are distinctly hooked with darker brown cross-lines. The antemedian line is strongly angled and the postmedian line is straight. It can be found in many habitats including woodland, damp pasture, suburban gardens and waste ground. Adults are attracted to light and flowers. It can occasionally be disturbed from low-growing vegetation during the day. The recorded flight period is prolonged, from 26 May to 6 November. The data indicates that there is just a single, annual generation. The larvae feed on Common Nettle. It overwinters as a larva.

White-line Snout
Schrankia taenialis (Hübner) SOCC (P) Category four.

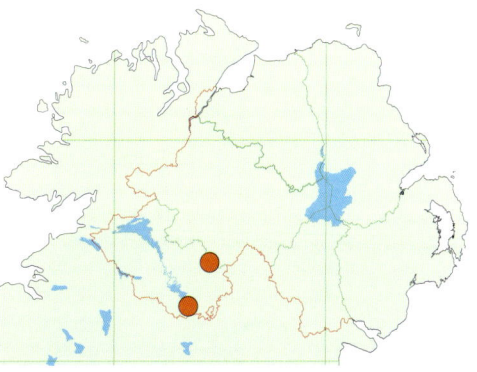

Rare. This species was discovered for the first time in Northern Ireland at Crom (Fermanagh) on 2 July 2000. Despite much trapping at Crom over many years, this remains the only record from this site. A single adult was trapped at Crockacleavan Lough (Tyrone) on 19 July 2000. It seems quite likely that it is under-recorded and may exist in other suitable localities, as has been the case in the past for other new species. This is a small micro-like species with pale brown forewings. The antemedian line is black. The postmedian line is white, finely edged on its inner side in black. There is also a small dark spot near the postmedian line which is visible in fresh specimens. The hindwings are brownish-white. It occurs in damp, broadleaved woodland and possibly heathland and moorland sites. Adults are not overly attracted to light which may account for its scarceness at seemingly suitable sites. In Britain, it flies from early July to mid-August. Little is known about its life cycle in the wild. The foodplants are unknown and it is thought that it overwinters in the larval stage.

Pinion-streaked Snout
Schrankia costaestrigalis (Stephens) Category two.

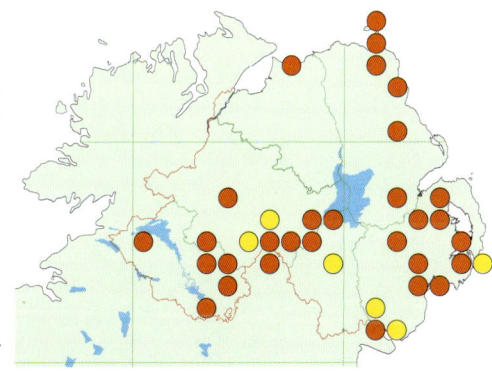

Widespread. This small, inconspicuous moth was discovered in Northern Ireland in 1997 at Peatlands Park (Armagh), but was undoubtedly overlooked in the past. Since then it has been found widely across southern counties. It is also found in north Antrim at Glenariff Forest and at a number of sites on Rathlin Island. In Londonderry, it is known from Duncrun and the Umbra. Adults resemble a micromoth. They have long narrow forewings which are pale straw in colour. There is a pale streak running obliquely from the apex to the dorsum of the wing. It is joined to a darker horizontal line in the median area of the forewing. The hindwings are pale white with a small distinctive discal spot. It has been recorded from a wide range of habitats, but it is particularly associated with damp woodland, bogs, fens and marshy places. Adults come to light and flowers, especially sedges and thistles. The recorded flight period is 22 June to 11 October. The larval foodplant is unknown in the wild. It overwinters as a larva.

Marsh Oblique-barred
Hypenodes humidialis (Doubleday) Category three.
Scarce. This is the smallest noctuid species in the region. It is known from Fermanagh, sites along the border of Tyrone and Armagh and Rathlin Island (Antrim). It was first discovered in Northern Ireland on 21 July 1998 when it was taken at light at Peatlands Park (Armagh). Subsequently it has been found in Fermanagh at Altaclanabryan, Correl Glen, Crom, Legatillida and Monmurry Bog. Other recorded sites are Crockacleaven Lough, Fivemiletown and Rehaghy Mountain in Tyrone. The small size of this moth makes it easy to be misidentified and overlooked as a micromoth, so it is probably more widespread than the few records show. The forewings are long and narrow and pale yellow-brown in colour. There are two well-marked lines running obliquely across the forewing, giving this moth its vernacular name. Its preferred habitats are bogs, fens, heaths and wet meadows. Adults are said to fly in the late afternoon and again at dusk. The recorded flight period is 22 June to 25 August. The larval foodplant is unknown in the wild.

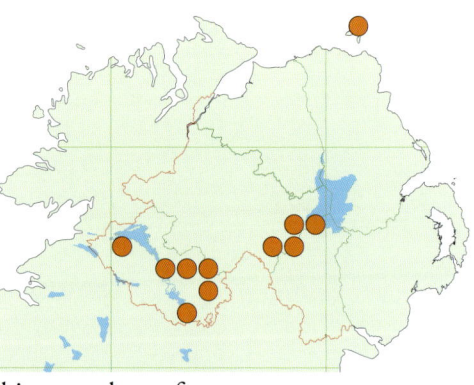

Fan-foot
Zanclognatha tarsipennalis (Treitschke) Category two.
Widespread. A widely distributed species in the southern counties especially Down but more local elsewhere especially in the north. In north Antrim it is recorded from Ballymena and Glenariff Forest. The only Londonderry localities are Duncrun and the Umbra. This species resembles Small Fan-foot. The forewings are yellowish-brown with three conspicuous cross-lines. The antemedian and postmedian lines are darker and the latter curves inwards towards the costa. The subterminal line is straight and clearly defined and there is a small crescent-like mark in the central area of the forewing. Fan-foot occurs in a wide range of habitats including woodland, bogs and gardens. Adults come to light in small numbers and are occasionally disturbed by day from vegetation. The recorded flight period is 26 May to 15 August. The larvae apparently feed on dead and withered leaves of Beech, oak and Bramble. It overwinters as a larva.

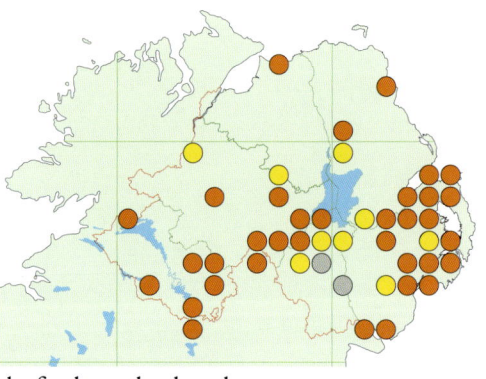

Small Fan-foot
Herminia grisealis (Denis & Schiffermüller) Category one.
Widespread. This species has a similar distribution to the closely-related Fan-foot. It is generally distributed across southern counties, but scarce north of Lough Neagh. Recently recorded sites in the north include Glenariff Forest, the oak woods of Breen Wood and Portglenone and also several suburban gardens in Antrim. Adults are normally slightly darker brown than Fan-foot. The antemedian and postmedian lines are darker than the ground colour and strongly marked. The subterminal line is distinctly curved towards the apex, unlike the Fan-foot which is straight. There is also a small crescent-like mark in the central area of the forewing above the postmedian cross-line. Small Fan-foot occurs mainly in woodland margins, clearings, bogs and some suburban gardens. Adults become active from dusk onwards and come to light in small numbers. They are occasionally disturbed from ground vegetation during the day. The recorded flight period is 15 May to 19 August. The larvae feed on both fresh and withered leaves of a variety of trees including oak, Hazel, birch, Hawthorn, and Alder. It overwinters as a pupa.

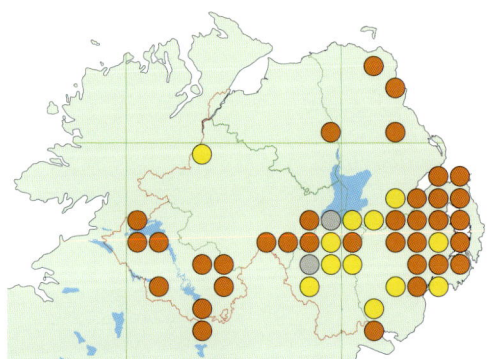

CHAPTER FOUR

Habitat gallery

List of sites illustrated

Derryleckagh Wood and Fen, County Down	357
Lackan Bog, County Down	358
Mill Bay, County Down	359
Brackagh Moss National Nature Reserve, County Armagh	360
Eshywulligan, County Fermanagh	361
Rehaghy Mountain, County Tyrone	362
Murlough National Nature Reserve, County Down	363
Rostrevor Wood National Nature Reserve, County Down	364
Murlough Bay, County Antrim	365
Peatlands Park, County Armagh	366
Breen Wood National Nature Reserve, County Antrim	367
The Argory, County Armagh	368
Rathlin Island, County Antrim	369
Glennasheevar, County Fermanagh	370
Killard Point National Nature Reserve, County Down	371
Correl Glen National Nature Reserve, County Fermanagh	372
Monawilkin, County Fermanagh	373
Crom, County Fermanagh	374
Loughgall, County Armagh	375

Habitat gallery

Derryleckagh Wood and Fen, County Down, grid ref. J1225.

BROADLEAVED WOODLAND and fen. Derryleckagh Wood, close to Newry in south Down, is an excellent example of native broadleaved woodland. The nearby fen is one of the largest and least damaged examples still in existence in Northern Ireland and notable for its plant communities and rare species of wetland mosses. Together the two habitats constitute the most interesting area of lowland habitat in this part of Northern Ireland. The fen occupies the flat floor of a small river valley and the wood, the steep, west-facing slope above it. The woodland is mainly Hazel with oak standards and it is considered significant for the lack of alien species. The floor of the woodland is strewn with boulders, typically covered in luxuriant growth of mosses. In spring there is an impressive display of woodland flowers including Wood Anemone, Bluebell and Lesser Celandine. Devil's-bit Scabious is locally abundant in the fen and the site supports a colony of Marsh Fritillary, the only extant population in the area. Other butterflies present include Small Copper, Common Blue and Silver-washed Fritillary. Noteworthy moth species include Lunar Hornet Moth, Little Emerald, Small White Wave, Chevron, Streak, Puss Moth, Lunar Marbled Brown, Vapourer, Ruby Tiger, Feathered Gothic, Large Wainscot, Anomalous, Haworth's Minor, Copper Underwing and Double Lobed.

357

Lackan Bog, County Down, grid ref. J2337.

CUTOVER BOG, lowland raised bog, birch woodland. County Down is the driest and most easterly county in Northern Ireland and has the least extent of lowland peat. This site, Lackan Bog, is the largest surviving block in the entire county. The bog is entirely cutover and comprises numerous flooded peat cuttings separated by drier ramparts. Bottle Sedge, Bogbean and Common Cottongrass are dominant in the wet areas, whilst the drier banks are covered in Heather and Blue Moor-grass. Scrubby woodland of birch with some willow exists around the margins of the site. In common with most relict cutover bogs, there has been no peat-cutting or other management for many years and consequently, scrub is encroaching on the open areas. Lackan Bog was trapped regularly in the early 1990s, but little recently and the current status of the more interesting species is uncertain. The fauna is dominated by woodland and particularly birch-feeding Lepidoptera; wetland and grassland species are also present. At least one significant species, Marsh Fritillary, is certainly lost from the site as the small area of habitat for it has become unsuitable. Noteworthy moths recorded are Gold Swift, Birch Mocha, Yellow Horned, Early Tooth-striped, Engrailed, Scallop Shell, Lesser Cream Wave, Autumn Green Carpet, Heath Rustic, Double Lobed and Silver Hook.

Habitat gallery

Mill Bay, County Down, grid ref. J2513.

SALTMARSH. Saltmarsh vegetation represents the transition between tidal mudflats and dry land. These marshes develop on the most inland parts of estuaries and at the heads of bays that are most sheltered from wave action and where fine sediments are deposited. The sediments are colonized by successive bands of specialised plants according to their tolerance of saltwater inundation. Saltmarsh is a naturally rare habitat in Northern Ireland as the conditions that encourage its development exist in only a few places around the coast. Many of the areas that would have supported the habitat have been reclaimed (for example, in Belfast Lough and Lough Foyle) and the marshes no longer exist. The photograph shows the saltmarsh at Mill Bay in Carlingford Lough which is the largest remaining intact example. The species living in this habitat are specialised as they have to cope with the tides and the salt-laden vegetation. Unfortunately, like the habitat, some species appear to have declined and several have not been recorded for many years, such as White Colon and Crescent Striped. Noteworthy moth species include Common Emerald, Lesser Cream Wave, Small Rivulet, Yellow-tail, Common Footman, Nutmeg, Dog's Tooth, Clay, Light Arches and Cloaked Minor.

Brackagh Moss National Nature Reserve, County Armagh, grid ref. J0251.

FEN AND RELICT BOG, willow and alder carr, birch and gorse scrub. Originally a raised bog beside the River Bann, Brackagh Moss is now an area of fen with associated wetland habitats. The presence of peat ramparts, standing one metre or so above the general level of the site, reveals the amount of peat that has been removed. The removal of so much peat means the wetland is now fed principally by enriched groundwater rather than acidic, nutrient-poor rainwater. This has changed the nature of the vegetation, allowing fen species to increase at the expense of the original raised bog species; only a small area of the original vegetation remains. This change has happened relatively slowly so the fauna to some extent has been able to adapt to the changing conditions; nevertheless, species requiring acid conditions have declined, many to local extinction. The moth fauna of Brackagh has been well studied and is outstanding and includes several species which are unknown elsewhere in Northern Ireland. The butterfly fauna is less interesting, although both Réal's Wood White and Green Hairstreak are present. Marsh Fritillary formerly occurred but habitat conditions no longer suit it. The moth list for the site includes Red-tipped Clearwing, Lunar Hornet Moth, Birch Mocha, Blue-bordered Carpet, Scallop Shell, Valerian Pug, Figure of Eight, Alder Moth and Silver Hook.

Eshywulligan, County Fermanagh, grid ref. H4838.

WET HEATH, blanket bog, conifer plantation, scrub and wet acid grassland. Marginal land in the hilly areas of Northern Ireland can be hard to categorise into a distinct habitat type. Variation in the vegetation is often subtle, reflecting slight changes in slope, aspect, drainage, soil type and management. This area of east Fermanagh is representative of this type of landscape that can be seen throughout the hilly areas of Northern Ireland. The predominant vegetation in this view is the Heather-dominated wet heath and blanket bog. It is interspersed with areas of rush-infested pasture which is promoted by the high rainfall, poor drainage and grazing. As agriculture is marginally productive in these upland areas, large parts have been converted to plantations of conifers. These woods provide an additional habitat and, with the removal of grazing, also allow development of birch and willow scrub. Further variety is often provided by stands of Gorse and Bog Myrtle. Little work has been done to record systematically the moths of these habitats, but this region of Fermanagh is proving to be a very interesting area, supporting a varied and rich assemblage of moths. Fox Moth, Narrow-bordered Bee Hawkmoth, Small Chocolate-tip, Red-necked Footman, Marbled Coronet, Glaucous Shears and Dark Brocade are noteworthy moths listed from this site.

Rehaghy Mountain, County Tyrone, grid ref. H7053.

MIXED WOODLAND, dry and wet heath. This site, like Eshywulligan, contains a mosaic of different habitats that is difficult to classify. These diverse sites seem to be particularly valuable habitat for moths throughout Northern Ireland and for its size, Rehaghy Mountain has proved to be remarkably rich in species. It is a low hill, located in the rolling countryside of south Tyrone, with open broadleaved woodland on the lower, rocky slopes. Birch is the dominant species, but small quantities of oak, Alder and Ash are present. Hazel and Holly are also common, while willows and stands of mature Blackthorn occur on the periphery of the woodland. The summit of the hill is open with ungrazed heath dominated in places by Bog Myrtle and Gorse. There is also a small lake, Black Lough, surrounded by wet heath, poor fen and acid grassland. Regular moth trapping started here in 1996 and the species list currently amounts to 240 species. This includes several rare birch-feeding species such as Scarce Prominent and Flounced Chestnut and the Blackthorn-feeding Figure of Eight, for which this is the only known site in Tyrone. Other noteworthy species include Birch Mocha, Small Seraphim, Northern Winter Moth, Satin Beauty, Scarce Umber, Pale Tussock, Miller, Sprawler, Dark Chestnut, Brick, Scarce Silver Y, and Beautiful Snout.

Habitat gallery

Murlough National Nature Reserve, County Down, grid ref. J3933.

SAND DUNE, coastal heath, acid grassland, scrub and woodland. Murlough National Nature Reserve is a sand dune system managed by the National Trust on the east coast of Down. It is a diverse site containing the largest extent of natural dune and heath habitat in Northern Ireland. The dunes are old and rather acid in nature, but still floristically rich. Unusual features of the site are the shingle flats covered in sparse low vegetation and carpets of lichens as illustrated in the photograph. There is some woodland on the reserve, principally of Scots Pine and Sycamore. Large areas of the dunes are covered in dense stands of introduced Sea Buckthorn. Murlough is the best documented coastal site in Northern Ireland and its insect fauna is very species-rich, including 20 resident butterflies and 263 moths. The butterfly population has been monitored by a transect established in 1979. In addition to many uncommon resident species, its east coast location means it receives many migrants. It is currently the richest butterfly site in Northern Ireland and includes significant populations of Réal's Wood White, Dark Green Fritillary, Marsh Fritillary, Grayling and Small Heath. Noteworthy moths include Dark Spinach, Clouded Magpie, Haworth's Pug, Plain Pug, Annulet, Archer's Dart, Black Rustic, Feathered Ranunculus and Marbled White Spot.

The Butterflies and Moths of Northern Ireland

Rostrevor Wood National Nature Reserve, County Down, grid ref. J1817.

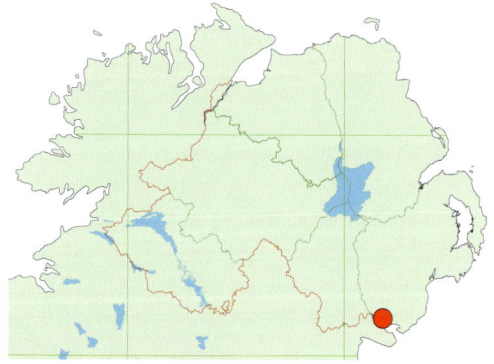

OAK WOODLAND. The region around Rostrevor in south Down was a favoured haunt of some of the early twentieth-century entomologists, especially Foster and Johnson. The area is one of the most wooded parts of Down, although the dominant woodland present today is coniferous plantation. This reserve protects a remnant area of broadleaved woodland on a steep west-facing slope above Rostrevor village and Carlingford Lough. The canopy in the woodland is closed and consists almost entirely of oak. Hazel is the dominant species in the understorey. The ground flora is dry and acidic in character with extensive carpets of Great Wood-rush. Little variation is apparent within the woodland as there are few glades and rides; consequently, many of the birch- and willow-feeding species found in more diverse and wetter woodland sites, such as Crom, are scarce or absent from the reserve. Nevertheless, the species list for the reserve is long and most common woodland species are present on it. The list includes the following noteworthy moths: Barred Hook-tip, Clay Triple-lines, Clouded Magpie, Scorched Carpet, Satin Beauty, Buff Footman, Brindled Green, Northern Deep-brown Dart, Slender Brindle and Oak Nycteoline. The butterfly fauna is not of special significance, although both Holly Blue and Silver-washed Fritillary occur in the area.

Habitat gallery

Murlough Bay, County Antrim, grid ref. D1842.

BROADLEAVED WOODLAND, heathland, unimproved calcareous grassland. Murlough Bay, lies on the eastern side of Fair Head, the prominent headland that dominates this part of the Northern Ireland coastline. The plateau behind Fair Head is covered mainly in wet heath and acid grassland. An interesting maritime heathland with Juniper is found in exposed areas along the top of the line of cliffs that extends south from Fair Head. These are initially composed of hard dolerite, but are replaced by softer chalk in Murlough Bay. Beneath the chalk cliffs, broad grassy slopes with planted shelter belts of trees stretch down to the sea in Murlough Bay. Native broadleaved woodland dominated by birch, Hazel and Rowan grows extensively on the steepest areas of undercliff at the northern end of Murlough Bay. Some of the most interesting vegetation at the site is found where there are flushes and seepages beneath the chalk cliffs; Mossy Saxifrage, one of the foodplants of the Yellow-ringed Carpet (a specialty of Murlough Bay), grows in these situations. The Lepidoptera list for this area includes several northern species of moth, some of which are unknown elsewhere in Northern Ireland. Some species may also be recent colonists from Scotland. Noteworthy species are Grayling, Yellow-ringed Carpet, Smoky Wave, Chimney Sweeper, Treble Bar, Anomalous and Centre-barred Sallow.

365

Peatlands Park, County Armagh, grid ref. H9061.

LOWLAND RAISED BOG, fen, native oak and birch woodland. Peatlands Park was set up to protect a small part of what was once an extensive area of lowland raised bog in north Armagh. Most of the bog has been cut over and the surface level lowered, but one small area of intact raised bog surface remains, protected within Mullenakill National Nature Reserve. A second national nature reserve at Annagarriff contains native oak woodland. Also present within the park boundaries are areas of fen and wet birch and willow woodland. Scots Pine is common and well established on the open bog. Invasive Rhododendron is a management problem, and much effort has been expended to remove it from the site. Peatlands Park is known to support one of the richest assemblages of bogland Lepidoptera in Northern Ireland. The woodland fauna is also diverse and birch-, willow-, oak- and pine-feeding moths are all well represented. The moth list for the site contains 274 species. The area was visited by several of the early entomologists, in particular Johnson, who called the area Churchill. Butterflies present include Green Hairstreak and Large Heath, the latter at its only Armagh site. Noteworthy moths include Fox Moth, Bordered Grey, Bordered White, Eyed Hawkmoth, Vapourer, Dark Tussock, Beautiful Yellow Underwing, Pine Beauty, Red Chestnut, Mother Shipton and Small Purple-barred.

Breen Wood National Nature Reserve, County Antrim, grid ref. D1233.

OAK WOODLAND and heath. Breen Wood is a semi-natural woodland nature reserve in north Antrim a few miles inland from Ballycastle. It is an example of the type of native oak woodland that is characteristic of impoverished, acid soils in western Britain and Ireland. The woodland grows on a series of small valleys and dry ridges of glacial deposits. The canopy is dominated by oak with significant amounts of birch. Holly and Rowan are the commonest components of the understorey and the ground flora is dominated by Bilberry and Great Wood-rush with a significant proportion of mosses. The woodland abuts conifer plantation on the east, whilst on the western side (illustrated in this photograph) it merges into heathland. There is a natural transition between the two habitats marked by the zone of birch saplings, shown in the photograph as they come into leaf. This sunny, sheltered habitat on the edge of the woodland is more attractive to many of the woodland Lepidoptera than the cooler conditions beneath the canopy. There have been records of just four common butterflies. One surprising omission is the Green Hairstreak as the habitat illustrated appears eminently suitable for it. Noteworthy moths listed from the site include Dotted Carpet, Scallop Shell, Northern Spinach, Buff Footman, Satin Beauty and Old Lady.

The Butterflies and Moths of Northern Ireland

The Argory, County Armagh, grid ref. H8757.

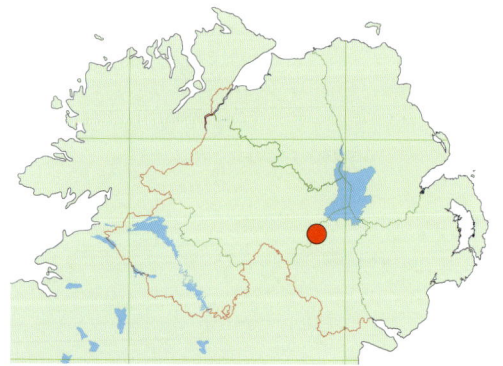

CUTOVER BOG, broadleaved woodland. The Argory is a National Trust property on the east bank of the River Blackwater. The estate has extensive areas of broadleaved estate woodland with much oak, small areas of wetland habitat beside the river and a lowland raised bog, the Argory Moss, which is shown in the photograph. The Moss has a total area of seventeen hectares, split into two sections — the high moss of some ten hectares with an uncut surface and the seven-hectare low moss. The low moss is cutover and suffers from encroachment by birch and pine trees. The insects of the Argory, particularly the moths, have been studied over the past 30 years at varying levels of intensity. There are 268 moth species on the site list, including most common woodland moths plus some bogland specialists. Changes in both the woodland and wetland habitats are apparent as some scarcer species appear to have been lost including Forester and Narrow-bordered Five-spot Burnet. Noteworthy moths recorded on the estate include Triple-spotted Pug, Early Tooth-striped, Oak Beauty, Grey Scalloped Bar, Lunar Marbled Brown, Dark Tussock, Figure of Eight, Barred Chestnut, Brindled Green, Poplar Grey, Old Lady, Green Silver-lines, Alder Moth, Oak Nycteoline, and Beautiful Snout. Green Hairstreak has been seen once on the Moss, but its current status is uncertain.

Habitat gallery

Rathlin Island, County Antrim, grid ref. D1547.

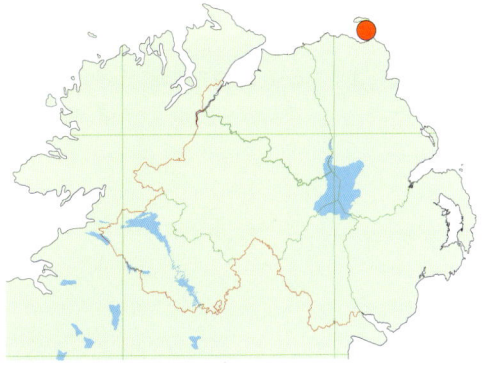

SEA CLIFFS, coastal grassland, heathland, acid grassland, poor fen. Situated ten kilometres off the north coast of Antrim, Rathlin is the largest and only permanently inhabited marine island in Northern Ireland. The north coast is almost completely cliff-bound with inaccessible undercliffs and hanging slopes. The coast of the southern arm of Rathlin is gentler and more low-lying. The photograph shows the low, rocky coast near Rue Point, the southernmost point of Rathlin. The productive land in the interior of the island has been improved for grazing, but large areas of heathland survive and there are also numerous small lakes and ponds. All the woodland on the island has been planted and it consists chiefly of conifers and Sycamore. The moths of Rathlin were first investigated by Dennis and Neal Rankin just before the Second World War. Recording then ceased until 1988 when it was resumed through the initiative of the RSPB wardens and has continued intermittently to the present day. Rathlin hosts a number of important species and has, in some cases, the only known Northern Ireland populations. Amongst these are Red Carpet, Chestnut-coloured Carpet, Annulet, Bordered Grey, Scarce Footman, Square-spot Dart, Northern Rustic, Vine's Rustic, Confused, Brindled Ochre, Dot Moth, Southern Wainscot, Tawny Shears and Pod Lover, Double Lobed, Marbled Coronet, Anomalous, and Feathered Ranunculus.

Glennasheevar, County Fermanagh, grid ref. H0554.

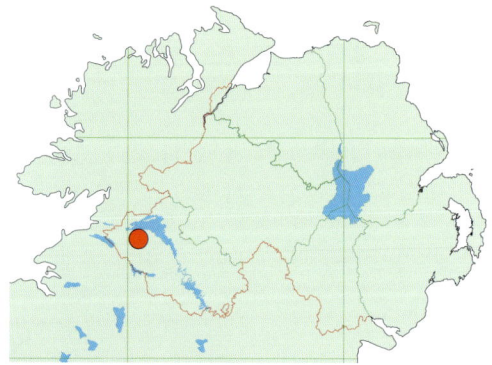

BLANKET BOG, wet and dry heath, flushed grassland. Glennasheevar is one of the most accessible and best examples of intact blanket bog in Northern Ireland. The site is located in the geologically varied district of west Fermanagh encompassed by Lough Navar and Conagher Forests. Conifer plantations cover a large proportion of this region, but many areas remain in natural condition. The unplanted land shown in the photograph is one of the most interesting of these open areas; it lies west of the exit of the Lough Navar Forest Drive. The site consists of acid grassland and wet heath that grades imperceptibly into blanket bog. Additional variety exists along the roadside which, like many forestry tracks, has been constructed with limestone and is covered in short herb-rich grassland. Glennasheevar came to the notice of lepidopterists with the discovery of a colony of Marsh Fritillary – currently the only extant population in Fermanagh. The site is also notable for day-flying moths including Emperor Moth, Argent and Sable, Grass Wave, Clouded Buff, (a blanket bog specialist) and Narrow-bordered Bee Hawkmoth. Other recorded species include Red-necked Footman and Small Chocolate-tip. Large Heath is also present, flying here with Small Heath, so providing a rare opportunity to see these two similar butterflies flying together.

Killard Point National Nature Reserve, County Down, grid ref. J6043.

UNIMPROVED GRASSLAND, dune. Killard Point is a prominent headland, jutting into the Irish Sea at the entrance to Strangford Lough in east Down. It is considered the most natural and intact coastal site in Northern Ireland. The point is covered in several types of grassland, including maritime grassland closest to the sea, dune grassland on areas of wind-blown sand, acid grassland (often with a significant heath element) on rocky outcrops and a neutral community on the higher and most inland parts. All the grassland is unimproved and of high floristic diversity. The shore is low and rocky with maritime lichen communities on the bare rock. There is very little scrub on the site. The community of ruderal species on the upper shore is well developed and undisturbed. The butterfly and moth assemblages of the grassland habitats are considered of high diversity, rivalling those found on Murlough NNR to the south; however, Killard is relatively inaccessible and, consequently, trapping has been less intensive than at Murlough NNR. Butterfly species include Dark Green Fritillary and huge populations of both Common Blue and Meadow Brown. Noteworthy moths include Blood-vein, Pimpinel Pug, Small Elephant Hawkmoth, Yellow-tail, Archer's Dart, Least Yellow Underwing, Broad-barred White, Pod Lover, Feathered Gothic, Straw Underwing and Bordered Sallow.

Correl Glen National Nature Reserve, County Fermanagh, grid ref. H0754.

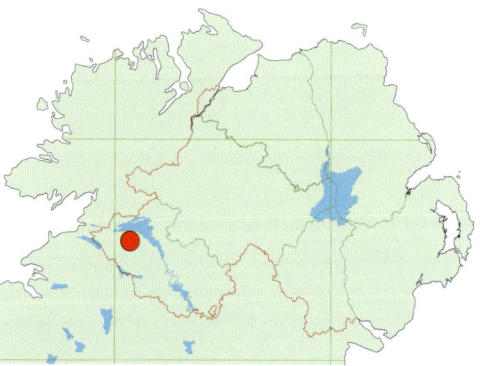

NATIVE OAK WOODLAND, heath. This nature reserve comprises narrow rocky valleys containing native broadleaved woodland separated by bands of wet heath. Correl Glen in west Fermanagh lies close to Monawilkin, but the vegetation of the two sites is very different, a reflection of their underlying geology. Monawilkin is on calcareous limestone whilst Correl Glen is underlain by acid sandstone. The woodland is oak-dominated with abundant Holly and birch and notable for its moss and lichen flora. The herb layer is dominated by Great Wood-rush, Bilberry and Wood Anemone, bryophytes and ferns. Common plants in the heathland are Bog Myrtle and Heather. The butterfly list for the reserve includes both Holly Blue and Purple Hairstreak. It is one of only two sites in Fermanagh for the Holly Blue. The status of the oak-feeding Purple Hairstreak is uncertain as it has rarely been seen, but the habitat appears eminently suitable. Green Hairstreak is found on the heathland, but is elusive. The woodland moth fauna is also notable and there is an especially good representation of woodland, birch and heathland species. Examples include Pale Eggar, Scalloped Hook-tip, Little Emerald, Plain Pug, Tissue, Clouded Buff, Heath Rustic, Barred Chestnut, Beaded Chestnut, Neglected Rustic, Black Rustic, Alder Moth, Large Ear, Scarce Silver Y and Beautiful Snout.

Habitat gallery

Monawilkin, County Fermanagh, grid ref. H0852.

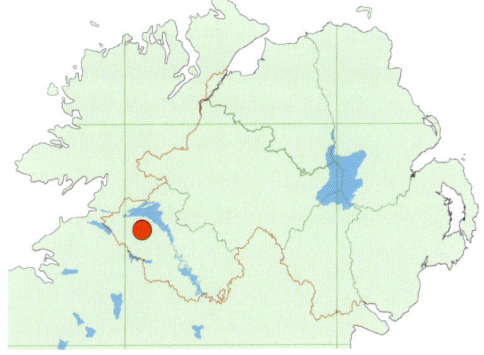

UNIMPROVED CALCAREOUS, neutral and acid grassland, broadleaved scrub. The calcareous grassland at Monawilkin is considered to be the best example of the habitat in Northern Ireland. The most valuable habitat is found on the warm, south-facing scarps, where the soils are thinnest, promoting a varied, short sward of fine grasses and herbs. The warm aspect of these scarps is of benefit to the butterfly and moth fauna. Other grassland types are also present, including acid grassland where sandstone replaces the limestone and neutral grassland in areas of deeper, poorly-drained soils. Additional interest is provided by patches of Hazel scrub and broadleaved woodland. The butterfly fauna of Monawilkin was formerly exceptional, with records of 20 resident species making it the richest inland site in Northern Ireland. However, several species may have become extinct, including Small Blue and Wall. The moth fauna is also notable, especially for species dependent on calcareous grassland. The site list so far includes over one hundred species of moth, some of which are restricted in Northern Ireland to this part of Fermanagh. Noteworthy species include Six-spot Burnet, Barred Carpet, Little Emerald, Small Engrailed, Narrow-bordered Bee Hawkmoth, Ruby Tiger, Wood Tiger, Reddish Light Arches, Anomalous, Mother Shipton, Burnet Companion and Small Purple-barred. Butterflies include Dingy Skipper, Réal's Wood White, Green Hairstreak, Common Blue and Grayling.

373

Crom, County Fermanagh, grid ref. H3524.

BROADLEAVED WOODLAND, fen, grassland. This large and very varied National Trust property is the most important inland locality for moths in Northern Ireland; there have been records of 289 species. The Crom estate is the most densely-wooded part of Northern Ireland. The woodland is structurally diverse including high oak forest, birch woodland, wet alder and willow carr and lake shore scrub with stands of Aspen. Buckthorn, Hazel and Spindle are all common components of the scrub and woodland understorey. Out-of-place plantations of conifers and poplars have been removed by the National Trust and the clearings created have been allowed to naturally revert to native woodland. The grassland on the estate has been improved for grazing except on the lake shores prone to winter flooding. Extensive fen and swamp communities grow intact on the shoreline of Upper Lough Erne and the smaller satellite lakes. Interesting moths are recorded from all the main habitats, several of which are unknown elsewhere in Northern Ireland; it is the woodland fauna that is most outstanding. The butterfly fauna is of lesser interest, although the Purple Hairstreak population is easily the largest in Northern Ireland. Other noteworthy moths are Lunar Hornet Moth, Poplar Lutestring, Clay Triple-lines, Brown Scallop, Scarce Prominent, Sprawler, Lead-coloured Drab, Dark Umber, Double Lobed, Rush Wainscot, Beautiful Snout and White-line Snout.

Loughgall, County Armagh, grid ref. H9152.

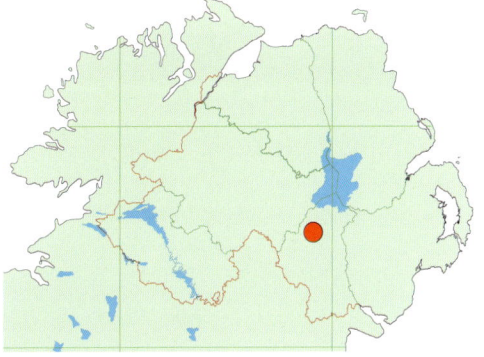

MATURE GARDEN. Gardens are the habitat that the majority will be most familiar with and, for the lepidopterist, this home patch will provide the most convenient trapping locality. Running a garden trap all year is straightforward, an especial advantage in the winter when the weather is unpredictable. The length and quality of a garden list will be dictated by geography, the garden attributes and the quality of the surrounding habitats. Suburban gardens on the edge of natural habitat are likely to be the most productive sites, but experience across Northern Ireland has shown that even tiny urban gardens can produce lists in excess of 200 species. Geographical location is also of significance. For instance, east coast gardens are more likely to be visited by migrant species; the experience of seeing a rare migrant moth is an incentive to many dedicated trappers. Most butterfly and moth species visiting a garden will be wanderers from an established population in a nearby habitat and only a proportion will be permanent residents. This proportion can be boosted by sympathetic management and appropriate planting to enhance its attractiveness to butterflies and moths as exemplified by this garden in Loughgall village. Over 240 species of moth have been recorded including Gold Swift, Scorched Carpet, Seraphim, Eyed Hawkmoth, Feathered Gothic, Brindled Green, Old Lady and Double Lobed.

The Butterflies and Moths of Northern Ireland

CHAPTER FIVE

Studying butterflies and moths in the field

MANY PEOPLE BECOME interested in butterflies and, to a lesser extent, moths through casual observation of common species in the field. Others develop an interest through studying other insect groups or indirectly through their professional capacity as land managers and conservationists. This chapter aims to provide the beginner with useful information on the common methods and items of equipment used to study Lepidoptera in the field. Publications which provide more detailed information on the techniques and the equipment are listed in the bibliography.

Butterflies
There are few naturalists who do not appreciate the beauty of butterflies. They are undoubtedly the most popular group of insects with the general public and probably the easiest group of insects for a beginner to study. There is only a small number of species of butterfly in Northern Ireland and, whilst a few are elusive and challenging to find, most are obvious, of general distribution and can be identified without difficulty. The ecology and life history of most of our resident species have been researched elsewhere in their range, but there have been few detailed studies undertaken on the species in Northern Ireland. This is unfortunate as the butterflies of Northern Ireland are undoubtedly worthy of study as the unravelling of the story of Réal's Wood White clearly demonstrates. There is still potential for making novel observations of behaviour and ecology, or for engaging in long-term monitoring projects. Amateur naturalists may be best placed to undertake such research as they are not constrained to publish results within a few years, as is frequently the case in academic research.

In the space of a single year, with the investment of time and travel, it is possible to see all our resident and common migrant butterflies. No special equipment is needed beyond a hand net and a pair of binoculars. These suffice also for the conspicuous day-flying moths, especially the burnets and hawkmoths. Butterfly identification is covered by numerous field guides and the field characteristics of the species are well described. The most comprehensive guides provide information on biology, behaviour and ecology of each species and include illustrations of all the stages of the life cycle from the egg to the adult. Some also introduce the reader to issues such as parasitism, colony structure, dispersal and migration and conservation. Information on the jizz of each species (those difficult to define characteristics of living butterflies that allow experienced observers to identify an individual to species with just a brief glimpse), is available in many of the most recent publications. This level of expertise is only gained through patient observation of the species in the field under all conditions.

Finding adult butterflies rarely presents difficulty for the observant naturalist; however, discovery of the other stages (egg, larva and pupa) provides a greater challenge. It may surprise some to learn that, even today, virtually nothing is known about the early stages of some species in the wild; knowledge of these species has come from captive breeding. This information is of great significance, as the ecology of the eggs and especially the larvae is often the key to understanding why species exist only in particular situations.

Regular recording of butterflies provides data on population levels and can reveal large-scale fluctuations and changes in abundance. The most valuable information on trends is provided by butterfly transects. These have become the established method of monitoring species and sites in Britain and in many European countries. Transects are timed counts performed once a week using a standard methodology along a set route in suitable weather conditions. All the butterflies seen in an imaginary box extending 2.5m either side and ahead of the observer are counted. The route may be walked throughout the season from April to September or just through the flight period of a single key species or group of species. Pooling of the data from

Page 376
Painted Lady.

Banbridge, County Down.

Preceding page
Kenny Murphy and Jonathan Thompson examining a Heath trap.

Rostrevor Wood National Nature Reserve, County Down.

Opposite page
Speckled Wood.

Montiaghs Moss National Nature Reserve, County Antrim.

The Butterflies and Moths of Northern Ireland

many transects allows the calculation of regional and national trends of the species. Standardisation of the technique is vital to ensure year-to-year compatibility for this analysis. Unfortunately, only a few transects have been operated in Northern Ireland for more than a few years and more are needed. Anyone considering setting up a transect should consult Butterfly Conservation Northern Ireland who can provide full instructions and advice on procedure and data submission.

Moths

Moths present a very different challenge to the naturalist than do butterflies because most are nocturnal and there are many more species to identify. There are some day-flying species which can be studied in a similar manner to butterflies. Other day-flying moths need specific techniques to locate and study them – for example, members of the clearwing family (Sesiidae), a very unusual group of mimetic moths resembling species of wasp. It is now possible to purchase synthetic versions of the pheromone attractants used by the female clearwing to signal to males and to use these to lure the insects into a trap. The use of these chemical lures has begun to revolutionise our knowledge of these species and, in particular, has helped in discovering how widespread and common they actually are.

The literature for moth identification is limited compared to that for butterflies, but there are several excellent books available. These are listed in the bibliography. Books and publications which deal with moths generally have sparser text and fewer illustrations than the field guides for butterflies; this is due to the greater number of moth species. There is also extensive literature on the use of light traps and other techniques for detecting the adults. In the next section we summarise the most common techniques, but first the study of the egg and larval stages is briefly discussed.

Opposite
Swallow-tailed Moth.

Peatlands Park, County Armagh.

Above
Orange Tip egg.

Newry Canal at Scarva, County Down.

Eggs

Locating the eggs of butterflies and moths in the wild requires considerable investment of time and experience as well as an intimate knowledge of species behaviour and foodplants. Even for experts, looking for eggs can be a laborious affair often with little reward, and thus is not practised widely. For some species, the egg may be easily found and may provide a simple way of locating the species.

Egg-laying strategies vary widely in the Lepidoptera. Most females lay their eggs on the foodplant so that the newly-hatched larvae have immediate access to it, but some species simply broadcast the eggs over the vegetation, for example, the swift moths and the Meadow Brown butterfly. Eggs may be laid singly or in large batches, concealed or in open, conspicuous places. The eggs of some common species of butterfly, such as the Orange Tip, are relatively easy to find, despite their small size. Examining the flower heads in May and early June of the Cuckoo-flower (its main foodplant in Northern Ireland), often reveals an orange egg placed conspicuously just behind the flower on the underside of the stem. Another species, the Marsh

The Butterflies and Moths of Northern Ireland

Marsh Fritillary egg mass.

Bentley Wood, Wiltshire

Fritillary, lays its eggs in large batches on the underside of the leaves of Devil's-bit Scabious, its principal foodplant. Egg batches may be found by direct observation of the females or by systematic searching of patches of the foodplant in known colonies. Locating the eggs of moths is more difficult; because almost all species are nocturnal, direct observation of egg-laying behaviour is rarely possible in the field. Females captured in a moth trap can be stimulated to produce eggs if they are placed in a container with some of the appropriate foodplant.

Larvae

One of the most interesting components of lepidopteran study is the search for the larval stage of species. This can be carried out on overcast days unsuitable for adult observation. It affords an additional aspect to recording, establishing proof of breeding and provides valuable information about preferred foodplants and habitat preferences (an important aspect of recording which is poorly covered in Northern Ireland). This information cannot be reliably obtained from studies of adults as they are often found well away from what is deemed to be their normal habitat. The most productive period for larval recording is late spring to early autumn.

The larval period is when most species suffer their highest mortality. They have numerous enemies including parasites and predatory insects, birds and even fungi. Birds are perhaps the most significant of the predators. As the overwhelming majority of insectivorous birds are diurnal, visual hunters, many Lepidoptera larvae become active only at night. This strategy was widely known to early entomologists who hunted for them after dark by torchlight. One of the major differences between the study of moths now compared with that of the Victorian

Blackcap feeding young.

Crom, County Fermanagh.

The young of many species of small insectivorous birds are raised on a diet of caterpillars.

era is in the emphasis placed in collecting larvae. This activity is not as popular with modern lepidopterists who are more reliant on light traps for obtaining adults. These were unavailable to Victorian naturalists, so a common alternative was to collect larvae and rear them through to adults. Those who practised this gained an intimate knowledge of their biology, especially the early stages of many moths – knowledge that, in some instances, has not been improved upon. For some species, it remains the only way of obtaining adults, as not all moths are attracted to light.

Finding larvae in the wild requires expertise and ingenuity as most are camouflaged and secretive in behaviour. Their size and structure varies immensely between families and species. Many are highly colourful or patterned, while others are smooth-skinned and rather mundane in appearance. The largest, and often the most conspicuous, larvae are those of the hawkmoths which can attain lengths of five and six centimetres or more, whereas the larvae of many of the geometrids, particularly the pugs, are only several millimetres in length when fully grown. The vast majority of larvae feed on the living leaves of higher plants. Other foods used by fewer species are lichens, algae and ferns. Some larvae, for example members of the clearwing family, live in the stems and trunks of trees, eating the living wood.

Eyed Hawkmoth larva.

Annagariff Wood National Nature Reserve, County Armagh.

This fully-grown larva was discovered resting low down on the central stem of a willow bush.

Feathered Thorn larva.

Peatlands Park, County Armagh.

The larvae of many species of tree-feeding geometrid moths resemble small twigs. They rely on this camouflage to escape detection.

The Butterflies and Moths of Northern Ireland

Vapourer larva.

Montiaghs Moss National Nature Reserve, County Antrim.

The long, irritant hairs covering this caterpillar provide an effective deterrent to predation by most birds. A notable exception is the Common Cuckoo which feeds especially on these hairy larvae.

Having a basic knowledge of the requirements and foodplants of the species you are targeting is essential in order to refine your search. Some species of plant appear regularly in the list of foodplants used by our butterflies and moths, so it is sensible to learn to identify these. With experience it will soon be apparent that larvae are easier to find on plants growing in particular conditions. In this part of the world, the importance of sunny conditions creating a warm microclimate is perhaps the overriding one. Early literature often cited the periphery of woodland and open areas with isolated trees as especially productive for a wide variety of species, but other promising areas which have the appropriate foodplants should not be neglected. The foliage of trees should be examined for the tell-tale signs of larval activity such as freshly-eaten leaves, and frass (larval droppings).

Species which have been collected can be identified, with practice, to family by specific characteristics. Hawkmoths, for example, have a horn-like projection at the end of their abdomen; this trait is a useful diagnostic feature and easily distinguishes them at any stage in the development from other similar species. Larvae of the tussocks have long hairs which are particularly conspicuous in most species belonging to this family. Geometrid larvae, which form the second largest family of moths in Northern Ireland, are generally long and thin, often with external features which resemble buds and withered leaves.

Not all lepidoptera larvae lead solitary lives. Some are gregarious, living communally with other larvae, especially in the early stages of development. The larvae of the Marsh Fritillary and Small Eggar construct a larval web or tent. The web protects the young larvae from the elements and potential predators. Most gregarious species, however, become solitary in the later instars prior to pupation. Many other butterflies, including the Peacock and Small

Studying butterflies and moths in the field

Marsh Fritillary larvae.

Lackan Bog, County Down.

The communal larvae of this protected butterfly are most frequently encountered in August and September and after hibernation in early spring.

Small Eggar larva.

The Umbra, County Londonderry.

The larvae of this species are gregarious and live in a larval tent until the final instar, then disperse.

Geometrid pupa.

When the adult moth is close to emergence, the pupal case softens and in some species the colour and outline of the forewings can be seen.

Tortoiseshell, also feed communally throughout the early stages of development. Most gregarious larvae betray their presence quite easily by the large areas of defoliation and frass on their foodplants.

The most popular method for finding larvae is beating, a straightforward technique of beating or shaking branches of trees and shrubs and dislodging the larvae which are then caught on a beating tray (basically a shallow bowl of white cloth held taut by a frame). These can be bought from suppliers of entomological equipment or improvised by placing a white sheet on the ground beneath the target plant. Some entomologists use an inverted umbrella. It is important to beat just a single plant species at a time so that the larvae are associated with the appropriate plant. This is necessary if the larvae are going to be reared for identification.

The technique of beating is best used on the lower branches of trees and low shrubs. In grassland and other habitats with stands of herbaceous plants and other non-woody vegetation, sweeping is the recommended technique. A sweep net has a bag made of strong canvas attached to a robust frame. The net has to be a balance between robustness to withstand coarse vegetation and lightness, so that it does not become unwieldy. The net is moved from side to side through the vegetation, knocking insects into it. Larvae obtained by sweeping are more difficult to associate with specific plants unless the vegetation is uniform which in reality tends to be unusual. Sweeping must only be done in dry conditions; it is inappropriate after rain or early in the day when dew is still present.

Both methods have advantages and disadvantages and are probably most effectively done at night. Species which are protected by legislation should not be taken without the appropriate permit from the relevant authority. Any larvae, once identified, should be returned to where they were found. For more detailed methods for finding larvae see Allen (1980), Porter (1997) and Tutt (1994).

Searching for pupae

The pupal stage of butterflies and moths is perhaps the least known part of the life cycle. In the wild they are often very difficult to find, except by chance, as they are usually extremely well hidden. Pupa-digging (as it was commonly referred to by early entomologists) is an activity carried out during the late autumn and winter months. It was commonly practised by Victorian naturalists, but is not as popular today. It can be monotonous and time-consuming; the majority of those found will inevitably be common species. The essential tools are a garden trowel and a small box lined with moss in which to place the pupae. The best areas to search are under trees and the most productive species are the oaks, birches and willows. Trees that have wide, spreading roots with mossy areas at the base of the trunk merit examination. Great care should be exercised when digging around the roots and base of the trunk. Any pupae that are present will be lying just under the surface and so might easily be damaged by the trowel.

The size of Lepidoptera pupae varies enormously, reflecting the dimensions of the larvae. The smallest are just a few millimetres in length in the case of pugs and up to 50 or 60 millimetres in the larger hawkmoths. The outward appearance of the pupa differs considerably across the order. Butterflies produce the most colourful and bizarrely-shaped pupae, incorporating a wide array of cryptic patterns; the pupae of moths are rather prosaic by comparison. Most are light to dark brown in colour and have a polished appearance. Butterflies, unlike many moths, do not construct cocoons and pupate near their foodplants, usually well hidden deep in the vegetation. The pupal structure in most species is highly cryptic and blends naturally with the surrounding vegetation, making it extremely difficult to detect. The Large White is an exception, as the pupae can often be seen in full view on walls or under ledges and windowsills. Fissures, crevices in bark on the trunk, or small cavities in the branches of trees make good pupation sites for some moths, including the Puss Moth and Sallow Kitten. In contrast, the conspicuous papery cocoons of the burnets and a few other day-flying moths are easy to find among the marram grass in early summer.

Caring for pupae is not difficult as long as the temperature and humidity around them can be controlled. It is vital to avoid high temperatures and low humidity and so they should be stored outside in a shed laid on damp, peat-free compost with a light covering of moss. A cold frame out of direct sunlight is a suitable alternative. The aim is to mimic in the box the natural conditions the pupae would experience in the wild.

Large White pupa.

Aughinlig, County Armagh.

Pupae of this butterfly are commonly found on the walls and under the window sills of houses with vegetable gardens.

Light traps

It has long been known that moths are attracted to artificial light. This has been widely illustrated and documented throughout the centuries in art and literature. This phenomenon has been witnessed at some time in all our homes during humid summer evenings, when moths congregate around outside lights or beat themselves on the windows of a lighted room.

Sallow Kitten larva.

Crom, County Fermanagh.

A larva in the process of constructing its cocoon at the base of a tree.

The Butterflies and Moths of Northern Ireland

Robinson trap and a portable generator.

Using light to attract moths is the most effective way of recording species. The best light source is one that emits part of its output as ultra-violet light which moths, unlike people, can see. The most popular light traps among moth enthusiasts are mercury vapour units (which use mains power as their source) and the portable fluorescent actinic tube type, which are powered either by mains or battery. Virtually all of the modern moth traps are designed to accommodate one of these sources.

The design of most modern moth traps is based primarily on the lobster pot principle, so the trap is easy to enter but difficult to exit. Moths are drawn towards the trap by the light which is placed above the entrance. The moths are usually guided into the body of the trap by vanes or baffles. On approaching the trap, the moths fly in ever-decreasing circles around the light source and strike one of the vanes and drop into the trap through the narrow entrance. Modern light traps have revolutionised the study of moths. The following is a brief description of the main types available, but Fry and Waring (2001) should be consulted for a more detailed account of the relative merits, effectiveness and construction of the different traps.

Robinson light trap

The design and name of this trap comes from the Robinson Brothers who developed the original concept in the late 1940s. Its commercial introduction in the 1950s revolutionised the study of moths and it remains the most effective moth trap currently available. The design used today has changed little since the original, the main difference being the use of plastic rather than metal for most of the components. The light source is a mercury vapour light bulb which is powered from the mains via a choke. The most commonly used bulbs are 125 and 80 watts. The base is lined inside with cardboard egg cartons which are normally placed vertically and rest against the walls of the trap. The egg cartons provide the moths with small individual hollows in which to settle without disturbing the others in the trap. Each carton can be removed individually for inspection the following morning.

A Robinson trap requires mains electricity, which restricts its use to areas with a supply. Many moth enthusiasts overcome this restriction by using portable generators which allow them to operate the moth trap in areas remote from buildings. There are a number of manufacturers who offer a good selection of portable generators in different sizes. Units between 350W and 650W are popular due to their lightness and ease of use, although one of the disadvantages of using a small generator with a standard 125W bulb is that it requires refilling periodically.

The Robinson trap is the recommended trap for garden use; it is not, however, without problems. The bright light emitted from the bulb may cause some irritation and annoyance to neighbours so it should be used with consideration. There is a black-blended bulb available which can be used as an alternative. This bulb emits mainly UV light which is more acceptable when trapping in the garden. Black bulbs are more expensive than standard MV bulbs and also more susceptible to cracking, if exposed to rain. Many enthusiasts also use time clocks which automatically switch the trap on and off at the appropriate times.

The Robinson is the most effective trap for attracting and retaining large numbers of moths; no other trap can compete with it in terms of the number of species and individuals captured.

The moths are not harmed in any way and after examination they can be released again. It can be used successfully in the garden; however, be mindful that some birds learn to associate the trap with a potential food source. Be particularly careful of this when releasing moths back into the vegetation as they become easy prey for hungry onlookers. In terms of portability, this trap is more bulky to carry any distance compared to the other traps.

Skinner trap

The concept and design of this trap centres on portability, speed and ease of use, although it does not seem to be as popular in Northern Ireland as it is in Britain. The trap folds up neatly making it easy to transport and assemble in the field. It utilizes the same bulb and electrical components used in the Robinson trap, so it can be powered either by the mains electricity or a portable generator. Some units are also fitted with actinic tubes. The overall size of the trap is about eighteen inches wide by twelve inches high. It is normally constructed out of aluminium sheeting, but exterior plywood is a cheaper and more readily available alternative. The lamp or tube is fixed to a long bar which spans two of the sides, thus providing additional support to the trap assembly. The central two transparent Perspex sheets where the moths enter the trap are angled down at about 30° towards the base and retained about an inch apart. The trap is normally placed on a sheet. Some moth enthusiasts also place egg cartoons inside for moths to rest on. Unlike the Robinson, the operator can inspect the catch by lifting either of the transparent sheets, without having to turn off the light. This trap can be operated in the rain with a suitable rain cover.

Skinner trap with a portable generator.

The Skinner trap is as effective as the Robinson trap in attracting moths, but its weakness lies in its ability to retain the catch successfully within its base. This is especially so at daybreak, when most of the catch can escape easily through the space between the transparent Perspex sheets. Although its design makes it easy to build and assemble, it requires the operator to remain with the trap while it is running. This trap is often used at public events as the moths can be inspected at any time without having to turn off the light, as with the Robinson trap.

Portable actinic or Heath trap

The Heath trap dates back to the mid-1960s when it was invented by the late John Heath, one of the most eminent English lepidopterists of the time. The trap is portable and battery operated and designed for use in locations remote from mains electricity. The traps produced today differ slightly from Heath's original design. The trap is based around a blue twelve-volt, six-watt fluorescent tube, which emits a percentage of its total light output as UV light.

The base and sides of the trap, constructed out of thin aluminium sheeting, are designed to slot neatly together forming a box; these can be disassembled easily and carried in a rucksack. The modern traps have a photocell switch which turns the light on at dusk and off at daybreak. This is a useful feature unavailable until recently and invaluable in saving battery power and when using a series of traps. These can safely be set up for operation in daylight, removing the need to switch them on at dusk. The actinic trap is usually powered from a twelve-volt motorcycle or small car battery, but it can also be run from the mains or a portable generator.

Heath trap.

The Heath trap is the most versatile and economic trap for beginners. It is light, extremely portable and can be assembled and dismantled quickly. More than one trap including batteries can be carried in a rucksack. They can be used in sites and localities that are difficult to access with a vehicle and it is generally safe to leave them in place overnight. Actinic bulbs produce little heat and there is no risk of the bulb cracking if it rains. They produce little light and so are visible over just a short distance and hence are unlikely to attract unwanted attention, although this low output also renders them much less effective than the mains-operated traps in attracting moths. This is a significant consideration in planning moth surveys, especially site inventories. The availability of small, more portable, generators at an affordable price does further restrict the need for Heath traps.

Rothamsted trap

The Rothamsted trap is a fixed trap developed by the Rothamsted Research Station, a government-funded, agricultural research organisation. The trap is a standardised design, based around a tungsten filament bulb. It is normally secured to the ground or a base and is operated every night of the year. The trap functions as a standardised sample unit in a number of sites including the general countryside, gardens, woodland and coastal localities; this is part of a long-running monitoring project which started in 1968. This survey has yielded some of the most comprehensive data ever published on the current population trends of many of our native species across Britain and Northern Ireland. At present there is only one operational trap in Northern Ireland, at Hillsborough Forest, but there are plans to introduce others in the near future.

Studying butterflies and moths in the field

The catches from a Rothamsted trap are relatively poor by comparison with a Robinson and similar units. Unlike other moth traps, this unit requires a killing agent. The only justification for using this kind of trap is long-term data analysis. The cumulative data since the 1960s has revealed some worrying trends and changes in abundance of many common species across a wide range of habitats and more recently, how climate and changes in land management have affected moth populations.

Home-made moth traps
Many moth enthusiasts who are DIY experts find it more economical to construct their own traps. The basic building components needed for the fabrication of a mains trap can be purchased from various shops and stores. The electrical units can be purchased from entomological dealers. The trap illustrated is the Lansdowne MV Trap designed by Maurice Hughes. Many enthusiasts in Northern Ireland currently use this design successfully as an alternative to the Robinson trap. Some believe the deeper base helps to retain the moths better than the standard Robinson unit. Because it uses readily available components, the trap can be built for considerably less than the cost of a manufactured Robinson. Details of the components used are given below.

Rothamsted trap.

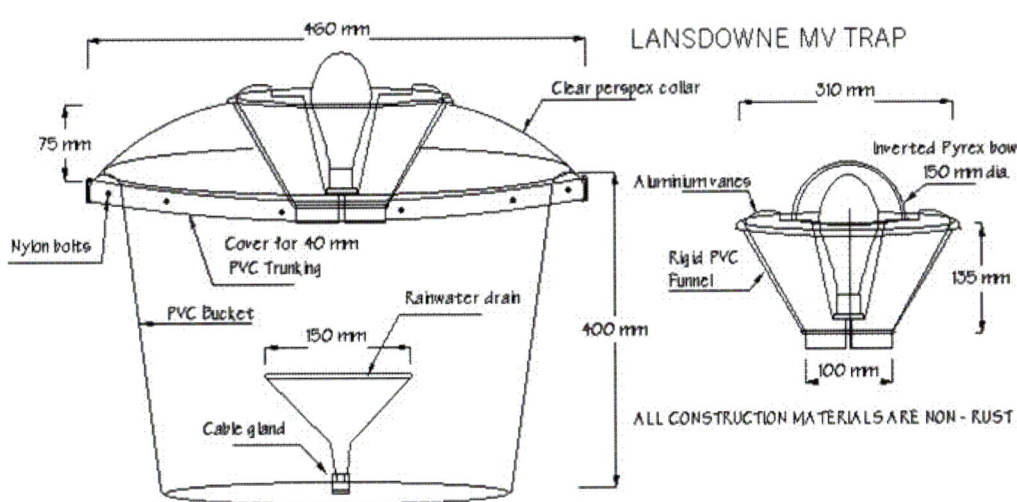

Top and bottom funnels

PVC trunking cover

Nylon bolts (numberplate bolts)

Pyrex bowl

PVC bucket

Perspex collar (1.5mm semi-rigid PVC)

Large cable gland

Aluminium for vanes

Bulb, bulb holder, waterproof control gear

391

Site selection

Having obtained a moth trap, the next stage is to decide where to use it. Choosing the right location for a trap is important as this can affect the number and abundance of species taken. Most enthusiasts begin trapping in their gardens which is an excellent way of becoming familiar with the common generalist species. Trapping over a year will reveal the changes in the seasonal abundance of species and show the effects of weather and trap location on the size of a night's catch. Gardens lend themselves to long-term studies of their moth fauna. If considering embarking on such a study, plan it carefully and consider getting advice on how best to run the trap to maintain consistency from year to year so that the data obtained can be analysed. Designing the study in this way will mean that the results will not be rendered meaningless due to variation in trapping effort or other factors; however, if the study is being done for enjoyment, then such considerations are irrelevant. The results of garden trapping can be truly impressive. A mature garden in north Down has been trapped practically every suitable night for the past fifteen years and has recorded over 215 species. Trapping over a four-year period at an old wooded estate in Seaforde, south Down has recorded 220 species. Also a small urban garden in Belfast has, rather surprisingly, produced 208 species over an eight-year period.

Whilst trapping in a garden is convenient and easy to do regularly, it can only ever produce a limited number of species. Many moths are habitat specialists and never venture beyond the confines of their habitat. The richest and most productive habitats in Northern Ireland for moths in terms of species diversity are generally natural and semi-natural broadleaved woodlands; bogs, coastal heaths, grasslands and dune slacks are also productive habitats, though generally less diverse than natural woodland. Sites with a mosaic of habitat types, for example, a bog with a woodland fringe, can be very productive sites, supporting a wide selection of species. These sites can be the most interesting for the moth enthusiast to explore.

Permission must always be obtained from the landowner before trapping at a site. This is usually easy to obtain and in some cases is actively encouraged. Environment and Heritage Service, the National Trust, RSPB and Ulster Wildlife Trust are usually very amenable to offers from moth trappers to survey sites. It is also prudent to inform the local police and anyone in the surrounding area about your activities, especially if you are running Robinson traps with generators; this will save any unnecessary bother and embarrassment with local people who may be suspicious about the noise and the bright lights.

Compiling a species inventory or site list requires trapping a site frequently, at least once every two weeks, and throughout the year. There is always some species of moth on the wing in Northern Ireland throughout the year, so the season truly never ends. The quietest period is late winter when there are few opportunities for trapping and when the species diversity is at its lowest. As spring progresses, the number of species on the wing increases noticeably. The peak period for moths is from late May through to late September.

Weather conditions

The most productive nights for moth trapping are warm, humid nights with good cloud cover and little or no wind. These conditions are vital in the wintertime to have any chance of success. Moths simply will not fly if there is a frost, strong winds or persistent rain so there is little point in trapping when these conditions are forecast to last all night. Clear nights with a full moon can also be unproductive as the light from the moon conflicts with the light from the trap and catches are often poor. A mild night with light intermittent rain seems to have less effect on moths and some excellent catches have been taken under these conditions.

Studying butterflies and moths in the field

Old fence post treated with a sugar solution.

Banffshire, Scotland.

Prolonged heavy rain during the day and into the night is not usually very productive and catches are usually poor.

Other techniques used for attracting moths
Although light traps are by far the most efficient and effective method for sampling moth populations, not all moths are attracted to light and males are more commonly seen at light than are females. These species may be attracted using one of the following methods.

Sugaring
Sugaring is the activity of painting a sweet, volatile liquid on a tree trunk, fence post or similar surface to attract moths. The basic ingredients of sugar mixture are treacle, brown sugar and stale beer. These are all boiled together and then rum and a small amount of amyl acetate are

The Butterflies and Moths of Northern Ireland

Wine rope placed across Bracken.

Peatlands Park, County Armagh.

An alternative moth attractant to sugar, most effective in the same weather conditions.

added to produce a volatile, sweet-smelling, black syrup. The sugar mixture has to be applied regularly for best results.

Sugaring was a popular method with nineteenth- and early twentieth-century lepidopterists, but it is less practised today although there are some species which are more easily found using this technique and it still has some strong adherents. There are several problems with the technique rendering it less popular. First, applying sugar regularly to trees is detrimental to other epiphytic organisms, in particular, lichens; sugaring is actively discouraged on many woodland sites for this reason. Anyone contemplating sugaring should therefore use old timber. The second reason is concern about the safety of amyl acetate (pear drops), one of the main ingredients of traditional sugar mixtures. Amyl acetate can only be obtained under licence. It is a dangerous substance and should be treated with extreme care. Anyone using it should heed the Health and Safety regulations regarding its use.

Some lepidopterists have developed their own variations of sugar mixture and are rather secretive about the recipe. The liquid is applied with a fairly wide brush in vertical strokes to fence posts and any other suitable structure and some of the surrounding vegetation. Weather conditions are an important factor to take into consideration when using this method. Warm, humid evenings are usually the most productive.

Wine ropes

This is a relatively new technique. Long lengths of absorbent rope are soaked in a solution of red wine and sugar and then laid across the branches and foliage of trees before dusk. The wine and sugar mixture presumably mimics fermenting fruit and attracts moths in to feed. It therefore works best in late summer and autumn. Wine ropes have been used rarely in Northern Ireland, but they have proved successful in Britain with species such as the Old Lady and Copper Underwing, which are rarely seen at light.

Studying butterflies and moths in the field

Pheromone lures
Pheromones are chemicals produced by female moths to attract males. Synthetic copies of these have become available in recent times and they have proved very effective in attracting the species of day-flying clearwing moths which are normally so elusive. They cannot be used

Oak Eggar larva.

Peatlands Park, County Armagh.

This fully-grown larva was found resting on a Heather stem. Recent light rain has left water droplets on the fine hairs.

for every type of clearwing, but do work for one of our species, the Currant Clearwing. The lures can be kept for several years if stored properly in a freezer. For a detailed account of their use and success, see Collins (2002).

Some of Northern Ireland's active lepidopterists.

Peatlands Park, County Armagh, July 2006.

From left to right
*Maurice Hughes, Trevor Boyd, Dave Allen, Brian Nelson, Kenny Murphy, Ted Rolston,
John Faulkner, Kerry Leonard, Rebecca Leonard and Ian Rippey.*

Buff-tip.

Peatlands Park, County Armagh.

Photography

PHOTOGRAPHY IS NOW widely recognised as an integral part of the study of any natural history group. It provides the means to accumulate photographic records of adults, their behaviour and habitats. Photographs can assist with the identification of closely related species and are often used as an educational aid to teach inexperienced observers. They can also reveal aspects of species behaviour which are impossible to ascertain from any set specimen. It is also a useful way to document natural changes in habitats. Images can be taken over a number of years and used to assist in the implementation of site management plans.

Photography has been described as the modern approach to collecting. It provides a visual record of a subject or habitat without inflicting any harm or damage to the organism or its environment. The plethora of cameras and equipment available today make it possible for anyone to obtain reasonably high-quality photographs of butterflies and moths and their environment.

The photographs of the adults, larvae and habitats for this publication, were taken over the last ten years, mostly on a Mamiya AFD medium format camera using Fuji Velvia Professional 120 roll film. The increased film size allows larger magnifications and increased resolution producing rich well saturated colours. Digital capture was used in the latter stages for a number of the adult photographs and to illustrate some of the other chapters. All digital images were taken either on a Kodak Pro SLR/n or a Nikon D2X. The adults and larvae were photographed using the following lenses: Nikon 105mm and 200 mm Micro Nikkors, Mamiya 80mm and 120mm Macro lenses. The habitat photographs were taken using a range

The Butterflies and Moths of Northern Ireland

Opposite page
Elephant Hawkmoth larva.

Annagariff Wood National Nature Reserve, Peatlands Park, County Armagh.

Right
Puss Moth larva.

Montiaghs Moss National Nature Reserve, County Antrim.

of lenses from 24mm to 110mm. Virtually all of the photographs were taken on a tripod. A monopod was occasionally used with some butterflies, or when the terrain made it impractical to use a tripod.

Good photographs are not necessarily the result of owning an expensive camera or an array of lenses. Having a sound photographic technique and some understanding of the behaviour and habits of the subjects is fundamental to achieving a high level of consistency of photographs. Digital capture has in the last few years been a serious contender to film and for many people the reasons for converting to this new technology are compelling. However, there are many issues to consider before abandoning film completely. Converting to digital capture is not cheap. There are many hidden costs, not to mention the time involved in learning a new technology and the post-processing time which is necessary to achieve the desired result. For an in-depth account on current cameras, equipment, and the methods and techniques used to photograph insects to professional standards, readers are referred to Thompson (2003 and 2005).

Checklist of Northern Ireland fauna

BUTTERFLIES

HESPERIIDAE
Dingy Skipper — *Erynnis tages* (Linnaeus, 1758)

PIERIDAE
Réal's Wood White — *Leptidea reali* Reissinger, 1989
Clouded Yellow — *Colias croceus* (Fourcroy, 1785)
Brimstone — *Gonepteryx rhamni* (Linnaeus, 1758)
Large White — *Pieris brassicae* (Linnaeus, 1758)
Small White — *Pieris rapae* (Linnaeus, 1758)
Green-veined White — *Pieris napi* (Linnaeus, 1758)
Orange Tip — *Anthocaris cardamines* (Linnaeus, 1758)

LYCAENIDAE
Green Hairstreak — *Callophrys rubi* (Linnaeus, 1758)
Purple Hairstreak — *Neozephyrus quercus* (Linnaeus, 1758)
Small Copper — *Lycaena phlaeas* (Linnaeus, 1761)
Small Blue — *Cupido minimus* (Fuessly, 1775)
Common Blue — *Polyommatus icarus* (Rottemburg, 1775)
Holly Blue — *Celastrina argiolus* (Linnaeus, 1758)

NYMPHALIDAE
Red Admiral — *Vanessa atalanta* (Linnaeus, 1758)
Painted Lady — *Vanessa cardui* (Linnaeus, 1758)
Small Tortoiseshell — *Aglais urticae* (Linnaeus, 1758)
Camberwell Beauty — *Nymphalis antiopa* (Linnaeus, 1758)
Peacock — *Inachis io* (Linnaeus, 1758)
Comma — *Polygonia c-album* (Linnaeus, 1758)
Dark Green Fritillary — *Argynnis aglaja* (Linnaeus, 1758)
Silver-washed Fritillary — *Argynnis paphia* (Linnaeus, 1758)
Marsh Fritillary — *Euphydryas aurinia* (Rottemburg, 1775)
Speckled Wood — *Pararge aegeria* (Linnaeus, 1758)
Wall — *Lasiommata megera* (Linnaeus, 1767)
Grayling — *Hipparchia semele* (Linnaeus, 1758)
Meadow Brown — *Maniola jurtina* (Linnaeus, 1758)
Ringlet — *Aphantopus hyperantus* (Linnaeus, 1758)
Small Heath — *Coenonympha pamphilus* (Linnaeus, 1758)
Large Heath — *Coenonympha tullia* (Müller, 1764)
Monarch — *Danaus plexippus* (Linnaeus, 1758)

MOTHS

HEPIALIDAE
Ghost Moth — *Hepialus humuli* (Linnaeus, 1758)
Gold Swift — *Hepialus hecta* (Linnaeus, 1758)
Common Swift — *Hepialus lupulinus* (Linnaeus, 1758)
Map-winged Swift — *Hepialus fusconebulosa* (DeGeer, 1778)

ZYGAENIDAE
Procridinae
Forester — *Adscita statices* (Linnaeus, 1758)

Zygaeninae
Six-spot Burnet — *Zygaena filipendulae* (Linnaeus, 1758)
Narrow-bordered Five-spot Burnet — *Zygaena lonicerae* (Scheven, 1777)

SESIIDAE
Sesiinae
Lunar Hornet Moth — *Sesia bembeciformis* (Hübner, 1806)

Paranthreninae
Currant Clearwing	*Synanthedon tipuliformis* (Clerck, 1759)
Red-tipped Clearwing	*Synanthedon formicaeformis* (Boisduval, 1840)

LASIOCAMPIDAE
December Moth	*Poecilocampa populi* (Linnaeus, 1758)
Pale Eggar	*Trichiura crataegi* (Linnaeus, 1758)
Small Eggar	*Eriogaster lanestris* (Linnaeus, 1758)
Lackey	*Malacosoma neustria* (Linnaeus, 1758)
Oak and Northern Eggar	*Lasiocampa quercus* (Linnaeus, 1758)
Fox Moth	*Macrothylacia rubi* (Linnaeus, 1758)
Drinker	*Euthrix potatoria* (Linnaeus, 1758)

SATURNIIDAE
Emperor	*Saturnia pavonia* (Linnaeus, 1758)

DREPANIDAE
Scalloped Hook-tip	*Falcaria lacertinaria* (Linnaeus, 1758)
Barred Hook-tip	*Watsonalla cultraria* (Fabricius, 1775)
Pebble Hook-tip	*Drepana falcataria* (Linnaeus, 1758)
Chinese Character	*Cilix glaucata* (Scopoli, 1763)

THYATIRIDAE
Peach Blossom	*Thyatira batis* (Linnaeus, 1758)
Buff Arches	*Habrosyne pyritoides* (Hufnagel, 1766)
Poplar Lutestring	*Tethea or* (Denis & Schiffermüller, 1775)
Common Lutestring	*Ochropacha duplaris* (Linnaeus, 1758)
Yellow Horned	*Achlya flavicornis* (Linnaeus, 1758)

GEOMETRIDAE

Alsophilinae
March Moth	*Alsophila aescularia* (Denis & Schiffermüller, 1775)

Geometrinae
Grass Emerald	*Pseudoterpna pruinata* (Hufnagel, 1767)
Large Emerald	*Geometra papilionaria* (Linnaeus, 1758)
Common Emerald	*Hemithea aestivaria* (Hübner, 1789)
Small Emerald	*Hemistola chrysoprasaria* (Esper, 1795)
Little Emerald	*Jodis lactearia* (Linnaeus, 1758)

Sterrhinae
Birch Mocha	*Cyclophora albipunctata* (Hufnagel, 1767)
Clay Triple-lines	*Cyclophora linearia* (Hübner, 1799)
Blood-vein	*Timandra comae* A. Schmidt, 1931
Mullein Wave	*Scopula marginepunctata* (Goeze, 1781)
Lesser Cream Wave	*Scopula immutata* (Linnaeus, 1758)
Cream Wave	*Scopula floslactata* (Haworth, 1809)
Smoky Wave	*Scopula ternata* (Schrank, 1802)
Small Fan-footed Wave	*Idaea biselata* (Hufnagel, 1767)
Single-dotted Wave	*Idaea dimidiata* (Hufnagel, 1767)
Riband Wave	*Idaea aversata* (Linnaeus, 1758)
Plain Wave	*Idaea straminata* (Borkhausen, 1794)
Vestal	*Rhodometra sacraria* (Linnaeus, 1767)

Larentiinae
Oblique Carpet	*Orthonama vittata* (Borkhausen, 1794)
Gem	*Orthonama obstipata* (Fabricius, 1794)
Flame Carpet	*Xanthorhoe designata* (Hufnagel, 1767)
Red Carpet	*Xanthorhoe decoloraria* (Esper, 1806)
Red Twin-spot Carpet	*Xanthorhoe spadicearia* (Denis & Schiffermüller, 1775)

Dark-barred Twin-spot Carpet	*Xanthorhoe ferrugata* (Clerck, 1759)
Silver-ground Carpet	*Xanthorhoe montanata* (Denis & Schiffermüller, 1775)
Garden Carpet	*Xanthorhoe fluctuata* (Linnaeus, 1758)
Shaded Broad-bar	*Scotopteryx chenopodiata* (Linnaeus, 1758)
Lead Belle	*Scotopteryx mucronata* (Scopoli, 1763)
July Belle	*Scotopteryx luridata* (Hufnagel, 1767)
Small Argent and Sable	*Epirrhoe tristata* (Linnaeus, 1758)
Common Carpet	*Epirrhoe alternata* (Müller, 1764)
Galium Carpet	*Epirrhoe galiata* (Denis & Schiffermüller, 1775)
Yellow Shell	*Camptogramma bilineata* (Linnaeus, 1758)
Yellow-ringed Carpet	*Entephria flavicinctata* (Hübner, 1813)
Grey Mountain Carpet	*Entephria caesiata* (Denis & Schiffermüller, 1775)
Shoulder Stripe	*Anticlea badiata* (Denis & Schiffermüller, 1775)
Streamer	*Anticlea derivata* (Denis & Schiffermüller, 1775)
Beautiful Carpet	*Mesoleuca albicillata* (Linnaeus, 1758)
Dark Spinach	*Pelurga comitata* (Linnaeus, 1758)
Water Carpet	*Lampropteryx suffumata* (Denis & Schiffermüller, 1775)
Purple Bar	*Cosmorhoe ocellata* (Linnaeus, 1758)
Striped Twin-spot Carpet	*Nebula salicata* (Denis & Schiffermüller, 1775)
Phoenix	*Eulithis prunata* (Linnaeus, 1758)
Chevron	*Eulithis testata* (Linnaeus, 1761)
Northern Spinach	*Eulithis populata* (Linnaeus, 1758)
Barred Straw	*Eulithis pyraliata* (Denis & Schiffermüller, 1775)
Small Phoenix	*Ecliptopera silaceata* (Denis & Schiffermüller, 1775)
Red-green Carpet	*Chloroclysta siterata* (Hufnagel, 1767)
Autumn Green Carpet	*Chloroclysta miata* (Linnaeus, 1758)
Dark Marbled Carpet	*Chloroclysta citrata* (Linnaeus, 1761)
Common Marbled Carpet	*Chloroclysta truncata* (Hufnagel, 1767)
Barred Yellow	*Cidaria fulvata* (Forster, 1771)
Blue-bordered Carpet	*Plemyria rubiginata* (Denis & Schiffermüller, 1775)
Pine Carpet	*Thera firmata* (Hübner, 1822)
Grey Pine Carpet	*Thera obeliscata* (Hübner, 1787)
Spruce Carpet	*Thera britannica* (Turner, 1925)
Chestnut-coloured Carpet	*Thera cognata* (Thunberg, 1792)
Juniper Carpet	*Thera juniperata* (Linnaeus, 1758)
Broken-barred Carpet	*Electrophaes corylata* (Thunberg, 1792)
Beech-green Carpet	*Colostygia olivata* (Denis & Schiffermüller, 1775)
Mottled Grey	*Colostygia multistrigaria* (Haworth, 1809)
Green Carpet	*Colostygia pectinataria* (Knoch, 1781)
July Highflyer	*Hydriomena furcata* (Thunberg, 1784)
May Highflyer	*Hydriomena impluviata* (Denis & Schiffermüller, 1775)
Ruddy Highflyer	*Hydriomena ruberata* (Freyer, 1831)
Slender-striped Rufous	*Coenocalpe lapidata* (Hübner, 1809)
Argent and Sable	*Rheumaptera hastata* (Linnaeus, 1758)
Scallop Shell	*Rheumaptera undulata* (Linnaeus, 1758)
Tissue	*Triphosa dubitata* (Linnaeus, 1758)
Brown Scallop	*Philereme vetulata* (Denis & Schiffermüller, 1775)
Dark Umber	*Philereme transversata* (Hufnagel, 1767)
Sharp-angled Carpet	*Euphyia unangulata* (Haworth, 1809)
November Moth	*Epirrita dilutata* (Denis & Schiffermüller, 1775)
Pale November Moth	*Epirrita christyi* (Allen, 1906)
Autumnal Moth	*Epirrita autumnata* (Borkhausen, 1794)
Small Autumnal Moth	*Epirrita filigrammaria* (Herrich-Schaeffer, 1846)
Winter Moth	*Operophtera brumata* (Linnaeus, 1758)
Northern Winter Moth	*Operophtera fagata* (Scharfenberg, 1805)
Barred Carpet	*Perizoma taeniata* (Stephens, 1831)
Rivulet	*Perizoma affinitata* (Stephens, 1831)
Small Rivulet	*Perizoma alchemillata* (Linnaeus, 1758)

Barred Rivulet	*Perizoma bifaciata* (Haworth, 1809)
Pretty Pinion	*Perizoma blandiata* (Denis & Schiffermüller, 1775)
Grass Rivulet	*Perizoma albulata* (Denis & Schiffermüller, 1775)
Sandy Carpet	*Perizoma flavofasciata* (Thunberg, 1792)
Twin-spot Carpet	*Perizoma didymata* (Linnaeus, 1758)
Slender Pug	*Eupithecia tenuiata* (Hübner, 1813)
Haworth's Pug	*Eupithecia haworthiata* Doubleday, 1856
Lead-coloured Pug	*Eupithecia plumbeolata* (Haworth, 1809)
Cloaked Pug	*Eupithecia abietaria* (Goeze, 1781)
Toadflax Pug	*Eupithecia linariata* (Denis & Schiffermüller, 1775)
Foxglove Pug	*Eupithecia pulchellata* Stephens, 1831
Mottled Pug	*Eupithecia exiguata* (Hübner, 1813)
Valerian Pug	*Eupithecia valerianata* (Hübner, 1813)
Marsh Pug	*Eupithecia pygmaeata* (Hübner, 1799)
Netted Pug	*Eupithecia venosata* (Fabricius, 1787)
Lime-speck Pug	*Eupithecia centaureata* (Denis & Schiffermüller, 1775)
Triple-spotted Pug	*Eupithecia trisignaria* Herrich-Schaeffer, 1848
Satyr Pug	*Eupithecia satyrata* (Hübner, 1813)
Wormwood and Ling Pug	*Eupithecia absinthiata* (Clerck, 1759)
Currant Pug	*Eupithecia assimilata* Doubleday, 1856
Bleached Pug	*Eupithecia expallidata* Doubleday, 1856
Common Pug	*Eupithecia vulgata* (Haworth, 1809)
White-spotted Pug	*Eupithecia tripunctaria* Herrich-Schaeffer, 1852
Grey Pug	*Eupithecia subfuscata* (Haworth, 1809)
Tawny-speckled Pug	*Eupithecia icterata* (Villers, 1789)
Bordered Pug	*Eupithecia succenturiata* (Linnaeus, 1758)
Shaded Pug	*Eupithecia subumbrata* (Denis & Schiffermüller, 1775)
Plain Pug	*Eupithecia simpliciata* (Haworth, 1809)
Thyme Pug	*Eupithecia distinctaria* Herrich-Schaeffer, 1861
Ochreous Pug	*Eupithecia indigata* (Hübner, 1813)
Pimpinel Pug	*Eupithecia pimpinellata* (Hübner, 1813)
Narrrow-winged Pug	*Eupithecia nanata* (Hübner, 1813)
Angle-barred and Ash Pug	*Eupithecia innotata* (Hufnagel, 1767)
Golden-rod Pug	*Eupithecia virgaureata* Doubleday, 1861
Brindled Pug	*Eupithecia abbreviata* Stephens, 1831
Oak-tree Pug	*Eupithecia dodoneata* Guenée, 1857
Juniper Pug	*Eupithecia pusillata* (Denis & Schiffermüller, 1775)
Larch Pug	*Eupithecia lariciata* (Freyer, 1841)
Dwarf Pug	*Eupithecia tantillaria* Boisduval, 1840
V-Pug	*Chloroclystis v-ata* (Haworth, 1809)
Green Pug	*Pasiphila rectangulata* (Linnaeus, 1758)
Bilberry Pug	*Pasiphila debiliata* (Hübner, 1817)
Double-striped Pug	*Gymnoscelis rufifasciata* (Haworth, 1809)
Streak	*Chesias legatella* (Denis & Schiffermüller, 1775)
Treble-bar	*Aplocera plagiata* (Linnaeus, 1758)
Chimney Sweeper	*Odezia atrata* (Linnaeus, 1758)
Welsh Wave	*Venusia cambrica* Curtis, 1839
Small White Wave	*Asthena albulata* (Hufnagel, 1767)
Small Yellow Wave	*Hydrelia flammeolaria* (Hufnagel, 1767)
Seraphim	*Lobophora halterata* (Hufnagel, 1767)
Early Tooth-striped	*Trichopteryx carpinata* (Borkhausen, 1794)
Small Seraphim	*Pterapherapteryx sexalata* (Retzius, 1783)
Yellow-barred Brindle	*Acasis viretata* (Hübner, 1799)

Ennominae

Magpie	*Abraxas grossulariata* (Linnaeus, 1758)
Clouded Magpie	*Abraxas sylvata* (Scopoli, 1763)
Clouded Border	*Lomaspilis marginata* (Linnaeus, 1758)

Scorched Carpet	*Ligdia adustata* (Denis & Schiffermüller, 1775)
Tawny-barred Angle	*Macaria liturata* (Clerck, 1759)
V-Moth	*Macaria wauaria* (Linnaeus, 1758)
Latticed Heath	*Chiasmia clathrata* (Linnaeus, 1758)
Brown Silver-line	*Petrophora chlorosata* (Scopoli, 1763)
Barred Umber	*Plagodis pulveraria* (Linnaeus, 1758)
Scorched Wing	*Plagodis dolabraria* (Linnaeus, 1767)
Brimstone Moth	*Opisthograptis luteolata* (Linnaeus, 1758)
Bordered Beauty	*Epione repandaria* (Hufnagel, 1767)
Lilac Beauty	*Apeira syringaria* (Linnaeus, 1758)
August Thorn	*Ennomos quercinaria* (Hufnagel, 1767)
Canary-shouldered Thorn	*Ennomos alniaria* (Linnaeus, 1758)
Early Thorn	*Selenia dentaria* (Fabricius, 1775)
Lunar Thorn	*Selenia lunularia* (Hübner, 1788)
Scalloped Hazel	*Odontopera bidentata* (Clerck, 1759)
Scalloped Oak	*Crocallis elinguaria* (Linnaeus, 1758)
Swallow-tailed Moth	*Ourapteryx sambucaria* (Linnaeus, 1758)
Feathered Thorn	*Colotois pennaria* (Linnaeus, 1761)
Pale Brindled Beauty	*Phigalia pilosaria* (Denis & Schiffermüller, 1775)
Brindled Beauty	*Lycia hirtaria* (Clerck, 1759)
Belted Beauty	*Lycia zonaria* (Denis & Schiffermüller, 1775)
Oak Beauty	*Biston strataria* (Hufnagel, 1767)
Peppered Moth	*Biston betularia* (Linnaeus, 1758)
Spring Usher	*Agriopis leucophaearia* (Denis & Schiffermüller, 1775)
Scarce Umber	*Agriopis aurantiaria* (Hübner, 1799)
Dotted Border	*Agriopis marginaria* (Fabricius, 1776)
Mottled Umber	*Erannis defoliaria* (Clerck, 1759)
Willow Beauty	*Peribatodes rhomboidaria* (Denis & Schiffermüller, 1775)
Bordered Grey	*Selidosema brunnearia* (Villers, 1789)
Ringed Carpet	*Cleora cinctaria* (Denis & Schiffermüller, 1775)
Satin Beauty	*Deileptenia ribeata* (Clerck, 1759)
Mottled Beauty	*Alcis repandata* (Linnaeus, 1758)
Dotted Carpet	*Alcis jubata* (Thunberg, 1788)
Brussels Lace	*Cleorodes lichenaria* (Hufnagel, 1767)
Engrailed	*Ectropis bistortata* (Goeze, 1781)
Small Engrailed	*Ectropis crepuscularia* (Denis & Schiffermüller, 1775)
Grey Birch	*Aethalura punctulata* (Denis & Schiffermüller, 1775)
Common Heath	*Ematurga atomaria* (Linnaeus, 1758)
Bordered White	*Bupalus piniaria* (Linnaeus, 1758)
Common White Wave	*Cabera pusaria* (Linnaeus, 1758)
Common Wave	*Cabera exanthemata* (Scopoli, 1763)
Clouded Silver	*Lomographa temerata* (Denis & Schiffermüller, 1775)
Early Moth	*Theria primaria* (Haworth, 1809)
Light Emerald	*Campaea margaritata* (Linnaeus, 1767)
Barred Red	*Hylaea fasciaria* (Linnaeus, 1758)
Annulet	*Charissa obscurata* (Denis & Schiffermüller, 1775)
Grey Scalloped Bar	*Dyscia fagaria* (Thunberg, 1784)
Grass Wave	*Perconia strigillaria* (Hübner, 1787)

SPHINGIDAE
Sphinginae

Convolvulus Hawkmoth	*Agrius convolvuli* (Linnaeus, 1758)
Death's Head Hawkmoth	*Acherontia atropos* (Linnaeus, 1758)

Smerinthinae

Eyed Hawkmoth	*Smerinthus ocellata* (Linnaeus, 1758)
Poplar Hawkmoth	*Laothoe populi* (Linnaeus, 1758)

Macroglossinae
Narrow-bordered Bee Hawkmoth — *Hemaris tityus* (Linnaeus, 1758)
Hummingbird Hawkmoth — *Macroglossum stellatarum* (Linnaeus, 1758)
Oleander Hawkmoth — *Daphnis nerii* (Linnaeus, 1758)
Striped Hawkmoth — *Hyles livornica* (Esper, 1779)
Elephant Hawkmoth — *Deilephila elpenor* (Linnaeus, 1758)
Small Elephant Hawkmoth — *Deilephila porcellus* (Linnaeus, 1758)
Silver-striped Hawkmoth — *Hippotion celerio* (Linnaeus, 1758)

NOTODONTIDAE
Notodontinae
Puss Moth — *Cerura vinula* (Linnaeus, 1758)
Alder Kitten — *Furcula bicuspis* (Borkhausen, 1790)
Sallow Kitten — *Furcula furcula* (Clerck, 1759)
Iron Prominent — *Notodonta dromedarius* (Linnaeus, 1758)
Pebble Prominent — *Notodonta ziczac* (Linnaeus, 1758)
Lesser Swallow Prominent — *Pheosia gnoma* (Fabricius, 1776)
Swallow Prominent — *Pheosia tremula* (Clerck, 1759)
Coxcomb Prominent — *Ptilodon capucina* (Linnaeus, 1758)
Scarce Prominent — *Odontosia carmelita* (Esper, 1799)
Pale Prominent — *Pterostoma palpina* (Clerck, 1759)
Lunar Marbled Brown — *Drymonia ruficornis* (Hufnagel, 1766)

Pygaerinae
Small Chocolate-tip — *Closera pigra* (Hufnagel, 1766)

Phalerinae
Buff-tip — *Phalera bucephala* (Linnaeus, 1758)

Dilobinae
Figure of Eight — *Diloba caeruleocephala* (Linnaeus, 1758)

LYMANTRIIDAE
Vapourer — *Orgyia antiqua* (Linnaeus, 1758)
Dark Tussock — *Dicallomera fascelina* (Linnaeus, 1758)
Pale Tussock — *Calliteara pudibunda* (Linnaeus, 1758)
Yellow-tail — *Euproctis similis* (Fuessly, 1775)

ARCTIIDAE
Lithosiinae
Round-winged Muslin — *Thumatha senex* (Hübner, 1808)
Muslin Footman — *Nudaria mundana* (Linnaeus, 1761)
Red-necked Footman — *Atolmis rubricollis* (Linnaeus, 1758)
Four-dotted Footman — *Cybosia mesomella* (Linnaeus, 1758)
Scarce Footman — *Eilema complana* (Linnaeus, 1758)
Buff Footman — *Eilema depressa* (Esper, 1787)
Common Footman — *Eilema lurideola* (Zincken, 1817)
Four-spotted Footman — *Lithosia quadra* (Linnaeus, 1758)

Arctiinae
Wood Tiger — *Parasemia plantaginis* (Linnaeus, 1758)
Garden Tiger — *Arctia caja* (Linnaeus, 1758)
Clouded Buff — *Diacrisia sannio* (Linnaeus, 1758)
White Ermine — *Spilosoma lubricipeda* (Linnaeus, 1758)
Buff Ermine — *Spilosoma luteum* (Hufnagel, 1766)
Muslin Moth — *Diaphora mendica* (Clerck, 1759)
Ruby Tiger — *Phragmatobia fuliginosa* (Linnaeus, 1758)
Cinnabar — *Tyria jacobaeae* (Linnaeus, 1758)

NOLIDAE
Least Black Arches — *Nola confusalis* (Herrich-Schaeffer, 1847)

NOCTUIDAE
Noctuinae

Square-spot Dart	*Euxoa obelisca* (Denis & Schiffermüller, 1775)
White-line Dart	*Euxoa tritici* (Linnaeus, 1761)
Garden Dart	*Euxoa nigricans* (Linnaeus, 1761)
Coast Dart	*Euxoa cursoria* (Hufnagel, 1766)
Light Feathered Rustic	*Agrotis cinerea* (Denis & Schiffermüller, 1775)
Archer's Dart	*Agrotis vestigialis* (Hufnagel, 1766)
Turnip Moth	*Agrotis segetum* (Denis & Schiffermüller, 1775)
Heart and Club	*Agrotis clavis* (Hufnagel, 1766)
Heart and Dart	*Agrotis exclamationis* (Linnaeus, 1758)
Crescent Dart	*Agrotis trux* (Hübner, 1824)
Dark Sword-grass	*Agrotis ipsilon* (Hufnagel, 1766)
Shuttle-shaped Dart	*Agrotis puta* (Hübner, 1803)
Flame	*Axylia putris* (Linnaeus, 1761)
Portland Moth	*Actebia praecox* (Linnaeus, 1758)
Flame Shoulder	*Ochropleura plecta* (Linnaeus, 1761)
Northern Rustic	*Standfussiana lucernea* (Linnaeus, 1758)
Large Yellow Underwing	*Noctua pronuba* Linnaeus, 1758
Lesser Yellow Underwing	*Noctua comes* (Hübner, 1813)
Broad-bordered Yellow Underwing	*Noctua fimbriata* (Schreber, 1759)
Lesser Broad-bordered Yellow Underwing	*Noctua janthe* (Borkhausen, 1792)
Least Yellow Underwing	*Noctua interjecta* Hübner, 1803
Double Dart	*Graphiphora augur* (Fabricius, 1775)
Autumnal Rustic	*Paradiarsia glareosa* (Esper, 1788)
True Lover's Knot	*Lycophotia porphyrea* (Denis & Schiffermüller, 1775)
Pearly Underwing	*Peridroma saucia* (Hübner, 1808)
Ingrailed Clay	*Diarsia mendica* (Fabricius, 1775)
Barred Chestnut	*Diarsia dahlii* (Hübner, 1813)
Purple Clay	*Diarsia brunnea* (Denis & Schiffermüller, 1775)
Small Square-spot	*Diarsia rubi* (Vieweg, 1790)
Setaceous Hebrew Character	*Xestia c-nigrum* (Linnaeus, 1758)
Double Square-spot	*Xestia triangulum* (Hufnagel, 1766)
Dotted Clay	*Xestia baja* (Denis & Schiffermüller, 1775)
Neglected Rustic	*Xestia castanea* (Esper, 1798)
Six-striped Rustic	*Xestia sexstrigata* (Haworth, 1809)
Square-spot Rustic	*Xestia xanthographa* (Denis & Schiffermüller, 1775)
Heath Rustic	*Xestia agathina* (Duponchel, 1827)
Gothic	*Naenia typica* (Linnaeus, 1758)
Great Brocade	*Eurois occulta* (Linnaeus, 1758)
Green Arches	*Anaplectoides prasina* (Denis & Schiffermüller, 1775)
Red Chestnut	*Cerastis rubricosa* (Denis & Schiffermüller, 1775)

Hadeninae

Beautiful Yellow Underwing	*Anarta myrtilli* (Linnaeus, 1761)
Nutmeg	*Discestra trifolii* (Hufnagel, 1766)
Shears	*Hada plebeja* (Linnaeus, 1761)
Grey Arches	*Polia nebulosa* (Hufnagel, 1766)
White Colon	*Sideridis albicolon* (Hübner, 1813)
Cabbage Moth	*Mamestra brassicae* (Linnaeus, 1758)
Dot Moth	*Melanchra persicariae* (Linnaeus, 1761)
Broom Moth	*Melanchra pisi* (Linnaeus, 1758)
Beautiful Brocade	*Lacanobia contigua* (Denis & Schiffermüller, 1775)
Pale-shouldered Brocade	*Lacanobia thalassina* (Hufnagel, 1766)
Dog's Tooth	*Lacanobia suasa* (Denis & Schiffermüller, 1775)
Bright-line Brown-eye	*Lacanobia oleracea* (Linnaeus, 1758)
Glaucous Shears	*Papestra biren* (Goeze, 1781)
Broad-barred White	*Hecatera bicolorata* (Hufnagel, 1766)
Campion	*Hadena rivularis* (Fabricius, 1775)

Tawny Shears and Pod Lover	*Hadena perplexa* (Denis & Schiffermüller, 1775)
Marbled Coronet	*Hadena confusa* (Hufnagel, 1766)
Lychnis	*Hadena bicruris* (Hufnagel, 1766)
Antler Moth	*Cerapteryx graminis* (Linnaeus, 1758)
Hedge Rustic	*Tholera cespitis* (Denis & Schiffermüller, 1775)
Feathered Gothic	*Tholera decimalis* (Poda, 1761)
Pine Beauty	*Panolis flammea* (Denis & Schiffermüller, 1775)
Small Quaker	*Orthosia cruda* (Denis & Schiffermüller, 1775)
Northern Drab	*Orthosia opima* (Hübner, 1809)
Lead-coloured Drab	*Orthosia populeti* (Fabricius, 1775)
Powdered Quaker	*Orthosia gracilis* (Denis & Schiffermüller, 1775)
Common Quaker	*Orthosia cerasi* (Fabricius, 1775)
Clouded Drab	*Orthosia incerta* (Hufnagel, 1766)
Twin-spotted Quaker	*Orthosia munda* (Denis & Schiffermüller, 1775)
Hebrew Character	*Orthosia gothica* (Linnaeus, 1758)
Brown-line Bright-eye	*Mythimna conigera* (Denis & Schiffermüller, 1775)
Clay	*Mythimna ferrago* (Fabricius, 1787)
Southern Wainscot	*Mythimna straminea* (Treitschke, 1825)
Smoky Wainscot	*Mythimna impura* (Hübner, 1808)
Common Wainscot	*Mythimna pallens* (Linnaeus, 1758)
Shore Wainscot	*Mythimna litoralis* (Curtis, 1827)
White-speck	*Mythimna unipuncta* (Haworth, 1809)
Shoulder-striped Wainscot	*Mythimna comma* (Linnaeus, 1761)

Cuculliinae

Chamomile Shark	*Cucullia chamomillae* (Denis & Schiffermüller, 1775)
Shark	*Cucullia umbratica* (Linnaeus, 1758)
Minor Shoulder-knot	*Brachylomia viminalis* (Fabricius, 1776)
Sprawler	*Asteroscopus sphinx* (Hufnagel, 1766)
Brindled Ochre	*Dasypolia templi* (Thunberg, 1792)
Northern Deep-brown Dart	*Aporophyla lueneburgensis* (Freyer, 1848)
Black Rustic	*Aporophyla nigra* (Haworth, 1809)
Pale Pinion	*Lithophane hepatica* (Clerck, 1759)
Grey Shoulder-knot	*Lithophane ornitopus* (Hufnagel, 1766)
Blair's Shoulder-knot	*Lithophane leautieri* (Boisduval, 1829)
Red Sword-grass	*Xylena vetusta* (Hübner, 1813)
Sword-grass	*Xylena exsoleta* (Linnaeus, 1758)
Early Grey	*Xylocampa areola* (Esper, 1789)
Green-brindled Crescent	*Allophyes oxyacanthae* (Linnaeus, 1758)
Merveille du Jour	*Dichonia aprilina* (Linnaeus, 1758)
Brindled Green	*Dryobotodes eremita* (Fabricius, 1775)
Dark Brocade	*Blepharita adusta* (Esper, 1790)
Grey Chi	*Antitype chi* (Linnaeus, 1758)
Feathered Ranunculus	*Polymixis lichenea* (Hübner, 1813)

Acronictinae

Satellite	*Eupsilia transversa* (Hufnagel, 1766)
Chestnut	*Conistra vaccinii* (Linnaeus, 1761)
Dark Chestnut	*Conistra ligula* (Esper, 1791)
Brick	*Agrochola circellaris* (Hufnagel, 1766)
Red-line Quaker	*Agrochola lota* (Clerck, 1759)
Yellow-line Quaker	*Agrochola macilenta* (Hübner, 1809)
Flounced Chestnut	*Agrochola helvola* (Linnaeus, 1758)
Brown-spot Pinion	*Agrochola litura* (Limmaeus, 1761)
Beaded Chestnut	*Agrochola lychnidis* (Denis & Schiffermüller, 1775)
Centre-barred Sallow	*Atethmia centrago* (Haworth, 1809)
Lunar Underwing	*Omphaloscelis lunosa* (Haworth, 1809)
Pink-barred Sallow	*Xanthia togata* (Esper, 1788)
Sallow	*Xanthia icteritia* (Hufnagel, 1766)

Poplar Grey	*Acronicta megacephala* (Denis & Schiffermüller, 1775)
Miller	*Acronicta leporina* (Linnaeus, 1758)
Alder Moth	*Acronicta alni* (Linnaeus, 1767)
Grey Dagger	*Acronicta psi* (Linnaeus, 1758)
Light Knot Grass	*Acronicta menyanthidis* (Esper, 1789)
Knot Grass	*Acronicta rumicis* (Linnaeus, 1758)
Coronet	*Craniophora ligustri* (Denis & Schiffermüller, 1775)

Bryophilinae

Marbled Beauty	*Cryphia domestica* (Hufnagel, 1766)

Amphipyrinae

Copper Underwing	*Amphipyra pyramidea* (Linnaeus, 1758)
Mouse Moth	*Amphipyra tragopoginis* (Clerck, 1759)
Old Lady	*Mormo maura* (Linnaeus, 1758)
Brown Rustic	*Rusina ferruginea* (Esper, 1785)
Straw Underwing	*Thalpophila matura* (Hufnagel, 1766)
Small Angle Shades	*Euplexia lucipara* (Linnaeus, 1758)
Angle Shades	*Phlogophora meticulosa* (Linnaeus, 1758)
Olive	*Ipimorpha subtusa* (Denis & Schiffermüller, 1775)
Angle-striped Sallow	*Enargia paleacea* (Esper, 1788)
Suspected	*Parastichtis suspecta* (Hübner, 1817)
Dingy Shears	*Parastichtis ypsillon* (Denis & Schiffermüller, 1775)
Dun-bar	*Cosmia trapezina* (Linnaeus, 1758)
Dark Arches	*Apamea monoglypha* (Hufnagel, 1766)
Light Arches	*Apamea lithoxylea* (Denis & Schiffermüller, 1775)
Reddish Light Arches	*Apamea sublustris* (Esper, 1788)
Crescent Striped	*Apamea oblonga* (Haworth, 1809)
Clouded-bordered Brindle	*Apamea crenata* (Hufnagel, 1766)
Clouded Brindle	*Apamea epomidion* (Haworth, 1809)
Confused	*Apamea furva* (Denis & Schiffermüller, 1775)
Dusky Brocade	*Apamea remissa* (Hübner, 1809)
Small Clouded Brindle	*Apamea unanimis* (Hübner, 1813)
Rustic Shoulder-knot	*Apamea sordens* (Hufnagel, 1766)
Slender Brindle	*Apamea scolopacina* (Esper, 1788)
Double Lobed	*Apamea ophiogramma* (Esper, 1794)
Marbled Minor	*Oligia strigilis* (Linnaeus, 1758)
Rufous Minor	*Oligia versicolor* (Borkhausen, 1792)
Tawny Marbled Minor	*Oligia latruncula* (Denis & Schiffermüller, 1775)
Middle-barred Minor	*Oligia fasciuncula* (Haworth, 1809)
Cloaked Minor	*Mesoligia furuncula* (Denis & Schiffermüller, 1775)
Rosy Minor	*Mesoligia literosa* (Haworth, 1809)
Common Rustic	*Mesapamea secalis* (Linnaeus, 1758)
Lesser Common Rustic	*Mesapamea didyma* (Esper, 1788)
Small Dotted Buff	*Photedes minima* (Haworth, 1809)
Small Wainscot	*Photedes pygmina* (Haworth, 1809)
Flounced Rustic	*Luperina testacea* (Denis & Schiffermüller, 1775)
Large Ear	*Amphipoea lucens* (Freyer, 1845)
Crinan Ear	*Amphipoea crinanensis* (Burrows, 1908)
Ear Moth	*Amphipoea oculea* (Linnaeus, 1761)
Rosy Rustic	*Hydraecia micacea* (Esper, 1789)
Frosted Orange	*Gortyna flavago* (Denis & Schiffermüller, 1775)
Haworth's Minor	*Celaena haworthii* (Curtis, 1829)
Crescent	*Celaena leucostigma* (Hübner, 1808)
Bulrush Wainscot	*Nonagria typhae* (Thunberg, 1784)
Rush Wainscot	*Archanara algae* (Esper, 1789)
Large Wainscot	*Rhizedra lutosa* (Hübner, 1803)
Treble Lines	*Charanyca trigrammica* (Hufnagel, 1766)
Uncertain	*Hoplodrina alsines* (Brahm, 1787)

Rustic	*Hoplodrina blanda* (Denis & Schiffermüller, 1775)
Vine's Rustic	*Hoplodrina ambigua* (Denis & Schiffermüller, 1775)
Small Mottled Willow	*Spodoptera exigua* (Hübner, 1808)
Mottled Rustic	*Caradrina morpheus* (Hufnagel, 1766)
Pale Mottled Willow	*Caradrina clavipalpis* (Scopoli, 1763)
Anomalous	*Stilbia anomala* (Haworth, 1809)

Heliothinae
Bordered Sallow	*Pyrrhia umbra* (Hufnagel, 1766)
Scarce Bordered Straw	*Helicoverpa armigera* (Hübner, 1808)
Bordered Straw	*Heliothis peltigera* (Denis & Schiffermüller, 1775)

Eustrotiinae
Marbled White Spot	*Protodeltote pygarga* (Hufnagel, 1766)
Silver Hook	*Deltote uncula* (Clerck, 1759)

Chloephorinae
Green Silver-lines	*Pseudoips prasinana* (Linnaeus, 1758)
Oak Nycteoline	*Nycteola revayana* (Scopoli, 1772)

Pantheinae
Nut-tree Tussock	*Colocasia coryli* (Linnaeus, 1758)

Plusiinae
Slender Burnished Brass	*Thysanoplusia orichalcea* (Fabricius, 1775)
Burnished Brass	*Diachrysia chrysitis* (Linnaeus, 1758)
Golden Plusia	*Polychrysia moneta* (Fabricius, 1787)
Gold Spot	*Plusia festucae* (Linnaeus, 1758)
Lempke's Gold Spot	*Plusia putnami* (Grote, 1873)
Silver Y	*Autographa gamma* (Linnaeus, 1758)
Beautiful Golden Y	*Autographa pulchrina* (Haworth, 1809)
Plain Golden Y	*Autographa jota* (Linnaeus, 1758)
Gold Spangle	*Autographa bractea* (Denis & Schiffermüller, 1775)
Scarce Silver Y	*Syngrapha interrogationis* (Linnaeus, 1758)
Dark Spectacle	*Abrostola triplasia* (Linnaeus, 1758)
Spectacle	*Abrostola tripartita* (Hufnagel, 1766)

Catocalinae
Clifden Nonpareil	*Catocala fraxini* (Linnaeus, 1758)
Mother Shipton	*Callistege mi* (Clerck, 1759)
Burnet Companion	*Euclidia glyphica* (Linnaeus, 1758)

Ophiderinae
Herald	*Scoliopteryx libatrix* (Linnaeus, 1758)
Small Purple-barred	*Phytometra viridaria* (Clerck, 1759)

Rivulinae
Straw Dot	*Rivula sericealis* (Scopoli, 1763)

Hypeninae
Beautiful Snout	*Hypena crassalis* (Fabricius, 1787)
Snout	*Hypena proboscidalis* (Linnaeus, 1758)

Strepsimaninae
White-line Snout	*Schrankia taenialis* (Hübner, 1809)
Pinion-streaked Snout	*Schrankia costaestrigalis* (Stephens, 1834)
Marsh Oblique-barred	*Hypenodes humidialis* Doubleday, 1850

Herminiinae
Fan-foot	*Zanclognatha tarsipennalis* Treitschke, 1835
Small Fan-foot	*Herminia grisealis* (Denis & Schiffermüller, 1775)

Species removed from the checklist or doubtfully recorded in Northern Ireland.

The following species were included on the Northern Ireland checklist but have been removed by the (NILRC) due to insufficient information, lack of credible evidence or in some cases voucher specimens.

Thrift Clearwing	*Bembecia muscaeformis* (Esper)
Satin Wave	*Idea subsericeata* (Haworth)
Heath Rivulet	*Perizoma minorata* (Treitschte)
Manchester Treble-bar	*Carsia sororiata* (Hübner)
Straw Belle	*Aspitates gilvaria* (Denis & Schiffermüller)
Poplar Kitten	*Furcula bifida* (Brahm)
Dotted Rustic	*Rhyacia simulans* (Hufnagel)
Fen Square-spot	*Diarsia florida* (Schmidt)
Sweet Gale	*Acronicta euphorbiae* (Denis & Schiffermüller)
Lunar Yellow Underwing	*Noctua orbona* (Hufnagel)

Potential addition of the September Thorn *Ennomos erosaria* (Denis & Schiffermüller) to the Northern Irish checklist

The fully mature larva illustrated in the photograph below was found on a small birch tree in Peatlands Park (Armagh) in July 2001. The caterpillar was retained but died during the pupal stage. The identification was confirmed from the photograph by Paul Waring and Jim Porter. However, in the absence of a certain adult record, the presence of September Thorn in Northern Ireland cannot be confirmed and so the species has not been included on the checklist. There have been only three Irish records, all from Killarney (Kerry). Adults closely resemble August Thorn which is common at Peatlands Park, and distinguishing the two species requires careful examination. September Thorn is found in broadleaved woodland. The larval foodplants include birch and oak.

Opposite page
Eyed Hawkmoth.

Annagarriff Wood National Nature Reserve, Peatlands Park, County Armagh.

This attractive hawkmoth is easily identified by the large eyespots on its hindwings, which are revealed when threatened.

Left
September Thorn larva.

Peatlands Park, County Armagh.

This final instar larva was discovered while searching a small birch tree for larvae. It was photographed in situ.

Bibliography

Allen, D. & Thompson, R. 2001. Significant records of Macro-moths in Northern Ireland (1990-February 2001). *Atropos* 14: 35-41.

Allen, P.B.M. 1980. *Leaves from a Moth-hunter's Notebooks.* E.W. Classey, Oxfordshire.

Asher, J., Warren, M., Fox, R., Harding, P., Jeffcoate G. & Jeffcoate, S. 2001. *The Millennium Atlas of Butterflies in Britain and Ireland.* Oxford University Press, Oxford.

Baynes, E.S.A. 1964. *A Revised Catalogue of Irish Macrolepidoptera (Butterflies and Moths).* E.W. Classey, Hampton.

Baynes, E.S.A. 1970. *A Supplement to a Revised Catalogue of Irish Macrolepidoptera (Butterflies and Moths).* E.W. Classey, Hampton.

Beirne, B.P. 1985. Irish Entomology: the first hundred years. *Irish Naturalists' Journal Special Entomological Supplement.*

Bond, K.G.M., Nash, R. & O'Connor, J.P. 2006. *An Annotated Checklist of the Irish Butterflies and Moths (Lepidoptera).* The Irish Biogeographical Society and the National Museum of Ireland, Dublin.

Bradley, J.D. 2000. *Checklist of Lepidoptera recorded from the British Isles.* Second edition (revised). D.J. Bradley & M.J. Bradley, Fordingbridge & Newent.

Campbell, D.C. 1893. The Macro-lepidoptera of the Londonderry District. *Irish Naturalist* 2: 19-22, 43-46, 72-74 & 147.

Campbell, D.C. 1906. Entomological notes from Londonderry. *Irish Naturalist* 15: 44-45.

Donovan, C. 1936. *A catalogue of the Macrolepidoptera of Ireland.* E.J. Burrow, Cheltenham and London.

Foster, G. 1932. Lepidoptera of County Down. *Proceedings of the Belfast Naturalists' Field Club* (Series 2) 9 (Appendix 5): 63-92

Fox, R., Asher, J., Brereton, T., Roy, D. & Warren, M. 2006. *The State of Butterflies in Britain and Ireland.* Pisces Publications, Newbury.

Fox, R., Conrad, K.F., Parsons, M.S., Warren, M.S. & Woiwood, I.P. 2006. *The State of Britain's Larger Moths.* Butterfly Conservation and Rothamsted Research, Wareham.

Fry, R. & Waring, P. 2001. *A guide to moth traps and their use. Amateur Entomologist* 24. (2nd ed.) Amateur Entomologists' Society, Orpington.

Greer, T. 1920-21. The Macro-Lepidoptera of County Tyrone. *Entomologist* 53: 217-221 & 274-277; 54: 31-35, 113-116, 206-208, 258-261 & 282-285.

Greer, T. 1922-1923. The Lepidoptera of the North of Ireland. *Proceedings of the Belfast Naturalists' Field Club* 8 (Appendix 4): 1-60.

Heal, H.G. 1965. The Wood White *Leptidea sinapis* and the railways. *Irish Naturalists' Journal* 15: 8-13.

Heath, J. & Emmet, A.M. (eds) 1976 to present. *The Moths and Butterflies of Great Britain and Ireland.* Volumes 1-10. Harley Books, Colchester.

Hickin, N. 1992. *The Butterflies of Ireland. A field guide.* Roberts Rinehart Publishers, Schull.

Johnson, W.F. 1902. Lepidoptera. In *A Guide to Belfast and the Counties of Down and Antrim.* pp. 199-204. Belfast Naturalists' Field Club and McCaw, Stevenson & Orr, Belfast.

Kane, W.F. de V. 1901. *A catalogue of the Lepidoptera of Ireland*. West, Newman & Co., London.

Leverton, R. 2001. *Enjoying Moths*. T&D Poyser, London.

Magurran, A.E. 1985. The diversity of Macrolepidoptera in two contrasting woodland habitats at Banagher, Northern Ireland. *Proceedings of the Royal Irish Academy* **85B**: 121-132.

Nairn, R.W.G. 1977. Lepidoptera. *Murlough Nature Reserve Scientific Report* **1977**: 104-109.

Nelson, B., Hughes, M., Nash, R. & Warren, M. 2001. *Leptidea reali* Reissinger 1989 (Lep.: Pieridae): a butterfly new to Britain and Ireland. *Entomologist's Record & Journal of Variation* **113**: 97-101.

O'Connor, J.P., Ashe, P. & Walsh, J. 2005. *First supplement to a Bibliography of Irish Entomology*. Irish Biogeographical Society and National Museum of Ireland, Dublin.

Pittaway, A.R. 1993. *The Hawkmoths of the Western Palaearctic*. Harley Books, Colchester.

Porter, J. 1997. *Colour Identification Guide to Caterpillars of the British Isles*. Viking, London.

Praeger, R.L. 1949. *Some Irish Naturalists*. Dundalgan Press, Dundalk.

Riley, A.M. & Prior, G. 2003. *British and Irish Pug Moths – a guide to their identification and biology*. Harley Books, Colchester.

Rippey, I. 1986. The Butterflies of Northern Ireland. *Irish Naturalists' Journal* **22**: 133-140.

Rippey, I. 1989. Butterflies (Lepidoptera) in Northern Ireland 1986-1987: additional records and corrigenda. *Irish Naturalists' Journal* **23**: 27-30.

Ryan, J.G., O'Connor, J.P. & Beirne, B.P. 1984. *A Bibliography of Irish Entomology*. The Flyleaf Press, Dublin.

Skinner, B. 1984. *Colour Identification Guide to Moths of the British Isles*. Viking, London.

Thomas, J. & Lewington, R. 1991. *The Butterflies of Britain and Ireland*. Dorling Kindersley, London.

Thompson, R. 2002. *Close-up on Insects – A Photographer's Guide*. Guild of Master Craftsman Publications, Lewes.

Thompson, R. 2005. *Close-up and Macro – A Photographer's Guide*. David and Charles, Newton Abbbot.

Tutt, J.W. 1994. *Practical Hints for the Field Lepidopterist*. Facsimile Edition. The Amateur Entomologists' Society, Orpington.

Waring, P. 1994. Moth traps and their use. *British Wildlife* **5**: 137-148.

Waring, P. & Townsend, M. 2003. *Field Guide to the Moths of Great Britain and Ireland*. British Wildlife Publishing, Gillingham.

Watts, C.W. 1894a. Lepidoptera in the Belfast District. *Entomologist's Monthly Magazine* **30**: 12-13.

Watts, C.W. 1894b. Lepidoptera in the Belfast District in 1893. *Irish Naturalist* **3**: 44-45.

Watts, C.W. 1894c. Lepidoptera in the Belfast District. *Proceedings of the Belfast Naturalists' Field Club* **4**, Appendix II (4): 115-131.

Wright, W.S. 1954. Some Macrolepidoptera from Co. Antrim, Northern Ireland. *Entomologist's Gazette* **5**: 237-242.

Wright, W.S. 1955. Some Macrolepidoptera from Co. Down, Northern Ireland. *Entomologist's Gazette* **6**: 228-230.

Wright, W.S. 1956. Some Macrolepidoptera from Co. Armagh, Northern Ireland. *Entomologist's Gazette* **7**: 83-85.

Wright, W.S. 1964. *The Macro-lepidoptera of Northern Ireland*. Publication No. 169. Ulster Museum, Belfast.

Young, M. 1997. *The Natural History of Moths*. T&D Poyser, London.

Appendices

Appendix 1   Plant Species Index

Plant species mentioned in the text with their scientific names.

Common name	Scientific name	Common name	Scientific name
Alder	*Alnus glutinosa*	Cow Parsley	*Anthriscus sylvestris*
Alder Buckthorn	*Frangula alnus*	Cowberry	*Vaccinium vitis-idaea*
Annual Meadow-grass	*Poa annua*	Crab Apple	*Malus sylvestris*
Apple	*Malus domestica*	Creeping Cinquefoil	*Potentilla reptans*
Ash	*Fraxinus excelsior*	Creeping Willow	*Salix repens*
Aspen	*Populus tremula*	Cross-leaved Heath	*Erica tetralix*
Beard lichen	*Usnea* species	Cuckoo-flower	*Cardamine pratensis*
Bedstraw	*Galium* species	Cypress	*Cupressus* species
Beech	*Fagus sylvatica*	Dandelion	*Taraxacum officinale*
Bell Heather	*Erica cinerea*	Deadly Nightshade	*Atropa belladonna*
Bilberry	*Vaccinium myrtillus*	Devil's-bit Scabious	*Succisa pratensis*
Birch	*Betula* species	Dock	*Rumex* species
Bird's-foot Trefoil	*Lotus corniculatus*	Dog lichen	*Peltigera* species
Bitter Vetch	*Lathyrus linifolius*	Dog Rose	*Rosa canina*
Black Currant	*Ribes nigrum*	Douglas Fir	*Pseudotsuga menziesii*
Blackthorn	*Prunus spinosa*	Eared Willow	*Salix aurita*
Bladder Campion	*Silene vulgaris*	Early Hair-grass	*Aira praecox*
Bluebell	*Hyacinthoides non-scripta*	Elder	*Sambucus nigra*
Blue Moor-grass	*Sesleria caerulea*	English Elm	*Ulmus procera*
Bog Myrtle	*Myrica gale*	European Larch	*Larix decidua*
Bogbean	*Menyanthes trifoliata*	Eyebright	*Euphrasia* species
Bottle Sedge	*Carex rostrata*	False Brome	*Brachypodium sylvaticum*
Bracken	*Pteridium aquilinum*	Fern	Pteridophyta
Bramble	*Rubus fruticosus*	Field Mouse-ear	*Cerastium arvense*
Broom	*Cytisus scoparius*	Field Scabious	*Knautia arvensis*
Buddleia	*Buddleja davidii*	Foxglove	*Digitalis purpurea*
Bugle	*Ajuga reptans*	Fuchsia	*Fuchsia magellanica*
Bulrush	*Typha latifolia*	Garden Marigold	*Calendula* species
Bur-reed	*Sparganium* species	Garlic Mustard	*Alliara petiolata*
Burnet Rose	*Rosa pimpinellifolia*	Goldenrod	*Solidago virgaurea*
Burnet-saxifrage	*Pimpinella saxifraga*	Gooseberry	*Ribes uva-crispa*
Buttercup	*Ranunculus* species	Goosefoot	*Chenopodium* species
Cabbage	*Brassica oleracea*	Gorse	*Ulex europaeus*
Chamomile	*Chamaemelum nobile*	Grapevine	*Vitis vinifera*
Cherry	*Prunus* species	Greater Plantain	*Plantago major*
Chickweed	*Stellaria* species	Greater Willowherb	*Epilobium hirsutum*
Clematis	*Clematis* species	Great Wood-rush	*Luzula sylvatica*
Clover	*Trifolium* species	Grey Willow	*Salix cinerea*
Cock's Foot	*Dactylis glomerata*	Ground-ivy	*Glechoma hederacea*
Common Club-rush	*Schoenoplectus lacustris*	Groundsel	*Senecio vulgaris*
Common Cottongrass	*Eriophorum angustifolium*	Harebell	*Campanula rotundifolia*
Common Couch	*Elytrigia repens*	Hare's-tail Cottongrass	*Eriophorum vaginatum*
Common Cow-wheat	*Melampyrum pratense*	Hawk's-beard	*Crepis* species
Common Hemp-nettle	*Galeopsis tetrahit*	Hawkweed	*Hieracium* species
Common Knapweed	*Centaurea nigra*	Hawthorn	*Crataegus monogyna*
Common Milkwort	*Polygala vulgaris*	Hazel	*Corylus avellana*
Common Nettle	*Urtica dioica*	Heather	*Calluna vulgaris*
Common Ragwort	*Senecio jacobaea*	Hedge Bedstraw	*Galium mollugo*
Common Reed	*Phragmites australis*	Hogweed	*Heracleum sphondylium*
Common Restharrow	*Ononis repens*	Holly	*Ilex aquifolium*
Common Sorrel	*Rumex acetosa*	Honeysuckle	*Lonicera periclymenum*
Common Toadflax	*Linaria vulgaris*	Hop	*Humulus lupulus*
Common Valerian	*Valeriana officinalis*	Horseshoe Vetch	*Hippocrepis comosa*
Couch	*Elytrigia* species	Horsetail	Equisetaceae

Hound's-tongue	*Cynoglossum officinale*	Rosebay Willowherb	*Chamerion angustifolium*
Ivy	*Hedera helix*	Rowan	*Sorbus aucuparia*
Juniper	*Juniperus communis*	Rush	*Juncus* species
Kidncy Vetch	*Anthyllis vulneraria*	Saxifrage	*Saxifraga* species
Knapweed	*Centaurea* species	Scentless Mayweed	*Tripleurospermum inodorum*
Knotgrass	*Polygonum aviculare*	Scots Pine	*Pinus sylvestris*
Lady's-mantle	*Alchemilla* species	Sea Aster	*Aster tripolium*
Larch	*Larix* species	Sea Bindweed	*Calystegia soldanella*
Large Bird's-foot Trefoil	*Lotus pedunculatus*	Sea Buckthorn	*Hippophae rhamnoides*
Lavender	*Lavandula* species	Sea Campion	*Silene uniflora*
Lesser Bulrush	*Typha angustifolia*	Sea Sandwort	*Honckenya peploides*
Lesser Celandine	*Ranunculus ficaria*	Sedge	*Carex* species
Lilac	*Syringa vulgaris*	Sheep's Fescue	*Festuca ovina*
Lime	*Tilia* species	Sheep's Sorrel	*Rumex acetosella*
Lousewort	*Pedicularis sylvatica*	Silver Fir	*Abies alba*
Marjoram	*Origanum majorana*	Sitka Spruce	*Picea sitchensis*
Marram	*Ammophila arenaria*	Smooth Hawk's-beard	*Crepis capillaris*
Marsh Ragwort	*Senecio aquaticus*	Sneezewort	*Achillea ptarmica*
Marsh Violet	*Viola palustris*	Sorrel	*Rumex* species
Marsh Willowherb	*Epilobium palustre*	Sow-thistle	*Sonchus* species
Mat-grass	*Nardus stricta*	Spindle	*Euonymus europaeus*
Meadow Vetchling	*Lathyrus pratensis*	Spruce	*Picea* species
Meadowsweet	*Filipendula ulmaria*	St. John's-wort	*Hypericum* species
Michaelmas-daisy	*Aster* species	Stitchwort	*Stellaria* species
Milkweed	*Asclepias* species	Stonecrop	*Sedum* species
Mossy Saxifrage	*Saxifraga hypnoides*	Swede	*Brassica napobrassica*
Mouse-ear	*Cerastium* species	Sycamore	*Acer pseudoplatanus*
Mugwort	*Artemisia vulgaris*	Teasel	*Dipsacus fullonum*
Nasturtium	*Tropaeolum majus*	Thistle	*Cirsium* and *Carduus* species
Noble Fir	*Abies procera*	Thrift	*Armeria maritima*
Norway Spruce	*Picea abies*	Traveller's-joy	*Clematis vitalba*
Oak	*Quercus* species	Trefoil	*Lotus* species
Oleander	*Nerium oleander*	Tufted Hair-grass	*Deschampsia caespitosa*
Orache	*Atriplex* species	Turnip	*Brassica rapa*
Pear	*Pyrus communis*	Vetch	*Vicia* species
Periwinkle	*Vinca* species	Violet	*Viola* species
Petunia	*Petunia* species	Viper's Bugloss	*Echium vulgare*
Pignut	*Conopodium majus*	Virginia Creeper	*Parthenocissus quinquefolia*
Pine	*Pinus* species	Water Mint	*Mentha aquatica*
Plantain	*Plantago* species	Wavy Hair-grass	*Deschampsia flexuosa*
Poplar	*Populus* species	White Campion	*Silene latifolia*
Potato	*Solanum tuberosum*	White Clover	*Trifolia repens*
Primrose	*Primula vulgaris*	Wild Angelica	*Angelica sylvestris*
Purging Buckthorn	*Rhamnus cathartica*	Wild Carrot	*Daucus carota*
Purple Loosestrife	*Lythrum salicaria*	Wild Cherry	*Prunus avium*
Purple Moor-grass	*Molinia caerulea*	Wild Marjoram	*Origanum vulgare*
Purple Saxifrage	*Saxifraga oppostifolia*	Wild Privet	*Ligustrum vulgare*
Purple Small-reed	*Calamagrostis canescens*	Wild Thyme	*Thymus polytrichus*
Ragged Robin	*Lychnis flos-cuculi*	Willow	*Salix* species
Ragwort	*Senecio* species	Willowherb	*Epilobium* species
Raspberry	*Rubus idaeus*	Wood Anemone	*Anemone nemorosa*
Red Bartsia	*Odontites vernus*	Wood Sage	*Teucrium scorodonia*
Red Campion	*Silene dioica*	Wood Small-reed	*Calamagrostis epigejos*
Red Clover	*Trifolium pratense*	Woodruff	*Galium odoratum*
Red Currant	*Ribes rubrum*	Wych Elm	*Ulmus glabra*
Red Valerian	*Centranthus ruber*	Yarrow	*Achillea millefolium*
Reed Canary Grass	*Phalaris arundinacea*	Yellow Iris	*Iris pseudacorus*
Rhododendron	*Rhododendron ponticum*	Yellow Mountain Saxifrage	*Saxifraga aizoides*
Ribwort Plantain	*Plantago lanceolata*	Yellow-rattle	*Rhinanthus minor*
Rock Sea-spurrey	*Spergularia rupicola*	Yew	*Taxus baccata*
Rose	*Rosa* species		

The Butterflies and Moths of Northern Ireland

Map showing the Irish counties.

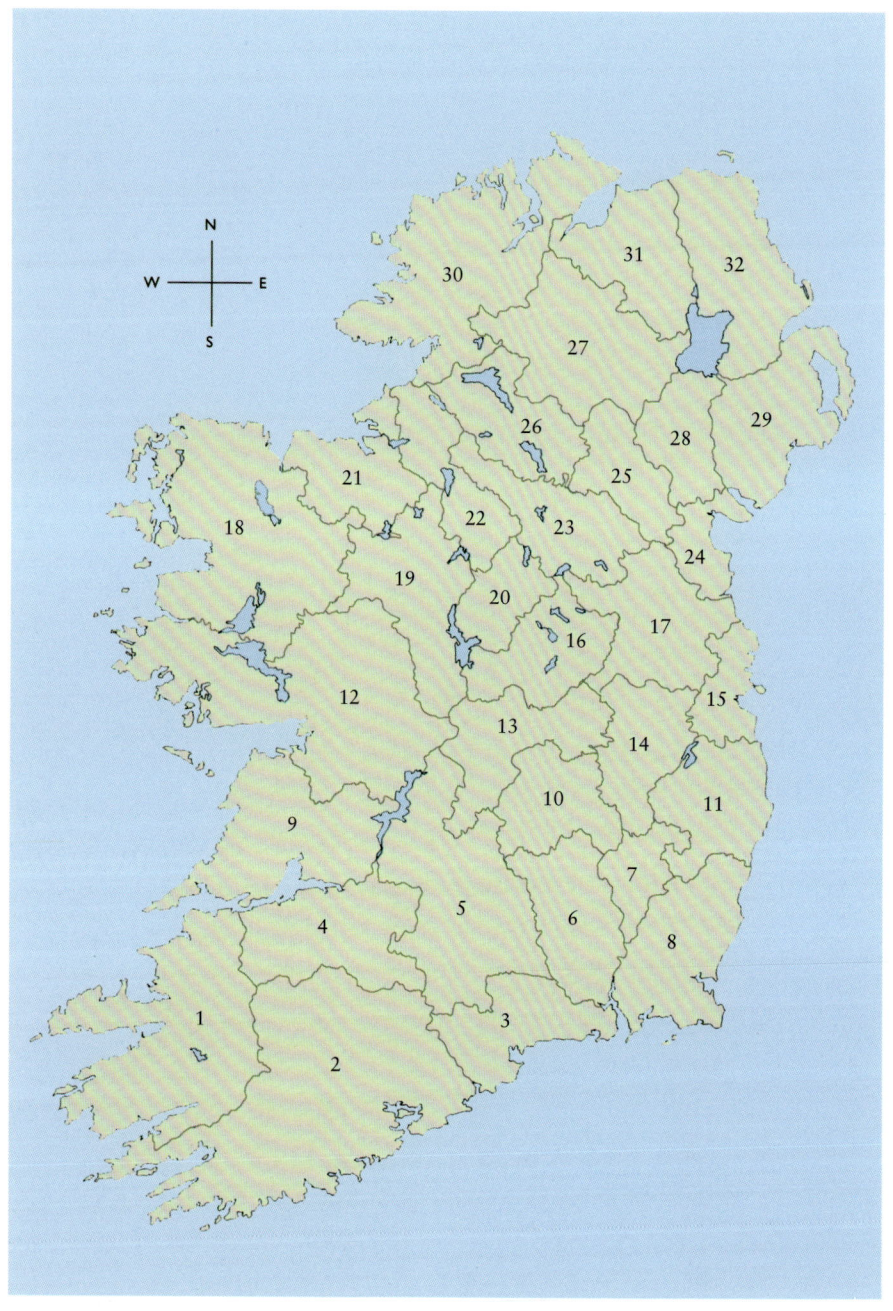

Key

1 Kerry	9 Clare	17 Meath	25 Monaghan
2 Cork	10 Laois	18 Mayo	26 Fermanagh
3 Waterford	11 Wicklow	19 Roscommon	27 Tyrone
4 Limerick	12 Galway	20 Longford	28 Armagh
5 Tipperary	13 Offaly	21 Sligo	29 Down
6 Kilkenny	14 Kildare	22 Leitrim	30 Donegal
7 Carlow	15 Dublin	23 Cavan	31 Londonderry
8 Wexford	16 Westmeath	24 Louth	32 Antrim

Appendix 2 Place Names Index

The list includes all the Northern Ireland place names referred to in the text with the county and four figure grid reference, where applicable, a guide to location.

Acton Glebe J0540	Armagh	Ballyteigue S9605	Wexford
Aghalane H3419	Fermanagh	Ballywalter J6269	Down
Aghalee J1265	Antrim	Banagher Glen C6606	Londonderry
Ahoghill D0401	Antrim	Banbridge J1244	Down
Altaclanabryan H4538	Fermanagh	Bangor J5182	Down
Altadavan H5949	Tyrone	Barmouth C7836	Londonderry
Altadiawan H5949	Tyrone	Bealnaslaght Bridge H7189	Londonderry
Altaveedan H4946	Tyrone	Beihy H1630	Fermanagh
Altdrumman H5677	Tyrone	Belcoo H0838	Fermanagh
Altnaheglish Forest C7003	Londonderry	Belfast Castle J3279	Antrim
Altnaveigh J0525	Armagh	Belfast Harbour Estate J3777	Down
Annagarriff Wood NNR H9061	Armagh	Belfast J3474	Antrim
Annalong J3719	Down	Bellaghy H9696	Londonderry
Annalong Valley J3423	Down	Belleek G9459	Fermanagh
Annaloughan Bog H5755	Tyrone	Belleisle H2835	Fermanagh
Ardess H2063	Fermanagh	Belmore Mountain H1440	Fermanagh
Ardglass J5537	Down	Belshaw's Quarry J2468	Antrim
Ardress H9155	Armagh	Belvoir Park J3469	Down
Ardress House H9155	Armagh	Benaughlin H1731	Fermanagh
Ardunshin H3742	Fermanagh	Benbradagh C7211	Londonderry
Argory Moss H8757	Armagh	Benburb H8152	Tyrone
Argory, the H8758	Armagh	Benone C7135	Londonderry
Armagh City H8745	Armagh	Bessbrook J0428	Armagh
Aughalun H3841	Fermanagh	Binevenagh C6930	Londonderry
Aughentaine H4651	Tyrone	Binevenagh Forest C6930	Londonderry
Aughinlig H8654	Armagh	Black Lough H4840	Fermanagh
Aughintober J0935	Armagh	Blackhill Bog H3647	Fermanagh
Aughnacloy H6652	Tyrone	Blaney H1652	Fermanagh
Aughnadarragh Lough J4459	Down	Bleanalung Bay H0261	Fermanagh
Aughrim J2818	Antrim	Bloody Bridge J3826	Down
Ballinderry J1267	Antrim	Boa Island H0862	Fermanagh
Ballycastle D1140	Antrim	Bohill Forest J3946	Down
Ballydorn J5263	Down	Boho H1244	Fermanagh
Ballygalley D3707	Antrim	Bracaghreilly C8000	Londonderry
Ballygalley Head D3807	Antrim	Brackagh Moss J0251	Armagh
Ballygowan J4163	Down	Breckenhill Dam J2495	Antrim
Ballykinler Dunes J4234	Down	Breen Wood D1233	Antrim
Ballylesson J3367	Down	Breesha Plantation D1941	Antrim
Ballymacashen Bog J4660	Down	Brookeborough H3840	Fermanagh
Ballymacilrany J1166	Antrim	Bushmills C9440	Antrim
Ballymaclary C7035	Londonderry	Caledon Estate H7543	Tyrone
Ballymacormick Point J5383	Down	Camlough J0326	Armagh
Ballymagorry C3601	Tyrone	Canary H8656	Armagh
Ballymena D1811	Antrim	Capanagh D3003	Antrim
Ballymoney C9425	Antrim	Carganamuck H8749	Armagh
Ballynacor J0358	Armagh	Carlingford Lough J11	Down
Ballynagilly H7484	Tyrone	Carmacullagh H4533	Fermanagh
Ballynakill Lake L6458	Galway	Carn J0356	Armagh
Ballynasollus H7383	Tyrone	Carnagat Forest H4552	Tyrone
Ballynewry Lough H9743	Armagh	Carndaisy H8286	Tyrone
Ballyquintin J6245	Down	Carne J0256	Armagh
Ballyquintin Point J6245	Down	Carnlough D2818	Antrim
Ballyrea H8444	Armagh	Carnmore H4735	Fermanagh
Ballyreagh Bog H3048	Fermanagh	Carnmore Lough H4735	Fermanagh
Ballyreagh H3839	Fermanagh	Carrick Lough H0953	Fermanagh
Ballyrobert J4481	Down	Carrickfergus J4187	Antrim

Location	County
Carrickfergus Marina J4187	Antrim
Carricknabrattoge H4234	Fermanagh
Carryduff J3665	Down
Castle Archdale H1860	Fermanagh
Castle Caldwell H0260	Fermanagh
Castle Coole H2643	Fermanagh
Castle Espie J4967	Down
Castlereagh J3771	Down
Castlerock C7736	Londonderry
Castleward J5749	Down
Castlewellan Forest J3237	Down
Castlewellan J3436	Down
Cavans Bog J4047	Fermanagh
Cave Hill J3179	Antrim
Churchill H8860	Armagh
Churchtown C8115	Londonderry
Clandeboye J4879	Down
Clare Glen J0244	Armagh
Clare Wood D0641	Antrim
Cleen H4446	Fermanagh
Cleggan South, Rathlin Island D1151	Antrim
Cleggan, Rathlin Island D1151	Antrim
Cliffs of Magho H0757	Fermanagh
Clogher H5451	Tyrone
Coagh H9079	Tyrone
Coalisland H8764	Tyrone
Colebrooke H4144	Fermanagh
Coleraine C8432	Londonderry
Colin Glen J2671	Antrim
Comber J4669	Down
Conagher Forest H0452	Fermanagh
Coneyglen H6087	Tyrone
Conlig J4976	Down
Cookstown H8178	Tyrone
Coolcush H8260	Tyrone
Cooneen H4542	Fermanagh
Cooraghy Bay D1050	Antrim
Copeland Bird Observatory J5985	Down
Copeland Island J5883	Down
Copeland Islands J58	Down
Corbet Fen J1846	Down
Corbet Lough J1844	Down
Corbet's Wood J4436	Down
Correl Glen H0854	Fermanagh
Cottage Farm H4380	Tyrone
Craigavon Lakes J0557	Armagh
Craigs H7780	Tyrone
Cranfield J2610	Down
Crawfordsburn Country Park J4682	Down
Crawfordsburn J4681	Down
Creeve Lough H7351	Tyrone
Creighton's Wood J4278	Londonderry
Crilly H6949	Tyrone
Crockacleavan Lough H4842	Tyrone
Crockanaver H2263	Fermanagh
Crockcor Quarry C7902	Londonderry
Crom H3524	Fermanagh
Cromore C8337	Londonderry
Crossgar J4552	Down
Cuilcagh Mountain H1228	Fermanagh
Culcrow C8923	Londonderry
Cullinane H4748	Tyrone
Curran Bog H8795	Londonderry
Cushendall D2327	Antrim
Cushendun D2432	Antrim
Cushleake Mountain D2236	Antrim
Dartry Lodge H8553	Armagh
Davagh Forest H7383	Tyrone
Derry City C4317	Londonderry
Derryadd Lough H9160	Armagh
Derryboye J4756	Down
Derrygonnelly H1251	Fermanagh
Derryhubbert H9060	Armagh
Derryleckagh Fen J1125	Down
Derryleckagh Wood J1225	Down
Derrylin H2827	Fermanagh
Derryola Bridge J0968	Antrim
Derryvore H3423	Fermanagh
Divis J2875	Antrim
Divis Mountain J2875	Down
Donard Demesne J3729	Down
Donard Wood J3629	Down
Donemana C4403	Londonderry
Doon Forest H4536	Fermanagh
Downpatrick J4844	Down
Dresternan Lough H5032	Fermanagh
Drum Lough J1932	Down
Drum Manor H7677	Tyrone
Drumadion J1931	Antrim
Drumask H8553	Armagh
Drumawhy J5475	Down
Drumbanagher J0536	Armagh
Drumbrughas North H3435	Fermanagh
Drumherriff H9152	Armagh
Drumkeeragh J3346	Down
Drumman More H8946	Armagh
Drumman More Lough H8946	Armagh
Drumnahavil (Drumcarn) H8128	Armagh
Drumquin J3274	Tyrone
Duncrun C8632	Londonderry
Dundonald J4273	Down
Dundrum J4038	Down
Dungannon H7962	Tyrone
Dunmurry J2968	Antrim
Dunseverick C9844	Antrim
East Light, Rathlin Island D1552	Antrim
Ecos Park D1203	Antrim
Ederney H2265	Fermanagh
Ely Lodge H1950	Fermanagh
Enniskillen Agricultural College H2447	Fermanagh
Enniskillen H2344	Fermanagh
Eshywulligan H4838	Fermanagh
Fair Head D1843	Antrim
Fardross Forest H4946	Tyrone
Favour Royal H6051	Tyrone
Finaghy J3069	Antrim
Fivemiletown H4447	Tyrone
Florencecourt H1834	Fermanagh
Fofanny Dam J2829	Down
Forkhill J0115	Armagh
Fox's Hill Bog H3049	Fermanagh
Garrison G9451	Fermanagh
Garron Point D2924	Antrim
Garry Bog C9430	Antrim

Garvary Wood H0161	Fermanagh	Knockaraven H3464	Tyrone
Gawleys Gate J0867	Antrim	Knockbarragh J1823	Down
Giant's Causeway C9545	Antrim	Knockmore H9050	Fermanagh
Gilford J0648	Down	Knocknalosset H4534	Fermanagh
Gilnahirk J4072	Down	Knockninny H2730	Fermanagh
Glasdrumman J3722	Down	Knocktarna C8729	Londonderry
Glen Wood H1733	Fermanagh	Lackan Bog J2337	Down
Glenariff Forest D2019	Antrim	Lagan Meadows J3369	Down
Glenarm D3015	Antrim	Larne D4002	Antrim
Glengad C9120	Antrim	Leathemstown Reservoir J2172	Antrim
Glengarriff V9057	Cork	Legacurragh H1530	Fermanagh
Glenkeen H7049	Tyrone	Legalough H0834	Fermanagh
Glenmakeeran D1636	Antrim	Legatillida H4539	Fermanagh
Glenmore H6561	Tyrone	Leitrim Lodge J2225	Down
Glennasheevar H0553	Fermanagh	Lemnagore Wood H7543	Tyrone
Glenshane Pass C7802	Londonderry	Light House Island J5985	Down
Glenshesk D1333	Antrim	Limavady C6723	Londonderry
Gobbins, the J4897	Antrim	Lisadian H8346	Armagh
Goragh Wood J0631	Armagh	Lisblake Bog H1637	Fermanagh
Gortatole H1136	Fermanagh	Lisburn J2566	Antrim
Gorteen H2823	Fermanagh	Lislasly H8753	Armagh
Gortin Glen H4985	Tyrone	Lisnabreeny J3670	Down
Gortmaconnell Rock H1330	Fermanagh	Lisnagunogue C9841	Antrim
Gosford Forest H9641	Armagh	Lissan Estate H8081	Tyrone
Greencastle J2411	Antrim	Lissan H8081	Tyrone
Grey Point J4583	Down	Lough Alaban H0743	Fermanagh
Hanging Rock H1036	Fermanagh	Lough Beg H9794	Antrim/Londonderry
Helen's Bay J4682	Down	Lough Bran C8303	Londonderry
Hillsborough Forest J2458	Down	Lough Corry H4636	Fermanagh
Hollymount J4743	Down	Lough Fea H7587	Tyrone
Holywood J4079	Down	Lough Gullion J0061	Armagh
Inish Doney H1853	Fermanagh	Lough Namanfin H0545	Fermanagh
Inishargy J6065	Down	Lough Navar Forest H0556	Fermanagh
Island Hill J4869	Down	Loughgall H9152	Armagh
Islandmagee J4699	Antrim	Loughry H8174	Tyrone
Jerrettspass J0633	Armagh	Luke's Mountain J3230	Down
Jordanstown J3584	Antrim	Lurgan J0657	Armagh
Kebble, Rathlin Island D0951	Antrim	Lusty More Island H1061	Fermanagh
Keel Point J4035	Down	Magherafelt H8990	Londonderry
Keeran Moss J3893	Tyrone	Magheramorne J4398	Antrim
Kilclief J5945	Down	Magheramorne Quarry J4398	Antrim
Kilkeel J3014	Down	Maghery H9262	Armagh
Killadeas H2053	Fermanagh	Maghery Bog J2515	Down
Killard Point NNR J6043	Down	Magho Cliffs H0557	Fermanagh
Killeter Bog H0882	Tyrone	Magilligan C63	Londonderry
Killeter Forest H0982	Tyrone	Magilligan Dunes C6638	Londonderry
Killinchy J5158	Down	Magilligan Field Centre C6733	Londonderry
Killowen Point J1815	Down	Magilligan Point C6638	Londonderry
Killybegs Bog J0899	Antrim	Maguiresbridge H3438	Fermanagh
Killycolpy Wood H9171	Tyrone	Maloon H8078	Tyrone
Killykeegan H0134	Fermanagh	Marble Arch H1234	Fermanagh
Killymoon H8276	Tyrone	Markethill H9639	Armagh
Killynick H3420	Fermanagh	Marlbank H1234	Fermanagh
Kilrea Bog C9211	Londonderry	Martinstown D1313	Antrim
Kilrea C9212	Londonderry	Megaberry J1763	Antrim
Kinarla H2045	Fermanagh	Milford H8542	Armagh
Kinnahalla J2528	Down	Mill Bay J2513	Down
Kinnegoe J0661	Armagh	Minnow Burn J3268	Down
Kinramer South D1050	Antrim	Monastir Gorge H1132	Fermanagh
Kinramer, Rathlin Island D1051	Antrim	Monawilkin H0852	Fermanagh
Kircubbin J5962	Down	Moneygal Bog H2388	Tyrone
Knockagh J3686	Antrim	Moneymore H8583	Londonderry

The Butterflies and Moths of Northern Ireland

Monlough J4064	Down	Six Road Ends J5278	Down
Monmurry H3842	Fermanagh	Sixmilecross H5667	Tyrone
Monmurry Bog H3842	Fermanagh	Skerry East D1420	Antrim
Montiaghs Moss J0965	Antrim	Slaghtfreeden H7586	Londonderry
Mount Stewart J5772	Down	Slieve Bearnagh J3128	Down
Mountgibbon Bog H3943	Fermanagh	Slieve Croob J3145	Down
Mountjoy H4178	Tyrone	Slieve Donard J3527	Down
Moy H8456	Tyrone	Slieve Gullion J0220	Armagh
Muck Island D4602	Antrim	Slieveanorra D1326	Antrim
Mullaghmore Bog H5854	Tyrone	Slievenacloy J2571	Antrim
Mullanary H8552	Antrim	Slisgarrow Quarry H0151	Fermanagh
Mullanawinna H0342	Fermanagh	Springhill H8782	Londonderry
Mullenakill Bog H9061	Armagh	Stewartstown H8571	Tyrone
Mullinure H8846	Armagh	Stillago H7956	Armagh
Muntober H7382	Tyrone	Stormont J3974	Down
Murley Mountain H4451	Tyrone	Strabane Glen H3599	Tyrone
Murlough Bay D1941	Antrim	Strabane H3497	Tyrone
Murlough NNR J3933	Down	Strand Lough J5337	Down
Navan Fort H8445	Armagh	Stranfeley H3838	Fermanagh
Navan Quarry H8545	Armagh	Strangford J5849	Down
Ness Wood C511	Londonderry	Strangford Narrows J6047	Down
Newcastle J3731	Down	Stranmillis J3470	Antrim
Newry Canal, Scarva J0642	Down	Tamnabrack D2711	Antrim
Newtownards J4974	Down	Tamnamore H8762	Tyrone
Newtownbreda J3569	Antrim	Tamniaran Bog H9294	Londonderry
Newtowncrommelin D1417	Antrim	Tandragee J0346	Tyrone
Oxford Island J0461	Armagh	Tannaghmore J0457	Armagh
Peatlands Park H8960	Armagh	Tardree Forest J1893	Antrim
Pettigo H1166	Fermanagh	Tattanellen Bog H4747	Tyrone
Pigeon Rock J2623	Down	Taylorstown J0494	Antrim
Portadown J0053	Armagh	Teal Lough H7590	Tyrone
Portaferry J5950	Down	Tempo H3548	Fermanagh
Portballintrae C9242	Antrim	Tempo Manor H3548	Fermanagh
Portglenone C9703	Antrim	Terryscollop H7954	Tyrone
Portglenone Wood C9802	Antrim	Thompson's Quarry H8346	Armagh
Portmore Lough J1168	Antrim	Tollymore Forest J3531	Down
Portora H2244	Fermanagh	Torr Head D2241	Antrim
Portrush C8540	Antrim	Traad Point H9587	Londonderry
Portstewart C8138	Londonderry	Trillick H3355	Tyrone
Portstewart Dunes C7936	Londonderry	Trory Glebe H2348	Fermanagh
Poyntzpass J0539	Armagh	Tullybrannigan J3630	Down
Quintin Bay J6350	Down	Tullychurry H0463	Fermanagh
Quoile Pondage J5047	Down	Tullyhogue H8273	Tyrone
Ranaghan Bridge C7901	Londonderry	Tullyhona H1535	Fermanagh
Randalshough H1550	Fermanagh	Tullylagan H7972	Tyrone
Randalstown Forest J0888	Antrim	Tullynacor H0541	Fermanagh
Raughlan J0631	Armagh	Tullyreagh Bridge H3744	Fermanagh
Ravella H6552	Tyrone	Turmennan Fen J4850	Down
Rea's Wood J1485	Antrim	Tyrella J4636	Down
Rehaghy Mountain H7053	Tyrone	Umbra, the C7335	Londonderry
Reilly Wood H3325	Fermanagh	Ushet, Rathlin Island D1448	Antrim
Richhill H9448	Armagh	Washing Bay H9064	Tyrone
Roe Valley C6621	Londonderry	Wellbrook H7579	Tyrone
Rostrevor J1818	Down	West Light, Rathlin Island D0951	Antrim
Rostrever Wood J1817	Down	White Mountain J3267	Antrim
Rue Point, Rathlin Island D1547	Antrim	White Park Bay D0244	Antrim
Saintfield J4059	Down	White Park H3742	Fermanagh
Seaforde J4043	Down	Whitehead J4791	Antrim
Selshion Moss H9854	Armagh	Willy Wood Island J5578	Down
Semples Bridge J3784	Tyrone	Wood Lough H7660	Tyrone
Shane's Castle J1188	Antrim	Woodford Bridge H3419	Fermanagh
Sharvogues Bog J1095	Antrim		

Appendix 3 List of Recorders

M. Abernethy, C. Acheson, G. Acheson, F. Adair, P. Adams, F. Agnew, E. Aiken, K.N. Alexander, B. Allen, D. Allen, J.E.R. Allen, M.D.B. Allen, R. Allen, R.L. Allen, D. Anderson, J. Anderson, M. Anderson, R. Anderson, C.J. Andrews, D. Andrews, J. Andrews, P. Andrews, A. Archdale, M. Archdale, J. Ard, Armagh Field Naturalists Society, G. Armstrong, M. Armstrong, R.E. Atkinson, C. Bailey, R. Bain, A. Baine, A. Baird, M. Ballagh, R. Ballentine, K.M. Bannon, F.L. Barr, H. Barton, R.L. Barton, E.S.A. Baynes, W. Beere, S. Beesley, Belfast Naturalists' Field Club, C. Bell, J. Bell, T. Bell, W. Belshaw, A. Bennington, N. Bingham, R.W. Bingham, H. Black, D. Blakely, D. Blamire, R.J. Bleakley, J. Bloomer, L.H. Bonaparte Wyse, G. Bond, K.G.M. Bond, M. Boston, D. Boyd, K.W.P. Boyd, M. Boyd, R.J.A. Boyd, T.D. Boyd, H.L.E. Boyd, M Bradley, S. Brady, I. Braund, C. Breen, D. Brennan, C.A. Brewster, J. Bristow, D. Broughton, D. Brown, J. Brown, M. Brown, P. Brown, R.A. Brown, A. Browne, D. Browne, V. Browne, E. Brownlow, W. Brush, Botanical Society of the British Isles, E. Buchanan, P. Buchanan, I.D. Bullock, J. Bullock, L. Burke, P. Burns, S. Burns, M. Bushby, E. Cahoon, J. Calladine, T.A. Cambell, C. Campbell, D.C. Campbell, G. Campbell, M. Campbell, O. Campbell, F. Carroll, P. Carruthers, L. Carson, N. Carson, V. Carter, W. Carvill, P. Casement, A. Chambers, S. Chambers, T. Chambers, W. Chambers, E.A. Chapman, P. Chester-Williams, J. Clarke, K. Clarke, M. Clarkson, C. Coates, G.N.L. Coates, B. Coburn, W. Cochrane, G.A. Cole, H. Cole-Baker, S. Cole-Baker, D. Coney, A.M. Coogan, A. Copeland, P. Corbett, P. Cormacain, P. Cornick, N. Corry, A.G.P. Cotter, D. Cotton, A. Craig, H. Craig, M. Craig, R. Craig, B. Crawford, W.M. Crawford, M. Crawley, A. Crooks, A. Crory, R. Crothers, P. Crowther, R. Cryer, W. Culbert, G. Darcy, S. D'Arcy Burt, P. Davidson, W. Davis, C. Dawson, N. Dawson, G.V. Day, C.D. Deane, J. Deery, L. Deery, E. Dempsey, M. Densley, M. Deverell, R. Devine, J. Devlin, H. Dick, G. Dickey, T. Dines, D. Doherty, H. Doherty, N. Doherty, P. Doherty, E. Donaghy, C. Donaldson, E. Donaldson, N. Donaldson, L. Donnelly, T. Donnelly, C. Donovan, E. Douglas, K. Douglas, N. Dowling, P. Dowling, R. Dowling, L. Duncan, P. Dunn, G. Dunneice, E. Dunwoody, R. Eagleson, D. Edgar, I. Enlander, J.S. Faulkner, V.M. Faulkner, R. Field, R.W. Field, A.F. Finegan, P. Finegan, J.D.F. Fisher, D. Fitzpatrick, P.M. Fogg, R.S. Forbes, A. Forde, P. Forrest-Jameson, P. Foss, A.P. Foster, Canon G. Foster, N. Foster, S. Foster, A. Fowles, D. Fox, E. Franklin, B. Fraser, J.F.D. Frazer, G. Freeburn, I. Frost, J.R. Frost, P. Fulton, R. Fulton, J.S. Furphy, J. Gamble, P. Gardiner, M. Garner, C.L. Garnett, C. Gates, J. George, F. Gibson, G. Gillespie, W. Gillespie, C. Gilmore, J. Gilpin, K. Glasgow, N. Glass, R. Glendinning, G.W. Gordon, K. Gordon, P.E. Gordon, A.V. Gore, V. Gotto, M. Graham, J.L. Graham, D. Gray, M. Gray, R. Gray, H. Greenlee, J. Greer, J. E. Greer, L. Greer, R.N. Greer, T. Greer, E. Greeves, L. Greeves, R.M. Guthrie, S.H. Guthrie, P. Hackney, R. Hadden, D. Hadrick, O. Hadrick, J.N. Halbert, P. Hale, L. Hall, B. Hamill, R.J.H. Hamill, E. Hamilton, M. Hampton, R. Hanna, P.T. Harding, D. Harrison, W.J. Harron, Y. Hart, T. Harvey, J. Haslett, D. Hawthorne, R.F. Haynes, H.G. Heal, E. Heath, G. Henderson, C. Henry, I. Herbert, M. Heyn, H. Higgins, A.G. Hill, P. Hillis, D. Hodges, P. Hodson, N.T. Holmes, A. Houston, R. Houston, D. Hughes, J. Hughes, M. Hughes, J. Hull, M. Hyndman, A. Irvine, E. Irvine, R. Irvine, A. Irwin, A.G. Irwin, I.M.C. Irwin, P. Irwin, I. Jackson, P.F. Jameson, C. Jeffs, C. Jenkins, S. Johnson, W.F. Johnson, G.A. Jones, T. Jones, R. Jordan, W.F. de V. Kane, H. Kee, L. Kennedy, W. Kennedy, A. Kernohan, L. Kernohan, P. Kernohan, C. Keys, M. Kinley, G. Kinney, P. Kirby, G. Knight, J. D. Lamont, C. Langham, D. Larmour, D. Lavery, J. Lavery, A. Lavin, M. Lee, T.A. Lee, K. Leonard, J. Leslie, R.M. Leslie, R. Liggett, J. Light, D. Looney, J. Lorimer, R.I. Lorimer, C. Loughran, E. Loughran, F. Lunny, J.W. Lyons, M. MacDougall, A. MacFadyen, P. Mackie, I. Maclaine, M. MacPolin, J. Magee, E. Maginnis, A.E. Magurran, F. Maitland, J. Malins, M. Malley, J. Manley, J.O. Mann, P. Marsh, G. Marshall, M. Marshall, M. Martin, C.J. Mason, L.J. Mason, R.G. Mathers, V. Mathers, F.C. Maxwell, D. McAllister, K. McAllister, D. McAnally, C. McAnerney, B. McAnuff, H. McBride, H. McCann, P. McCausland, J. McCleery, E. McClelland, P. McClelland, T. McCloy, S. McClure, A. McComb, D. McConnell, V. McCrae, D. McCready, P. McCrossan, L. McCulloch, J. McCurdy, W.I. McCutcheon, V. McDonagh, I.C. McDonald, D. McDonnell, R. McDowell, W. McDowell, P. McEvoy, L. McFaul, A.B McFerran, D.M. McFerran, E. McFerran, F.C. McFerran, K.M. McFerran, L.K.B. McFerran, A. McGeehan, G. McGeehan, E. McGonagle, J. McGregor, R. McGuffin, P. McHaffie, D. McIlwrath, A. McIntosh, N. McIntyre, C. McIvor, B. McKee, I. McKee, J.B. McKee, N.D. McKee, S. McKee, J.B. McKeown, C. McLarnon, J. McLauchlan, V. McLoughlin, T. McLucas, G.E. McMichael, T. McMillan, P.B. McMullan, A.S. McMullin, J. McMurray, M. McNeelly, D. McNeill, I. McNeill, P. McNulty, T. McNulty, S. McRobert, G. McRoberts, J. Medhurst, E. Meharg, M. Meharg, C. Mellon, J.L. Messenger, J. Milburne, A. Miller, R. Milligan, R.J. Milligan, R. Milliken, D. Mitchell, I. Mitchell, R. Moffit, R. Moles, B. Montgomery, J. Montgomery-Watson, D. Moody, C. Moore, P. Moore, D. Moran, R. Morell, L. Morgan, M. Morgan, D. Morris, T. Moseley, R. Mulholland, C. Mullan, C. Murphy, J. Murphy, K. Murphy, M. Murphy, S. Murphy, R.G.W. Nairn, R. Nash, National Trust, B. Nelson, P.I. Nelson, R. Nelson, L. Nesbitt, S. Nesbitt, A. Nicholson, H.D. Nicholson, J. Nimmons, D. Nixon, H. Noble, H.A.H. Northridge, H.J. Northridge, R.H. Northridge, R.M.H. Northridge, J.D. Nunn, J. O'Boyle, Omagh Wild Bird Society, J. Orr, M. Orr, M. O'Sullivan, D.F. Owen, Oxford Island NNR Staff, P. Pace, N. Parker, F. Parkinson, J. Parsons, C.E. Partridge, R.L. Patterson, E. Peel, E.C. Pelham-Clinton, J.V. Peters, G. Phillips, J.C.L. Phillips, M. Phillips, P. Phillips, R.A. Phillips, J. Piper, R. Piper, L. Playle, R. Pollen, S. Porter, W. Potts, J Preston, C.R. Price, D. Price, R. Price, A. Prior, I. Quaile, S. Rainey, M. Ramsey, E. Randall, P. Randall, D.H. Rankin, N. Rankin, F. Ratniaks, D. Rayner, S. Reeve-Smith, C.A.M. Reid, D. Reid, L. Reid, R. Reid, H. Richter, A. Riley, D.H. Riley, B. Rippey, I. Rippey, S. Rippey, T. Rippey, G. Roberts, B. Robinson, G. Robinson, B. Robson, A. Robson, E. Rolston, C. Ronayne, M. Rose, A. Ross, M. Ruddock, B. Russell, J. Rutherford, R. Rutherford, J. Samuel, G. Saunders, D. Scott, G. Scott, R. Scott, P. Scovell, R. Sellar, J. Semple, J.W.D. Semple, N. Semple, C. Sharpe, C. Shaw, L. Sheppard, R. Sheppard, G. Sheridan, M. Shields, S. Shortell, M. Simms, R. Simpson, F.W. Sinclair, B. Skinner, M. Slater, R. Smartt, J. Smith, M. Smith, U.A.M. Smith, M. Smyth, M. Snowdon, F. Somerville, C. Stamp, K. Stanfield, D. Stanley, D.K. Stanley, J.A. Steele, R. Steele, R.D. Steele, A.W. Stelfox, R. Stephenson, A. Sterritt, I. Stewart, J. Stewart, T. Stinson, W. Strahan, L. Stronge, J. Symington, R. Tate, M.R. Thomas, P. Thomlinson, D. Thompson, K. Thompson, R. Thompson, H.C. Thurgate, M. Tickner, K. Tomkins, A. Towers, P. Townsend, A. Travers, M. Turner, F. Turtle, R.S. Twist, Ulster Wildlife Trust, A. Upton, N. Veale, W.H. Veale, J. Vinycomb, W. Voles, M. Walker, N. Walsh, M. Warden, P. Waring, J.S. Warnock, F.B. Warren, A. Waterman, F. Watson, F. Watson, G. Watson, G.A. Watson, P. Watson, C.W. Watts, I. Watts, R. Weatherup, D. Weir, B. West, R.S. Weyl, J. Whatmough, J. White, J. Whitla, H. Wiggins, J. Wilde, C.G. Wilkinson, R. Williams, T.R. Williams, A. Wilson, J. Wilson, R. Wilson, I. Winters, S.A. Wolfe-Murphy, L. Wolsey, R. Wood, B. Woodall, F. Woods, S. Woods, G. Woulahan, M. Wright, W.S. Wright, J. Wyatt, R. Young.

Appendix 4 Index

Index entries are organised alphabetically. Species on the Northern Ireland checklist are indicated in bold. Page numbers in bold denote the species account and if the species is illustrated the entry is italicised.

Abraxas grossulariata see Magpie
Abraxas sylvata see Clouded Magpie
Abrostola triplasia see Dark Spectacle
Abrostola tripartita see Spectacle
Acasis viretata see Yellow-barred Brindle
Acherontia atropos see Death's Head Hawkmoth
Achlya flavicornis see Yellow Horned
Acronicta alni see Alder Moth
Acronicta euphorbiae see Sweet Gale
Acronicta leporina see Miller
Acronicta megacephala see Poplar Grey
Acronicta menyanthidis see Light Knot Grass
Acronicta psi see Grey Dagger
Acronicta rumicis see Knot Grass
Actebia praecox see Portland Moth
Adonis Blue *Polyommatus bellargus* 10
Adscita statices see Forester
Aethalura punctulata see Grey Birch
Aglais urticae see Small Tortoiseshell
Agriopis aurantiaria see Scarce Umber
Agriopis leucophaearia see Spring Usher
Agriopis marginaria see Dotted Border
Agriopis marginaria f. *fuscata see* Dotted Border
Agrius convolvuli see Convolvulus Hawkmoth
Agrochola circellaris see Brick
Agrochola helvola see Flounced Chestnut
Agrochola litura see Brown-spot Pinion
Agrochola lota see Red-line Quaker
Agrochola lychnidis see Beaded Chestnut
Agrochola macilenta see Yellow-line Quaker
Agrotis cinerea see Light Feathered Rustic
Agrotis clavis see Heart and Club
Agrotis exclamationis see Heart and Dart
Agrotis ipsilon see Dark Sword-grass
Agrotis puta see Shuttle-shaped Dart
Agrotis segetum see Turnip Moth
Agrotis trux see Crescent Dart
Agrotis vestigialis see Archer's Dart
Alcis jubata see Dotted Carpet
Alcis repandata see Mottled Beauty
Alcis repandata f. *conversaria see* Mottled Beauty
Alder Kitten *Furcula bicuspis* 10, **207**, *208*, 405
Alder Moth *Acronicta alni* **302**, *303*, 360, 368, 372, 408
Allen, J.E.R. (fl. 1897-1930) 4
Allophyes oxyacanthae see Green-brindled Crescent
Alsophila aescularia see March Moth
American Painted Lady *Vanessa virginiensis* 45
Amphipoea crinanensis see Crinan Ear
Amphipoea (Ear Moths) 314-316
Amphipoea lucens see Large Ear
Amphipoea oculea see Ear Moth
Amphipyra berbera see Svensson's Copper Underwing
Amphipyra pyramidea see Copper Underwing
Amphipyra tragopoginis see Mouse Moth
Anaplectoides prasina see Green Arches
Anarta myrtilli see Beautiful Yellow Underwing
Angle Shades *Phlogophora meticulosa* **310**, *311*, 408
Angle-barred Pug *see* Angle-barred and Ash Pug
Angle-barred and Ash Pug *Eupithecia innotata* **152**, 403

Angle-striped Sallow *Enargia paleacea* **312**, 408
Annulet *Charissa obscurata* **196**, 363, 369, 404
Anomalous *Stilbia anomala* **334**, *335*, 357, 365, 369, 373, 409
Anthocaris cardamines see Orange Tip
Anticlea badiata see Shoulder Stripe
Anticlea derivata see Streamer
Antitype chi see Grey Chi
Antler Moth *Cerapteryx graminis* **267**, 407
Apamea crenata see Clouded-bordered Brindle
Apamea crenata f. *combusta see* Clouded-bordered Brindle
Apamea epomidion see Clouded Brindle
Apamea furva see Confused
Apamea lithoxylea see Light Arches
Apamea monoglypha see Dark Arches
Apamea oblonga see Crescent Striped
Apamea ophiogramma see Double Lobed
Apamea remissa see Dusky Brocade
Apamea scolopacina see Slender Brindle
Apamea sordens see Rustic Shoulder-knot
Apamea sublustris see Reddish Light Arches
Apamea unanimis see Small Clouded Brindle
Apeira syringaria see Lilac Beauty
Aphantopus hyperantus see Ringlet
Aplocera plagiata see Treble-bar
Aporophyla lueneburgensis see Northern Deep-brown Dart
Aporophyla nigra see Black Rustic
Archanara algae see Rush Wainscot
Archer's Dart *Agrotis vestigialis* **236**, *237*, 363, 371, 406
Arctia caja see Garden Tiger
Arctiidae 222-232, 405
Argent and Sable *Rheumaptera hastata* 16, *21*, **122**, 370, 402
Argynnis aglaja see Dark Green Fritillary
Argynnis paphia see Silver-washed Fritillary
Ash Pug *see* Angle-barred and Ash Pug
Asher et al. (2001) 12
Aspitates gilvaria see Straw Belle
Asteroscopus sphinx see Sprawler
Asthena albulata see Small White Wave
Atethmia centrago see Centre-barred Sallow
Atolmis rubricollis see Red-necked Footman
August Thorn *Ennomos quercinaria* **174**, 404, 411
Autographa bractea see Gold Spangle
Autographa gamma see Silver Y
Autographa jota see Plain Golden Y
Autographa pulchrina see Beautiful Golden Y
Autumn Green Carpet *Chloroclysta miata* 109, **110**, *111*, 358, 402
Autumnal Moth *Epirrita autumnata* **128**, 402
Autumnal Rustic *Paradiarsia glareosa* **246**, *247*, 406
Axylia putris see Flame

Ballintra (Donegal) 13
Barred Carpet *Perizoma taeniata* **132**, 373, 402
Barred Chestnut *Diarsia dahlii* **248**, *249*, 368, 372, 406
Barred Hook-tip *Watsonalla cultraria* 10, *11*, **74**, 364, 401
Barred Red *Hylaea fasciaria* **195**, *196*, 404
Barred Rivulet *Perizoma bifaciata* **134**, 403
Barred Straw *Eulithis pyraliata* **109**, 402
Barred Umber *Plagodis pulveraria* **170**, 404
Barred Yellow *Cidaria fulvata* **112**, 402
Baynes, Edward Stuart Augustus (1889-1972) 6
Beaded Chestnut *Agrochola lychnidis* **298**, 372, 407

Beautiful Brocade *Lacanobia contigua* **260**, 406
Beautiful Carpet *Mesoleuca albicillata* **104**, *105*, 402
Beautiful Golden Y *Autographa pulchrina* **344**, *345*, 409
Beautiful Snout *Hypena crassalis* **352**, 362, 368, 372, 374, 409
Beautiful Yellow Underwing *Anarta myrtilli* **256**, *257*, 366, 406
Bedstraw Hawkmoth *Hyles galii* 198
Beech-green Carpet *Colostygia olivata* 15, **119**, 402
Beirne, Bryan Patrick (1918-1998) 7
Belmullet (Mayo) 14
Belted Beauty *Lycia zonaria* 3, 4, *14*, 15, 16, **181**, 404
Bembecia muscaeformis see Thrift Clearwing
Bilberry Pug *Pasiphila debiliata* **157**, 403
Birch Mocha *Cyclophora albipunctata* **86**, *87*, 358, 360, 362, 401
Birchall, Edwin (1819-1884) 2
Biston betularia see Peppered Moth
Biston betularia f. *carbonaria see* Peppered Moth
Biston stratria see Oak Beauty
Black Rustic *Aporophyla nigra* **284**, 363, 372, 407
blackcap feeding young 382
Blair's Shoulder-knot *Lithophane leautieri* **286**, *287*, 407
Bleached Pug *Eupithecia expallidata* **145**, 403
Blepharita adusta see Dark Brocade
Blood-vein *Timandra comae* **88**, 371, 401
Blue-bordered Carpet *Plemyria rubiginata* **114**, *115*, 360, 402
Boloria euphrosyne see Pearl-bordered Fritillary
Bond et al. (2006) checklist 25
Bordered Beauty *Epione repandaria* **171**, *172*, 404
Bordered Grey *Selidosema brunnearia* 5, 16, **185**, *186*, 366, 369, 404
Bordered Pug *Eupithecia succenturiata* **147**, *148*, 403
Bordered Sallow *Pyrrhia umbra* **335**, 371, 409
Bordered Straw *Heliothis peltigera* **336**, *337*, 409
Bordered White *Bupalus piniaria* **192**, *193*, 366, 404
Brachylomia viminalis see Minor Shoulder-knot
Bradley checklist (2000) 25
Brakey, Reverend S.L. (fl. 1872) 3
Bray Head (Wicklow) 183
Brenan, Samuel Arthur (1837-1908) 2
Brick *Agrochola circellaris* **296**, 362, 407
Bright-line Brown-eye *Lacanobia oleracea* **262**, 406
Brimstone *Gonepteryx rhamni* 16, 17, 28, *37*, 400
Brimstone Moth *Opisthograptis luteolata* **170**, 404
Brindled Beauty *Lycia hirtaria* **180**, *181*, 404
Brindled Green *Dryobotodes eremita* **292**, *293*, 364, 368, 375, 407
Brindled Ochre *Dasypolia templi* **283**, 369, 407
Brindled Pug *Eupithecia abbreviata* **153**, 154, 403
Bristow, James (fl. 1872-1902) 4
Broad-barred White *Hecatera bicolorata* **264**, 371, 406
Broad-bordered Bee Hawkmoth *Hemaris fuciformis* 10
Broad-bordered Yellow Underwing *Noctua fimbriata* **244**, 406
Broken-barred Carpet *Electrophaes corylata* **118**, 402

Broom Moth *Melanchra pisi* 260, 406
Brown, E.W. *see* Partridge and Brown
Brown Hairstreak *Thecla betulae* 41
Brown Rustic *Rusina ferruginea* 309, 408
Brown Scallop *Philereme vetulata* 125, 374, 402
Brown Silver-line *Petrophora chlorosata* 169, 404
Brown-line Bright-eye *Mythimna conigera* 275, 407
Brown-spot Pinion *Agrochola litura* 298, 407
'browns' *see* Nymphalidae
Brussels Lace *Cleorodes lichenaria* 189, 404
Buff Arches *Habrosyne pyritoides* 78, 401
Buff Ermine *Spilosoma luteum* 229, 405
Buff Footman *Eilema depressa* 226, 364, 367, 405
Buff-tip *Phalera bucephala* 216, *397*, 405
Bulrush Wainscot *Nonagria typhae* 329, *330*, 319, 408
Bupalus piniaria see Bordered White
Burnet Companion *Euclidea glyphica* 35, 349, 373, 409
burnets *see* Zygaenidae
Burnished Brass *Diachrysia chrysitis* 341, *342*, 409
Burren Green *Calamia tridens* 7, 10

Cabbage Moth *Mamestra brassicae* 259, 406
Cabera exanthemata see Common Wave
Cabera pusaria see Common White Wave
Calamia tridens see Burren Green
Callistege mi see Mother Shipton
Calliteara pudibunda see Pale Tussock
Callophrys rubi see Green Hairstreak
Camberwell Beauty *Nymphalis antiopa* 3, 28, 47, 400
Campaea margaritata see Light Emerald
Campbell, David Callender (1860-1926) 3
Campbell, W.H. (fl. 1878-1894) 3
Campion *Hadena rivularis* 264, 406
Camptogramma bilineata see Yellow Shell
Canary-shouldered Thorn *Ennomos alniaria* 174, *175*, 404
Caradrina clavipalpis see Pale Mottled Willow
Caradrina morpheus see Mottled Rustic
Carsia sororiata see Manchester Treble-bar
Catocala fraxini see Clifden Nonpareil
Celaena haworthii see Haworth's Minor
Celaena leucostigma see Crescent
Celastrina argiolus see Holly Blue
Centre-barred Sallow *Atethmia centrago* 299, 365, 407
Cerapteryx graminis see Antler Moth
Cerastis rubricosa see Red Chestnut
Cerura vinula see Puss Moth
Chamomile Shark *Cucullia chamomillae* 280, 407
Charanyca trigrammica see Treble Lines
Charissa obscurata see Annulet
Chesias legatella see Streak
Chestnut *Conistra vaccinii* 294, 407
Chestnut-coloured Carpet *Thera cognata* 116, 369, 402
Chevron *Eulithis testata* 107, *108*, 357, 402
Chiasmia clathrata see Latticed Heath
Chimney Sweeper *Odezia atrata* 159, 365, 403
Chinese Character *Cilix glaucata* 74, 76, 401
Chloroclysta citrata see Dark Marbled Carpet
Chloroclysta miata see Autumn Green Carpet
Chloroclysta siterata see Red-green Carpet
Chloroclysta truncata see Common Marbled Carpet
Chloroclystis v-ata see V-Pug

Cidaria fulvata see Barred Yellow
Cilix glaucata see Chinese Character
Cinnabar *Tyria jacobaeae* 231, *232*, 405
Clay *Mythimna ferrago* 276, 359, 407
Clay Triple-lines *Cyclophora linearia* 86, 364, 374, 401
Clearwing Moths *see* Sesiidae
Cleora cinctaria see Ringed Carpet
Cleorodes lichenaria see Brussels Lace
Clifden Nonpareil *Catocala fraxini* 348, 409
Cloaked Minor *Mesoligia furuncula* 323, 359, 408
Cloaked Pug *Eupithecia abietaria* 16, 137, *138*, 403
Clonbrock, Lord *see* Dillon, R. E.
Clostera pigra see Small Chocolate-tip
Clouded Border *Lomaspilis marginata* 166, *167*, 403
Clouded Brindle *Apamea epomidion* 317, *318*, 408
Clouded Buff *Diacrisia sannio* 222, 229, 230, 370, 372, 405
Clouded Drab *Orthosia incerta* 272, 273, 407
Clouded Magpie *Abraxas sylvata* 31, 165, 363, 364, 403
Clouded Silver *Lomographa temerata* 194, *195*, 404
Clouded Yellow *Colias croceus* 28, 36, 400
Clouded-bordered Brindle *Apamea crenata* 316, *317*, 408
Coast Dart *Euxoa cursoria* 236, 406
Cockayne, Edward Alfred (1880-1956) 5
Coenocalpe lapidata see Slender-striped Rufous
Coenonympha pamphilus see Small Heath
Coenonympha tullia see Large Heath
Colias croceus see Clouded Yellow
Colias croceus f. *helice see* Clouded Yellow
Colocasia coryli see Nut-tree Tussock
Colostygia multistrigaria see Mottled Grey
Colostygia olivata see Beech-green Carpet
Colostygia pectinataria see Green Carpet
Colotois pennaria see Feathered Thorn
Comma *Polygonia c-album* 28, 48, 400
Common Blue *Polyommatus icarus* 18, 22, 34, 43, *44*, 357, 371, 373, 400
Common Carpet *Epirrhoe alternata* 99, 402
Common Emerald *Hemithea aestivaria* 84, 359, 401
Common Footman *Eilema lurideola* 224, *225*, 226, 359, 405
Common Heath *Ematurga atomaria* 192, 404
Common Lutestring *Ochropacha duplaris* 79, 401
Common Marbled Carpet *Chloroclysta truncata* 112, *113*, 402
Common Pug *Eupithecia vulgata* 143, 145, 403
Common Quaker *Orthosia cerasi* 272, 407
Common Rustic *Mesapamea secalis* 324, 408
Common Swift *Hepialus lupulinus* 59, 400
Common Wainscot *Mythimna pallens* 278, 407
Common Wave *Cabera exanthemata* 193, 404
Common White Wave *Cabera pusaria* 192, 404
Confused *Apamea furva* 16, 318, 319, 369, 408
Conistra ligula see Dark Chestnut
Conistra vaccinii see Chestnut
Convolvulus Hawkmoth *Agrius convolvuli* 198, 404
Coole Lough (Galway) 125
Coolorta (Clare) 125
Copper Underwing *Amphipyra pyramidea* 306, 357, *394*, 408
Corofin (Clare) 161

Coronet *Craniophora ligustri* 306, *307*, 408
Cosmia trapezina see Dun-bar
Cosmorhoe ocellata see Purple Bar
Coxcomb Prominent *Ptilodon capucina* 213, 405
Craniophora ligustri see Coronet
Crawford, William Monod (1872-1941) 5, 68, 69
Cream Wave *Scopula floslactata* 15, 89, 401
Crescent *Celaena leucostigma* 329, 408
Crescent Dart *Agrotis trux* 15, 239, 406
Crescent Striped *Apamea oblonga* 14, 15, 316, 359, 408
Crinan Ear *Amphipoea crinanensis* 327, 408
Crocallis elinguaria see Scalloped Oak
Cryphia domestica see Marbled Beauty
Cucullia chamomillae see Chamomile Shark
Cucullia umbratica see Shark
Cupido minimus see Small Blue
Currant Clearwing *Synanthedon tipuliformis* 14, 15, 66, *396*, 401
Currant Pug *Eupithecia assimilata* 144, 403
Cybosia mesomella see Four-dotted Footman
Cyclophora albipunctata see Birch Mocha
Cyclophora linearia see Clay Triple-lines

Danaidae (monarchs) 45
Danaus plexippus see Monarch
Daphnis nerii see Oleander Hawkmoth
Dark Arches *Apamea monoglypha* 314, 408
Dark Brocade *Blepharita adusta* 292, 361, 407
Dark Chestnut *Conistra ligula* 295, 362, 407
Dark Green Fritillary *Argynnis aglaja* 28, 49, *50*, 363, 371, 400
Dark Marbled Carpet *Chloroclysta citrata* 112, *113*, 402
Dark Spectacle *Abrostola triplasia* 347, 409
Dark Spinach *Pelurga comitata* 104, 363, 402
Dark Sword-grass *Agrotis ipsilon* 239, 406
Dark Tussock *Dicallomera fascelina* 16, 219, 366, 368, 405
Dark Umber *Philereme transversata* 125, 374, 402
Dark-barred Twin-spot Carpet *Xanthorhoe ferrugata* 95, 402
Dasypolia templi see Brindled Ochre
Death's Head Hawkmoth *Acherontia atropos* 199, 404
December Moth *Poecilocampa populi* 67, 401
Deilephila elpenor see Elephant Hawkmoth
Deilephila porcellus see Small Elephant Hawkmoth
Deileptenia ribeata see Satin Beauty
Deltote uncula see Silver Hook
Diachrysia chrysitis see Burnished Brass
Diacrisia sannio see Clouded Buff
Diaphora mendica see Muslin Moth
Diarsia brunnea see Purple Clay
Diarsia dahlii see Barred Chestnut
Diarsia florida see Fen Square-spot
Diarsia mendica see Ingrailed Clay
Diarsia rubi see Small Square-spot
Dicallomera fascelina see Dark Tussock
Dichonia aprilina see Merveille du Jour
Dillon, Robert Edward, later Lord Clonbrock (1869-1926) 3, 4
Diloba caeruleocephala see Figure of Eight
Dingle (Kerry) 166
Dingy Shears *Parastichtis ypsillon* 313, 408
Dingy Skipper *Erynnis tages* 4, 16, 28, *35*, 350, 373, 400
Discestra trifolii see Nutmeg
Dog's Tooth *Lacanobia suasa* 262, 359, 406
Donovan, Charles (1863-1951) 4

Dot Moth *Melanchra persicariae* 259, *260*, 369, 406
Dotted Border *Agriopis marginaria* 183, **184**, 185, 404
Dotted Carpet *Alcis jubata* 188, 367, 404
Dotted Clay *Xestia baja* 251, 406
Dotted Rustic *Rhyacia simulans* 411
Double Dart *Graphiphora augur* 245, 406
Double Lobed *Apamea ophiogramma* 320, *321*, 357, 358, 369, 374, 375, 408
Double Square-spot *Xestia triangulum* 251, 406
Double-striped Pug *Gymnoscelis rufifasciata* 157, 403
Drepana falcataria see Pebble Hook-tip
Drepanidae 74-77, 401
Drinker *Euthrix potatoria* **71**, 401
Dromore Forest (Clare) 296
Dromore Wood (Clare) 87
Drymonia ruficornis see Lunar Marbled Brown
Dryobotodes eremita see Brindled Green
Dublin City (Dublin) 66
Dun-bar *Cosmia trapezina* 313, *314*, 408
Dusky Brocade *Apamea remissa* 319, 408
Dwarf Pug *Eupithecia tantillaria* 155, 403
Dyscia fagaria see Grey Scalloped Bar

Ear Moths see Amphipoea
Ear Moth *Amphipoea oculea* **327**, *328*, 408
Early Grey *Xylocampa areola* 289, *290*, 407
Early Moth *Theria primaria* 194, 404
Early Thorn *Selenia dentaria* 175, *176*, 404
Early Tooth-striped *Trichopteryx carpinata* 119, *162*, **163**, 358, 368, 403
Ecliptopera silaceata see Small Phoenix
Ectropis bistortata see Engrailed
Ectropis crepuscularia see Small Engrailed
Eggars see Lasiocampidae
Eilema complana see Scarce Footman
Eilema depressa see Buff Footman
Eilema depressa f. *unicolor* Banks see Buff Footman
Eilema lurideola see Common Footman
Electrophaes corylata see Broken-barred Carpet
Elephant Hawkmoth *Deilephila elpenor* 204, *205*, 405
Elephant Hawkmoth larva *399*
Ematurga atomaria see Common Heath
Emperor *Saturnia pavonia* **71**, *72*, *73*, 370, 401
Enargia paleacea see Angle-striped Sallow
Engrailed *Ectropis bistortata* 189, *191*, 358, 404
Ennomos alniaria see Canary-shouldered Thorn
Ennomos erosaria see September Thorn
Ennomos quercinaria see August Thorn
Entephria caesiata see Grey Mountain Carpet
Entephria flavicinctata see Yellow-ringed Carpet
Epione repandaria see Bordered Beauty
Epirrhoe alternata see Common Carpet
Epirrhoe galiata see Galium Carpet
Epirrhoe tristata see Small Argent and Sable
Epirrita species 126-130
Epirrita autumnata see Autumnal Moth
Epirrita christyi see Pale November Moth
Epirrita dilutata see November Moth
Epirrita filigrammaria see Small Autumnal Moth
Erannis defoliaria see Mottled Umber
Erebia aethiops see Scotch Argus
Erebia epiphron see Mountain Ringlet
Eriogaster lanestris see Small Eggar
ermine moths see Arctiinae
Erynnis tages see Dingy Skipper
Euclidea glyphica see Burnet Companion
Eulithis populata see Northern Spinach

Eulithis prunata see Phoenix
Eulithis pyraliata see Barred Straw
Eulithis testata see Chevron
Euphydryas aurinia see Marsh Fritillary
Euphyia unangulata see Sharp-angled Carpet
Eupithecia abietaria see Cloaked Pug
Eupithecia abbreviata see Brindled Pug
Eupithecia absinthiata see Wormwood and Ling Pug
Eupithecia absinthiata f. *goossensiata* see Wormwood and Ling Pug
Eupithecia assimilata see Currant Pug
Eupithecia centaureata see Lime-speck Pug
Eupithecia distinctaria see Thyme Pug
Eupithecia dodoneata see Oak-tree Pug
Eupithecia exiguata see Mottled Pug
Eupithecia expallidata see Bleached Pug
Eupithecia haworthiata see Haworth's Pug
Eupithecia icterata see Tawny-speckled Pug
Eupithecia indignata see Ochreous Pug
Eupithecia innotata see Angle-barred and Ash Pug
Eupithecia innotata f. *fraxinata* see Angle-barred and Ash Pug
Eupithecia lariciata see Larch Pug
Eupithecia linariata see Toadflax Pug
Eupithecia nanata see Narrow-winged Pug
Eupithecia pimpinellata see Pimpinel Pug
Eupithecia plumbeolata see Lead-coloured Pug
Eupithecia (Pugs) 136-165, 386
Eupithecia pulchellata see Foxglove Pug
Eupithecia pusillata see Juniper Pug
Eupithecia pygmaeata see Marsh Pug
Eupithecia satyrata see Satyr Pug
Eupithecia simpliciata see Plain Pug
Eupithecia subfuscata see Grey Pug
Eupithecia subumbrata see Shaded Pug
Eupithecia succenturiata see Bordered Pug
Eupithecia tantillaria see Dwarf Pug
Eupithecia tenuiata see Slender Pug
Eupithecia tripunctaria see White-spotted Pug
Eupithecia trisignaria see Triple-spotted Pug
Eupithecia valerianata see Valerian Pug
Eupithecia venosata see Netted Pug
Eupithecia virgaureata see Golden-rod Pug
Eupithecia vulgata see Common Pug
Euplexia lucipara see Small Angle Shades
Euproctis similis see Yellow-tail
Eupsilia traversa see Satellite
Eurois occulta see Great Brocade
Euthrix potatoria see Drinker
Euxoa cursoria see Coast Dart
Euxoa nigricans see Garden Dart
Euxoa obelisca see Square-spot Dart
Euxoa tritici see White-line Dart
Eyed Hawkmoth larva *383*
Eyed Hawkmoth *Smerinthus ocellata* 200, 203, 366, 375, 404, *410*

Falcaria lacertinaria see Scalloped Hook-tip
Fan-foot *Zanclognatha tarsipennalis* 354, 409
Feathered Gothic *Tholera decimalis* 268, 357, 371, 375, 407
Feathered Ranunculus *Polymixis lichenea* 294, 363, 369, 407
Feathered Thorn *Colotois pennaria* 179, 404
Feathered Thorn larva *383*
Fen Square-spot *Diarsia florida* 411
Figure of Eight *Diloba caeruleocephala* 216, *217*, 360, 362, 368, 405
Flame *Axylia putris* 240, *241*, 406
Flame Carpet *Xanthorhoe designata* 94, 401

Flame Shoulder *Ochropleura plecta* 242, 406
Flounced Chestnut *Agrochola helvola* 297, 362, 407
Flounced Rustic *Luperina testacea* 326, 408
footman moths see Notodontidae
Forester *Adscita statices* 16, **60**, *61*, 368, 400
Foster, Canon George (1870-1935) 5
Fountainstown (Cork) 321
Four-dotted Footman *Cybosia mesomella* 224, 405
Four-spotted Footman *Lithosia quadra* 226, 405
Fox Moth *Macrothylacia rubi* **70**, 361, 366, 401
Foxglove Pug *Eupithecia pulchellata* 138, *139*, 403
Frosted Orange *Gortyna flavago* 328, 408
Furcula bicuspis see Alder Kitten
Furcula bifida see Poplar Kitten
Furcula furcula see Sallow Kitten
Furphy, Joe 7

Galium Carpet *Epirrhoe galiata* 99, 402
Garden Carpet *Xanthorhoe fluctuata* 96, 402
Garden Dart *Euxoa nigricans* 235, 406
Garden Tiger *Arctia caja* 227, *228*, 405
Gatekeeper *Pyronia tithonus* 45
Gem *Orthonama obstipata* 93, 401
Geometra papilionaria see Large Emerald
Geometrid pupa *386*
Geometridae 15, 82-197, 384, 401-404
Ghost Moth *Hepialus humuli* 58, *59*, 400
Glanville Fritillary *Melitaea cinxia* 10
Glaucous Shears *Papestra biren* 263, 361, 406
Glenageary (Dublin) 6, 116, 243
Glenveigh National Park (Donegal) 187
Gold Spangle *Autographa bractea* 345, 409
Gold Spot *Plusia festucae* 342, *343*, 409
Gold Swift *Hepialus hecta* 58, *59*, 358, 375, 400
Golden Plusia *Polychrysia moneta* 341, 409
Golden-rod Pug *Eupithecia virgaureata* 147, *152*, 403
Gonepteryx rhamni see Brimstone
Gortyna flavago see Frosted Orange
Gothic *Naenia typica* 253, 406
Graphiphora augur see Double Dart
Grass Emerald *Pseudoterpna pruinata* 83, 401
Grass Rivulet *Perizoma albulata* 134, 403
Grass Wave *Perconia strigillaria* 197, 370, 404
Grayling *Hipparchia semele* 29, **52**, *53*, 363, 365, 373, 400
Great Brocade *Eurois occulta* 254, 406
Green Arches *Anaplectoides prasina* 254, *255*, 406
Green Carpet *Colostygia pectinataria* 119, *120*, 402
Green Hairstreak *Callophrys rubi* 16, 28, **41**, *42*, 360, 366, 367, 368, 372, 373, 400
Green Pug *Pasiphila rectangulata* 157, 403
Green Silver-lines *Pseudoips prasinana* 338, 368, 409
Green-brindled Crescent *Allophyes oxyacanthae* 289, *290*, 407
Greene, Joseph (1824-1906) 2, 348
Greene and Hogan catalogue (1854) 2
Green-veined White *Pieris napi* 33, **39**, 400
Greer, Thomas (1874-1949) xii, 5
Grey Arches *Polia nebulosa* 258, 406
Grey Birch *Aethalura punctulata* 15, 190, 404
Grey Chi *Antitype chi* 293, 407
Grey Dagger *Acronicta psi* 303, *304*, 408
Grey Mountain Carpet *Entephria caesiata* 101, *102*, 402
Grey Pine Carpet *Thera obeliscata* 114, **115**, 116, 402

Grey Pug *Eupithecia subfuscata* **146**, 153, 403
Grey Scalloped Bar *Dyscia fagaria* **197**, 368, 404
Grey Shoulder-knot *Lithophane ornitopus* **286**, 287, 407
Gymnoscelis rufifasciata see Double-striped Pug

Habrosyne pyritoides see Buff Arches
Hada plebeja see Shears
Hadena bicruris see Lychnis
Hadena confusa see Marbled Coronet
Hadena perplexa see Tawny Shears and Pod Lover
Hadena perplexa capsophila see Tawny Shears and Pod Lover
Hadena rivularis see Campion
hawkmoths see Sphingidae
Haworth's Minor *Celaena haworthii* **329**, 357, 408
Haworth's Pug *Eupithecia haworthiata* **136**, 363, 403
Heal, Henry George (1920-1986) *1*, 7
Heart and Club *Agrotis clavis* **238**, 406
Heart and Dart *Agrotis exclamationis* **239**, 406
Heath, John 344, 389
Heath Rivulet *Perizoma minorata* 411
Heath Rustic *Xestia agathina* **252**, 253, 358, 372, 406
Hebrew Character *Orthosia gothica* **274**, 407
Hecatera bicolorata see Broad-barred White
Hedge Rustic *Tholera cespitis* **268**, 407
Helicoverpa armigera see Scarce Bordered Straw
Heliothis peltigera see Bordered Straw
Hemaris fuciformis see Broad-bordered Bee Hawkmoth
Hemaris tityus see Narrow-bordered Bee Hawkmoth
Hemistola chrysoprasaria see Small Emerald
Hemithea aestivaria see Common Emerald
Hepialidae (Swift moths) 58-60, 381, 400
Hepialus fusconebulosa see Map-winged Swift
Hepialus fusconebulosa f. *gallicus* see Map-winged Swift
Hepialus hecta see Gold Swift
Hepialus humuli see Ghost Moth
Hepialus lupulinus see Common Swift
Herald *Scoliopteryx libatrix* **350**, *351*, 409
Herminia griseolis see Small Fan-foot
Hesperiidae (Skippers) 12, 35-36, 400
'highflyers' 121, 402
Hillsborough Forest Rothamsted trap 13, 132, 154, 390
Hipparchia semele see Grayling
Hippotion celerio see Silver-striped Hawkmoth
Hogan, Arthur Riky (fl. 1850-1860) 2
Holly Blue *Celastrina argiolus* 16, **44**, *45*, 364, 372, 400
Hoplodrina alsines see Uncertain
Hoplodrina ambigua see Vine's Rustic
Hoplodrina blanda see Rustic
Hughes, Maurice 391
Hummingbird Hawkmoth *Macroglossum stellatarum* **202**, 405
Hydraecia micacea see Rosy Rustic
Hydrelia flammeolaria see Small Yellow Wave
Hydriomena furcata see July Highflyer
Hydriomena impluviata see May Highflyer
Hydriomena ruberata see Ruddy Highflyer
Hylaea fasciaria see Barred Red
Hyles galii see Bedstraw Hawkmoth
Hyles livornica see Striped Hawkmoth
Hypena crassalis see Beautiful Snout
Hypena proboscidalis see Snout
Hypenodes humidialis see Marsh Oblique-barred

Idaea aversata see Riband Wave
Idaea biselata see Small Fan-footed Wave
Idaea dimidiata see Single-dotted Wave
Idaea straminata see Plain Wave
Idea subsericeata see Satin Wave
Inachis io see Peacock
Ingrailed Clay *Diarsia mendica* **248**, 406
Ipimorpha subtusa see Olive
Irish Annulet *Odontognophos dumeteta* 10
Iron Prominent *Notodonta dromedarius* **209**, 405
Issoria lathonia see Queen of Spain Fritillary

Jodis lactearia see Little Emerald
Johnson, William Frederick (1852-1934) **3**
July Belle *Scotopteryx luridata* **98**, 402
July Highflyer *Hydriomena furcata* **121**, 402
Juniper Carpet *Thera juniperata* **116**, 117, 118, 402
Juniper Pug *Eupithecia pusillata* **154**, 403

Kane, William Francis de Vismes (1840-1918) **2**
Kenmare (Kerry) 213
Kenmare Demesne (Kerry) 125
Killarney (Kerry) 161, 214
Killarney National Park (Kerry) 166, 214
Kingston, Dun Laoghaire (Dublin) 348
Kitten Moths see Notodontidae
Knot Grass *Acronicta rumicis* **302**, 305, 408

Lacanobia contigua see Beautiful Brocade
Lacanobia oleracea see Bright-line Brown-eye
Lacanobia suasa see Dog's Tooth
Lacanobia thalassina see Pale-shouldered Brocade
Lackey *Malacosoma neustria* **4**, 15, **69**, 401
Lampropteryx suffumata see Water Carpet
Landsdowne MV trap (fig. 1) 391
Langham, Charles (1870-1951) **4**
Laothoe populi see Poplar Hawkmoth
Larch Pug *Eupithecia lariciata* **155**, 403
Large Copper *Lycaena dispar* 7
Large Ear *Amphipoea lucens* **327**, 372, 408
Large Emerald *Geometra papilionaria* **83**, 84, 401
Large Heath *Coenonympha tullia* 16, 29, **55**, *56*, 366, 370, 400
Large Wainscot *Rhizedra lutosa* **331**, 357, 408
Large White *Pieris brassicae* **38**, 387, 400
Large White pupa *387*
Large Yellow Underwing *Noctua pronuba* **234**, 243, 406
Lasiocampa quercus see Oak and Northern Eggar
Lasiocampa quercus callunae see Oak and Northern Eggr
Lasiocampidae (Eggars) 66-72, 401
Lasiommata megera see Wall
Latticed Heath *Chiasmia clathrata* **168**, *169*, 404
Lead Belle *Scotopteryx mucronata* **97**, 402
Lead-coloured Drab *Orthosia populeti* **271**, 374, 407
Lead-coloured Pug *Eupithecia plumbeolata* **137**, 403
Least Black Arches *Nola confusalis* **233**, 405
Least Yellow Underwing *Noctua interjecta* **244**, 371, 406
Lempke's Gold Spot *Plusia putnami* **344**, 409
Leptidea reali see RÈal's Wood White
Leptidea sinapis see Wood White
Lesser Broad-bordered Yellow Underwing *Noctua janthe* **244**, *245*, 406
Lesser Common Rustic *Mesapamea didyma* **324**, *325*, 408
Lesser Cream Wave *Scopula immutata* **88**, 358, 359, 401

Lesser Swallow Prominent *Pheosia gnoma* **210**, 211, 405
Lesser Yellow Underwing *Noctua comes* **243**, 406
Ligdia adustata see Scorched Carpet
Light Arches *Apamea lithoxylea* **315**, 316, 359, 408
Light Emerald *Campaea margaritata* **195**, 404
Light Feathered Rustic *Agrotis cinerea* 14, 15, **236**, 406
Light Knot Grass *Acronicta menyanthidis* **305**, 408
Lilac Beauty *Apeira syringaria* **172**, *173*, 404
Lime-speck Pug *Eupithecia centaureata* **142**, 403
Ling Pug see Wormwood and Ling Pug
Lithophane hepatica see Pale Pinion
Lithophane leautieri see Blair's Shoulder-knot
Lithophane ornitopus see Grey Shoulder-knot
Lithosia quadra see Four-spotted Footman
Lithosiinae (footman and allied moths) 222, 405
Little Emerald *Jodis lactearia* **85**, *86*, 357, 372, 373, 401
Lobophora halterata see Seraphim
Lomaspilis marginata see Clouded Border
Lomaspilis marginata f. *diluta* see Clouded Border *167*
Lomographa temerata see Clouded Silver
Lough Bunny (Clare) 125
Lough Caragh (Kerry) 256
Lough Coole (Clare) 125
Lough Creag Dubh, Connemara, (Galway) 117
Lunar Hornet Moth pupa and exit holes *65*
Lunar Hornet Moth *Sesia bembeciformis* **63**, *64*, 357, 360, 374, 400
Lunar Marbled Brown *Drymonia ruficornis* **214**, *215*, 357, 368, 405
Lunar Thorn *Selenia lunularia* **177**, 404
Lunar Underwing *Omphaloscelis lunosa* **299**, 407
Luperina testacea see Flounced Rustic
lutestring moths see Thyatiridae
Lycaena dispar see Large Copper
Lycaena phlaeas see Small Copper
Lycaenidae 41-45, 400
Lycia hirtaria see Brindled Beauty
Lycia lapponaria see Rannoch Brindled Beauty
Lycia zonaria see Belted Beauty
Lycophotia porphyrea see True Lover's Knot
Lymantriidae 218-221, 384, 405

Macaria liturata see Tawny-barred Angle
Macaria liturata f. *nigrofulvata* see Tawny-barred Angle
Macaria wauaria see V-Moth
Macroglossum stellatarum see Hummingbird Hawkmoth
Macrothylacia rubi see Fox Moth
Magpie *Abraxas grossulariata* **165**, 403
Magurran, Anne 101, 116
Malacosoma neustria see Lackey
Mamestra brassicae see Cabbage Moth
Manchester Treble-bar *Carsia sororiata* 411
Maniola jurtina see Meadow Brown
Map-winged Swift *Hepialus fusconebulosa* **60**, 400
Marbled Beauty *Cryphia domestica* **306**, *307*, 408
Marbled Coronet *Hadena confusa* **265**, *266*, 361, 369, 407
Marbled Minor *Oligia strigilis* **322**, 408
Marbled White Spot *Protodeltote pygarga* **337**, 363
March Moth *Alsophila aescularia* **82**, *83*, 401
Marsh Fritillary egg mass *382*

Marsh Fritillary *Euphydryas aurinia* 15, 16, *17*, 20, 22, 29, 51, 357, 360, 363, 370, 382, 400
Marsh Fritillary larvae 384, *385*
Marsh Oblique-barred *Hypenodes humidialis* 354, 409
Marsh Pug *Eupithecia pygmaeata* 15, 140, 403
May Highflyer *Hydriomena impluviata* 121, 402
Meadow Brown *Maniola jurtina* 54, 371, 381, 400
Melanchra persicariae see Dot Moth
Melanchra pisi see Broom Moth
Melitaea cinxia see Glanville Fritillary
Merveille du Jour *Dichonia aprilina 291*, 407
Mesapamea didyma see Lesser Common Rustic
Mesapamea secalis see Common Rustic
Mesoleuca albicillata see Beautiful Carpet
Mesoligia furuncula see Cloaked Minor
Mesoligia literosa see Rosy Minor
Middle-barred Minor *Oligia fasciuncula* 323, 408
Miller *Acronicta leporina* 302, 362, 408
Minor Shoulder-knot *Brachylomia viminalis* 282, 407
Monarch *Danaus plexippus* 28, 56, 400
monarchs *see* Danaidae
Morell, Roy (d. 1991) 8
Mormo maura see Old Lady
Mother Shipton *Callistege mi* 35, *348*, *349*, 366, 373, 409
Mottled Beauty *Alcis repandata* 185, 187, 404
Mottled Grey *Colostygia multistrigaria* 119, 162, 402
Mottled Pug *Eupithecia exiguata* 139, 403
Mottled Rustic *Caradrina morpheus* 333, 409
Mottled Umber *Erannis defoliaria* 185, 404
Mountain Ringlet *Erebia epiphron* 45
Mouse Moth *Amphipyra tragopoginis* 308, 408
Mullein Wave *Scopula marginepunctata* 88, *89*, 401
Murphy, Kenny *377*
Muslin Footman *Nudaria mundana* 222, 405
Muslin Moth *Diaphora mendica* 229, 231, 405
Mythimna comma see Shoulder-striped Wainscot
Mythimna conigera see Brown-line Bright-eye
Mythimna ferrago see Clay
Mythimna impura see Smoky Wainscot
Mythimna litoralis see Shore Wainscot
Mythimna pallens see Common Wainscot
Mythimna straminea see Southern Wainscot
Mythimna unipuncta see White-speck

Naenia typica see Gothic
Nairn, R. 349
Narrow-bordered Bee Hawkmoth *Hemaris tityus* 16, *201*, 361, 370, 373, 405
Narrow-bordered Five-spot Burnet *Zygaena lonicerae* 62, 368, 400
Narrow-winged Pug *Eupithecia nanata* 150, *151*, 403
Nebula salicata see Striped Twin-spot Carpet
Neglected Rustic *Xestia castanea* 251, 372, 406
Nelson, Phyllis Ismay (1907-1979) 6, 344
Neozyphyrus quercus see Purple Hairstreak
Netted Pug *Eupithecia venosata 141*, 403
Newbridge (Kildare) 87
Noctua comes see Lesser Yellow Underwing
Noctua fimbriata see Broad-bordered Yellow Underwing
Noctua interjecta see Least Yellow Underwing
Noctua janthe see Lesser Broad-bordered Yellow Underwing
Noctua pronuba see Large Yellow Underwing

Noctuidae 234-354, 406
Nola confusalis see Least Black Arches
Nolidae 233, 405
Nonagria typhae see Bulrush Wainscot
Northern Deep-brown Dart *Aporophyla lueneburgensis* 284, 364, 407
Northern Drab *Orthosia opima* 270, 407
Northern Eggar *see* Oak and Northern Eggar
Northern Rustic *Standfussiana lucernea* 242, 369, 406
Northern Spinach *Eulithis populata* 108, 367, 402
Northern Winter Moth *Operophtera fagata* 10, *131*, 362, 402
Notodonta dromedarius see Iron Prominent
Notodonta ziczac see Pebble Prominent
Notodontidae (Prominent and kitten moths) 206-218, 405
November Moth *Epirrita dilutata* 126, *127*, 128, 402
Nudaria mundana see Muslin Footman
Nutmeg *Discestra trifolii* 256, 359, 406
Nut-tree Tussock *Colocasia coryli 340*, 409
Nycteola revayana see Oak Nycteoline
Nymphalidae 45-56, 400
Nymphalis antiopa see Camberwell Beauty

Oak Beauty *Biston stratara* 32, 182, 368, 404
Oak and Northern Eggar *Lasiocampa quercus* 69, 401
Oak Eggar larva *395*
Oak Eggar *Lasiocampa quercus* 70
Oak Eggar *see* Oak and Northern Eggar
Oak Nycteoline *Nycteola revayana* 338, *339*, 364, 368, 409
Oak-tree Pug *Eupithecia dodoneata* 154, 403
Oblique Carpet *Orthonama vittata* 92, 401
Ochreous Pug *Eupithecia indignata* 149, 403
Ochropacha duplaris see Common Lutestring
Ochropleura plecta see Flame Shoulder
Odezia atrata see Chimney Sweeper
Odontognophos dumetata see Irish Annulet
Odontopera bidentata see Scalloped Hazel
Odontosia carmelita see Scarce Prominent
O'Farrell, A.F. 240, 315
Old Lady *Mormo maura* 308, *309*, 367, 368, 375, 394, 408
Oleander Hawkmoth *Daphnis nerii 203*, 405
Oligia fasciuncula see Middle-barred Minor
Oligia latruncula see Tawny Marbled Minor
Oligia strigilis see Marbled Minor
Oligia versicolor see Rufous Minor
Olive *Ipimorpha subtusa* 16, 312, 408
Omphaloscelis lunosa see Lunar Underwing
Operophtera brumata see Winter Moth
Operophtera fagata see Northern Winter Moth
Opisthograptis luteolata see Brimstone Moth
Orange Tip *Anthocaris cardamines* 9, *39*, *40*, 400
Orange-Tip egg *381*
Orgyia antiqua see Vapourer
Orthonama obstipata see Gem
Orthonama vittata see Oblique Carpet
Orthosia cerasi see Common Quaker
Orthosia cruda see Small Quaker
Orthosia gothica see Hebrew Character
Orthosia gracilis see Powdered Quaker
Orthosia incerta see Clouded Drab
Orthosia munda see Twin-spotted Quaker
Orthosia opima see Northern Drab
Orthosia populeti see Lead-coloured Drab
Ourapteryx sambucaria see Swallow-tailed Moth

Painted Lady *Vanessa cardui* 28, 47, 400
Pale Brindled Beauty *Phigalia pilosaria* 180, 404
Pale Eggar *Trichiura crataegi* 16, 67, *68*, 372, 401
Pale Mottled Willow *Caradrina clavipalpis* 334, 409
Pale November Moth *Epirrita christyi* 126, *127*, 402
Pale Pinion *Lithophane hepatica* 284, *285*, 407
Pale Prominent *Pterostoma palpina 214*, 405
Pale Tussock *Calliteara pudibunda* 220, 362, 405
Pale-shouldered Brocade *Lacanobia thalassina 261*, *262*, 406
Panolis flammea see Pine Beauty
Papestra biren see Glaucous Shears
Paradiarsia glareosa see Autumnal Rustic
Paranthreninae (Clearwing moths) 66, 401
Pararge aegeria see Speckled Wood
Parasemia plantaginis see Wood Tiger
Parastichtis suspecta see Suspected
Parastichtis ypsillon see Dingy Shears
Partridge, C.E. and Brown, E.W. (fl. 1893-1897) 4
Pasiphila debiliata see Bilberry Pug
Pasiphila rectangulata see Green Pug
Peach Blossom *Thyatira batis* 77, *78*, 401
Peacock *Inachis io* 48, *49*, 384, 400
Pearl-bordered Fritillary *Boloria euphrosyne* 45
Pearly Underwing *Peridroma saucia* 248, 406
Pebble Hook-tip *Drepana falcataria 76*, 401
Pebble Prominent *Notodonta ziczac* 210, *211*, 405
Pelurga comitata see Dark Spinach
Peppered Moth *Biston betularia* 182, 404
Perconia strigillaria see Grass Wave
Peribatodes rhomboidaria see Willow Beauty
Peridroma saucia see Pearly Underwing
Perizoma affinitata see Rivulet
Perizoma albulata see Grass Rivulet
Perizoma alchemillata see Small Rivulet
Perizoma bifaciata see Barred Rivulet
Perizoma blandiata see Pretty Pinion
Perizoma didymata see Twin-spot Carpet
Perizoma flavofasciata see Sandy Carpet
Perizoma minorata see Heath Rivulet
Perizoma taeniata see Barred Carpet
Petrophora chlorosata see Brown Silver-line
Phalera bucephala see Buff-tip
Pheosia gnoma see Lesser Swallow Prominent
Pheosia tremula see Swallow Prominent
Phigalia pilosaria see Pale Brindled Beauty
Philereme transversata see Dark Umber
Philereme vetulata see Brown Scallop
Phlogophora meticulosa see Angle Shades
Phoenix *Eulithis prunata* 107, 109, 402
Photedes minima see Small Dotted Buff
Photedes pygmina see Small Wainscot
Phragmatobia fuliginosa see Ruby Tiger
Phytometra viridaria see Small Purple-barred
Pieridae 36-40, 400
Pieris brassicae see Large White
Pieris napi see Green-veined White
Pieris rapae see Small White
Pimpinel Pug *Eupithecia pimpinellata* 150, *151*, 371, 403
Pine Beauty *Panolis flammea* 268, *269*, 366, 407
Pine Carpet *Thera firmata* 114, 116, 402
Pinion-streaked Snout *Schrankia costaestrigalis* 353, 409
Pink-barred Sallow *Xanthia togata 300*, 407
Plagodis dolabraria see Scorched Wing
Plagodis pulveraria see Barred Umber

Plain Golden Y *Autographa jota* **345**, 409
Plain Pug *Eupithecia simpliciata* 149, 363, 372, 403
Plain Wave *Idaea straminata* **91**, 401
Plemyria rubiginata see Blue-bordered Carpet
Plusia festucae see Gold Spot
Plusia putnami see Lempke's Gold Spot
Pod Lover *see* Tawny Shears and Pod Lover
Poecilocampa populi see December Moth
Polia nebulosa see Grey Arches
Polia nebulosa f. *pallida see* Grey Arches
Polychrysia moneta see Golden Plusia
Polygonia c-album see Comma
Polymixis lichenea see Feathered Ranunculus
Polyommatus bellargus see Adonis Blue
Polyommatus icarus see Common Blue
Poplar Grey *Acronicta megacephala* **301**, 368, 408
Poplar Hawkmoth *Laothoe populi* 23, **200**, 404
Poplar Kitten *Furcula bifida* 209, 411
Poplar Lutestring *Tethea or* 79, *80*, 374, 401
Porter, Jim 411
Portland Moth *Actebia praecox* 241, *242*, 406
Powdered Quaker *Orthosia gracilis* 272, *273*, 407
Praeger, Robert Lloyd 2, 3
Pretty Pinion *Perizoma blandiata* 16, **134**, 403
Privet Hawkmoth *Sphinx ligustri* 4
Prominent and Kitten Moths *see* Notodontidae
Protodeltote pygarga see Marbled White Spot
Pseudoips prasinana see Green Silver-lines
Pseudoterpna pruinata see Grass Emerald
Pterapherapteryx sexalata see Small Seraphim
Pterostoma palpina see Pale Prominent
Ptilodon capucina see Coxcomb Prominent
Pug Moths *see* Eupithecia
Purple Bar *Cosmorhoe ocellata* **106**, 402
Purple Clay *Diarsia brunnea* 249, 406
Purple Hairstreak *Neozephyrus quercus* 4, 16, 18, 41, *42*, 372, 374, 400
Puss Moth *Cerura vinula* 206, *207*, 357, 387, 405
Puss Moth larva *398*
Pyronia tithonus see Gatekeeper
Pyrrhia umbra see Bordered Sallow

Queen of Spain Fritillary *Issoria lathonia* 45

Rankin, Denis Henderson (1919-1944) 6
Rankin, Matthew Neal (1918-1999) 6
Rannoch Brindled Beauty *Lycia lapponaria* 10
Rathdrum (Wicklow) 87
Réal's Wood White *Leptidea reali* 10, 28, **36**, 360, 363, 373, 400
Red Admiral *Vanessa atalanta* 28, *46*, 400
Red Carpet *Xanthorhoe decoloraria* **94**, 369, 401
Red Chestnut *Cerastis rubricosa* 255, 366, 406
Red Sword-grass *Xylena vetusta* 286, **288**, 289, 407
Red Twin-spot Carpet *Xanthorhoe spadicearia* **95**, 401
Reddish Light Arches *Apamea sublustris* 315, 373, 408
Red-green Carpet *Chloroclysta siterata* 109, *110*, 111, 402
Red-line Quaker *Agrochola lota* **296**, 407
Red-necked Footman *Atolmis rubricollis* **223**, 361, 370, 405
Red-tipped Clearwing *Synanthedon formicaeformis* 16, 19, **65**, *66*, 360, 401
Rheumaptera hastata see Argent and Sable
Rheumaptera undulata see Scallop Shell
Rhizedra lutosa see Large Wainscot

Rhodometra sacraria see Vestal
Rhyacia simulans see Dotted Rustic
Riband Wave *Idaea aversata* **91**, 401
Ringed Carpet *Cleora cinctaria* 4, 15, **186**, 404
Ringlet *Aphantopus hyperantus* 54, 400
Rivula sericealis see Straw Dot
Rivulet *Perizoma affinitata* **132**, 402
Rosy Minor *Mesoligia literosa* 324, 408
Rosy Rustic *Hydraecia micacea* 328, 408
Round-winged Muslin *Thumatha senex* **222**, 405
Ruby Tiger *Phragmatobia fuliginosa* **231**, 357, 373, 405
Ruddy Highflyer *Hydriomena ruberata* **121**, 402
Rufous Minor *Oligia versicolor* 322, 408
Rush Wainscot *Archanara algae* 331, 374, 408
Rusina ferruginea see Brown Rustic
Rustic *Hoplodrina blanda* 332, *333*, 409
Rustic Shoulder-knot *Apamea sordens* 319, 408
Ryan *et al.* 2

Sallow *Xanthia icteritia* **301**, 407
Sallow Kitten *Furcula furcula* 208, 387, 405
Sallow Kitten larva *387*
Sandy Carpet *Perizoma flavofasciata* **135**, 403
Satellite *Eupsilia traversa* **294**, 407
Satin Beauty *Deileptenia ribeata* **187**, 362, 364, 367, 404
Satin Wave *Idea subsericeata* 411
Saturnia pavonia see Emperor
Saturniidae (silkmoths) 72-73, 401
Satyr Pug *Eupithecia satyrata* **143**, 403
Savage, W. 67
Scallop Shell *Rheumaptera undulata* **123**, 358, 360, 367, 402
Scalloped Hazel *Odontopera bidentata* **177**, 404
Scalloped Hook-tip *Falcaria lacertinaria* 74, *75*, 372, 401
Scalloped Oak *Crocallis elinguaria* 177, *178*, 404
Scarce Bordered Straw *Helicoverpa armigera* **336**, 409
Scarce Footman *Eilema complana* 224, *225*, 226, 369, 405
Scarce Prominent *Odontosia carmelita* **212**, 213, 362, 374, 405
Scarce Silver Y *Syngrapha interrogationis* **346**, 362, 372, 409
Scarce Umber *Agriopis aurantiaria* 183, *184*, 362, 404
Schrankia costaestrigalis see Pinion-streaked Snout
Schrankia taenialis see White-lined Snout
Scoliopteryx libatrix see Herald
Scopula floslactata see Cream Wave
Scopula immutata see Lesser Cream Wave
Scopula marginepunctata see Mullein Wave
Scopula ternata see Smoky Wave
Scorched Carpet *Ligdia adustata* 166, *167*, 364, 375, 404
Scorched Wing *Plagodis dolabraria* 170, *171*, 404
Scotch Argus *Erebia aethiops* 10
Scotopteryx chenopodiata see Shaded Broad-bar
Scotopteryx luridata see July Belle
Scotopteryx mucronata see Lead Belle
Seapoint (Dublin) 341
Selenia dentaria see Early Thorn
Selenia lunularia see Lunar Thorn
Selidosema brunnearia see Bordered Grey
Selidosema brunnearia f. *tyronensis* 5, 186
September Thorn *Ennomos erosaria* 411
September Thorn larva *411*
Seraphim *Lobophora halterata* 161, *162*, 375, 403
Sesia bembeciformis see Lunar Hornet Moth

Sesiidae (Clearwing Moths) 12, 15, 63-66, 381, 383, 395, 400
Setaceous Hebrew Character *Xestia c-nigrum* **250**, 406
Shaded Broad-bar *Scotopteryx chenopodiata* **97**, 402
Shaded Pug *Eupithecia subumbrata* **148**, 403
Shark *Cucullia umbratica* 280, *281*, 407
Sharp-angled Carpet *Euphyia unangulata* **125**, 402
Shears *Hada plebeja* **256**, 406
Shore Wainscot *Mythimna litoralis* 278, *279*, 407
Shoulder Stripe *Anticlea badiata* 102, *103*, 402
Shoulder-striped Wainscot *Mythimna comma* 280, 407
Shuttle-shaped Dart *Agrotis puta* 240, *241*, 406
Sideridis albicolon see White Colon
silkmoths *see* Saturniidae
Silver Hook *Deltote uncula* **338**, 358, 360, 409
Silver Y *Autographa gamma* **344**, 409
Silver-ground Carpet *Xanthorhoe montanata* **95**, *96*, 402
Silver-striped Hawkmoth *Hippotion celerio* 5, *8*, 205, 405
Silver-washed Fritillary *Argynnis paphia* 29, 49, *50*, *51*, 357, 364, 400
Single-dotted Wave *Idaea dimidiata* **91**, 401
Six-spot Burnet *Zygaena filipendulae* **61**, *62*, 63, 373, 400
Six-striped Rustic *Xestia sexstrigata* **252**, 406
Skinner, Bernard 90, 137
skippers *see* Hesperiidae
Slender Brindle *Apamea scolopacina* 320, *321*, 364, 408
Slender Burnished Brass *Thysanoplusia orichalcea* **341**, 409
Slender Pug *Eupithecia tenuiata* **136**, 403
Slender-striped Rufous *Coenocalpe lapidata* 4, 14, 15, **122**, 402
Small Angle Shades *Euplexia lucipara* 310, *311*, 408
Small Argent and Sable *Epirrhoe tristata* **98**, 402
Small Autumnal Moth *Epirrita filigrammaria* **129**, 402
Small Blue *Cupido minimus* 3, *13*, 14, 16, 17, 20, 28, **43**, 373, 400
Small Chocolate-tip *Clostera pigra* 216, *217*, 361, 370, 405
Small Clouded Brindle *Apamea unanimis* **319**, 408
Small Copper *Lycaena phlaeas* **43**, 357, 400
Small Dotted Buff *Photedes minima* **325**, 408
Small Eggar *Eriogaster lanestris* 5, 16, 19, 22, **67**, *68*, 401
Small Eggar larvae 384, *385*
Small Elephant Hawkmoth *Deilephila porcellus* **204**, 371, 405
Small Emerald *Hemistola chrysoprasaria* 14, 15, **85**, 401
Small Engrailed *Ectropis crepuscularia* 190, *191*, 373, 404
Small Fan-foot *Herminia grisealis* **354**, 409
Small Fan-footed Wave *Idaea biselata* **90**, 401
Small Heath *Coenonympha pamphilus* 55, 363, 370, 400
Small Mottled Willow *Spodoptera exigua* **333**, 409
Small Phoenix *Ecliptopera silaceata* **109**, 402
Small Purple-barred *Phytometra viridaria* **350**, 366, 373, 409
Small Quaker *Orthosia cruda* **270**, 407

Small Rivulet *Perizoma alchemillata 133*, 359, 402
Small Seraphim *Pterapherapteryx sexalata 163*, *164*, 362, 403
Small Square-spot *Diarsia rubi* 249, 406
Small Tortoiseshell *Aglais urticae* 47, 386, 400
Small Wainscot *Photedes pygmina* 326, 408
Small White *Pieris rapae* 38, 400
Small White Wave *Asthena albulata* 160, 357, 403
Small Yellow Wave *Hydrelia flammeolaria* 161, 403
Smerinthus ocellata see Eyed Hawkmoth
Smoky Wainscot *Mythimna impura* 277, 407
Smoky Wave *Scopula ternata* 10, 90, 365, 401
Snout *Hypena proboscidalis* 353, 409
Southern Wainscot *Mythimna straminea* 276, 277, 369, 407
Speckled Wood *Pararge aegeria* 51, *378*, 400
Spectacle *Abrostola tripartita* 348, 409
Sphingidae (Hawkmoths) 198-205, 384, 386, 404
Sphinx ligustri see Privet Hawkmoth
Spilosoma lubricipeda see White Ermine
Spilosoma luteum see Buff Ermine
Spodoptera exigua see Small Mottled Willow
Sprawler *Asteroscopus sphinx* 16, 21, *282*, 283, 362, 374, 407
Spring Usher *Agriopis leucophaearia* 183, 404
Spruce Carpet *Thera britannica* 115, 116, *117*, 402
Square-spot Dart *Euxoa obelisca* 16, 234, 369, 406
Square-spot Rustic *Xestia xanthographa* 252, 406
Standfussiana lucernea see Northern Rustic
Stelfox, Arthur Wilson (1883-1972) 3, 215
Stilbia anomala see Anomalous
Straw Belle *Aspitates gilvaria* 411
Straw Dot *Rivula sericealis* 351, 409
Straw Underwing *Thalpophila matura* 310, 371, 408
Streak *Chesias legatella* 158, *159*, 357, 403
Streamer *Anticlea derivata* 103, *105*, 402
Striped Hawkmoth *Hyles livornica* 204, 405
Striped Twin-spot Carpet *Nebula salicata* 106, 402
Suspected *Parastichtis suspecta* 15, 312, 408
Svensson's Copper Underwing *Amphipyra berbera* 308
Swallow Prominent *Pheosia tremula* 210, 211, 405
Swallow-tailed Moth *Ourapteryx sambucaria* 178, *380*, 404
Sweet Gale *Acronicta euphorbiae* 411
Swift moths *see* Hepialidae
Sword-grass *Xylena exsoleta* 15, 16, 288, 407
Synanthedon formicaeformis see Red-tipped Clearwing
Synanthedon tipuliformis see Currant Clearwing
Syngrapha interrogationis see Scarce Silver Y

Tawny Marbled Minor *Oligia latruncula* 322, 323, 408
Tawny Shears *see* Tawny Shears and Pod Lover
Tawny Shears and Pod Lover *Hadena perplexa* 264, 265, 369, 407
Tawny-barred Angle *Macaria liturata* 168, 404
Tawny-speckled Pug *Eupithecia icterata* 147, 403
Tethea or see Poplar Lutestring
Thalpophila matura see Straw Underwing
Thecla betulae see Brown Hairstreak

Thera britannica see Spruce Carpet
Thera cognata see Chestnut-coloured Carpet
Thera firmata see Pine Carpet
Thera juniperata see Juniper Carpet
Thera obeliscata see Grey Pine Carpet
Theria primaria see Early Moth
Tholera cespitis see Hedge Rustic
Tholera decimalis see Feathered Gothic 'thorns' 174
Thrift Clearwing *Bembecia muscaeformis* 411
Thumatha senex see Round-winged Muslin
Thyatira batis see Peach Blossom
Thyatiridae (lutestring moths) 77-82, 401
Thyme Pug *Eupithecia distinctaria* 149, 403
Thysanoplusia orichalcea see Slender Burnished Brass
tiger and ermine moths *see* Notodontidae
Timandra comae see Blood-vein
Timoleague (Cork) 4
Tissue *Triphosa dubitata 124*, 372, 402
Toadflax Pug *Eupithecia linariata* 139, 403
Treble Lines *Charanyca trigrammica* 331, 408
Treble-bar *Aplocera plagiata* 159, 365, 403
Trichiura crataegi see Pale Eggar
Trichopteryx carpinata see Early Tooth-striped
Triphosa dubitata see Tissue
Triple-spotted Pug *Eupithecia trisignaria* 142, 368, 403
True Lover's Knot *Lycophotia porphyrea 246*, 406
Turner, Henry Jerome (1856-1951) 5
Turnip Moth *Agrotis segetum* 238, 406
tussock moths *see* Lymantriidae
Twin-spot Carpet *Perizoma didymata* 135, 403
Twin-spotted Quaker *Orthosia munda* 273, 407
Tyria jacobaeae see Cinnabar
Tyria jacobaeae ab. *flavescens see* Cinnabar

Uncertain *Hoplodrina alsines* 332, 333, 408

Valerian Pug *Eupithecia valerianata* 19, 140, 360, 403
Vanessa atalanta see Red Admiral
Vanessa cardui see Painted Lady
Vanessa virginiensis see American Painted Lady
Vapourer larva *384*
Vapourer *Orgyia antiqua* 218, *219*, 357, 366, 405
Venusia cambrica see Welsh Wave
Vestal *Rhodometra sacraria* 92, *93*, 401
Vine's Rustic *Hoplodrina ambigua* 333, 369, 409
V-Moth *Macaria wauaria* 14, 15, **168**, 403
V-Pug *Chloroclystis v-ata* 156, 403

Wall *Lasiommata megera* 16, 18, 20, 29, 52, *53*, 373, 400
Waring, Paul 333, 411
Waring and Townsend (2003) 12, 25, 92
Water Carpet *Lampropteryx suffumata* 106, 402
Watsonalla cultraria see Barred Hook-tip
Watts, Charles W. 3
Welsh Wave *Venusia cambrica* 160, 403
White Colon *Sideridis albicolon* 14, 15, 258, 359, 406
White Ermine *Spilosoma lubricipeda* 229, *230*, 231, 405
White-line Dart *Euxoa tritici* 235, 406
White-line Snout *Schrankia taenialis* 16, 353, 374, 409
White-speck *Mythimna unipuncta* 278, 407
White-spotted Pug *Eupithecia tripunctaria* 145, 146, 403

Willow Beauty *Peribatodes rhomboidaria* 185, 188, 404
Winter Moth *Operophtera brumata 130*, 132, 402
Wood Tiger *Parasemia plantaginis* 16, *57*, 222, 227, 373, 405
Wood White *Leptidea sinapis* 7, 36
Wormwood Pug *see* Wormwood and Ling Pug
Wormwood and Ling Pug *Eupithecia absinthiata* 143, 403
Wright, William Stuart, Captain (fl. 1946-1976) 7

Xanthia icteritia see Sallow
Xanthia icteritia f. *flavescens see* Sallow
Xanthia togata see Pink-barred Sallow
Xanthorhoe decoloraria see Red Carpet
Xanthorhoe designata see Flame Carpet
Xanthorhoe ferrugata see Dark-barred Twin-spot Carpet
Xanthorhoe f. *unidentaria see* Dark-barred Twin-spot Carpet
Xanthorhoe fluctuata see Garden Carpet
Xanthorhoe montanata see Silver-ground Carpet
Xanthorhoe spadiceria see Red Twin-spot Carpet
Xestia agathina see Heath Rustic
Xestia baja see Dotted Clay
Xestia castanea see Neglected Rustic
Xestia c-nigrum see Setaceous Hebrew Character
Xestia sexstrigata see Six-striped Rustic
Xestia triangulum see Double Square-spot
Xestia xanthographa see Square-spot Rustic
Xylena exsoleta see Sword-grass
Xylena vetusta see Red Sword-grass
Xylocampa areola see Early Grey

Yellow Horned *Achlya flavicornis* 80, *81*, 358, 401
Yellow Shell *Camptogramma bilineata 100*, 402
Yellow-barred Brindle *Acasis viretata* 164, 403
Yellow-line Quaker *Agrochola macilenta* 297, 407
Yellow-ringed Carpet *Entephria flavicinctata* 10, *11*, 16, **101**, 365, 402
Yellow-tail *Euproctis similis* 221, 359, 371, 405
Youghal (Cork) 333

Zanclognatha tarsipennalis see Fan-foot
Zygaena filipendulae see Six-spot Burnet
Zygaena lonicerae see Narrow-bordered Five-spot Burnet
Zygaenidae (Burnets) 60-63, 387, 400

Additional photographic credits
Opposite acknowledgement page
Stephen Dalton
Small Blue page 25 Brian Nelson
Marsh Fritillary eggs page 380 Maurice Hughes
Sugaring page 391 Roy Leverton